D1217677

Nature Knowledge

Nature Knowledge
Ethnoscience, Cognition,
and Utility

Edited by
Glauco Sanga and Gherardo Ortalli

Berghahn Books
New York • Oxford

Istituto Veneto di Scienze, lettere ed Arti
Venezia

Published in 2003 by

Berghahn Books

www.berghahnbooks.com

©2004 Istituto Veneto di Scienze, lettere ed Arti

Library of Congress Cataloging-in-Publication Data

Nature knowledge : ethnoscience, cognition, and utility / edited by Glauco Sanga and Gherardo Ortalli.
  p. cm.
Includes bibliographical references (p.   )
ISBN 1-57181-822-7 (alk. paper)--ISBN 1-57181-823-5 (pbk. : alk. paper)
  1. Ethnobiology. 2. Ethnobotany--Nomenclature. 3. Ethnozoology--Nomenclature.
4. Philosophy of Nature. 5. Conservation of natural resources. I. Sanga, Glauco. II. Ortalli, Gherardo.

British Library Cataloguing in Publication Data

A Catalogue record for this book is available from the British Library.

Printed in the United States on acid-free paper

ISBN 1-77181-822-7 (hardback)

Contents

List of Illustrations

List of Tables

Preface and Acknowledgements

This book presents the materials proposed and the results that emerged during the International Conference *Nature Knowledge / Saperi naturalistici*, which took place in Venice, from 4 to 6 December 1997, organized by the Istituto Veneto di Scienze, Lettere ed Arti and the Centro Interuniversitario di Studi sulla Trasmissione del Sapere of the Department of Historical Studies of Ca' Foscari University of Venice.

The purpose of the conference was to compare the main lines of research into nature knowledge over the last few years: the folk, traditional, and local forms of knowledge and uses of nature; and the way nature can be protected and conserved.

This field of study is now very wide and involves anthropologists, historians, economists, linguists, and biologists. It ranges from theoretical problems relating to forms of perception and the classification of the real, to concrete procedures of economic and political intervention in developing countries. The time was thus ripe for a moment of reflection and real comparison, to allow us to assess what has been done and what remains to be done, and to discuss the problems that have emerged from different research projects.

The proceedings were organized in thematic sessions, during which a coordinator posed questions and directed the discussion. The participants were already acquainted with both the questions proposed and the replies sent in by the various speakers (distributed in pre-prints before the conference), since the intention was to leave ample room for discussion among the scholars present, all from different backgrounds and orientations.

The structure of this volume reflects the way in which the conference was organized. The studies and materials presented are ordered according to the same five sessions in which the individual talks were grouped: each session opened with a series of questions, which are reflected in the introductory chapters of each part; these were followed by the papers of the guest-speakers, now presented as the numbered chapters of the volume; the sessions closed with broad ranging and organic discussions, which have been summarized and presented as the conclusion to each part of the volume.

- Session 1. *Classification.* Ethnotaxonomies: the ways of perceiving, knowing and classifying nature. Ethnotaxonomies are discussed, i.e., the forms of folk and local classification of nature, and new hypotheses and perspectives in the field of classification (polythetical classifications, fuzzy classifications).
  Coordinator: Marta Maddalon.
  Participants: Brent Berlin, Roy Ellen, Oddone Longo, and John Trumper.

- Session 2. *Naming.* Ethnolinguistics: the ways of naming nature and by means of nature. Linguistic mechanisms (metonymy, metaphor, derivation) through which names are given to animals, plants, parts of the body, atmospheric phenomena etc. are discussed, as are the transfers of systems of denomination

from one sphere to another (animal names to name plants, names of parts of the body to denominate land, etc.).
Coordinator: Glauco Sanga.
Participants: Mario Alinei, Brent Berlin, Maurizio Gnerre, Jane Hill, Giovan Battista Pellegrini, Nicole Revel, John Trumper.

- Session 3. *Thought*. Nature as symbol, as model and as metaphor. The symbolic uses of nature are discussed, with particular attention to the theme of universals.
Coordinator: Daniel Fabre.
Participants: Jean-Pierre Albert, Marlène Albert-Llorca, Giulio Angioni, Jack Goody, and Francis Zimmermann.

- Session 4. *Use*. Domestication: modification and practical use of nature. Particular attention is paid to the theoretical problem of the practical application of local nature knowledge (Local Technical Knowledge, Indigenous Knowledge Systems) in the contexts of development processes.
Coordinator: Antonino Colajanni.
Participants: Giulio Angioni, Roy Ellen, Tim Ingold, Pier Giorgio Solinas, and Michael Warren.

- Session 5. *Conservation*. Biodiversity, environmentalism, museums and parks. The discussion concerns the problems raised by conservation of the natural environment: dynamics and conflicts between local populations; conservation of resources and farmers'"rights"; inter-relations between conservation and the transformation of cultures and natural resources.
Coordinator: Cristina Papa.
Participants: Mauro Ambrosoli, Laurence Bérard, Stephen Brush, Philippe Marchenay, Diego Moreno, Gherardo Ortalli.

This book is published several years after the conference, since the editorial work has been long and difficult; nevertheless we trust the proceedings are still relevant and that the conference remains a milestone in the study of folk nature knowledge.

Acknowledgements

As the volume is sent to press, a special thanks is due to all those who helped to make the initiative such a success: for the scientific organization, the anthropologists Nadia Breda, Giovanni Dore, Lidia Sciama; for the technical organization, Gianfranco Bonesso and the other students of Ethnology and Cultural Anthropology of the Faculty of Humanities and Philosophy of Ca' Foscari University of Venice. A major contribution was made by Alessandro Franchini and Giovanna Palandri of the Istituto Veneto di Scienze, Lettere ed Arti of Venice, for the general organization, and by Janine A. Treves-Habar of Berghahn Books for the excellent editing work.

List of Abbreviations

AEPS	Arctic Environmental Protection Strategy
AgREN	Agricultural Research and Extension Network
AIDS	Acquired Immune Deficiency Syndrome
AION	Annali dell'Istituto Universitario Orientale di Napoli
AKIS	agricultural knowledge information systems
ALE	*Atlas Linguarum Europae* by Mario Alinei
AMT	Alonso Méndez Ton (ch. 1)
AOC	Appellation d'origine contrôlée
APFT	Avenir des Peuples de Forêts Tropicales
ASEMI	Asie du Sud-Est et Monde Insulindien
ASLEF	*Atlante Storico Linguistico Etnografico Friulano* by Giovan Battista Pellegrini
CGIAR	Consultative Group on International Agricultural Research
CIAD	Center for Integrated Agricultural Development
CIDA	Canadian International Development Agency
CIKARD	Center for Indigenous Knowledge for Agriculture and Rural Development
CIMMYT	Centro Internacional para Mejoramiento de Maiz y Trigo
CIP	Centro Internacional de la Papa
CIRAN	Center for International Research and Advisory Network
CN	*Chicham Nekatai* (ch. 7)
CSC	Certificate of Specific Character
CTHS	Comité des Travaux Historiques et Scientifiques
CTTA	communication for technology transfer in agriculture
CUP	Cambridge University Press
DHS	distinction, homogeneity, and stability
DOC	Denominazione di Origine Controllata
FAO	Food and Agriculture Organization
FEW	*Französisches Etymologisches Wörterbuch* by Walther von Wartburg
GARAE	Groupe Audois de Recherche et d'Animation Ethnographique
ICMs	idealized cognitive models
IDG	International Development Group
IDRC	International Development Research Centre
IE	Indo-European
IEW	*Indogermanisches Etymologisches Wörterbuch* by Julius Pokorny
IGM	Istituto Geografico Militare
IIRR	International Institute of Rural Reconstruction
IITAP	International Institute for Theoretical and Applied Physics
IK	indigenous knowledge
IKS	indigenous knowledge systems
ILEIA	Information Centre for Low External Input and Sustainable Agriculture

INAO	Institut National des Appellations d'Origine
INRA	Institut National de la Recherche Agronomique
IPGRI	International Plant Genetic Resources Institute
IRRI	International Rice Research Institute
IUCN	International Union for Conservation of Nature
L.	Linnaeus
LBs	language builders
LDC	least developed country
LTK	local traditional knowledge (ch. 19)
MIT	Massachusetts Institute of Technology
MSH	Maison des Sciences de l'Homme
MTK	modernist traditional knowledge (ch. 19)
NDDC	*Nuovo Dizionario Dialettale della Calabria* by Gerhard Rohlfs
NGOs	non-governmental organizations
NIE	New Indo-European
ODA	UK Overseas Development Administration
ORSTOM	Institut français de recherche scientifique pour le développement en coopération
OUP	Oxford University Press
PDO	Protected Designation of Origin
PGI	Protected Geographical Indication
PN	*Prison Notebooks* by Antonio Gramsci (ch. 17)
PUF	Presses Universitaires de France
REPPIKA	Regional Program for the Promotion of Indigenous Knowledge in Asia
SELAF	Société d'Etudes Linguistiques et Anthropologiques de France
SLARCIK	Sri Lanka Resource Centre for Indigenous Knowledge
sp.	species
SRI	Silsoe Research Institute
STDs	Sexually Transmitted Diseases
TOGA	Tropical Ocean Global Atmosphere
TOT	transfer-of-technology (ch. 21)
UK	United Kingdom
UNDP	Unites Nations Development Programme
UNEP	Unites Nations Environment Programme
UNESCO	United Nations Educational, Scientific and Cultural Organization
UNRISD	United Nations Research Institute for Social Development
USAID	United States Agency for International Development
WHO	World Health Organization
WWF	World Wildlife Fund

List of Contributors

Jean-Pierre Albert — Ecole des Hautes Etudes en Sciences Sociales, Toulouse, France

Marlène Albert-Llorca — University of Toulouse, France

Mario Alinei — University of Utrecht, Netherland

Mauro Ambrosoli — University of Udine, Italy

Giulio Angioni — University of Cagliari, Italy

Laurence Bérard — Centre National de la Recherche Scientifique, Paris, France

Brent Berlin — University of Georgia, Athens, United States

Stephen Brush — University of California, Davis, United States

Antonio Colajanni — «La Sapienza» University, Rome, Italy

Roy Ellen — University of Kent, Canterbury, United Kingdom

Daniel Fabre — Ecole des Hautes Etudes en Sciences Sociales, Toulouse, France

Maurizio Gnerre — Istituto Universitario Orientale, Naples, Italy

Jack Goody — St. Johns College, Cambridge, United Kingdom

Holly Harris — University of Kent, Canterbury, United Kingdom

Jane H. Hill — University of Arizona, Tucson, United States

Tim Ingold — University of Aberdeen, United Kingdom

Oddone Longo — University of Padua, Italy

Marta Maddalon — University of Calabria, Cosenza, Italy

Philippe Marchenay — Centre National de la Recherche Scientifique, Paris, France

Diego Moreno — University of Genoa, Italy

Gherardo Ortalli — Ca' Foscari University, Venice, Italy

Cristina Papa — University of Perugia, Italy

Giovan Battista Pellegrini — University of Padua, Italy

Nicole Revel — Centre André-Georges Naudricourt, Centre National de la Recherche Scientifique, Paris, France

Glauco Sanga — Ca' Foscari University, Venice, Italy

Pier Giorgio Solinas — University of Siena, Italy

John B. Trumper — University of Calabria, Cosenza, Italy

†D. Michael Warren — University of Iowa, Ames, United States

Francis Zimmermann — Ecole des Hautes Etudes en Sciences Sociales, Paris, France

Glauco Sanga

Introduction

Human Nature

Commenting on *The Giaour* by Byron, in 1818 Ludovico di Breme affirmed the romantic belief in the "continuous action of omnigenous nature," and defined its scope. "If I speak of exposing the spirit to the action of nature, I mean its moral frames as well as its physical ones, and I hold man as the first of the objects to contemplate, and a knowledge of times and customs as an essential part of this nature." Giacomo Leopardi answered him in the same year, disagreeing on everything, except on the nature of nature: "not that nature which not only surrounds and presses on every side, but is within us alive and shouting, can ever become extraordinary for men."

Indeed, we are nature too; and this is a good starting point for an understanding of the special relationship that man has created with nature—with ourself and at the same time with what is other; a relationship that starts with *human nature* and broadens out progressively towards the construction of a nature exterior to us, a knowable, controllable and exploitable nature.

If we are part of nature, human culture can be understood as the means by which our species relates to nature, through the transference of biological functions into the socialized external objects of tools and language. Viewed in this light, culture is in fact part of our biology, part of human nature: tools are cultural extensions of biological organs, and our relationship with the world and with others is a symbolic relationship in that it is mediated by language.

Nature knowledge—relations in nature—which every living being implicitly possesses, in humans *also* constitutes an exteriorized and objectified heritage, on account of the specific form that the particular biological strategy we call culture assumes in us. Culture is not exclusive to humans, because other animals that use tools and symbols can be found, but in us it is necessary, because no human activity exists that is not cultural, which does not have a symbolic "form."

All human activity, practical and symbolic, has a direct relationship with nature: from the economic sphere of the vital to the symbolic sphere, which is to say from modes of subsistence and reproduction to modes of relating, organizing, and reflecting. Nature knowledge is a point of encounter and amalgamation—not of opposition—between categories that are at the basis of human action, such as practical and symbolic, functional and formal, where the abstract is not considered in opposition to, but rather a specific "form" of the concrete.

Studying the origin and evolution of tools (that run alongside the origin and evolution of language), André Leroi-Gourhan has demonstrated that function assumes a form, because every form is the expression of a specific functionality, and this is true on the biological level even before the cultural level. Thus form cannot be arbitrary, it is always necessary—both the form of a natural organism and the form of a cultural object (just like language, which, far from being arbitrary, is simply necessary).

Elaborating the concept of "ethos of transcendence in value," Ernesto de Martino has shown us that in human culture the economic takes on meaning, since it is impossible to conceive of a purely biological *human* activity: every human action, even the most basically vital, is always symbolic *as well*. See the deep analysis of Pasquinelli (1984).[1]

In human beings the "economic"—or the "vital," if one prefers—must be seen as referring to the activity of subsistence in the widest sense of the term, including reproduction, because there is no difference between *human* production and reproduction, both having been symbolically identified since the origins of mankind. We have become human, distinguishing ourselves culturally from the other primates, with the development of hunting activities, for which tools were fabricated, and, at the same time, with the development of language. Hunting transforms subsistence-activities from animal to human, from biological to cultural, because the Paleolithic hunter, sexualizing hunting, identifying his weapon with his penis, production with reproduction, gave it a sense that transcended the need of nourishment: it went beyond the biological scope of the practical to enter the cultural scope of the economic, which is practical and symbolic at the same time. And here it is clear that culture is not opposed to nature, but is a way of living in nature—the human way.

For this reason the nature-culture dichotomy is unsatisfactory, in the same way that the many other dichotomies of which we regularly make use are insidious and misleading, such as human-animal, sacred-profane, religion-magic, local-global, *langue-parole* (competence-performance), objective-subjective, right up to the most radical and unsuspected, life-death. Dichotomy is an efficacious but rudimentary tool; it is an initial analytical approximation, based on the principle of maximum differentiation, which, while it allows us to distinguish, does not permit us to regain the solidarity that is at the origin of distinction—distinction petrified into a binary opposition which, by its polar nature, makes a simple difference into something radical and extreme.

The conference would go beyond summarizing the "state of the art" of the studies on *Nature Knowledge*—American, French, English and Italian (in particular the work of the late Giorgio Raimondo Cardona)—and really act as a forum for the useful comparison of tendencies, methods, and results.

In the following, I provide a succinct account of the contributions, notably the theoretical positions expressed by the participants.

Classification

After pointing out that folk classifications have mainly been studied in the three highly structured domains of the natural world (plants and animals), color terms, and kinship systems, Marta Maddalon at once raised the theme that proved central to the whole conference, namely that of the cóntrast between universal models—proposed by the scholars of a cognitivist orientation—and culturally determined models. From this main theme stems a series of connected issues, such as the innate versus the acquired (innate taxonomic capacities with respect to culturally acquired capacities), perception versus utility (formal classifications with respect to functional classifications), realism versus idealism (categories as entities with real existence in the external world with respect to categories as mere sociocultural constructions). According to Maddalon, one can hypothesize a universal level in the perception of nature that favors a morphological form of classification; in modern postagricultural societies there is a greater development of lower taxonomic ranks (varietal, cultivars) to the detriment of higher ranks (generic).

Maddalon also raises a number of other questions, notably: (i) the importance, from the taxonomic point of view, of the intermediate level, the space between the generic and life-forms levels, (ii) the concepts of metonymy, metaphor, and prototype (with examples taken from the classification of birds in Latin and in Italian dialects), (iii) the "interplay of similitudes" between the various spheres of the natural world, and (iv) the suggestion that we should examine just how much scientific classifications owe to ethnoscientific classifications.

In chapter 1, Brent Berlin points to two universal principles of ethnobiological classification: not all generic and specific taxa are named, and those that are named are done so on the basis of the perception of affinities independent of culture, which reflect the intrinsic organization of nature and not a human, economic or symbolic order. Berlin presents an exceptional example: the folk botanical system of classification of the informant Alonso Méndez Ton, a Tzeltal Maya Indian from the Highlands of Chiapas, Mexico. This folk system, representing some 900 scientific genera and nearly 3000 valid botanical species, is both natural and comprehensive: an unfamiliar species is identified on perceptive grounds and is classified by similarity (a known taxon is defined as "genuine," "true," a taxon assimilated to a known one is described in such terms as: "it's like an," "it's similar to"). Alonso Méndez Ton's system is natural because it is perceptually based on psychological principles that underlie what we know today as scientific systematic botany. "When his classification of the flora does not conform with currently recognized phylogenetic boundaries of Western botany, it is generally the case that the organisms in question nonetheless share many perceptual features in common which justify his grouping them as members of the same conceptual category."

According to Roy Ellen, in chapter 2, it is undoubtedly legitimate to seek classifications that can operate independently of cultural inputs or context, but these cognitive propensities are so general and abstract that they can tell us little about the concrete classifications of people, at lower and more functional levels. "Noncultural input operates in terms of the process of categorization, rather than underpinning particular categories. And certainly regularities may be the product of general mechanisms operating across different and very varied domains, constrained by the data being organized." In particular, "in the domain of living kinds these tendencies converge in a particular way, not obviously because of features of the mind that does the classifying, but because of regularities of the objective world which is classified and to that the mind responds." The universality of the ethnotaxonomies proposed by Berlin actually functions only if we clearly separate the general-purpose models (logical and 'natural') from the special-purpose models (those which have arisen for cultural reasons). It is only these latter that are found in ethnographic practice; indeed, the more profound the research, the more the taxonomies tend to collapse, since they work well only with simplified data, as is demonstrated, for example, by the classifications of domesticates.

Ellen also emphasizes the difficulty of drawing a clear distinction between the symbolic and the mundane, the social and the non-social: "symbolic things are in an important sense practical, and practical classifications of the non-social world often rely on metaphors that are ultimately social, as in the use of the terms 'genus' and 'family' to organize plants and animals." Therefore, "it is impossible, for example, to make sense of Austronesian terms and categories for 'bird' and for 'tree' without considering utilitarian and symbolic criteria."

In chapter 3, Oddone Longo points out that Aristotle's zoological classification is not intended to be systematic and does not introduce new names, but uses the common ones. It is based on traditional popular knowledge, in particular on that of professional specialists: shepherds, hunters, and fishermen. While the criteria adopted are perceptive (shape, color, size) and ecological (habitat), the background behind Aristotelian classifications is certainly utilitarian, since the nucleus of knowledge was provided by the professional techniques of breeders and hunters. Thus we are not dealing with abstract knowledge, but with concrete knowledge. There is no coherent taxonomic system; animals are grouped according to the point of view of their treatment (anatomical, ecological, ethological etc.) in different and often contrasting classifications, which are not ends in themselves and general, but functional and special.

In his chapter 4, John Trumper presents the two great models of classification that have been present in our culture for 2500 years, Aristotle's "ontological" model and Plato's "henological" model. The ontological model presents a unitarian vision of the world, with a substantial symmetry between the parts of the natural continuum and "precise symmetries between individual 'souls,' in which genus/species etc. are analytical tools belonging to a particular model; individuals are characterized by 'sums' of accidents or properties, fuzzy classes are either accommodated or relegated to the 'wondrous,' perhaps the older system." This archaic model, which formalizes a previous vision of the world, is formulated by Aristotle and taken up by Theophrastus, Galen, Dioscorides, Lucretius, and Lucan. The henological model presents a hierarchic vision of the world, in which the Anima Mundi reveals itself in individuals; it has hierarchies in tree-form, uses binary traits with absolute definitory value, and has no room for fuzzy classes. It is Plato's model: the One generates All-Soul which generates the souls. Already present in Homer (the "golden cord," the chain of being of the *Iliad*), it is principally developed by Plotinus: the One generates a series of dyadic principles, above all the Nous. Since Plotinus entered the Christian Middle Ages, via Arabian transmission, as an Aristotelian, and his pupil Porphyry was translated by Boetius, a syncretism was created between the two models.

In the Christian west the ontological model prevailed (with Anselm of Canterbury, Thomas Aquinas), but the Neoplatonic model is also strongly present, propagated by Johannes Scotus Eriugen, by mystic monachism, and by the Franciscans (above all Raymond Lull). With Nicola Cusano it entered the University of Padua and prevailed over the Aristotelian model. When the hierarchic ladder model, with differences that function dichotomously in binary fashion, was adopted and perfected by Descartes, Newton, Leibniz, Linnaeus, and Darwin, a breach was created between popular and scientific classifications. Thus folk systems, whether European or extra-European, derive from the first model, and scientific ones from the second. The difference is not so much in the hierarchy, which in the sense of "ranking" and "relations" is also present in the first model, as in the algebraic approach to hierarchy.

In Indo-European denominations there are precise symmetries between humans, other animals, birds, fish and even plants, which have nothing to do with similitude, metonymy, or metaphor, but which reflect an ancient worldview: when the "shoulder muscle" is called "mouse" (Latin *musculus*), and the "thigh muscle" is called "lizard" (Latin *lacertus*), they "are small animals that move under the skin of the man-animal as a composite higher animal who exists conjunctively with other animals in an animal continuum."

Naming

Glauco Sanga sets forth a set of questions about nature naming: (i) linguistic mechanisms (phonetic, morphological, semantic, rhetorical), such as onomatopoeia, derivation, extension, specialization, metonymy, metaphor; (ii) organizing principles: physical, structural, utilitarian, social, magic-religious; (iii) transfer of naming systems from one domain to another; (iv) totemic naming; (v) taboo names, hunting names; (vi) arbitrariness and motivation in the naming process.

In chapter 5, Mario Alinei believes that motivation is the most general naming mechanism. The most frequent types of motivation are, in order of importance: (a) metonymy, (b) metaphor, (c) expressive phonosymbolism, usually connected with baby talk, and (d) onomatopoeia. According to Alinei, the arbitrariness of the sign is an inescapable principle. The question that arises is the following: since words are arbitrary, how can they also be motivated? Motivation is indispensable to socialize the word, to make it comprehensible. The sign is intrinsically arbitrary, opaque, it is the motivation that renders it transparent and comprehensible. "Motivation is a basic component of the genesis of the word, but not of its function. Moreover, arbitrary is also the *choice* of motivation." This raises an intriguing question: "How could the first words uttered by *Homo loquens* be motivated, in the absence of pre-existing words?" An adequate answer can be supplied by the phonosymbolism of infantile language, much more than by onomatopoeia. "Significantly, thus, at the glottogonic level, motivation and arbitrariness appear to be inextricably interwoven, as if showing the importance of both aspects."

On the other hand, in chapter 6, Brent Berlin believes that phonosymbolism calls into question the arbitrariness of the sign as decreed by Ferdinand de Saussure. In chapter VI of *Ethnobiological Classification* (1992), Berlin proposes that the names applied to living things, especially the names for animals, often reflect some aspect of the inherent qualities of the organisms being named. The productive use of sound symbolism is particularly relevant for animal names: the names for "bird" and "fish" in Huambisa, a Jivaroan language of Amazonian Peru, are, for example, strongly marked by sound-symbolic properties: bird names show high frequency segments, in contrast with the lower frequency segments of fish names. In addition, names of small birds and fish commonly show high frequency vowel [i] stems, while larger birds and fish are referred to by names made up of the low frequency vowels [a] and [u].

The analysis of the names of the "tapir" and the "squirrel" in languages of nineteen distinct linguistic families of South American Indian languages confirms the hypothesis of size-sound symbolism: terms for "tapir" show the central vowel [a] (low acoustic frequency, hence "large/slow") and terms for "squirrel" exhibit the high front vowel [i] (high acoustic frequency, hence "small/quick"). "Tapirs are natural kinds that are large in appearance and slow (relative to squirrels) in their behavior; their names show a preference for the vowel [a]. Squirrels are small and quick (relative to tapirs), and their names show, in contrast, a preference for the high front vowel [i]." Berlin concludes that "the study of ethnobiological sound symbolism may shed light on the evolution of humans' representation of the perceived structure of the natural world in speech."

In chapter 7, Maurizio Gnerre studies the names of streams of the Shuar (Jivaroan, Upper Amazon, Eastern Ecuador), made up mostly of the male names of animals and the female names of plants (useful or cultivated). The animal or the plant name does not directly refer to the stream, since the stream is named generally through the mediation of the name of a person. It is in fact the person that denominates the stream (often the name of an eminent person), humanizing the "space" by transforming it into "place": "Persons are already 'given' as physical beings and necessarily named and culturally shaped, while places are 'created' through naming. People exist in (come from, traverse and go to) places, and 'places' arise because of human activities." Most Shuar people names are connected with an animal or a plant, but, more than to the person, the names refer itself to his *wakan'* (usually translated as "soul," is means rather "self-consciousness"): animals and plants are animated by male or female *wakan'*, and consequently human beings have the corresponding names. A person is a unique individual and gendered realization of a *wakan'*, occupying a place "vacated" by a dead person: "therefore, *wakan'* is the hidden link between persons and animals or plants, and it is the referent of many names."

The *wakan'* circulate as names among humans, animals, plants and some minerals, and name transparency favors this circulation. The Shuar language privileges name transparency. "Transparency, as a general linguistic feature, is strongly related to the issue of intralanguage relations and of relative arbitrariness; as a characteristic of names it makes the lexicon 'lighter.'" Transparency must nonetheless be distinguished from motivation; the former term covers a much larger conceptual area than the latter: many names are transparent, which is to say analyzable, but at the same time they are not clearly motivated; no motivation precludes opacity. "For most fully transparent Shuar river names there is no obvious motivation, and we can only guess at one. When, as in most Shuar streams' names, metonymy or synecdoche is present, motivation is only a sort of *ex-post* construction," as in the case of a name such as *Pamá éntsa*, "stream of the tapirs," for which we can only advance the purely hypothetical motivation that there were tapirs there.

Jane Hill, in chapter 8, deals with the problem of lexical loss among the Tohono O'odham (formerly "Papago") of the Sonoran Desert in south-central Arizona and the northern border of Sonora, Mexico. Her conclusions add to the common proposition "that when knowledge is lost, names are lost with it" its reverse, which is equally true: when names are lost, knowledge is lost with them. Empirical data find their theoretical basis in the reaffirmation of the intrinsic and necessary link between form and significance. Jane Hill also criticizes Saussure's notion of the arbitrariness of the sign, but from a point of view within structuralism. Adopting the position of Emile Benveniste, she emphasizes the necessity of the relationship between signifier and signified, all within language, thus excluding the third term (the external referent) introduced by Saussure: "meanings reside only in the language and signifiers project their signifieds rather than somehow matching up to categories in the world." For this reason Hill does not agree with Berlin, at least on the level of the practical; practical knowledge, although it must include universal components, is profoundly local and specific.

In chapter 9, Giovan Battista Pellegrini underlines the importance of metaphor and metonymy in the process of naming. Through deep etymological analysis and comparison, he provides various examples, taken from Italian dialects and romance languages, of the use of animal names for plants and tools, and he also analyzes metonymies among parts of the body.

In chapter 10, Nicole Revel analyzes in detail the processes of lexicalization of natural objects in the Palawan language of the Philippines. The mechanism of lexicalization of plants is distinct from that of birds and insects. Derivation is one of the six main lexemic types. It is interesting to note that a particular prefix, *mäg-*, is used in the vocabulary of wild plants to designate an analogy, a similitude: "related with," "looking alike" (e.g., *mäg-mamaqan* means "like areca nut"). In the lexicalization of birds and insects in Palawan, as well as in many languages of the Philippines and South-East Asia, ideophones are widespread and productive, both "sound icons" and "visual icons." Revel points out that "the link between sound and meaning is not an absolute one, it is rather to be considered within the respective constraints of the language where it appears. As R. Jakobson has analyzed, 'symbolism, although conditioned by the neuropsychological laws of synesthesia—and according to these very laws—is not the same for all.'" To name singing birds and stridulant insects, the root-word is made from an onomatopoeia. For the names of non-singing birds and non-stridulant insects, the ideophone-making turns to visual perception, as in the naming of plants, mushrooms and shells.

In chapter 11, John Trumper tackles three topics: taxonomic levels, ono-matopoeia, and metonymy and metaphor. What is the pertinent level at which naming processes begin? Whether at the taxonomic and perceptual generic/spe-cific level, or the intermediate or life-form level, and whether from a particular level lexical spreading is then directionally bottom-up or top-down as a produc-tive process. He discusses a few examples in depth: we can observe in Latin "orig-inally a bottom-up movement, involving the naming of a life-form built from lexical substance existing at the intermediate level, i.e., *quercus* (generic) $\subset$ *arbor* (intermediate) $\subset$ *arbusta* (life-form: a collective neuter pl., derived originally from *arbor*) $\subset$ *sata* (unique-beginner), whereas in medieval Latin we have the converse, top-down expansion in which the life-form becomes a new intermediate, aligned with *arbor*, leaving a covert life-form: *quercus* (generic) $\subset$ *arbor* vs. *arbustum* (inter-mediate) $\subset$ X (life-form) $\subset$ *planta* (unique-beginner). Thus we note that parto-nomic *planta* (originally tree-cutting, vine-cutting, slip, shoot) has taken the place of partonomic *sata*." "Seed" in Italian dialects reveals two different treatments: "lexical creation and movement appear to begin at the unique-beginner stage, in some cases, with top-down movement partonomic > unique-beginner > life-form > generic (derivatives of Latin *sēmen* 'seed'), at others with a bottom-up movement partonomic> generic > life-form (derivatives of Latin *cibus* 'food'). In the first case the movement is from 'sowing,' a typically agricultural concept, in the second, from 'seed' = 'food,' which would seem still characteristic of a plant- or seed-gathering society."

As regards the linguistic role of onomatopoeia, Trumper agrees with Whitney that "if we admit there ever was an imitative stage in human proto-linguistic development, then it must be a short-term phase in evolution: 'the onomapoetic stage was only a stepping-stone to something higher and better.'" He adds fur-thermore that this stage is not a social or anthropologically significant state, but is determined by what humans biologically have in common. From the linguistic point of view, much seems to depend on language typology: an agglutinative or isolative language may, because of its very morphology, keep associations with onomatopœic bases longer, while flexional languages, more especially if stress-timed, may destroy initial onomatopœic symmetries in their normal development. Since there is a continuous remotivation of sound representation, onomatopœa is decidedly culture-specific.

Trumper argues (see also chapter 4) "that in the case of natural phenomena (animals, birds, fish, plants) naming processes involve either straightforward simil-itude or the application of particular symmetrical schemes between animals and plants, animals and fish, fish and birds, etc. in terms of a particular cosmic vision": for example, "wolf" in Latin and Italian dialects presents a network of correspon-dences that include (a) mammals, (b) plants, (c) fish, (d) birds. He emphasizes that "none of these cases represent instances of classical metonymy or metaphor, rather the way in which particular Cognitive Models are interrelated with each other in a particular vision of the natural world"; whereas metonymy and metaphor are at work in trade jargon, which represents a necessity for naming non-natural objects, artifacts. In the tinker jargon of Calabria, southern Italy, there is "a terminology for metals and metal parts that is far richer than that of any dialect, richer even that of some standard languages, in which the primary element in naming seems to be synecdoche and other mechanisms of classical metonymy, e.g., *campanaru* 'lead' is 'object made of lead' (bell) > 'substance lead.'" Trumper puts forward the interesting hypothesis that specialization, whether in agriculture, or in industry or rather proto-industry, "creates an overall situation that brings utilitarian criteria to the fore and imposes them on and above all other possible criteria. Man's pure classificatory instinct in specialized societies or groups seems to give way to a series of utilitarian criteria which then determine intricate metonymic chains, therefore the creation of metaphors."

Thought

Daniel Fabre proposes a frank confrontation between anthropology and cognitivism, posing three crucial questions. (1) Should thought be collocated inside man, inside the spirit, the mind, or outside, in the public use of signs, in symbolic exchange? "For the anthropologist there is no spirit that is not objective. Thoughts are filled with content in a historic context, in the greater or lesser *hic and nunc* of a tradition, of customs, of institutions." (2) Ethnoscience has been constructed on the basis of the empirical approach to objects of science. However the object is not the botany or the zoology of a group, but the contents of thought, of which these areas of practical knowledge are merely occasions and supports, which is to say, in the last analysis, man himself, who uses and produces knowledge and thought. Therefore "we should examine the theoretical status of the micro-specialties produced by the plural notion of 'knowledges.'" (3) What has happened, today, to the universalistic program of anthropology of passing from the local to the general, from particular societies to man as such? To define universals as logical constrictions inherent to the human spirit means naturalizing or desocializing thought, relocating it inside the spirit. In this case there would be more direct methods than the ethnographic one to observe thought, and anthropology would either prove superfluous or lose its specificity.

Jean-Pierre Albert, in chapter 12, observes that the debate between ethnoscience and symbolic anthropology has an exact parallel in the debate concerning the specificity of the religious. The evolutionist argument in favor of the thesis of universals in knowledge of natural realities claims that certain forms of behavior are adaptive, and certain cognitive abilities are the outcome of the evolution of the species and are genetically innate. But adaptation, remarks Albert, is not only biological, but also social; thus we may suppose that the more general constraints on social life introduce an aspect of universality. In addition, similar societies produce similar rules and solutions.

The religious dimension is usually implicit in social rules: "religion may well be a socio-cognitive arrangement intended to accredit the unbelievable, i.e., statements in contrast with the world to which we spontaneously surrender our individual cognitive abilities." If we consider that the realities postulated by religion, although distinct from natural realities, tend towards the form of objectivity, "this means that religion gives an objective form (or origin) to principles of action, obligations, rules." Religion provides the social with an "objective" and "natural" form: "Nothing seems more common in the most varied mythological and ritual systems than the idea of continuity, or a community of essence between the order of nature (cosmic order) and social order. Indeed, the two areas of rules can be joined simply by exchanging some of their features: the social order gains in terms of presuming necessity and universality; nature, on the other hand, becomes a moral entity, a world of rule and not laws (in the sense of deterministic laws). In plainer words, nature tends to become a projection of society, while its supposed laws at the cosmic level are a reflection of the more general constraints of social life. We thus understand how cosmological universals may exist, not as expressions of innate mindsets, but as correlatives of the more general constraints of social life." The very idea of "nature in general" is never a naturalist representation; equally the *naturalization* of principles of a social order refers to a representation of nature as the place of the immutable, of order and of necessity.

According to Marlène Albert-Llorca, in chapter 13, one cannot distinguish ethnoscience from symbolic thought, because folk knowledge of nature is not always the simple empirical record of observable data, as it may also be the outcome of the play of symbolic thought itself. The notion of ethnoscience is problematic, because it suggests, on the one hand, that people have a relationship with nature that is not only utilitarian but also speculative; on the other hand, it implies the universality of the knowledges of nature. Berlin, totally rejecting the Whorf's relativistic vision, states: (a) that ethnoscientific classifications are based on observable affinities (morphological or ecological or ethological) of the species themselves, independent of the cultural importance of these species; and (b) that these classifications always take the form of taxonomy, i.e., a hierarchical classification so that at a given level of hierarchy all the categories are mutually exclusive. But in this way only one part of the indigenous knowledge of nature is considered. Alongside the formal criteria (morphological, perceptive), there also exist the functional (utilitarian) criteria and the symbolic criteria. Of an animal or a plant people know the form, the practical use and also the symbolic value. Ethnoscience runs the risk of isolating arbitrarily one form of knowledge from the others; classifications cannot be separated into "symbolic" and "mundane," because symbolism is not outside the world.

In chapter 14, Giulio Angioni observes that in the study of nature knowledge too much emphasis has been placed on an approach that privileges linguistic means of knowing, almost as if they were the only ones, reducing folk knowledge to the operations of classifying and naming. Naturally language is important, but, as Piaget teaches, it is not the basis of thought, nor can every thought be expressed linguistically. On this subject, Angioni points out the importance of the traditional knowledge implicit in "doing." This is an operative knowledge, a set of experiences, abilities and knowledges incorporated in the individual and in the group; it is pretheoretical, not verbalized and not easily verbalizable, and present in primitive societies as in our own society: not abstract notions and oral memory, but embodied knowledge and body memory, which are not "normally the subject of reflection and explicit and conscious speech but, since they have become mechanical concatenations of trains of thought and gestures, they are a kind of second nature, almost a part of our instinctive side and outcrop in our conscious minds almost exclusively in the case of accidents, difficult situations or something that disturbs their normal utilization, which does not require lucid behavior or special attention" (think, for example, of the operation of tying one's shoes or tie). "Even today, as in pre-industrial societies, this patrimony of abilities and elementary notions is learned and becomes incorporated mostly by impregnation, inference and experience."

Man "does not perform only semiotic operations such as naming and classifying: together with these, and before these, he acts to satisfy needs; he thinks of nature so as to 'do' and not only and not principally so as to say.'" Therefore, it is perhaps easier to find cultural universals in the very field of thought-for-doing, in that of practical abilities, than in the field of thought-for-saying, or in any case of knowledge formulated and organized through language: "break, cut, pound, tear, throw and dig, besides techniques of the body (Mauss): how to run or jump, get dressed or inhabit, are cultural universals (besides being precultural universals) more than to name, classify, categorize or elaborate other forms of semiotic knowledge concerning nature."

For Jack Goody, in chapter 15, the topic of nature must be considered in terms of systems of knowledge, influenced by specific cultural constellations, like the means of communication (e.g., literacy) or the means of production (e.g., pastoralism). Cognitive scientists should provide some satisfactory proof of the existence of universals, which can have some other explanation, like transcultural experiences (which are not always universal but widespread). Taxonomies of natural objects should also be accounted for in historical terms, because they change over time (even though the so-called traditional cultures are often seen as self-perpetuating in their conceptualizations). In addition, taxonomies are not always as explicitly formalized as written accounts suggest: any form of written knowledge tends to generalize and make sense of human behavior. "To establish order out of the varied and abundant material which we are faced with, we select a limited number of aspects of the phenomena to deal with, to re-present."

The same problems present themselves when one wishes to fit thought into taxonomies. Following Lévi-Strauss, one is forced to categorize concepts into oppositional binary pairs, analogical schemes that exclude ambivalence. Even where taxonomies remain the same, the cultural context of reference can change completely: a cow, says Goody, is not the same thing for me and for my Anglo-Saxon ancestors; the position of the cow in my cognitive space has changed: the term has remained but with a quite different practical and symbolic meaning. According to Goody it is difficult to distinguish between symbolic and pragmatic classifications of nature, since every linguistic use is symbolic. One has to work on a plurality of taxonomies, which vary according to the context, "so that the tomato is a vegetable in one context and a fruit in another."For example, the LoDagaa of northern Ghana do not know families, genres or species: they classify animals implicitly into edible and inedible, explicitly into wild and domesticated, and into black and white for ritual uses.

In chapter 16, Francis Zimmermann enters directly into the dispute between symbolic anthropology and cognitive anthropology, the biologistic naturalism of which he criticizes. Zimmermann defends a symbolic anthropology and a comparative ethnology, which are situated on an intermediate level on the scale of abstraction: "Neither too concrete nor too abstract. Neither specific (related to one particular language and culture) to the point of their being untranslatable, nor abstract to the point of their being unable to refer specifically to one particular semantic domain. Bodily humors like bile and phlegm in Galenic and Ayurvedic medicine seem to perfectly exemplify this kind of concept and names that are endowed with ontological implications." Zimmermann, in addressing the question of universals in the context of comparative ethnology with special reference to nature knowledge, tries "to find a way out of the dilemmas of cultural relativism. In the field of medical anthropology, for example, cultural relativism, which has been the dominant paradigm for the last two decades, resulted in our juxtaposing all the medical systems of the world with one another, on an equal footing, as so many pieces of local knowledge good for transmission, and so many ethnic commodities good for consumption. It is high time we broke with the prejudices of relativism."But we must not for this reason seek universal or primary concepts, as one used to say in the realm of cognitive sciences, but rather conditional or implicate universals.

Use

Antonino Colajanni emphasizes the dynamic and adaptive nature of indigenous knowledge systems that often have—it must not be forgotten—a long history behind them, of which it is important to know about. "Native societies possess their own forces of response and reaction: they are, in short, active and not passive subjects, also in the field of technique and natural knowledge." It is essential to reflect on the process of *technical domestication* that the West has exercised over native systems, and on the local forms of resistance, "technical syncretism," and functional re-adaptation of elements from outside. "This progressive recognition of the potentiality of knowledge systems and of indigenous action is producing—and will increasingly continue to produce—a beneficial effect on the processes of creation and re-creation of local *social and cultural identities*, which, as is well-known, constitute an important aspect of social change."

In chapter 17, Giulio Angioni observes that, compared to official knowledge, traditional knowledges and skills are not only tacit, implicit, intuitive and informal, but also and above all politically subordinate. The"bearers" of traditional knowledge know what to do but they do not have the power to do it. Angioni reminds us that Antonio Gramsci has some important things to say on these themes in his

Prison Notebooks. Angioni goes on to examine the propagation, in the context of contemporary globalization, of a form of mild racism, founded on the sense of the superiority of the West. Angioni claims that anthropologists have the unwitting responsibility for having spread the idea that ethnocentrism (understood as the inferiority of the other) is universal, shared by other animals and almost genetic. But ethnocentrism is neither universal nor inevitable. There is not only the attitude of considering others as inferior, but also of considering them as superior (as in the case of the Aztecs, the Incas, and of Sardinia).

Roy Ellen and Holly Harris—whose contribution is given in a summarized form in chapter 18—point out the changing and often contradictory scientific and moral attitudes towards indigenous knowledge. "Much Western science and technology emanates from European folk knowledge (e.g., herbal cures) and knowledge acquired in a colonial context. The nomenclature and classificatory schema employed by Linnaeus, for example, depended extensively on Asian folk knowledge as this was absorbed into the writings of colonial naturalists working in the seventeenth and eighteenth centuries." According to Ellen and Harris, indigenous knowledge, which is a tacit, intuitive, experiential, and informal, will always be necessary as an interface between real-world situations and literate expert knowledge, the latter of which will always have to be translated and adapted to local situations.

Tim Ingold, in chapter 19, deals with two separate arguments: (i) the man-environment relationship, and (ii) traditional knowledge. He proposes "to replace the conventional idea of organisms and persons as distinct, substantive entities with a view of the organism-cum-person as a position or nexus—situated within an unbounded field of relations—where growth is going on." Ingold considers that for social anthropology the mycological model of continuity among individuals (fungi) is more suitable than the zoological model of discrete individuals (animals); he thus suggests the model of the *fungal person*, because the person is not a substantive entity, but is a point of growth or emergence within a wider field or network of social relationships.

Ecological relations are not really between two entities "given" independently, organism and environment. It is commonly thought that an environment surrounds, and therefore presupposes something—an organism—to be surrounded, with an "inside" (the organism) and an "outside" (the environment). But we can, instead, think of the environment as a network of lines, branching out and coming together at various points, and ramifying indefinitely, without hierarchies. An ecological relationship "cannot be an interaction between one thing and another, for that would be to suppose that they existed, as discrete entities, in advance of their mutual engagement. If organisms, in general, 'issue forth' along the lines of their relationships, then each organism must be coextensive with the relationships issuing from a particular source. It is not possible, therefore, for any relationship to cross a boundary separating the organism from the environment. If the concept of environment is to mean anything at all, it must refer to the interpenetration of organisms. This is perhaps easier to see in the case of persons, where we are used to using the word 'social' to denote the condition of interpenetrability. But just as we need to be careful not to reify the social as an exclusive, higher-order domain going by the name of 'society,' we also have to avoid reifying the interpenetrability of organisms as a domain that exists apart from them, and with which they can interact, namely 'the environment.' In short, organisms no more interact with the environment than do individuals with society. Rather, ecological relations—like social relations—are the lines along which organisms-persons, through their processes of growth, are mutually implicated in each others' coming into being."

Ingold observes that traditional knowledge is generally associated with a genealogical model (that of the kinship diagrams), based on the idea that the elements that go together to constitute a person are passed down, from that person's ancestors, along lines of descent, which represent channels for the transmission of substance, which may be in part material, providing the recipient with a component of biology ("blood"), and in part mental, providing a complementary component of culture (a corpus of ideal rules, recipes and prescriptions). In this view, however, the environment is simply the backdrop of nature and it plays no part in the constitution of persons. "So long as the stuff of tradition could be passed along, like a relay baton, from generation to generation, it made no difference where the people were." Local traditional knowledge, however, "is not really 'passed down' at all. Rather, it is continually generated and regenerated within the contexts of people's skilled, practical engagements with significant components of the environment." It is not cognitive; it does not lie "inside people's heads." "It lies, rather, in the mutually constitutive engagement between persons and environment in the practical business of life." Tradition is not a kind of substance, it is a type of process.

Cognitive scientists think of the person in terms of container and content. "Equipped by nature with universal capacities, human beings are viewed as containers for the culturally variable, substantive content which specifies traditional knowledge in its diverse spheres of application." According to Ingold, however, local traditional knowledge might be better denoted by the concept of skill. Skills are not properties of the individual body; they are rather properties of the whole system of relations constituted by the presence of the agent in a richly structured environment. Moreover skills are refractory to codification in the programmatic form of rules and representations; they are learned through a mixture of imitation and improvisation in the settings of practice. "It would be wrong, then, to say of local traditional knowledge that it is 'cultural' rather than 'biological,' or in the head rather than in the body. It is rather a property of the whole human organism-person, having emerged through the history of his or her involvement in an environment."

Pier Giorgio Solinas, in chapter 20, notes that the relationship between knowledge and application is generally seen as hierarchic: first one knows and then one does. In fact, however, historic examples do not appear to confirm this presupposition. Solinas in any case declares himself pessimistic as regards the possibility of recovering indigenous knowledge. In the cases of critical transitions or revolutions, the new technical systems cannot tolerate the presence of two different kinds of knowledge, and the local one is swept away. "Of course after, but only after, complete deculturation has taken place, those born into the new society may acknowledge and research their lost patrimony. However, the results are often prolix or artificial: a laborious revival ritual performed by a mimical mind." Solinas points out that nature knowledges do not only concern animals and plants, but also man and his reproduction, and he presents an example taken from his research into the Santal tribe, India, which reveals that it is not possible to separate the ideology of reproduction from the practice. "Ideology is not a cognitive premise to which we add a pragmatic extension. That which is between one and the other is perhaps a firm link of analogical coincidence, but not necessarily a stringent consistency. When a Santal has to explain why blood transmitted on a paternal line prevails over that of the maternal line, he will attempt to defend his own supremacy of belief rather than demonstrate the validity of his thesis. The fact is that the thesis in itself is proposed neither as an axiom nor as a provable statement, and in all probability is not limited to the circle of things that can be affirmed or negated, but is rather an accessory, an ethical and mental institution at the same time."

In chapter 21, late Michael Warren affirms that indigenous knowledge is a powerful ally in moving away from a transfer of top-down approach technology to a more participatory one. Indigenous knowledge "represents localized—sometimes ethnically-based or community-based—knowledge that has evolved within a micro-environmental context." Its counterpart is the global knowledge produced by the universities and by the research laboratories. Warren emphasizes that indigenous knowledge is always dynamic, reflecting indigenous approaches to changing sets of problems faced by any local community. It was for this reason that the term "indigenous knowledge" was introduced instead of the earlier "traditional knowledge," in order to avoid the underlying stereotypes of simple, static, and primitive that the term "traditional" carried with it.

Conservation

Cristina Papa gives a dynamic interpretation of conservation, placing the emphasis on the human use of environmental resources, which with controlled forms of manipulation, transformation, and domestication can be conserved and renewed rather than exhausted and destroyed: therefore, according to the definition adopted in 1991 by the International Union for Conservation of Nature (IUCN), conservation is "the management of the human use of organisms and ecosystems capable of ensuring that such use is sustainable." In opposition to this view, linking conservation with use, is the particularly significant adoption of a concept of conservation excluding human use: conservation as the preservation of a wild humanless environment, in a world where man and nature are basically considered to be antagonistic. Of course conservation is not a neutral category, but generates conflicts that oppose social subjects with different interests in the management of natural resources.

As regards the conservation of genetic resources, Papa defends the sustainable development projects that envisage *in situ* and on-farm conservation, which implies the involvement of local populations. "In pursuing conservation their objective is controlled change aimed at increasing income and yield, without replacing the local genetic resources. Obviously the cultures (techniques, representations, and forms of learning), and the social and productive systems that have made the conservation of these varieties possible cannot be seen as immobile situations in the projects but rather as agents subject to change as part of more general transformations. Thus cultural change, instead of being exorcised, must be oriented towards a sustainability so that nature knowledge is developed and its transmission encouraged."

According to Papa, in the West the conservation of on-farm biodiversity by local producers can be targeted to the needs of the Western urban market, in the form of "typical," "quality" food products; thus local varieties become "traditional products," reflecting the transition from direct consumption and local markets to global markets. But it must be borne in mind that "the mechanisms to protect these local productions, however, lead to deep changes in the product itself and the local culture, by influencing their variability on the one hand, and their changeable nature as living material on the other," through a process of standardization of the food product and of the local culture itself that produces it, cutting them off from variability and from becoming historical and fixing them in a rigid, compact "typicalness.""In other words, a reduced simplified tradition is constructed in order to present a purported unchanging 'overall' continuity. This is then attributed with potential legitimacy and social recognition."

"Change is the rule of nature": according to Mauro Ambrosoli, this point of view better addresses human relations towards the environment. In chapter 22, Ambrosoli indicates the very dilemma of specialization: selection weakens the system. The botanical history of European agriculture is played out through the contrast between species and variety. The deliberate selection of seeds led to the elimination of the worst cultivars. Selected crops have advantages but are ambiguous: high-quality seed is more expensive and only suitable to the best soils; moreover this man-made selection causes a loss of germoplasm. This loss was balanced out by a continuous exchange of seeds from distant regions, from poor lands to fertile lands, which brought mountain seeds to the plains, and seeds from open fields to enclosed ones. The conscious exchange of seeds was practiced on a European and then on a world scale from the sixteenth to the nineteenth centuries thanks to the work of individual botanists, academic botanical gardens, and privately owned gardens. The peasant sector maintained its function of providing great quantities of local seeds for agricultural purposes and acted as a veritable bank of germplasm, which was necessary for any future development.

In chapter 23, Laurence Bérard and Philippe Marchenay point out the importance of the study of local agricultural products and foodstuffs. "Local food products are situated in a complex world of relations involving the biological and the social. An animal race, a cultivated plant, and a product such as a sausage or cheese are the outcome of an accumulation of knowledge, practice, observation and adjustments that must be seen in relation to the way they are represented. In short, they are objects heavily invested with many processes and meanings." A policy of protection and conservation poses cultural problems that must be carefully assessed. Protection procedures end up restricting diversity, simplifying it, stabilizing it, standardizing it, all terms that are in contrast with the very notion of diversity. Furthermore, "by generating new technical, biological and cultural references and bringing new players onto the scene, regulation has been overlaid on the existing complexity. How and which players can best protect products in this context of variability and diversity?"

Stephen Brush, in chapter 24, proposes a criticism of the current idea of genetic erosion and of conservation biology, which "aims to save species' diversity by salvaging key fragments of wilderness. The intent of conservation biology is to save a domain for nature so that it can re-conquer the Earth's surface if and when human disturbance ceases, whether this be a century, a millennium, or longer." The theme of the loss of genetic resources of crops illustrates some of the challenges and conflicts of alloying social science and conservation. By the 1960s the idea of the destruction of local crop diversity by global processes was widespread. But Brush's research, carried out on potatoes in Peru, on maize in Mexico, and on wheat in Turkey, reveals that "improved varieties easily root themselves in peasant production without displacing local varieties or dramatically reducing their diversity." Biological diversity on farms persists for three different reasons: (i) environmental advantages of different types of cultivars with regard to local microclimates, (ii) risk management of crop failure, by providing a form of biological insurance against pests, pathogens, or bad weather; (iii) cultural value of local varieties, because of their taste and quality or symbolic meanings. Nor must we forget the role of the market, or rather the lack of markets: "our research on crop variety choice in Peru, Mexico, and Turkey, revealed that peasant households produce more diversity than is necessary or optimal given environmental and risk conditions. Overproduction of diversity may be explained by the cultural value of local varieties, especially taste and cooking qualities, but why haven't peasant households discovered the benefits from specialization and exchange, so that not all households need to produce a whole array of varieties? In fact, markets for local varieties at the village level don't seem to operate, and households which consume a particular variety must also grow it."

In chapter 25, Diego Moreno states that the "natural" resources are always historically conditioned in their own ecology by the practices adopted by the previous societies that have settled on the site over time. Environmental archaeology has demonstrated the historical nature and finiteness of environmental resources. "There is no primordial natural Eden in the European history of the last 10,000 years." Biodiversity too, on a local scale, reveals its nature as a "historical product," obtained, for example, through particular strategies of selective harvesting. In fact "the *local plant heritage* (this category allows us to deal not only with the domestic flora but also the putatively wild flora) is managed according to definite production strategies. Today they appear to be documented historically more by the environmental mechanisms introduced by production practices (and their previous effects), than by the sources conventionally referred to by historians of agriculture." A rich cultural, environmental and productive legacy (i.e., environmental resources, practices, forms of knowledge and local plants and animal production) has been "unwittingly" conserved. "Only by adopting the historical ecology approach and the local history scale of observation will these environmental aspects of the European rural heritage be identified, recovered and correctly developed."

Finally, in chapter 26, Gherardo Ortalli reminds us that in the Middle Ages the very concept of conservation may have appeared decidedly anachronistic. Then, "the dialectical relation between man and the natural environment was thus determined by a powerful theological premise: since nature was created for man, there were no limits to its exploitation. These doctrinal bases paved the way to interpretations of the man-environment relationship whereby Christian anthropocentrism is associated with a consequent arrogance towards nature, considered to be completely subordinate."After the recovery of the year 1000 the issue of conserving resources began to be posed. With the twelfth century we register a growing number of objective forms of environmental protection, due to the needs of practical knowledge rather than any pressure from new ideologies: control of woods, protection of particularly valuable trees, obligation to plant new trees, limits on hunting, dumping dangerous or toxic materials, controls on polluting manufacturing processes. However "there had been a change of attitude towards the environment: the natural heritage was no longer seen as being inexhaustible," although the mediaeval conservation of resources followed a logical development which had little in common with safeguarding the environment: it was essentially a question of pursuing practical protection whenever required by contingent needs immediately perceived by a society with a very direct dependence on natural resources.

Ortalli invites us to reflect on the highly instructive case of Venice, which is primarily a city but strongly marked by natural elements (water, lagoon, canals, tides and islands)."Venice has always been characterized by its fragile equilibrium of water and land, built on a refined system of knowledge founded on the continuous and exact measurement of the tide, the relation between salt and fresh water, the influence of rivers and their deposits, the necessary and functional co-existence of islands, *barene* (the flat emerged grassy mud banks in the lagoon only ever covered by very high tide) and *ghebi* (narrow vein-like channels). These forms of knowledge have always played a key role in the survival of the lagoon equilibria. Without them, today the lagoon would not be what it is." These equilibria have been governed for centuries through the integration of different kinds of knowledge: both theoretical and experimental, and erudite and popular, organized over the centuries in various organs of control which gradually came together in the institution of the Magistrato alle Acque, which until just a few years ago decided on issues concerning the lagoon, "listened to the opinion of nine fishermen, something that shows how these different forms of knowledge were interwoven. "This would seem to be an excellent example of how knowledge produced by research institutes dialogued with empirical information from people with daily experience of life in the lagoon and fully aware of all its vital rhythms."

In the 1980s the state entrusted the monopoly management of the lagoon to a consortium (the Consorzio Venezia Nuova), bringing together some of the major private operators in Italy in the field of large public works. "The leap in the logic of the various forms of knowledge may be illustrated in the transition from experimental knowledge, including that of the fishermen, to the mathematical models used by the consortium." Naturally the interventions proposed to control the phenomenon of "acqua alta" are radically different: the scientific culture based on the local knowledges tends to put forward "gentle" solutions of minimal but continuous adjustments to the natural balances aimed at accompanying and guiding the ongoing evolutional processes; while the scientific culture expressed by the private consortium frames the problem in an engineering perspective of large-scale works and incomparably higher (and more costly) levels of technical intervention. We thus see once again the "contrast between traditional widespread forms of knowledge developed in the local context and those produced by general epistemological processes and introduced to the city from outside through procedures with considerable social and economic implications."

Conclusion

One theme dominated the conference, that of universals, authoritatively proposed by Brent Berlin. The idea of innate cultural universals, localized in the mind, aroused suspicion and reservations, especially in its most rigid formulations characteristic of cognitive scientists. The theoretical existence of universal features is not so much disputed as is their practical relevance; most anthropologists are more interested in culture-specific features, of varying import, whether local or transcultural—features determined by means of transmission (orality, literacy), modes of production (hunting-gathering, pastoralism, agriculture), and linguistic families. In any case, just as the belief in the theory of universal models does not mean a return to structural anthropology, so too its rejection in favor of culturally conditioned models has not meant a return to cultural relativism.

Another widely debated theme was that of the arbitrariness of the linguistic sign, as opposed to phonosymbolism and other hypotheses of "external" motivation of language.

The theme of indigenous, local, and traditional knowledges, and their rediscovery and re-use, was examined in depth; there emerged a certain uneasiness, both theoretical and practical, as regards the actual import of these concepts and the effective viability of projects linked to them. Nonetheless, there was much insistence on the historical and dynamic nature of local knowledge and of the natural resources themselves, and on the difficulties that this implies with respect to conservation and protection projects, which must avoid the contrasting dangers of rigidity and mummification on the one hand, and on the other that of a dynamic conservation that may simply be a disguise for a development artificially induced from without. But if nature is historical, what is a "natural" conservation? A *laissez-*

faire or a guided transformation? And what is the relationship between transformation and development?

If nature is a historical construction due to human activity, to the interaction of man with his environment, this opens the great question of rights over nature: legally nature would no longer be *res nullius*, and its exploitation would thus not be the exploitation of a thing, but of the men and cultures that have constructed it over time. In this case other major legal problems would seem to arise: a nature that is "historical"and no longer "natural"would cease to belong to everybody (to the whole world) and would become the property of *someone*—the collective property of only the local community that has "brought up"this nature. These are the themes that are becoming central to the debate on the rights to biodiversity, its exploitation and its conservation.

References

Berlin, Brent. 1982 *Ethnobiological Classification*. Princeton: Princeton University Press.

Cardona, Giorgio Raimondo, ed. 1981. Antropologia simbolica. Categorie culturali e segni linguistici. *La ricerca folklorica* 4: 3–98.

_____ 1985. *La foresta di piume. Manuale di etnoscienza*. Bari: Laterza.

_____ 1985. *I sei lati del mondo. Linguaggio ed esperienza*. Bari: Laterza.

_____ 1990. *I linguaggi del sapere*, edited by Corrado Bologna. Bari: Laterza.

de Martino, Ernesto. 1977. *La fine del mondo. Contributo all'analisi delle apocalissi culturali*, edited by Clara Gallini. Torino: Einaudi.

Leroi-Gourhan, André. 1964. *Le geste et la parole. Technique et langage*. Paris: Albin Michel.

_____ 1964. *Les religions de la préhistoire (Paléolithique)*. Paris: Presses Universitaires de France.

Pasquinelli, Carla. 1984. Trascendenza ed ethos del lavoro. Note su"La fine del mondo"di Ernesto de Martino. *La ricerca folklorica* 9: 29–36.

PART I

Classification

Marta Maddalon

Recognition and Classification of Natural Kinds

General Remarks

The "semantic fields"[1] usually and most frequently invoked in research and theoretical speculation on categorization/classification belong to: (a) the natural world
(plants and animals), (b) color terms, (c) kinship systems. This particular choice of
fields may well depend on apriori analytical approaches that automatically privilege highly structured and hierarchized domains; the domains chosen will thus
highlight intrinsic organization, clearly structured levels and the internal hierarchy
evidenced between levels. Another reason for this choice may well be that these
three domains, because they are central cultural, symbolic and practical issues in
the everyday life of human communities, are sensitive to cultural relativism. At the
same time, their everyday importance also makes them obvious candidates for
testing both linguistic and anthropological universals.

Such domains consequently became the testing ground not only for a host of
theoretical studies on meaning but also for the anthropological theory and practice which sprang from first enthusiasms on the possible applications of structuralist theory to culture and social behavior patterns. We refer of course to an
approach originally based on the famous analogy drawn between language and
culture (Sapir 1921). Both are organized in patterns that tend to form a unique
configuration. The originally linguistically oriented analogy is the basis of much of
Claude Lévi-Strauss' work (1947, 1951, 1953 etc.); it characterizes throughout what
is commonly called the New Ethnography School, whose initial steps are heavily
influenced by Componential Analysis and problems posed by distinctive features
and related topics. Slightly later adjuncts to the theory involve the simultaneous
presence of different levels of analysis, as well as the presence of covert categories
or structures, i.e., non-lexicalized ones, which have some sort of debt to Generative Semantics. It is within such an overall panorama that the particular semantic
fields referred to are focused, used and, we might even say abused.

These first considerations lead to reflections on a series of problems that need
to be tackled.

The Problem of the Ethnolinguistic Classification of Nature

The type of domains chosen answered classical objections that the lexicon of natural languages was relatively unstructured. This search for such domains showing

a highly structured linguistic organization, linked to a parallel search for universals, played a historically important role in the evolution of semantic theory, though now outdated. It played a similar role in ethnoanthropology, where the possibility of applying structuralist principles to cultural organization, whether myth, religion or symbol, is still at the heart of the most recent debates on the subject.

Instrumentalizing these three "fields" does not necessarily mean that the "direction" of research moves from usually limited or partial studies, based on specific domains, institutions or beliefs, towards universal implications. This would make them mere litmus paper for testing the goodness of theories. More recent research shows, however, that the careful study of natural knowledge triggers off meaningful reflection with a spin-off on the current debate in semantic theory and is linked to the study of cognitive processes.

We might fruitfully discuss whether it is useful for the study and analysis of folk classifications be confined within its self-imposed limits, which consisted in confirming or denying—usually confirming, however—extremely general principles of categorization and classification. At the same time, starting from possible principles that make up the basis of folk classification opens up a series of useful questions: what is the relationship between scientific representation and folk culture, what is that between natural phenomenon and artifact, that between various analytical levels etc. They cannot be overlooked.

Differences in Category Classification

It has already been said that the above three "fields" can be represented as well-structured sets.[2] What is of interest, however, is their belonging to different universes of discourse. Plants and animals, like colors, are primarily physical, have an undeniably symbolic role, their use is bound to the "scientific" knowledge of a particular period or particular cultural surrounds. On the other hand, plants and animals are essential to survival, whereas the knowledge of chromatic distinctions is not, though research on color perception can tell us something meaningful on how human perception functions in terms of physiological differences which, unlike cultural ones, cannot be relativized. Kinship terms differ in that they are intrinsically institutional: their only physical constituent is the act of procreation. One might also consider that the above domains are differently constituted vis-à-vis elements and artifacts; it might not be legitimate to consider the latter on a par with natural elements.

The Case of Folk Classification

Folk taxonomy is generally opposed to scientific classification. This oversimplification is unconsciously accepted more often than not in both linguistic and anthropological studies. Most discussions of the topic usually involve cultural aspects of the question. The whole debate, from classical Relativism up to the recent *prises de position* taken by the movement known as "New Ethnography," opposes the existence of crosscultural classificatory principles to the practical and theoretical unreality of such generalizations, which seem too dependent on the observer's theoretical position.

This is less understood at the scientific end of the debate because the tendency is for even the most painstaking of scholars to postulate as a *condicio sine qua non* the singleness of scientific representations or paradigms, underestimating how the origin and development of the history of science has epistemologically and practically conditioned the approach to knowledge, as also how a particular theoretical apparatus has in time prevailed over others.[3]

A propos particular semantic fields, keeping to ethnobiological ones as our main topic, we have already mentioned those factors that make them appear to be "good examples" and shown that they are an essentially good starting point for a discussion on the theoretical and practical problems posed by folk taxonomies' classification of "reality." For example, we have not really come to terms- though most realize it is a kernel problem- with the collecting, sifting and interpretation of data on folk taxonomies, often underestimated because the concrete referent, whether plant, animal, rock, watercourse or whatever, *seems* to shield us from wrong interpretations of data. This is because seemingly obvious set relations, inclusion/exclusion relationships etc., often volunteered by speakers, appear to point to utilitarian criteria, therefore apparently easier to interpret. We usually think such a naïve vision belongs to our scientific past, although, unfortunately, it is still with us and underpins much ethnoanthropological field work, a large chunk of even modern dialect investigation as well as traditional lexicography, often the only available and uncritically used source materials for many who write about folk taxonomies!

At a still more meaningful theoretical level, folk taxonomy forces us to reflect on classification as a primary human activity: this brings us to theoretical considerations on universals. A corollary of this brings us to the question of whether this activity, though primary, is necessarily simple. Secondly, if this activity is concerned with "ordering" the extralinguistic real world by linguistic means, is only the denotative aspect involved?

These questions do not, obviously, exhaust the topic, as is demonstrably well known. As a first point I propose limiting the initial discussion to the consideration that to interpret classificatory models, even from a linguistic standpoint, we must take into account the "interplay of similitudes" between the various spheres of the natural world, as also the complexity involved in projecting domains or parts of domains on to others. It is difficult to restrict the scope of taxonomic processes to that of more or less structured systems with suitable linguistic labels attached to most of the nodes that have been individuated.

The Different Levels of Folk Classification

We have already mentioned the analogy between language and culture and its influence; let us now focus our attention more on anthropology than on linguistics, taking as starting point the Relativism debate, passing through its various stages and reaching very differentiated positions, which have given rise to radically different theoretical choices.[4] I feel it useful in this case to separate out criticisms leveled at the inadequacy of models used in ethnobiological classification, the subject of much debate over the last few years, from more radical criticisms on the theoretical validity of the taxonomic approach in itself. This is, in turn, a very different question from that of raising legitimate doubts on the by-products of a particular analysis, which often give the appearance of lacking taxonomic depth. In such a case it is not so much analytical presuppositions that have to be questioned but the goodness of the data and interpretation.

Avoiding any such position that would claim that some people's formal models used to represent folk taxonomies, whether trees, Venn diagrams or other, have "imprisoned" and "bridled" analysis, I would like to underscore some precise problems for debate. First, it must be kept in mind that biotaxonomies were the starting point for most of the recent debate on basic categories; they were key points in a discussion of root problems such as (a) physical and psychological perception, (b) categories as entities with real existence in the external world, whose existence has been biologically and unequivocally proved, (c) acquisition as a new mainline theme. According to the cognitive approach human beings *do*, in fact, have an innate categorizing capacity and what they *develop* are *particular* conceptual systems dependent on the *particular* environment in which they live.

The General Problem of Categorization
Biotaxonomies as a Springboard for Discussion

Any future debate on semantics is bound to meaningful reflection on the nature that categories as such must possess. Taking as starting point the theory of natural kind terms, which states that the world is made up, on one hand, of sets of natural kinds, on the other, of languages that must possess lexicon to be associated with such kinds, and testing such a postulate by means of natural domains such as plants and animals, is a very decisive and unequivocal stand to take.

We have already mentioned radical criticisms that deny any validity whatsoever to the folk taxonomy approach. We might more profitably discuss "partial" criticisms such as Scott Atran's 1990, inter al., study which is the product of a great deal of research on biotaxonomies, as well as on the history of scientific classification as a topic belonging the history of science as such. Based on the postulate that there are undeniable regularities that go beyond single cultures in the very way in which humans classify the external world, one brings doubt to bear on the possible extension of such classificatory criteria from the cognitive point of view, (Atran 1990: 47): "[...] whether there are domain-specific cognitive universals that account for the peculiar kinds of regularities, apparent in folk systems of knowledge and belief world-over, or whether those regularities are the product of general processing mechanisms that cross such domains as living kind, artifacts and substances."

Here we find, in a nutshell, one of the most relevant points in a critique on categories, which has meaningful repercussions on classification. The postulate that not all concepts are theoretically equal, i.e., that there are universal domain-specific cognitions, is strengthened by its corollary that basic knowledge is domain-specific. The proof is furnished by research on the cultural transmission of folk-knowledge, some of it little influenced by social change—and natural categories are included here—some of it heavily conditioned, in the transmission chain, by specific social institutions. Such criticism inevitably goes to the root of the problem posed by classificatory processes, especially when determining "necessary features" in order to identify a plant or animal as itself.[5] Too general a debate on such all-pervading themes is perhaps outside the limits of our discussion, we must perforce limit discussion to topics that can hopefully be dealt with here in a more or less exhaustive fashion. Accordingly, I propose for debate the postulate whether "hierarchical ranking of living kinds is apparently unique to that domain" (Atran 1990: 57). In other words, whether it is useful or not to insist on universals of taxonomic classification, given empirically observed variability, even large-scale variability, in the study of particular, concrete situations.

Schematically, one might also say that taking as starting point entities in nature to test whether mental categories, as manifested in language, tell us something about real world categories, and how much, has implied postulating the existence of a classificatory level that seems to play a multifaceted central role, i.e., the Generic as Basic Level. First of all we propose bypassing problems involving how the debate about a basic level has evolved, and how far it has evolved, with all the implications, i.e., cognitive, linguistic, psychological, cultural etc., because insisting on this aspect of the problem will bring us back to a discussion of the difference between natural categories and artifacts that we have already raised. It would also mean bringing up the problem of the acquisition of basic levels or categories, given the amount of research that exists on the evolution of classificatory systems in children and category impairment in pathological cases. This, also, cannot be dealt with here in any exhaustive fashion.

Often quite opposite positions on the theme arise from a confusion of levels. This, however, is not always the case: it is certainly not in work that gets to grips with epistemological problems linked to the "nature" of categories: in other words, research that has as its object the repercussions of category definitions on accounts of the human mind, on accounts of the existence and validity of cognitive processes, right up to the discussion of Idealized Cognitive Models (ICMs), as also research that gets to grips with the problem of universals and the relationship between the capacity to conceptualize and that of learning conceptual systems, capacities that may often be quite different.[6]

*

Sticking to the problem of the Generic Level's existence, i.e., its reality at the psychological, cognitive, linguistic and taxonomic levels, we might make the following observations:

(a) postulating the presence of similarities in classificatory behavior that go beyond single cultures—that is, within ethnobiology—is a hypothesis that involves more the conceptualizing capacity level than the development of particular classificatory systems. What varies in the second case, and in a striking way, is the result, i.e. the creation of systems whose diversity, at certain levels of analysis, is a function of differences in their ecological and cultural niches.[7] That this occurs in phonological systems has been recognized for a long time in phonology and phonetics. For example, as Ferrari Disner (1983) showed quite convincingly, the apparently identical seven-vowel systems of Italian and Yoruba have a quite different acoustic and perceptual symmetry. Even in genetically related Romance dialects, seven-vowel basic systems show a typologically different use of perceptual space, as in the comparative study of acoustic data for Neapolitan, Paduan, Tuscan and North Calabrian (Trumper, Romito and Maddalon 1991). The operation of a different type of symmetry among apparently identical structures and contrast patterns is not often recognized in folk taxonomy studies. This forcefully contradicts a theoretical vision based on the primary role of utilitarian considerations in natural category distinctions, a point of view now largely and thankfully abandoned. On the other hand, I have the impression that putting the stress on differences in classificatory strategies, *at the lowest level,* often largely owing to the specific linguistic possibilities offered by a particular code, just to remain at the linguistic level, means underestimating one of the most common and powerful strategies in linguistic differentiation, at any level, that consists in maximizing the distinctive contrasts between sets. So perhaps it would be better to look for such contrast strategies rather than for the same type of expression in apparently the same semantic slot in every language.

(b) On the contrary, pointing out the limits of certain postulates in folk taxonomy and submitting them to debate does not mean complete rejection of a given model's axioms, especially those regarding the Basic Level. Most folk knowledge is structured at this level, whether it is a question of category learning or from the point of view of taxonomic depth. It may mean, however, questioning an aspect that is the object of current debate, starting with concrete examples of folk taxonomy. The possible existence of a further level whose nature seems more problematic than what is usually called the "intermediate," at least as originally defined,[8] is widely discussed in Atran 1990: 41 et seq. (Family Fragments), and by many others. It seems that in this "portion" of taxonomic space we more frequently find so-called residual or covert categories and, on the other hand, we meet the same lexeme used at different levels in a folk-taxonomy (with prototypical status?), referring as well to crosscutting categories, on the basis of discontinuities that are not strictly biological (Berlin,1992: 141 et seq.). The point is extremely relevant. It cannot be dismissed too lightly as a question involving only the model's formal apparatus. A European plant example of the case referred to would be *carduus* ('thistle" or some such equivalent) and its derivates in all Romance languages and dialects. As a further example of a problem internal to the model I would emphasize that of the nature of the Life Form level, both for its late insertion into the tax-

onomic schema and its rather complex role in the formation of biological categories, an ambiguity evidenced from the ancient world on and a topic I would like
to go back to.

Bringing to the fore the problem involved in a discussion of classificatory capacity development vis-à-vis, say, learning processes, which we have already
referred,[9] I think it useful to highlight research on Gestalt and more generally on
the relationship between wholes and parts in categorization. Experimental proof
brought to the debate bears on the fact that, inter alia, the Basic Level is largely
organized on such principles: this may be interpreted as psychological confirmation of a fundamentally "morphological" hypothesis that underpins the reasoning
and uses of Linnean taxonomy when it refers to differences that are easily seen
without technical aids. However, one might also rebut that recognition, in psychological terms- a general consideration that springs from perception studies (B.
Tversky- K. Hemenway 1984, B. Tversky 1986, etc.)-, does not exclude the possibility that differing systems may develop and even co-exist in the speaker's cultural experience. This derives from the rejection of a purely objectivist approach in
favor of one in which the relationship between elements in nature and their categorization according to the knowledge humans have of the real world is central to
the model.

Metonymy and Metaphor

In the light of both the complex theoretical apparatus that surrounds the prototype
concept, including recent modifications to Rosch problematically presented in
Lakoff 1987, and the enormous quantity of studies, not only from a linguistic point
of view, on the prototypical effects produced by metonyms and metaphors[10] within
a general approach to cognitive processes, the two categories might well be discussed in a perspective that both linguistically and culturally includes contempo-
raneously a recognition level (denotative) and an encyclopedic knowledge level,
i.e., a knowledge of all relevant practical levels to which determined biological categories are related.

We might first consider the following diachronic implications. A meaningful
role is played by similitudes in the criteria used in categorization and classification
in the ancient world. What relationship exists between the network of correspon-
dences that springs from viewing plants, animals and humans as part of a natural
continuum vis-à-vis the beginnings of a theory of discrete distinctions which are
central to classificatory model?

Using the linguistic level, which is readily analyzable, the "hypothetical
diachronic sequences" discussed in Berlin (1992: 274 sq., inter al.), might be seen
as a consequence of Metonymic Models. The creation of higher levels such as Life
Form or Kingdom from Generics might be interpreted as an example of the passage of the central member of a precise cultural configuration to designate higher
forms or vice versa. Obvious Indo-European cases are generic trees (oak, fir etc.)
and the general "tree" concept, "salmon" and "trout" vs. "fish" (cf. Anglo-Saxon leax,

German Lachs with Tokharian, Old Irish fír iasc "true fish" = "salmon," etc), interchangeable "water," ""sea, ""lake" and so on.

Keeping to the diachronic level, one might debate both the conceptual and quantitative relevance of crystallized metaphors. Secondly, some metaphors have lost their original status, becoming hazy because speakers no longer know the cultural and linguistic referents that gave them their original value,[11] with two possible consequences: (1) lexemes are accepted with their sheer denotative value, owing to the incomprehension of processes that led to the metaphor, (2) lexemes exist only as a similitude or metaphor no longer with a clearly understood referent.

The important point seems not so much to emphasize the complete substitution of category theory based on classical models, as crystallized in many epistemological approaches,[12] but to prop it in a critical manner with many of the points already brought up.

An example might be the capacity to distinguish between biological entities, with their linguistic correlates, on the basis of, say, Linnean morphological elements, recognizing biological discontinuity, does not preclude or exclude the coexistence of another sort of recognition grounded on common properties, uses or practices, etc. We can take the Linnean scientific taxonomy and compare it first with the late imperial Latin folk taxonomy Pliny's Naturalis Historia and other less technical authors commented in André (1967), and Capponi (1979), then with the present-day folk taxonomy in NeoVenetian Central dialects (PD–VI).

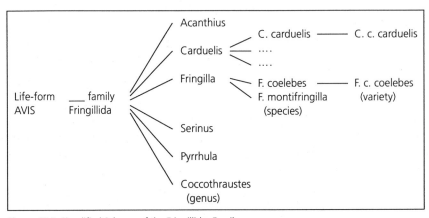

Figure M.1 Simplified Schema of the Fringillidæ Family

avis

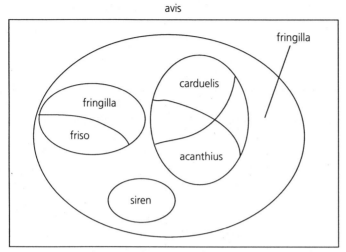

Figure M.2 Simplified Schema of Fringillidæ in Late Spoken Latin

oselo

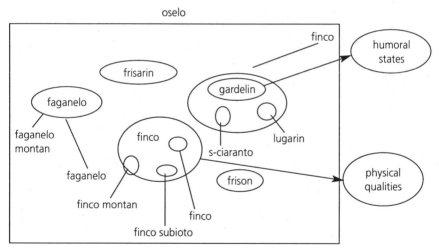

Figure M.3 Folk-taxonomic Representation of Finco (Finch, Chaffinch, Goldfinch etc.) in Veneto Dialect, with its Metaphorical Projections on to Other Cognitive Domains

Legenda:

Finco= *Fringilla coelebs coelebs* L.; Finco Montan= *Fringilla montifrangilla* L.; Finco subiòto=*Pyrrhula pyrrhula europaea* Viell.; Gardelìn= *Cardueslis carduelis carduelis* L.; Lugarìn= *Carduelis spinus* L.; S-ciaranto= *Chloris chloris mühlei* Parr.; Faganèlo= *Carduelis cannabina cannabina* L.; Faganèlo de montagna= *Carduelis linaria linaria* L.; Frisón= *Coccothrauster Coccothrauster* L.; Frisarìn= *Serinus canarius serinus* L. (The indications follows E. Arrigoni Degli Oddi (1929) on Italian birds.)

In the first instance schema 3, which should be derived (in a linguistically genetic sense) from schema 2, presents problems, not brought to the fore in the first two schemes, regarding the level at which to insert some of the taxa, e.g. gardelín > gardelín + lugarín + s-ciaranto, with relexicalization, vs. finco > finco + finco montán + finco subiòto, with modification. A further problem is created by the opposition finco subiòto ~ finco, finco montán, internal to the genus FINCO at the ethnobiological level, which corresponds to an opposition between biological genera (*Pyrrhula* vs. *Fringilla*). This kind of hypodifferentiation highlights two aspects: (1) it gives weight to Berlin's (1992) criticism that ornithologists/zoologists hyper-differentiate on the basis of few, not clearly visible morphological points with respect to botany or ethnotaxonomy (culture-specific); (2) it underscores a finch prototype, i.e., the inclusion within the category of various ornithological genera or species indicates the importance of such a prototype. With regard to"practice"or "use"it might be said that venatory practice and songbird breeding (gardelín subsumes the 3 European songbirds of the family) determine the centrality in folk classification of FINCO and GARDELIN as opposed to biologically based classification.

Besides such considerations and not as alternatives to them, we have lexical fossilization originating from no longer understood practices from which metaphors have been created, with complete loss of the original denotatum. Examples from NeoVenetian dialect begin as similitudes, e.g. el canta come un gardelín "he's singing/chirping happily away," el xe òrbo come un finco "he's as blind as a bat/a mole" [="finch"] The connection between"finches"and"bad sight" lies in venatory practices, where finches were first blinded and then left on perches to attract other birds. *Carduelis carduelis* and cheerfulness are connected because, before the advent of canaries, they were typically caged as songbirds. The metaphor consists in the passage: el canta come un gardelìn "he sings like a goldfinch" > el par un gardelìn "he seems a goldfinch" > el xé un gardelìn "he is a goldfinch."The same path is taken in el xe un finco "he's blind," as well as in CIUÍN *Phylloscopus sp.*, in the similitude: el xe magro come un ciuín "he's as thin as a rake" [="wood-wren"], becams a metaphor: el xe un ciuín "he's very thin."Town people who use the metaphor usually have no idea what the referents are.

This change in perspective goes beyond the positivist (and neopositivist) postulate whereby what you"believe about nature" is in a relatively simple manner connected to what you"know about nature,"and, we might add, to what you"call something in nature."Though frowned on in theory, this point of view seems to spring out, perhaps unconsciously, in much modern fieldwork.

Possible prototypical effects on classificatory systems, as part of a theoretical model of human categorization of nature, as well as metaphorical and metonymic creations and their consequences on internal/external taxonomic configurations, must be taken into account. One of the desirable results of such an approach would be to include part of lexico-semantic variability in the model, in order to guarantee analytical depth, including both denotative (taxonomic) and connotative (experiential, cultural) levels.

Proposals for Discussion

Linguistic Models

The weight attached to the linguistic level in models such as those proposed by Berlin et al. is significant, notwithstanding *some* criticism, on the more practical side, leveled at their conclusions, especially from the standpoint of lexical formation. It seems to be even more important to insert the linguistic *amongst* the other levels considered, which is different from setting it *before* or *above* these others. If one maintains that language belongs to cognition as such and uses general cognitive mechanisms, then a typical way of looking at language understood not so much as a guideline to world interpretation (both actively and passively, i.e.,"perceived">"classified,'"'perceived">"interpreted") but as a precondition both to the formation of a Weltanschauung and later to its interpretation, ceases implicitly to have any sense. Proposals for a first discussion might be along the following lines. Can recourse to the linguistic level be exploited to plan "depth" in natural world representations?

This latter would have the dual purpose of demonstrating the validity of the basic level as the starting point for forming projections—linguistic development would confirm or deny this—either in a more abstract sense (higher taxonomic levels) or at lower taxonomic levels, or, on the other hand, that of testing the validity of prototypes and metonymic models. In the first case, if such diachronic linguistic processes were shown to be general, then they would constitute proof of a universal level in the perception of nature that favors a form of classification in which recognition occurs more naturally and more frequently at the level of certain morphological, easily visible discontinuities based on dimensions etc., which may then be projected in a *number* of directions. A paradox is observed in the development of lower taxonomic ranks (varietal, cultivars), as an effect of a fairly recent modification of natural category knowledge in modern postagricultural societies, quite frequent in every day life, to the detriment of higher ranks (Generic), which often only remain in a residual metaphorical use.

Secondarily, the debate on divisions,[13] in the first place between the animal and plant kingdoms, to keep to the most obvious, but also between genera and species in a particular classification, up to the more "refined" models of modern scientific taxonomy, literally recapitulates the history of science, even -to be more precise-the entire history of culture in an anthropological and not an institutional sense.

Careful reflection on classificatory models and, once again, on their linguistic correlates, especially when compared in their oldest formulations (at least from the Third century BC on in Europe), are an unavoidable guide to the complex labyrinth that separates holistic folk world views from those supposedly more objective views of so-called "specialists," in the positivist sense. Discovering, or rather rediscovering, how much so-called "scientific" classifications owe older "ethnoscientific" representations is a way of introducing another dimension into our reflections on categories. This is obvious from the strictly linguistic standpoint, even if not always in a direct or simple manner. The influence of scientific terminology in our Western tradition is heavily connected to the Greco-Latin tradition of which, in part, it is the expression. Even supposedly original folk names are often relatable -not always in a simple and direct manner- with folk and specialist nature knowledge that the Ancient World and later the Middle Ages, with its own slant, had filtered and spread around. Folk knowledge, philosophical speculation and "specialist" knowledge went hand in hand in Europe up to at least the seventeenth century. Of course, this is an aspect of the problem related, in a broad fashion, to the limits of Western culture. However, these are the necessary premises to a debate on the complicated relations between types of "natural knowledge" and consequently classificatory models that only apparently belong to different, rigidly separated spheres, but are evidently connected both diachronically and synchronically. If and how this can be extended to other cultures is an open topic for debate.

Numerous other reflections, even of a more methodological nature, can be linked in an obvious manner with the more general problem of assigning natural objects to categories. With regard to the possibility of determining category limits, Ellen notes that such a task is made easier "by imposing culturally agreed boundaries, or indeed by creating these inadvertently or deliberately through genetic and other physical manipulations of the natural world." The initial part of this statement may especially be understood as referring to cultural limits superimposed in the actual reading of data. Obviously I do not mean, at the linguistic level, the brutal and naïve search for dialect items that correspond to biological categories, listed alphabetically, nor, at the anthropological level, the acceptance of names to be associated with a preconstituted taxonomy, normalizing any eventual incongruences. I refer above all to the intrinsic variability of extremely discordant answers given by informants belonging to the same speech community, which obviously cannot be explained in terms of cultural differences, or from differing informants with specialized, "technical" knowledge or I.K., that can be linked to sex or occupational or age differences and so on, the fruit, I feel, of "guiding" answers beyond legitimate limits in a way that is difficult to grasp. The apriori postulation of universal or so-called natural categories on the part of the researcher, or even admitting the simple possibility of creating classes of elements which the native speaker does not necessarily associate, implies, as far as I can see, that one obtains answers that are often quite unrealistic, in terms of folk organization and classification, precisely because, by virtue of a phenomenon that we might call overcollaboration, speakers may force their classificatory possibilities beyond and even in a contrary sense to the gamut of variability that is intrinsic and normal and even culturally permitted and perhaps encouraged.

Metonymic Models

Some criticisms of the application of the prototype concept to the analysis of natural categories are based on the uncontrovertible fact that amongst the components defining a category there are certain distinctions that are essentially biological and thus cannot directly depend on individuals'idiosyncratic physical or cultural perception. Apart from the fact that prototype theory does not aprioristically exclude the existence of clear-cut distinctions between categories, the main point is not whether biological discontinuities exist or not, but how they are perceived and categorized in folk taxonomies. More interesting for discussion are some of the axioms developed in more recent debate and the effects they have on our approach to the general problem of categorization.

The first criticism is that neither classification as such nor its explicit linguistic expression are purely denotative at the distinctive level and remain so until appropriate symbolic usage later intervenes. Structural anthropology assumes, instead, that symbolic activity as a thought process is activated by violating norms, by an act of transgression. It implies, in other words, the prior existence of a code. On this basis Tornay (1966), for example, proposed initiating the study of symbolism and its processes from well-structured fields like those of the taxonomic classifications of the physical world's "kingdoms." The present proposal for discussion starts from the contrary assumption that denotation and connotation develop and proceed linguistically at the same time and at the same rate, just as a strictly"distinctive"view of things at the experiential level lives and interacts with our symbolic world.

A second criticism is that metonymic processes may well be the basis for further in-depth developments of biological taxonomies. Inversely, reconsidering the role and weight, often quantitatively assessable, that complex networks of metaphorical projection have in lexicalization processes, at the ethnobiological level, serves more often than not as a basic guide for following through underlying processes that produce classification systems, processes simultaneously bound to knowledge and beliefs about nature.

Although naming problems in the strict sense will be dealt with in part 2 of these Proceedings, I would like to consider the importance of similitude in classificatory processes that create"cognitive models"in which plants, animals, the earth itself, do have veins, sinews, etc., just to make a very simple and common example. On the other hand, the merger of prototypical exemplars and the importance of metonymic and metaphorical effects on category formation and classificatory depth cannot be ignored.

Notes

1. One might object that the term is"dated"; however, its use avoids committing ourselves, in what are fundamentally the premises to a debate, to an in-depth discussion on linguistic and cognitive categories that will become, later on, the central point of our discussion.

2. It is useless to insist on the relativity of this supposition here, since it will become obvious in the debate.

3. The opposite has also to be considered, viz. how the cultural climate of a particular epoch conditions the formation and the development of scientific representations and paradigms.

4. See the very concrete discussion dedicated to the topic in Berlin 1992 (passim) where this counterposition is central to the debate. It is not randomly that Berlin makes cites Tyle's (1987) statement (p.

172) "I speak of the world by means of language," followed by the very strong claim that "The world is what we say it is," in other words, a claim that springs from the assumption, expressed in its most extreme form, that "species" etc. are a product of human imagination, pure creations of the mind.

5. Cf. Sperber 1975: 22 "If an animal does not actually possess a feature ascribed to it by its definition, then it possesses it virtually: not in its appearance but in its nature. In such conditions it would be hard for empirical evidence to contradict the definitions of folk taxonomies."

6. Cognitive Science provides the type of answer that Lakoff (1987) suggests to questions as to why human beings have conceptualizing systems, what their extension might be, how they are acquired or later modified by children etc. "The idea that people are born with a conceptualizing capacity seems to provide answers for all these questions." (p. 335).

7. This involves primary and distinct uses of plants and animals, an important aspect of the problem at hand, though not the main one.

8. Cf. Berlin's (1992: 16) comment on the categories in Berlin, Breedlove and Raven (1973): "The rank dubbed 'intermediate' was proposed cautiously and tentatively, its validity to be determined only by future research." Cf. also p. 17: "Intermediate taxa as members of the category 'intermediate' usually include taxa of generic rank, are rare in folk taxonomy, and are seldom named, leading Berlin, Breedlove and Raven to refer to them as "covert categories" (1968)." Further research has in fact partly altered original axioms.

9. Since Cognitive Psychology and new developments in Neurology constitute a sort of New Frontier in which many place their trust with the complete faith one usually associates with movements that promise to unveil at long last all the mind's secrets, discussion of this topic will necessarily be limited and hedged.

10. This is very fashionable at the moment, though much confusion seems to reign between what is true metaphor and what are usually considered normal elements of a Metonymic Model: a rediscussion would perhaps be opportune.

11. In the European tradition, e.g., the crosscultural passage from Greek to Latin to many European languages from the Ancient World to the early Middle Ages, intensified later in the Renaissance and in the eighteenth century, also has very drastic linguistic repercussions, especially, say, in Romance languages, but not only in these.

12. "Crystallized" in the sense that Aristotelian postulates about categories and the millenary discussion based on them are often "quoted" without reference to the problems they posed, even to the historical ones raised by Theophrastus during Aristotle's life and Aristotle's "corrections" based on discussion with Theophrastus and his other "students" (see, e.g., R. French, 1994). What is generally used from classical theory is usually the antithesis between "necessary" and "characteristic" features, between "qualities" and "quantities," etc.

13. We must be careful here: a similar expression may well be ambiguous if we only mean the search, in historical terms, for those sole criteria that are used to separate class from class, individual from individual.

References

Albert-Llorca, M. 1991. *L'ordre des choses. Le récits d'origine des animaux et des plantes en Europe*. Paris Editions du Comité des Travaux Historiques et Scientifiques.

Andrè, J. 1967. *Les noms d'oiseaux en latin*. Paris: Klincksieck.

Atran, S. 1990. *Cognitive foundation of natural history*. Cambridge: Cambridge University Press.

Barsanti, G. 1992. *La scala, la mappa, l' albero. immagini e classificazioni della natura fra Sei e Ottocento*. Firenze: Sansoni.

Belardi, W. 1996. *Tassinomie, Taxonomie e Metafore, Opuscola III* 3. Roma: Il Calamo: 209–27.

Berlin, B., D.E. Bredlove and P. H. Raven. 1968. Covert categories and Folk Taxonomies. *American Anthropology* 70. 290–99.

_____ Bredlove and P. H. Raven. 1973. General Principles of Classification and Nomenclature in Folk Biology, *American Anthropology* 75: 214–42

Berlin, B. 1992. *Ethnobiological classification*. Princeton: Princeton University Press.

_____ 1997. Tapir and Squirrel. Further Nomenclature Meanderings Toward a Universal Sound-Symbolic Bestiary. Paper presented at the Intenational *Conference on Nature Knowledge*, 4–6 December at the Itstituto Veneto di Scienze Lettere ed Arti, Venice.

Bulmer, R. N. H. 1974. Memoirs of a small game hunter: on the track of unknown animal categories in New guinea. *J. d'Agricul. Tropicale Botan. Appliquée* 21: 79–99.

Capponi, F. 1979. *Ornitologia latina*. Genua: Università di Genova, Facoltà di Lettere.

Ellen, R. 1979a. Introductory essay. In Classificatios in their social context edited by R. F. Ellen and D. A. Reason. London: London Academic Press, 1–31.

_____ 1979b. Omniscience and ignorance: variation in Nuaulu knowledge, identification and classification of animals. *Language and Society* 8, 337–64.

_____ 1993 *The cultural relations of classification*. Cambridge: Cambridge University Press.

Ferrari Disner, S. 1983. *Vowel Quality: the relation between universal and language-specific factors*. Los Angels: UCLA WPP, 58.

Foucault, M. 1966. *Les Mots et les Chose*. Paris: Gallimard.

French, R. 1994. *Ancient Natural History*. London and New York: Routledge.

Ingold, T. 1997. Two reflections on ecological knowledge, Paper presented at the Intenational Conference on *Nature Knowledge* 4–6 December at the Itstituto Veneto di Scienze Lettere ed Arti, Venice.

Kay, P. 1979. The Role of Cognitive Schemata in Word Meaning: Hedges Revisited. Department of Linguistic, University of California: Bercheley.

Keil, F. 1986. The acquisition of natural kind and artifact terms. In *Conceptual Chang*: ed. A. Marrar and W. Demopoulos: Norwodd N.J. Ablex Publishing Corp.

La Vergata, A. 1979. *L'evoluzione biologica da Linneo a Darwin*. Torino: Loescher.

Lakoff, G. 1987. *Women fire, and dangerous things*. Chicago and London: The University of Chicago Press.

Lévi-Bruhl, L. 1960. *La mentalité primitive*. Paris: Presses Universitaires de France.

Lévi-Strauss, Claude. 1958. *Anthropologie structurale*. Paris: Plon.

Liljencrants, J. and B. Lindblom. 1972. Numerical similation of vowel quality systems: the role of perceptual contrast. *Language* 48: 839–62.

Linné, C. 1972. *L'équilibre de la nature*, franc. trans. B. Jasmin. Paris: Vrin (originally published in Latin in 1744).

Lovejoy, A. 1964. *The great chain of being*, Cambridge Mass., Harvard University Press, 1961 (originally published in 1936).

Putnam, H. 1975. Mind, Language, and Reality, Cambridge Eng. New York: Cambridge University Press.

Sapir, E. 1921. *Language*. New York: Harcourt & Brace.

_____ 1929. A study in phonetic symbolism. *Journal of Experimental Psycology*, XII: 225–39.

Sperber, D. 1975. *Rethinking symbolism*. Cambridge, Cambridge University Press.

Tornay, S. 1980. *Percezione dei colori e pensiero simbolico. La ricerca folklorica* n. 4, 87–98.

Trumper, John L. Romito and Marta Maddalon. 1991. *Vowel systems and areas compared: definitional problems*. Conference on "*L'interfaccia tra Fonologia e Fonetica*" E. Magno Caldognetto e Paola Benincà eds. Padova: Unipress, 43–72.

Trumper, John and M.T. Vigolo. 1995. *Il Veneto Centrale. Problemi di classificazione dialettale e di fitonimia*. Padova: Edom.

Tversky, B. and K. Hemenway. 1983. Categories of environmental scenes. *Cognitive Psychology*, 15, 121–49.

Wittgenstein, L. 1975. *Osservazioni sopra il "Ramo d'oro" di Frazer*. Milano: Adelphi.

CHAPTER 1

How a Folk Botanical System can be both Natural and Comprehensive: One Maya Indian's View of the Plant World[1]

Brent Berlin

One of the first universal principles of ethnobiological classification states that no folk system is comprehensive in that taxa at its lower ranks, say folk genera and folk species, partition the full range of plant diversity for a local habitat. After the most prominent botanical species[2] are named in some particular system, there will always remain hundreds of others that are not given linguistic recognition. A second general principle claims that the conceptual ordering of those taxa that *are* recognized will be based on the affinities that humans observe among the species themselves, independently of these species' actual or potential cultural importance (see Berlin 1992). The application of this principle implies that when the outlines of an ethnobiological system of classification are known, they will reflect in great part the order inherent *in nature*, not an order arbitrarily laid *on nature* by humans' economic or symbolic concerns.

In this chapter, I present data on the folk botanical system of classification of Alonso Méndez Ton, a Tzeltal Maya Indian from the Highlands of Chiapas, Mexico. Analysis of the data supports the second principle, naturalness, but requires qualification of the strict application of the first principle, comprehensiveness. The data show that *comprehensiveness* is achieved when species not afforded standard linguistic recognition are nonetheless stated to be *similar to* (or *related to*) some set of folk taxa in the system that are themselves named. The data show that naturalness is achieved when it is observed that those plant species said to be "similar to" named species are also found to share readily observable morphological characteristics with those species to which they are said to be close. Often, this morphological similarity reflects taxonomic affinity, as well (i.e., perceptually related species are often members of the same genus or same family as the target species). Both findings are supported by data drawn from Alonso's classification of more than 5000 botanical collections representing some 3000 species carried out as part of a general botanical collecting program over a period of several years. It is proposed that Alonso's folk system of classification is based on psychological principles that underlie what we know today as scientific systematic botany.

Comprehensiveness

If it is generally true that only a subset of the organisms in any given region are usually extended conceptual recognition in folk systems of biological classification, one is led to ask, "Just how are species that are *not* recognized linguistically actually treated conceptually?" Most Americans, when asked for the name of an unfamiliar plant, might simply respond with causal remarks such as "I don't know, it's just a bush" or "it's some kind of tree, I don't have a name for it." But thirty years ago, when Dennis Breedlove and I began our general ethnobotanical work among the Tzeltal Maya, we soon noted that these types of "I don't know" or "it's just a (tree), (grass)… etc." responses were extremely rare among our Tzeltal collaborators. In *Principles of Tzeltal Plant Classification*, we wrote that:

> General botanical collecting quickly revealed that the Tzeltal lacked true plant names for much of the local flora. On the other hand, when presented with a particular plant [species that was unfamiliar to them] informants would rarely respond that the species had no name. Instead, they would systematically attempt to relate it to one of the categories in their named taxonomy. For example, an informant might state that such and such a plant was "*kol pahaluk sok X*,'" "it's similar to X,' where X represents the Maya name of a known, lower-level category… These classificatory responses allowed us to determine [what we called the *basic ranges* of each recognized category (i.e., all "genuine exemplars") and the *extended ranges* of these genuine categories (i.e., all the "it's similar to X" exemplars)]. (Berlin, Breedlove, and Raven 1974: 53)[3]

When our Maya collaborators made such statements as "it's like an x," or "it's similar to x," they were engaging in what might be called first-approximation botanical classification. The process is something like the following: first, a specimen of some unfamiliar species is examined and its perceptual characteristics are evaluated, a process that we can call *identification*. Once identified, the species is then conceptually assigned to a place "near to" a known species to which it is seen to be most similar, one already given linguistic recognition in the folk system, a process that we can call *classification*.[4] This kind of basic, first-order classificatory behavior, of course, is no different than that of the field botanist who makes a preliminary determination of a particular species as "near to x," as in *Smilax* aff. *bonanox*, a designation that the species will bear until a more definitive determination can be made at leisure, normally on the basis of comparison with herbarium material. If applied in a general, encompassing fashion, classifying plants in this manner allows for a folk botanical system of classification to be nearly comprehensive in that most plant species of a local habitat can either be assigned to one of the recognized lower-level folk categories of the folk system or, if not recognized as a legitimate member of a named category, nonetheless seen as "conceptually related to" one of the named lower-level folk taxa.

Knowledge of the extent to which such a system of classification actually works depends ideally on data comprised of large numbers of general botanical collections that have been comprehensively classified by numerous native collaborators. Barring large numbers of collaborators, an initial glimpse into the nature of such a system can be gained by the classification system that emerges from the remarkable botanical collections of Alonso Méndez Ton (hereafter, AMT), a Tzeltal Maya botanical collaborator who worked continually with Breedlove and me in our already mentioned ethnobotanical research in the 1960s and 1970s (see Berlin 1984 for a brief description of AMT's botanical contributions).

Shortly after beginning our ethnobotanical work among the Tzeltal, Breedlove and I quickly recognized AMT's talents as a botanical collector and in a few short months, he was in the field working alone, making his own fine botanical collections. At the completion of our project, AMT continued collecting, both as an assistant in Breedlove's *Flora of Chiapas* project and later as a brigade leader for one of the National Autonomous University of Mexico's *Flora of Mexico* collecting teams. Over a period of about five or six years, AMT had more than 5000 botanical collection numbers to his credit, (most in sheets of five), taken from numerous, widely separated localities throughout the state of Chiapas.[5]

The large majority of AMT's botanical collections were made in areas far removed from *kulak'tik*, a small hamlet of the municipality of Tenejapa where AMT grew up. Many of the species he collected, pressed, and dried were only slightly familiar to him, and it is certain that he had never seen a major portion of them. Nonetheless, only forty-four of the full set 5552 numbers are given the entry "no name" in AMT's collection books, less than one percent of the total inventory.

Linguistically, AMT's Maya plant names (*sensu laxu*) can be analyzed as one of two types: the first set is indicated by a noun accompanied by the qualifier *batz'il* (abbreviated as "*b.*") meaning "true" or "genuine," e.g., *batz'il atz'am te'* "genuine *atz'am te*" (*Rapanea myricoides*). Another set is comprised of nouns modified by the expression *kol pahaluk sok*, (abbreviated as "kps"), literally glossed as "it is somewhat the same as" < pahal "same," -uk subjunctivizing suffix "as it were," thus pahaluk x "likened to x." One can freely gloss the descriptive phrase *kps atz'am te'* as "similar to *atz'am te*'" (*Rapanea juergensenii*).

Both of these types of expressions can be seen on any a page of AMT's collection notebooks, in which each name is followed by a short Maya text. The text is a brief botanical description of the specimen (e.g., collection number 3429 reads "*b. ichil ak'* (genuine *ichil ak'*) [later determined as *Clematis grossa*, a clambering vine in the Ranunculaceae and described by AMT as] *lom nahtik yak'ulya xmo ta te,' sakik xnich* "[a] very long vine climbing in trees, with white flowers.")

The entry for the immediately following collection, 3430, is *kps ichil ak'* (similar to *ichil ak'*). AMT describes this specimen as *lom nahtik yak'ul ya xmo ta te,' tzajik sit* "[a] very long vine climbing in trees, with red fruits." This collection was ultimately determined as *Serjania hispida*, a sprawling vine in the Sapindaceae. While in different botanical families, *Clematis* and *Serjania* share many common characters of gross morphology and overall aspect. Furthermore, although there a number of species of *Serjania*, none of them receives a single, stable, distinctive "genuine name." This genus is always incorporated into AMT's classification system as "a vine similar to *ichil ak,' Clematis grossa.*

As a result of AMT's long-term collecting efforts in Chiapas, it is possible to state with some confidence the actual number of basic folk generic taxa that comprise his system of ethnobotanical classification. As we have come to expect from the ethnobotanical systems of traditional horticulturalists, AMT shows an inventory of approximately 500 folk genera (485 to be exact). He employs several hundred names and only these names—either in the form "genuine X" or "related to X"—to classify all of the botanical diversity with which he has come in contact during his collecting efforts in Chiapas, a diversity representing some 900 scientific genera and nearly 3000 valid botanical species (this number, by the way, comes close to representing the full set of species for the Chiapas central plateau). Conceptually, AMT's ethnobotanical system is comprehensive of the region's botanical diversity.[7]

Naturalness

If AMT's ethnobotanical system is close to comprehensive, is it natural? What are the patterns revealed in his conceptual organization of the considerable botanical diversity with which he is confronted? How consistent is his system of classification? Are the patterns revealed understandable in terms of a perceptually based theory of botanical classification based on over-all morphological similarity, which in the main also reflect the natural affinities of botanical species?

To fully answer these questions will require a long monograph, but the general outlines of the structure and content of AMT's view of the plant world can be seen by examining in some detail a number of typical examples from the massive ethnobotanical database created as a result of his efforts.

One of the most straightforward examples that can be discussed can be seen in AMT's conceptual treatment of the several species of *Lantana* in his collections. Table 1 indicates that AMT employs the Maya term *ch'ilwet* to refer to thirty-one botanical collections. The term ranges over a number of species of *Lantana*, as well as *Phyla stoechadifolia* and *Lippia graveolens*. AMT provided the name *batz'il ch'ilwet* to nine of the ten collections of *L. camara*, five of the six collections of *L. hispida* and two of the four collections of *L. hirta*. The remaining species in the set receive "*kol pahaluk sok ch'ilwet*" designations.

Botanical species	AMT's names		Total collections
	batz'il ch'ilwet	*kps ch'ilwet*	
Lantana camara	9	1	10
Lantana hispida	5	1	6
Lantana hirta	2	2	4
Lantana velutina		3	3
Lantana scorta		2	2
Lantana costaricensis		1	1
Lantana frutilla		1	1
Lantana trifolia		1	1
Phyla stoechadifolia		2	2
Lippia graveolens		1	1
Total	**16**	**15**	**31**

Table 1.1
Tabular Representation of AMT's Concept of *ch'ilwet*

The distribution of AMT's responses allow one to make fairly clear inferences about his concept of *ch'ilwet*. First, it will be noted that all of the plants in the set are closely related species of the same family, the Verbenaceae. *L. camara* and *L. hispida* are seen as the best exemplars or prototype of the category, most likely due to their relative frequency of occurrence in the area vis-à-vis other species of the genus. They are highly common, quite similar species with brightly-colored orange and white flowers found on exposed rocky slopes throughout much of Highland Chiapas. (Sometimes the genus is subdivided into folk species where flower color is marked by the specific epithet, e.g., tzajal *ch'ilwet* "red ch'ilwet" and *sakil ch'ilwet* "white ch'ilwet").

Several other species of the genus (*L. hirta, L. velutina, L. scorta, L. costaricensis, L. frutilla,* and *L. trifolia*) are considered to be conceptually related to the prototype. These species are less common and show fewer characters in common with the two major species. *Phyla stoechadifolia* and *Lippea graveolens* are related genera in the same family. *Lippia graveolens* is represented by but a single collection in the dataset. In addition to its designation as "kps ch'ilwet," the only other collection of *Phyla stoechadifolia* was said to be "related to" *yakan k'ulub wamal,* a folk genus with a botanical range of *Verbena litoralis,* also perceptually similar to *ch'ilwet* and in the same family. The single collections of *L. camara* and *L. hirta* identified by AMT as "like-ch'ilwet," as well as the two collections of *L. hirta* identified as "genuine ch'ilwet" are likely errors. Such misidentifications on AMT's part are rare and clearly explainable as part of dealing with normal species variation associated with botanical identifications in the field context, as any field botanist will readily admit.

A slightly more complex example can be seen in AMT's concept of the folk genus *ajoj*. *Ajoj* is the folk generic name afforded several species of the genus *Saurauia* (Actinidiaceae) (see Table 2). These plants are soft-wooded, low-branched trees and shrubs with rather large (20–30 cm) leaves covered with dense rusty-pubescent intricately branched hairs. It is one of the most common trees in the pine-oak-liquidambar forests of Chiapas.

Botanical species	batz'il ajoj	kps ajoj	Total collections
Saurauia scabrida	11		11
Saurauia comitis-rossei	3	1	4
Saurauia oreophila	2	5	7
Saurauia angustifolia	1		1
Saurauia conzattii		1	1
Saurauia hegeliana		2	1
Saurauia villosa		1	1
Kohleria elegans		3	3
Kohleria deppeana		1	1
Palicourea		1	1
Palicourea tridiocarpa		1	1
Achimenes grandiflora		1	1
Comocladia guatemalensis		1	1
Sommera grandis		1	1
Total	**17**	**19**	**36**

Table 1.2

Tabular Representation of AMT's Concept of *ajoj*

Table 2 shows that the prototypical species in this set is *S. scabrida,* represented by 11 of AMT's collections. In addition, three of the four collections of S. comitis-rossei are considered to be "genuine members" of this category, as is a single collection of S. angustifolia

Five other species of *Saurauia* are included in the extended range of the concept and treated as "similar to genuine ajoj." The most prominent of these is *S. orephila* (with five collections). In addition, two species of *Kohleria* and two species of *Palicourea* are also recognized as perceptually similar to *ajoj,* as are single collections of *Achimenes grandiflora Comocladia guatemalensis Sommera grandis.* While these genera represent different families (Gesneriaceae, Rubiaceae, and Rubiaceae), they are all generally similar to *Saurauia* in their over-all aspect.

A final example from AMT's folk system can be seen in his classification of the two conceptually related folk genera, *atz'am te'* and *tilil ja,'* two prominent small trees with a primary botanical range of *Rapanea myricoides* and *Parathesis chiapensis,* both members of the Myrsinaceae. *R. myricoides* is a delicately branched shrub with narrowly elliptic leaves , 5–9 cm long. Fruit appears in small clusters along the stems, and often appear like glistening beads of salt (hence the name *atz'am te'* a "salt tree"). *P. chiapensis* is also a small tree with somewhat broadly elliptic leaves. Both species are common and conspicuous understory shrubs throughout the region.

Table 1.3 Tabular Representation of AMT's Concepts of *atz'am te'* and *tilil ja'*

Botanical species	AMT's names				Total collections
	batz'il atz'am te'	*kps atz'am te'*	*batz'il tilil ja'*	*kps tilil ja'*	
Rapanea myricoides	13		1		14
Rapanea juergensenii		7			7
Ardisia escallonioides		3		3	6
Wimmeria acuminata		1			1
Parathesis chiapensis			11	3	14
Parathesis donnell-smithii			3	5	8
Parathesis melanosticta				3	3
Parathesis microcalyx				2	2
Parathesis leptopa				1	1
Gentlea micrantha				8	8
Ardisia compressa				7	7
Ardisia liebmannii				2	2
Ardisia alba				1	1
Ardisia karwinskyana				1	1
Dendropanax arboreus				1	1
Synardisia venosa				1	1
Total	**13**	**11**	**15**	**38**	**77**

Table 3 shows that "genuine *atz'am te'*" to be finds a clear focus on *Rapanea myricoides* (thirteen collections). Species that are considered conceptually close to *atz'am te'* include *R. juergensenii* (seven collections), *Ardisia escallonioides* (three collections) and a single collection of *Wimmeria acuminata.*

The folk taxon *tilil ja'* is centered on the closely related tree *Parathesis chiapensis* (eleven collections). This taxon's conceptual relatives include other species of *Parathesis* (namely, *P. donnell-smithii, P. melanostica , Parathesis microcalyx,* and *P. leptopa*), as well as *Gentlea micrantha* and several species of *Ardisia* (*A. compressa, A. liebmannii, A. alba,* and *A. karwinskyana,* as well as a single collection of *Dendropanax*). Both *atz'am te'* and *tilil ja'* are linked perceptually via *Ardisia escallonioides,* a species similar in overall aspect to both *Rapanea* and *Parathesis.* The basic and extended ranges of these two folk genera nearly cover the full set of all species in the Myrsinaceae as attested in AMT's collections, and only *Wimmeria* (Celastraceae) and *Dendropanax* (Araliaceae), both assigned "similar to x" designations, are in distinct botanical families.

Discussion

Each of these examples, representing species from families as diverse as the Verbenaceae, Actinidiaceae, and Myrsinaceae, are multiplied many times in AMT's collections, ultimately exhausting the several thousand species found in some of Chiapas' most varied ecological zones. In almost all of his classificatory decisions dealing with the placement of each of his 5000 collections, this Maya Indian's view of the plant world proves to be perceptually based, remarkably consistent, and, for the most part, botanically informed as to the natural affinities of the species involved.

When his classification of the flora does not conform with currently recognized phylogenetic boundaries of Western botany, it is generally the case that the organisms in question nonetheless share many perceptual features in common which justify his grouping them as members of the same conceptual category. AMT is not a trained botanist, guided by the conceptual framework found in evolutionary theory, full of notions concerning the likely phylogeny of the Chiapas flora. He lacks any basic formal education, having completed only the second grade in Mexico's notoriously bad rural schools. Breedlove's and my influence on him could hardly account for his performance, our contributions to his training being nothing more than preparing him in standard botanical collecting methods and instruction in the techniques of modern linguistic transcription.

What we see here is the systematic application of an ethnobotanical system of classification comprised of no more than 500 folk genera which conceptually organizes the biodiversity of one of the most botanically complex regions of Mexico. AMT's recognition of "genuine exemplars" of basic, folk generic botanical categories, as well as the perceptually "closely related" species seen to be conceptual affiliates of these basic categories, allows him organize his regional botanical reality in a highly comprehensive manner. Few professional botanists could excel the accuracy of his field determinations, all made on the spot without benefit of comparative herbarium materials (Breedlove has marveled at this on countless occasions). Thus, as the ethnobiological principle of naturalness is confirmed by AMT's system of ethnobotancial classification, our general principle on comprehensiveness must be modified, or at least clarified, so as to recognize the all-encompassing properties that AMT's view of the plant world so clearly reflects.

One might speculate that the folk system demonstrated by Alonso Méndez Ton is an early, though uncodified, formulation of the cognitive-linguistic ordering of nature that we know today as Western systematic biology. Perceptually distinctive species are first given linguistic recognition, the remaining plant world being largely ignored ("it's just a tree", "it's some kind of vine"). With enlarging experience, additional species are next incorporated into the older system by being recognized as similar to those with already codified names ("it's similar to x", "it's y's companion", "it's a relative of z"). Finally, with the development of classical botany, and especially with Linnaeus' nomenclatural standarizations, species said to be "similar to x" are recognized, if slightly different, as new species of established genera, or, if they are distinctive enough, become established as new genera and are given their own distinctive generic names.[7] In AMT's natural and comprehensive system of classification we see the unfolding of the first stages of this universal conceptual process.

Notes

1. I appreciate the comments of Ben Blount, David Porter, and Douglas Medin on an earlier version of this paper.

2. E. Hunn proposes several factors that relate to what counts in assessing prominence, size of organism being one of the most important (see Hunn n.d.).

3. An identical process has been noted by Breedlove and Laughlin in their description of Tzotzil ethnobotany. "When evoking botanical names, two descriptive terms dominated our conversations—*batz'i*, "genuine" or "true" and *yit'ix*, "bastard" or "false." With such a handy term [as *yit'ix*] any unknown plant can be easily [classified] (1993: 110).

4. Several competing psychological models have been proposed for this cognitive process. Medin (1989) discusses the most prominent ("overall similarity," exemplar-based," or some combination of the two. An exemplar-based strategy seems most likely to account for the behavior described here.

5. AMT's efforts during this period were devoted to making general botanical collections, that is, collections of "anything and everything in flower or fruit." This included, of course, many hundreds of species for which he lacked Maya names. His collections represent a significant contribution to our knowledge of the flora of Chiapas and, in partial recognition for his efforts, some 20 new species have been named in his honor (e.g., *Ilex tonii* Lundell, *Calypthrantes tonii* Lundell). His collections are currently housed at the California Academy of Sciences, San Francisco, and at the Herbarium of the National Autonomous University of Mexico in Mexico City.

6. My intuitive impressions concerning AMT's remarkable performance suggests something about what Medin calls "nontrivial exposure to the natural world" (personal communication), coupled with the thoughtful observation of an expert. The latter, I think, plays a minor role if my personal interaction with indigenous ethnobiologists over the last 30 years is any indication. Of course, maybe nontrivial exposure to the natural world makes one an expert by definition. And, as with everything else, sheer native intelligence is surely a major although unquantifiable factor in this whole story.

7. For a discussion of the issues involved in the linguistic processes involved in this nomenclatural elaboration, see Atran 1990, Bartlett 1940, Berlin 1972, 1986, Brown 1986, Cain 1956, 1958, 1959a, 1959b, Greene 1983 [1909], Hunn and French 1984, Walters 1986.

References

Atran, S. 1990. *Cognitive Foundations of Natural History.* Cambridge: Cambridge University Press.

Bartlett, H. H. 1940. The concept of the genus. I. History of the generic concept in botany. *Bull. Torrey Bot. Club* 67: 349–62.

Berlin, B. 1972. Speculations on the growth of ethnobotanical nomenclature. *Language and Society* 1: 51–86.

_____ 1986. Comment on "The growth of ethnobiological nomenclature by C. H. Brown." *Current Anthropology* 27: 12–13.

_____ 1992. *Ethnobiological classification: Principles of categorization of plants and animals in traditional societies.* Princeton, N. J.: Princeton University Press.

_____ , D. E. Breedlove, and P. H. Raven. 1974. *Principles of Tzeltal plant classification.* New York: Academic Press.

Brown, C. H. 1986. The growth of ethnobiological nomenclature. Current Anthropology 27: 1–18.

Breedlove, D. E. and R. M. Laughlin. 1993. The flowering of man: A Tzotzil botany of Zinacantán. Washington, DC: Smithsonian Institution Press.

Cain, A. J. 1956. The genus in evolutionary history. *Sys. Zool.* 5: 97–109.

_____ 1958. Logic and memory in Linnaeus's system of taxonomy. Proc. Linn. Soc. London Session 169: 144–63.

_____ 1959a. Taxonomic concepts. Ibis 101:302–18.

_____ 1959b. The post-Linnaean development of taxonomy. *Proc. Linn. Soc.* London Session 170: 234–44.

Greene, E. L. 1983 [1909]. *Landmarks of botanical history* 2 vols. edited by F. N. Egerton. Stanford, California: Stanford University Press.

Hunn, E. n.d. For factors governing the cultural recognition of biological taxa. Unpublished ms. University of Washington, Seattle.

_____ and D. French. 1984. Alternatives to taxonomic hierarchy: the Sahaptin case. *Journal Ethnobiology* 3: 73–92.

Medin, D. 1989. Concepts and conceptual structure. *American Psychology* 12: 1469–1481.

Walters, S. M. 1986. The name of the rose: A review of the ideas on the European bias in angiosperm classification. Tansley Review Paper 6, *New Phytologist* 104: 527–46.

CHAPTER 2

Arbitrariness and Necessity in Ethnobiological Classification: Notes on some Persisting Issues

Roy Ellen

The Role of Non-cultural Factors in the Human Configuration of the Natural World

In her paper in this collection, Marta Maddalon distinguishes three different semantic domains: color, kinship, and the natural world. All are qualitatively different in terms of the "things" classified. Colors are not really "things" at all, but rather properties of things, while kinship classes refer to the properties of the relations between things. Of the three domains, only natural kinds map directly onto real things in an objective world.

In an important sense, then, the objective "thinginess" of nature sets it apart from many other semantic domains (Ellen 1996), and what separates it from other domains that classify objects (say, cultural objects) is the degree to which we can organize it according to its plausibly conjectured evolution. Thus, classifying natural objects *a* and *b* together is more likely to indicate (though not always) natural historical affinities (common origin) than, say, a classification of furniture. I agree with Maddalon that we underestimate the difficulties of categorizing the natural world precisely because it consists of concrete entities with utilitarian referents. But to speak of the thinginess of the natural world is simply to acknowledge the universal human imperative to turn the natural world into things and to think of the things so identified in terms of their essential qualities. This is not to say that such a capacity is innate in the sense of springing into action from the first moment of postpartum development; it is simply to recognize the existence of a process that takes place over time, a consequence of interaction between normal developmental processes and environmental stimuli.

Much hinges on the extent to which the classifying of the natural world is indeed "spontaneous." It is clearly spontaneous in the sense that it is conducted in the context of a large number of previous classifying acts,[1] is subject to much learned cultural knowledge, and is rule-bound. However, Berlin (1992) and others would go beyond this and say that the degree of spontaneity is hugely constrained, not simply by cognitive mechanisms (the character of sense perception), but by innate abilities to recognize "natures plan." Whatever the case, we would be advised, when speaking of the origins of classifying behavior, to employ the language of ontogeny rather than that of predisposition.

I accept that folk classifications co-evolve with the plants and animals that are their subject, and in the most general sense agree with Boster (1996) that at the level of clearly-discriminated prototypes of natural kinds, humans "carve nature at the joints," that there are certain discontinuities that are so protean, so much a part of the experience of so many human populations, that they can be said to be universal. I believe this to be true for natural kinds as a phenomenal type, but also "unique beginners" such as plant or animal. One is reminded here of the position adopted by Reed (1988), that "animacy" or "animality" is not simply an end-product of classification based on multiple cognitive discriminations, but relates to a wired-in ability of the brain to distinguish an organic form that registers a particular kind of saliency which matches objective phylogenetic features. Since hominids have evolved in environments that display a particular phylogenetic and phenomenal discontinuity, it is not entirely surprising that they should demonstrate a capacity to (a) utilize a notion of natural kind that assists the management of diversity, and (b) recognize more diffuse prototypes in non-cultural ways (e.g., "animal,"'"plant,"perhaps"tree,"'"bird,"'"fish"). However, such artifacts of cognition are logically different from "life-forms" in the sense developed by Berlin (e.g., Berlin, Breedlove, and Raven 1973). These latter vary cross-culturally (Atran 1998: 568 n4), but do not always partition "the living world into broadly equivalent divisions." The notion of "life-form" relates to linguistic and categorical discrimination (and to "rank").[2]

However, much emphasis (e.g., Boster 1996; Brown 1984) has been placed on the roots of natural-kind classification in evolutionary psychology when there is equal reason to believe that classifications that cut across morphologically "natural" classifications, such as edible nonedible and dangerous nondangerous, may be in part a consequence of non-cultural recognition abilities, in this case those which seem to be reinforced through the limbic-frontal-cortical system (Fox 1989). Thus, humans may not see "stones," but they may well perceive objects in their environment with the properties of stones which can serve a particular purpose, and be grouped accordingly (Ingold 1992). We certainly need to investigate further the extent to which "affordance-based" classification can operate independent of cultural inputs or contexts. However, my own view is that these cognitive propensities are so abstract as to tell us relatively little about how people classify in their everyday lives, at lower (and more functional) levels of discrimination. On the whole, non-cultural input operates in terms of the process of categorization, rather than underpinning particular categories, while certain regularities may be the product of general mechanisms operating across different and very varied domains, constrained by the data being organized.

Theoretical Validity of the Taxonomic Approach

The most discussed specificity of classifications of the natural world concerns the extent to which they are necessarily rendered taxonomically, and the extent to which there is a clear relationship between cognition and the very stuff of classification. I have stated my own views on this point most recently in *The Cultural Relations of Classification (1993)*. In brief, I agree that the principle of taxonomy is a powerful one available as a universal classifying strategy. There can be little doubt also that people classify living things into increasingly inclusive groups, and that this provides a useful inductive framework for making systematic inferences about the properties of organisms. But this need not imply taxonomy in the formal or domain-specific sense. Systematic contrast and class inclusion are present across a number of domains. It is particularly striking in the case of plants and animals because of their "thinginess" and because they are the outcome of an evolutionary process that is reflected in patterned physical and behavioral resemblances (as discussed above). In the domain of living kinds these tendencies converge in a particular way, not obviously because of the features of the mind that does the classifying, but because of regularities in the objective world which is classified and to which the mind responds.

It would seem that some cultural profiles encourage taxonomic thinking as a way of representing relationships between things more than others (see e.g., Lancy and Strathern 1981), and some sub-cultural contexts encourage it more than others (e.g., formal literary-based operations in classroom contexts). Moreover, because of the propensity of most anthropological researchers to rely heavily on a taxonomic approach embedded in Western science, it is easy to yield taxonomies in patterns of data collected from non-literate informants. In asserting a universal "abstract taxonomic structure," the approach seems all too often to be to delete features of peoples classifying behavior of living organisms which do not fit the expected pattern, until such a pattern is obtained.

Brent Berlin (Berlin, Breedlove and Raven 1973: Berlin 1992) has consistently argued in favor of the universality of taxonomy for ethnobiological schemes, but this only really works if we also assert the clear separation of *general-purpose* from *special-purpose* schemes, that is those that are logical and "natural," from those that arise to meet the needs of particular cultural requirements. However, the effective demonstration of the empirical primacy of taxonomy depends on the extent of linkage between categories (in often flexible ways), ways that undermine implicit taxonomic levels and contrasts and the general-purpose/special-purpose distinction. It also depends upon the ease with which ethnographers can elicit transitivity statements (of the kind *a* is a *b* and *b* is a *c*, therefore *a* is a *c*). It is, then, basically an appeal to our common (cultural) sense.

Atran (1998: 563) no longer thinks that folk taxonomy defines the inferential character of folk biology as suggested in *Cognitive Foundations of Natural History*, and his recent findings do not uphold the customary distinction between general-purpose and special purpose classifications. This is consistent with the results of my own ethnobiological ethnography (e.g., Ellen 1993: 123–24). Nuaulu, like Itzaj Maya, do not "essentialize ranks," which would violate their primary concern with "ecological and morpho-behavioral relationships" in favor of abstract properties. The development of worldwide scientific systematics has until recently explicitly required rejecting such relationships (Atran 1998: 561–2) with their crosscutting classifications. However, as Professor Minelli pointed out in our discussions, the needs of modern taxonomy direct us much more to an "un-ranked systematics." I believe that one of the problems central to the methodology that we use to generate so much of our ethnobiological data is not knowing quite how independent the system of ranks that we discover may be from the kinds of concepts with which we start. On the whole, it is my experience that empirical ethnographic reality is rather of a single dynamic conception of the relations between categories, which allows for the generation of particular "classifications" depending on context. Thus, the variable position of palms in ethnobotanical schemes is an excellent example of the preeminence of local and cultural considerations, but also of some general fundamental ambiguity (Ellen 1998).

I agree with what I understand to be Atran's current position, that the more dense our knowledge the more we deviate from the general model, and that in a very real sense taxonomies are the result of "degenerate knowledge," that is they only become possible by simplifying experiential complexity in ways which makes knowledge less useful. The failure to integrate the classification of domesticates into general accounts of the working of classifications of the natural world, given the practical importance of such classificatory knowledge for most humans, is a major problem; it is not just a "special case." I also find the idea expressed by Maddalon that cultural selection of domesticates makes taxonomy possible by heightening the differences between categories of cultivars a neat and fertile one, and one that reinforces the interpretation of other current work (e.g., Shigeta 1996).

Symbolic and Mundane, Social and Non-social

Humans classify the world about them by matching perceptual images, words and concepts (Ohnuki-Tierney 1981: 453). The operations work equally in terms of unmodified sense data or their cultural representations. In this sense, the cognitive and cultural tools available to do this do not distinguish between the social world and the non-social world, although this has become a conventional distinction in the analysis of classification. Similarly, classification can treat its subject in a pragmatic and mundane way or by using various symbolic allusions. Since so much of what we sense and experience is mediated by social consciousness, and since the boundary between the mundane and the symbolic is often unclear, it has sometimes been difficult, in practice, to know where to divide these two axes. It seems to me that there is more consensus on the principles of categorization than on the status of patterns of categories that are the outcome.

The distinctions between symbolic and mundane classifications and between those of the social and non-social worlds cannot always be neatly drawn: symbolic things are in an important sense practical, and practical classifications of the non-social world often rely on metaphors that are ultimately social, as in the use of the terms "genus" and "family" to organize plants and animals. Attempts to bring these aspects of classifying behavior together have met with varying degrees of success. Those who espouse extreme formulations of the universalist (formal) relativist (symbolic) divide sometimes claim that they are engaged in separate kinds of endeavor, and that one body of work should not invalidate the other. This is, I think, the view that Mary Douglas (e.g. 1993: 161–65) has defended for the symbolists, and Brent Berlin for the formalists. However, although I notice no inclination on the part of Douglas to shift ground in the face of recent evidence and arguments, Berlin does appear to present a moderated version of his early views in *Ethnobiological Classification*.

Rather differently, some (including myself) have stressed the intrinsic empirical connections between the mundane and the symbolic. I find support for this view in Maddalon's discussion of metonymy and metaphor in the evolution of natural-kind classification. It is impossible, for example, to make sense of Austronesian terms and categories for "bird" and for "tree" without considering utilitarian and symbolic criteria. I also find support for this view in what we can discover about the historical development of particular natural-kind classifications in Europe and elsewhere (as evidenced by Trumper's and Maddalon's examples of the relationship between early Latin and dialect Italian classifications of different kinds of organisms), and their remarks on the interplay of similitudes between various spheres of the natural world, and the impact of culture contact and history on ethnobiological classification.

It is self-evident that we generate classifications, think about nature, and articulate knowledge about the environment in social contexts. Sometimes, even, we use forms of intelligence which appear to have evolved to cope with social interaction between humans to make sense of the natural world. In other words, we "anthropomorphize" nature through what Mithen (1996: 164–84) describes as "cognitive fluidity," the merging of different kinds of thought processes. As human beings, as opposed to say sticklebacks or even chimpanzees, it would seem, we can do little else.

However, when we do engage in classificatory acts as humans, we systematically repress or forget or ignore certain characteristics and associations of particular natural things, and exaggerate and foreground others. Any one species presents too complex an aggregation of traits to take into account in routine practical memory storage and information handling. This is why, for example, numerical taxonomy does not provide a good model for understanding how human minds process data-it is just too multidimensional. Sometimes this simplification results in more naturalistic classifications, sometimes it results in more symbolic ones, or a combination of the two. This is very clear when we look at graphic icons for natural species in different aesthetic and writing traditions. I think that on the whole I am rather suspicious of theories that claim that we should try to conflate or aggregate all meanings of nature and natural things all of the time in order to achieve some inductive understanding of the whole. When we do, we often generate cognitive contradictions that pose spurious interpretative problems for those scholars seeking an overarching synthesis. Maybe it is the ability to cope with these contradictions, to separate out potentially awkward representations of the same perceptual reality, that is itself some kind of universal mechanism of the mind.

The difficulties of assigning things to categories may be made easier, then, by imposing culturally agreed boundaries, or indeed by creating these inadvertently or deliberately through genetic and other physical manipulations of the natural world, e.g., breeding varieties of plants that emphasize phenotypic difference for aesthetic reasons or planting trees individually to display their architecture in ways that are often occluded in natural settings. Because parts of our experience of the world are complexly continuous, it is occasionally necessary to impose boundaries to produce categories at all. Sometimes these can be quite arbitrary, and even in such an apparently technically precise area as engineering design, it is now apparent that the scope for cultural arbitrariness over technical necessity is considerable (e.g., Lemonnier 1992).

With analytic (that is partonymic) classifications of material things (e.g., the human body), there is a large degree of cross-cultural conformity, as one might expect. But with biodiversity, it is different; some gaps are bigger and more salient than others, in most environments, and therefore serve as more widespread (even universal) markers in classifying behavior. What is it that makes a tree a convincing life-form? Our experience, in many diverse environments, does not make it automatic that we recognize it as a clearly separate bounded kind of thing, as we can see in any photograph of a stretch of forest. Trees often merge imperceptibly into bushes (Ellen 1998). They are often polythetic in definition, single features being neither essential to group membership nor sufficient to allocate an item to a group. It may seem, therefore, that categories vary according to the complexity of their definition, rather than simply the scope of their content.

Classifying as a Cognitive Process and Classifications as Cultural Artifacts

Early models of ethnobiological classification were heavily constrained by adherence to linguistically defined entities and a language-based interpretation of how classification worked, even if formal recognition was given to the separation of category and label. This model has been described by some (e.g., Bloch 1991) as the "linear-sentential" model of culture. With a shift away from the use of distinctive features, emphasis on core-periphery models and cognitive prototypes, and a growth in the use of psychological at the expense of linguistic approaches, greater recognition has been given to how we might classify and engage with objective differences in the natural world without necessarily using language as an intermediary.

Problems arise when the *process of classifying* (the cultural and cognitive mechanisms by which the assignation of objects, concepts and relations to categories is achieved) is conflated with *classifications* (the linguistic, mental and other cultural representations which result). This reifies schemes as permanent cultural artifacts or mentally-stored old knowledge, when they are more often properly understood as the spontaneous and often transient end product of underlying processes in an individual classifying act. We might call such a misinterpretation "the classificatory fallacy" (Ellen 1997).

To extend this distinction, and make it more productive, it is useful to employ the model of agency and structure (structuration), which we owe in its sociological form to Anthony Giddens (e.g., 1986). Thus, the relationship between classifying as a cognitive and cultural process and "a classification" as a representation is recursive and dialectical: you cannot have one without the other. The classifying process is always situated in and assumes some context of previous classifications, while itself modifying the context for the next classifying act. As this largely operates within the constraints of human culture and memory, it is clearly a matter of degree, depending on the knowledgeability of the classifier, the variability of the contexts, and the entities being classified. I believe this model to be well adapted to an understanding of the classification of natural kinds (see figure 2.1).

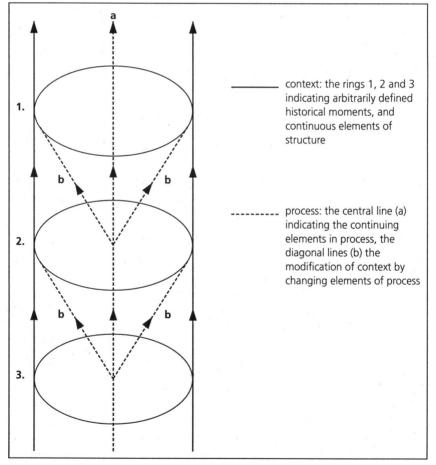

context: the rings 1, 2 and 3 indicating arbitrarily defined historical moments, and continuous elements of structure

process: the central line (a) indicating the continuing elements in process, the diagonal lines (b) the modification of context by changing elements of process

Figure 2.1 The Classificatory Relationship between Context and Process

Classifications of all kinds connect culture, psychology, and perceptual discontinu-ities of the concrete world. Confusion has arisen in the past from failure to distin-guish clearly between individual instruments of cognitive process and the collective medium in which these operate, comprising belief, cultural representations, and social practice. It is also crucial to distinguish information storage from representa-tion, abstract knowledge of the world, and the pragmatic schemata we use to nego-tiate our way through it. Our propensity to classify in the ways that we do results from the possession of certain innate cognitive skills, plus an ability to organize our perceptions through culture (aided by language) based on models drawn from somatic experience (such as right and left and bodily rhythms), and from social and perceptual experience of the material world. The form that a particular classification takes will sometimes be a culturally defined whole, but often as not will be the out-come of interaction in particular circumstances: the interplay of past knowledge, material context, and social inputs. Classifications as things, therefore, are not the inventions of individuals, but arise through the historically contingent character of cultural transmission, linguistic constraints, metaphorical extensions, and shared social experience in relation to individual cognitive practice.

Classificatory Knowledge and Applied Knowledge

One problem that has emerged in recent years is a consequence of the growth of interest in practical "indigenous knowledge." We can now see more clearly than in early discussions of the distinction between special-purpose and general purpose classifications that the way people use knowledge of the natural world to under-stand the world or to modify it, in real situations and the organization of such knowledge, does not always fit the kinds of models of classification which have emerged in ethnobiological work, and that we sometimes refer to as "classificatory knowledge." The former is often about abstract representation, the latter about knowledge for use.

Ethnobiological knowledge is the consequence of practical engagement in everyday life, and is constantly reinforced by experience; its adaptiveness stem-ming from the multiplicity of ways it can be organized (including classified) and the redundancy that is built into this process. Such classifications must always be fluid and negotiable, produced as well as reproduced. Although ethnobiological knowledge may focus on particular individuals and may achieve a degree of coherence in rituals and other symbolic constructs, its distribution is always frag-mentary: it does not exist in its totality in any one place or individual, despite the extraordinary oral encyclopedism of the likes of Alonso Ton Mendez (see Berlin, in this volume) or Saem Majnep (Marcus 1991). Such knowledge is socially distrib-uted. Indeed, to a considerable extent it is devolved not in individuals at all, but in the practices and interactions in which people themselves engage. We must expect this to be reflected in classifications.

The things we call ethnobiological *classifications* are, therefore, an emergent product of the application or core folk biological knowledge. I have described this as *prehension* (Ellen 1993: 229): those empirical processes determined by the interaction between knowledge, context, purpose, and innate cognitive tools which give rise to particular classificatory outcomes.

Conclusion

Looking at classifying behavior more generally, and seeing the connection between the ethnobiological classifications of others, the classifying behavior of ordinary people in Western contexts, and the classifications that have been sanctified by the growth of science, I agree with Maddalon that the use of the concept "folk" has arguably closed-off lines of enquiry and focused heavily on particular human groups. It is important to recognize the cognitive and cultural similarities at work in the "indigenous" knowledge of others and the "tacit" knowledge of ourselves. Such knowledge operates in relation to even the most complex and advanced of modern technologies (Ellen and Harris 2000).

Ethnobiological classifications generally organize knowledge that is orally transmitted, or transmitted through imitation and demonstration. The corollary of this is that writing it down changes some of its fundamental properties. Writing, of course, also makes it more portable and permanent, reinforcing the dislocation that arises when knowledge that is rooted in a particular place and set of experiences (local or indigenous), and generated by people living in those places, is transferred to other places. And although there has been a constant interaction between folk and scientific classifications throughout history, I am gratified that Atran now acknowledges that he might have been hasty in identifying taxonomy in science and in folk science as a simple manifestation of some common pan-human hard-wiring.

Notes

1. The phrases "classifying act" or "an act of classification" are used purely as a rhetorical device here, and I fully accept that in real life acts of classification are embedded in real situations and hardly separable from what goes on before and what comes afterwards. Indeed, the "act" may evolve, be reinforced or rescinded, over a period of time, as in, for example, drawing a person's attention to an object.

2. Thus, although the basic image prototype of "tree" may have existed for millions of years, the life-form category and term seem relatively recent (Witkowski, Brown, and Chase 1981), while its earliest labeling appears to have involved functional considerations reflected in tree/wood polysemy. Some life-forms, it appears, are more natural than others.

References

Atran, S. 1990. *Cognitive foundations of natural history: Towards an anthropology of science.* Cambridge: Cambridge University Press.

———— 1998. Folk biology and the anthropology of science: Cognitive universals and cultural particulars. *Behavioural and Brain Sciences* 21, 547–609.

Berlin, Brent. 1992. *Ethnobiological classification: Principles of categorisation of plants and animals in traditional societies*. Princeton, New Jersey: Princeton University Press.

_____, D. E. Breedlove and P. H. Raven. 1973. General principles of classification and nomenclature in folk biology. *American Anthropologist*. 75, 214–42.

Bloch, M. 1991. Language, anthropology and cognitive science. *Man* 26 (2): 183–98.

Boster, J. 1996. Human cognition as a product and agent of evolution. In *Redefining nature: Ecology, culture and domestication*, edited by R. F. Ellen and K. Fukui. Oxford: Berg.

Douglas, M. 1993. Hunting the pangolin [correspondence]. *Man* 28 (1): 161–65.

Ellen, R. F. 1993. *The cultural relations of classification: An analysis of Nuaulu animal categories from central Seram*. Cambridge: Cambridge University Press.

_____ 1996. The cognitive geometry of nature: A contextual approach. In *Nature and society: Anthropological perspectives*, edited by P. Descola and G. Palsson. London: Routledge.

_____ 1997. Classification. *Encyclopedic dictionary of social and cultural anthropology*, edited by A. Barnard and J. Spencer. London: Routledge.

_____ 1998. Palms and the prototypicality of trees: Some questions concerning assumptions in the comparative study of categories and labels. In *The social life of trees*, edited by Laura Rival Oxford: Berg.

_____ and H. Harris. 2000. Introduction. In *Indigenous environmental knowledge and its transformations: a critical anthropological perspective*, edited by Roy Ellen, Peter Parkes and Alan Bicker. Amsterdam: Harwood academic publishers, Studies in Environmental Anthropology, vol. 5.

Fox, Robin. 1989. The passionate mind: Brain, dreams, memory, and social categories. In *The search for society: Quest for a biosocial science and morality*, Robin Fox. New Brunswick, London: Rutgers University Press.

Giddens, A. 1986. *The constitution of society: Outline of the theory of structuration*. London: Polity.

Harris, P. and P. Heelas. 1979. Cognitive processes and collective representations, *European Journal of Sociology* 20: 211–41.

Hull, D. 1992. Biological species: An inductivist's nightmare. In *How classification works: Nelson Goodman among the social sciences*, edited by Mary Douglas and David Hull. Edinburgh: Edinburgh University Press.

Ingold, T. 1992. Culture and the perception of the environment. In *Bush base: Forest farm, culture, environment and development*, edited by E. Croll and D. Parkin. London: Routledge.

_____ 1996. Hunting and gathering as ways of perceiving the environment. In: *Redefining nature: Ecology, culture and domestication*, edited by R. F. Ellen and K. Fukui. Oxford Berg.

Karim, W.-J. 1981. *Mah Betisék concepts of living things*. London Athlone.

Lancy, D. F. and A. J. Strathern. 1981. Making twos: Pairing as an alternative to the taxonomic mode of representation. *American Anthropologist* 83: 773–95.

Lemonnier, P. 1992. *Elements for an anthropology of technology*. Anthropological Papers, Museum of Anthropology 88, Ann Arbor: University of Michigan.

Marcus, George E. 1991. Notes and quotes concerning the further collaboration of Ian Saem Majnep and Ralph Bulmer: Saem becomes a writer. In *Man and a half: Essays in Pacific Anthropology and Ethnobiology in honour of Ralph Bulmer*, edited by Andrew Pawley Auckland: The Polynesian Society.

Mithen, S. 1996. *The prehistory of the mind: A search for the origins of art, religion and science*. London: Thames and Hudson.

Ohnuki-Tierney, E. 1981. Phases in human perception/cognition/symbolisation processes: Cognitive anthropology and symbolic classification *American Ethnologist* 8 (2): 451–67.

Reed, Edward S. 1988. The affordances of the animate environment: Social science from the ecological point of view. In *What is an animal?*, edited by Tim Ingold London: Unwin Hyman.

Shigeta, M. 1996. Creating landrace diversity: The case of the Ari people and ensete (Ensete ventricosum) in Ethiopia. In *Redefining nature: Ecology, culture and domestication*, edited by R. F. Ellen and K. Fukui Oxford: Berg.

Witkowski, S. R., Cecil H. Brown and Paul K. Chase. 1981. Where do tree terms come from? *Man* (N.S.) 16: 1–14.

CHAPTER 3

Tackling Aristotelian Ethnozoology

Oddone Longo

Why just Aristotle?

One might wonder why a paper dealing with Aristotelian zoological classification should be considered relevant to the concerns of the present conference: after all, it seems rather extraneous to its (ethno)linguistic and anthropological focus. The response is that the Aristotelian "system," despite all incongruities and inadequacies we discover in it today with the benefit of hindsight, remains the first organized classification of living beings in the history of Western science, or at least the only one that survived the loss of most of ancient Greek scientific works. It long provided the model for subsequent classifications, though in actual fact, from Pliny down to Gesner and beyond through the centuries, and with a few exceptions like Albertus Magnus, systematic vision tended through the centuries to be sacrificed to an episodical, merely erudite, or even marvellous pattern of description of single animals. When scholars began to reconsider, with Linnaeus and his successors, the systematic organization of animal species, it was Aristotle that they were obliged to use as their starting point, and it was the progressive superseding of the Aristotelian conception, as a whole and in detail, that provided in later times the platform for the launch of the new zoological science.

A survey of some aspects of the Aristotelian classification practice, and particularly some problems of a linguistic nature, might therefore appear justified in this occasion.

At the same time, and with a view toward recovering and reusing the legacy of knowledge and issues we have received from the ancient world, the inclusion, however anomalous, of that legacy as a subject for consideration in a conference of primarily anthropological and (ethno)linguistic matters may suggest a role for the sciences of antiquity that is different and more significant than the one commonly assigned to them. Reciprocally, an updated, critically and historically oriented recovery of Greek scientific and speculative tradition may turn out to be of some benefit to present-day logico-linguistic and anthropological sciences. This general view was already taken up twenty years ago by Scott Atran in a paper devoted to aspects of Aristotelian definition and classification of animals in a logico-epistemological perspective, a perspective however different from the one I am pursuing here (Atran 1985).

Classifying and Naming

In arranging the various animal types in a more or less consistent system, the two acts of naming and classifying (which in the conference and in this volume are kept separated in two different sessions / parts) would appear to be, as a matter of fact, closely interconnected. It is particularly the identification of new species that automatically involves the creation of names for them; a glaring instance of this is the current proliferation of names in the zoological sciences, since nowadays new species are constantly being discovered, or known species are included in new taxonomic schemes.

On the opposite side of the story, old father Adam had to start from square one in naming individual animals or couples, as Jahweh introduced them to him (whether he was not only *naming* but also *classifying* them remains a matter of controversy). In Aristotle's case, on the contrary, the problem of naming new animal species did not arise; new names were not needed because no new, recently discovered species asked to be introduced into the established zoological nomenclature. There were no newly discovered animals, nor was any effort made to discover them; just a few "exotic" species stand out among the Aristotelian fauna (the ostrich, the crocodile, the elephant), but these already had their own traditional names.

The matters are rather different when it is a question of denoting "taxa" (to use a term unknown to the Greeks) of a higher rank than the species (*eîdos*), like the genus (*génos*) or the groupings in even higher categories, for which no name was at hand. In such cases, Aristotle might follow two different paths, either looking to tradition for already used classifying names, or holding it necessary to resort to linguistic innovation. The prevalent Aristotelian tendency is to approach the question of possibly coining new names for zoological "classes" with a dual perspective: (1) to maintain in every case the linguistic conventions already in use, and (2) to respect whenever possible the lack of denomination of taxa of higher rank that are, as Aristotle calls them, *anónyma*.

There is indeed also a tendency (3), for cases in which the creation of a new term was unavoidable, and inseparable from the identification of a particular animal grouping. An instance of tendency (3) is provided by the case of flying animals, such as birds, bats, and insects. Here Aristotle observes that, while for animals equipped with feathered wings there exists in current language a general name, *órnis* ("bird"), no such general denomination is at hand for the two other groups. In this case, the philosopher was obliged to coin two new names of taxa, which did not exist before him, and which were destined to different fortunes: these two new names were *éntoma* for insects (the Latin word *insecta* is but a calque of the Greek word), and *Dermoptera* for bats (*Historiae animalium* 490a5). The later are called *Chiroptera* in the actual nomenclature, while *Dermoptera* has been assigned to embrace another genus of little, quasi-flying mammals from the Far East.

It must not in any case be forgotten that Aristotle himself was not a "systematist" as a matter of *parti pris*, nor did he ever set out to construct a general classification of the living word. An Aristotelian taxonomic scheme is a not wholly justified modern extrapolation from his works, in that the philosopher only resorted, if ever, to classificatory procedures not as an end in and of themselves, but for their employment as an useful tool for the comparative description of animal species and their anatomo-physiological features and behaviours, which was the core of his scientific program.

Classifying

"To classify" means to put together groups of objects that share an entire range of characters (morphological, functional, etc.) into higher units that share only a limited set, rather than all of those characters. This is true also for the case of animal species (for Aristotle and for Plato before him, the "species" or *eîdos* is the ultimate taxon, below which there are only single individuals). As one proceeds from "species" to "genera," and then (resorting to modern taxa, which do not exist in Aristotle) to "families," "orders," and "classes," the number of the shared characters diminishes at the higher levels to the point where they are reduced to the most general, or even just one. For example, the highest Aristotelian taxonomic subdivision distinguishes between animals "with blood" (*énaima*) and those "without blood" (*ánaima*), a grouping that corresponds approximately to our distinction between Vertebrates and Invertebrates (although the taxonomic meaning of this categorization has diminished considerably since the time it was introduced).

The process of classifying requires the availability of suitable instruments of classification, or "classifiers," and it is of the utmost importance, in a historico-cultural inquiry, to ensure correct identification of the provenance of these classifiers and recognition of how they work. In the case of Aristotle (as of Plato and the Academics before him), we can state that he ranged living beings into classes by superimposing and adapting to the zoological context categories that had been developed, not for zoology, but for logics and discourse analysis. This remains, as we shall see later, the principal difficulty with a consideration of Aristotle's taxonomic system from the point of view of present-day ethnozoology.

There is a further problem: the two, or even the three basic classifiers, "genus" (*génos*), "species" (*eîdos*) and "variety" (*diaphorá*), already in use before Aristotle, never achieved an unambiguous semantic nor logic status of their own. The use and the meaning of *eîdos* and *génos*, in particular, often appear to be interchangeable, and sometimes the same classifier (*génos*) is even used, in the zoological works, to designate both a set of taxa and the greater taxon that includes them.

There are no more comprehensive designations of animal taxa than *eîdos* or *génos*, and Aristotle confines himself to speaking of "greatest" or "highest genera," which is also a clear indication of Aristotle's reluctance to innovate classifications and introduce new signifiers.

From Naming to Classifying. The Case of Coleoptera

Recourse to linguistic innovation in order to name something—the creation of a new signifier—is however inevitable when it becomes necessary to refer a single species of animal, already in possession of a commonly used name, to a higher grouping that does not have commonly perceived "visibility," or not enough to enable it to be useful for classifying purposes. In such cases—an instance of which we have already considered—and despite his preference for the "anonimity" of higher genera or groupings, Aristotle could be forced to introduce new signifiers, which would nevertheless remain perfectly clear to Greek speakers of average education.

It does not follow, however, that the new signifier already, always, and unequivocally denotes a genus as would a true classifier: it may happen that it begins by having a merely descriptive and not yet a truly classifying function. There are, to put it otherwise, different degrees of classificatory strength to obtain before the new term acquires a really taxonomic status.

In this connection we may quote the example of beetles (Greek *koleóptera*), a taxon designation actually used in zoology, where it forms a major order of the class of Insects (*phylum* Arthropoda), featuring a pair of elytra in place of the two front wings. The order Coleoptera includes today more than 370.000 known species, and it is therefore the most numerous order in the entire animal realm. Aristotle shows to be acquainted with only three species of beetles, the ladybird, the scarab, and the stag beetle, or these are the only species he quotes as samples of the genus, while we may suppose that more of them were known to him and to the Greek common people.

Aristotle, as we have seen, divided all "winged" animals into two large classes, the "sanguineous winged animals" (birds and bats) and "non-sanguineous winged animals" (insects); but while in the last resort he identified beetles as a proper genus of insects, in the actual use of the newly coined compound, he oscillated between a merely descriptive meaning ("insects that have wings (*pterá*) covered by a protective sheath (*koleós*)," or rather by two elytra), and an openly classificatory use. In the first case we have, as it were, a classification "in the nascent state," where the compound has not yet acquired a truly technical value. Later on, even the entire class of insects could be divided by Aristotle into the two groups of the *koleóptera*, with their pair of elytra, and, we dare say, the "non-*koleóptera*," or insects "without elytra" (*anélytra*).

Linguistic Pre-conditionings

The standard classificatory terminology in modern zoological nomenclatures is highly indebted to the Greek language, and markedly benefits from its well-known tendency to form nominal compounds. With the exception of the limited set of terms derived from Latin, or from other sources, the major amount of zoological entries is (or rather was!) based on Greek, exploiting its ductility. It would be of some interest to speculate—and we have here perhaps a less naive question than it would seem—what the effects might have been on modern zoological nomenclature if it had not been allowed to draw on the inexhaustible lexical mine of the Greek language.

The first traces of a specialist lexicon in the zoological field occurred with the authors of the 5th century b. C., with Plato and Aristotle at their head; we would know much more, were it not that the works of such authors as Democritus, Speusippus, and many others were lost to nothing more than scanty fragments. But to speak of a"specialist lexicon"does not necessarily mean that a complete set of"scientific" terms were at these authors' disposal, and it has been objected by French scholars such as Manquat and Pellegrin, that Aristotle in particular never attempted to introduce into usage what we would call a "scientific term." The "specialist"terms that he uses in his zoological works were taken, as for the names of single animal species, from the"professional"language of practitioners such as hunters, fishers, and livestock farmers. These were also the witnesses that provided Aristotle with information about the animals he was considering in his zoological treatisies.

Aristotle, we are told by these scholars, could not attain a rigorous taxonomy (in the modern sense) without breaking with generally accepted language. If the break did not happen, and if the philosopher was so cautious in introducing new zoological terms, we ought to deduce that he certainly had no such taxonomy in his mind.

In its extreme form it is stated that we are the victims of an"illusion of perspective"when faced with the apparently "scientific"quality of the language used in Aristotelian zoological treaties:"when the names used by Aristotle appear to us to be perfectly 'scientific,' this in fact happens because the philosopher spoke Greek and it is on Greek that our naturalistic terminology is based" (Manquat 1932). Aristotelian language may then only be scientific "by reflection": the opinion may be debatable, but discussion of the point might be profitable, and the topic may anyhow usefully introduce us to the following pages dealing with a possible meeting-point between Aristotelian zoology and modern ethnobiological views.

Aristotelian Ethnozoology?

The question we have to ask at this point is whether we can legitimately analyze Aristotelian zoology and its systematics using the conceptual tools of ethnobiology, and of ethnozoology in particular. In my opinion, there can be no *a priori* discriminant against this operation: the Greek culture, although rightly considered as one of the highest points ever rejoined by human civilization, was itself an ethnoculture like many others, despite the enormous advances it achieved in scientific and philosophical thinking as well as in other fields. There is nothing to justify the maintenance of the traditional discriminations of the Classicist model and the use of different methods, and different principles, in the study of Greek culture in comparison with other historical or even "ethnological" cultures.

As concerns Aristotle's zoological works, which are our principal, if not our only source of Greek "scientific" zoology, we have to bear in mind that the material presented therein was drawn from a repertoire of traditional knowledge that the author considered sufficiently reliable. Aristotle designates the possessors of this traditional knowledge (that we could define as 'folkloristic') using very general terms such as "men" (*hoi ánthropoi*), "people" (*hoi pollòi*), and describes the taxonomic denominations as "common names handed down" to the present time (*paradedoména koinà onómata*). Aristotle never takes any preconceived negative stance against these "traditions"; if anything, he merely questions their single tenets.

As we have seen, the depositaries of traditional zoological knowledge are to be sought ámong the practitioners in this area: breeders and shepherds, hunters and fishermen. Aristotelian zoology relies much more on these professional categories—for the information, the names of the various animal species, and also (though only in part) the taxonomic divisions proper—than on direct observation of the living world. It has been rightly pointed out that this choice led Aristotle to expand the sources of science, drawing information from outside the "illustrious" cultural tradition and turning towards a world that had hitherto been considered unworthy of the scientist's attention (Vegetti 1976). We ought however not to overemphasize the "professional" character of this knowledge, as if it were limited to the aforesaid categories of persons: knowledge about the animal, as well as about the plant world, was indeed largely diffused in Greek society, which lived in straight contact with the natural environment, and where a city life as we could suppose it was never reached, or was only in single, particular cases.

Many and Few, Large and Small

A noteworthy aspect in relation to the ethnozoological perspective proposed here is the matter of quantity. As we have already seen, Aristotle does not proceed with any identification or discovery of new species in addition to those already known: his "zoological garden" is that of Greek folklore. The total number of animal species identified in the *Historiae animalium* amounts to about 480 (and no more are added in subsequent treatises). Now, starting from a suggestion from Lévi-Strauss in his *Pensée Sauvage*, Berlin came to the conclusion that the mean number of "generic taxa" included in the known ethnobotanical or ethnozoological classifications amounted to 520 for the plant species and 390 for the animal species. The range of values fluctuates, in any case, around approximately 500 units, almost as if there existed an upper numerical limit of "specific" perceptibility or memory; this limit is higher, albeit only slightly—and for obvious reasons—in the case of the plant species. Berlin himself observes that the known ethnobotanical data coincide perfectly with those of ancient botany: 550 species of plants were named by Theophrastus and 537 by Dioscorides (as mentioned earlier, the mean value emerging from the "traditional cultivators" of ten societies was 520 species) (Berlin 1992). The 480 animal species of Aristotle's zoology fit perfectly into this dimension.

A second aspect we wish to consider here is that of the (possible) existence of common criteria or parameters for description and classification between the ethnozoological systems and their ethnotaxonomies and their Aristotelian counterparts.

Drawing again from Berlin, and from other authors, what governs the identification and the related naming and classification of the various animal types in the ethnozoologies is the level of immediate perception (the "perceptual givens"); the most common perceptual parameters primarily include the color, relative size, and the form of the living "objects" (we shall limit ourselves to these three in the present context). "Most of these dimensions correspond to some of the most readily and immediately apprehended sense impressions that human beings have as they interact with the physical and natural world" (Berlin 1992).

It is not difficult to see how similar criteria are also actively involved in Aristotelian zoology. At the beginning of the *Historiae animalium*, Aristotle clearly and concisely formulates the criteria on which to rely in describing the different animal species (he will keep to these criteria in subsequent biological works too). With a drastic simplification, he brings the numerous visible differences down to fit into a picture of mainly quantitative opposition or graduation: the species differ from each other in their "greater" or "lesser" degree of possession of certain features.

First of all, this scheme considers the size of the animal as a whole and in its component parts, and the number of parts absent or present, then its form (*schêma, eîdos*) and color (*chrôma*), the more visible characteristics of a living being for common eyes. The last two are also condensed, following a logical line of reasoning that we cannot pause to discuss here, into a quantitative contrast (e.g. "gray" is considered by the philosopher to be "blacker than white" and "whiter than black").

A few examples will clarify what may appear obscure in this exposition. Generally speaking, at the level of the higher "classes," what distinguishes the «sanguineous» animals (mammals, birds, fish, etc.) from the "bloodless" species (insects, crustaceans, etc.) is primarily their size; sanguineous animals are generally "larger" than bloodless creatures (*Historiae animalium* 490a21); from another taxonomic standpoint, viviparous species are generally "much larger" than ovoviviparous ones.

But even in the description of genera and species, and in their precise distinction, the criterion of size remains prevalent. A fairly typical example concerns various species of small birds. The criterion of a dimensional grading between the single varieties of the same species remains a constant, combined with the criterion of comparison between the dimensions of different species (or varieties). For example, there are three "varieties" (*eíde*) of titmice, i.e. the "fringilline," the "mountain" and a nameless (*anónymos*) third variety. The first is the largest and takes its name from the fact that it has the size of a finch; the second, which lives however in a different habitat, is of medium size and has a long tail; the third is the smallest of all. The situation is similar for the varieties of woodpecker, two of which are entirely similar, even in terms of their voice, but one is larger than the other. The woodpecker called *keleós* in Greek has the size of a turtle dove and is green (this is, in fact, the green woodpecker). There is also mention of three "genera" (or perhaps, simply "kinds," *géne*) of woodpecker, listed in ascending order according to their size: the first species is "smaller than a blackbird," the second is "larger than a blackbird" and the third is "slightly smaller than a hen" (*Historiae animalium* 592b–93b).

Cross-classifications

Another of the criteria for classification mentioned by Berlin (1992), the one concerning habitat or the ecological-ethological parameter, returns punctually in Aristotle, where it interferes with and is superimposed on the criteria of a morphological and physiological type. To suggest an example still relating to the avifauna, the ecological criterion may become dominant in the classification of birds. In fact, the species of fowl are distinguished as terrestrial or aquatic, and the aquatic birds as river, marsh, and marine species (*Historiae animalium* 593a25). The identification of their habitat is not to satisfy an abstract taxonomic need, but because birds, like other animals, draw their food (*trophé*) from the environment in which they live. In the case of water fowl, the precise definition of the different habitats combines with the morphological classification of certain species, distinguished as species with "joined toes" (*steganópodes* or web-footed) and with "separate toes" (*schizópodes*); the web-footed species have their habitat in water, and are water fowl proper, while the others live near courses or bodies of water.

There is no question that this crosswise combination of classification criteria is indicative of a taxonomic view that transcends the simple ethnozoological level; but it is equally certain that the observational data on which these taxonomic arrangements are constructed were drawn precisely from the folkloristic heritage of knowledge, as this could be transmitted to Aristotle by practitioners in this area (the hunters or fowlers, in this case), or even by common people living in the countries haunted by those kinds of fowl.

Practical Knowledge ...

A further aspect emphasized by Berlin and others concerns the "utilitarian," "functional" or "pragmatic" nature of classificatory distinctions, which would contrast with the "perceptual" character mentioned earlier. Berlin seems to prefer the "perceptual" to the "utilitarian" motivation, but as far as Aristotle is concerned, I believe the latter cannot be ruled out of consideration. On the contrary, there are several cases in which—behind a taxonomic adjustment or a description of different species—there are clear signs of a utilitarian motivation referring to the exploitation of animals for the purpose of their use as food or whatever else (principally as food).

For example, the text in *Historiae animalium* 595a13 concerning the two great classes of "herbivorous" (*poephága*) and "fructivorous" (*karpophága*) animals includes ample sections devoted to the description of techniques for fattening the animals in view of their final slaughtering. (It is worth noting that the animals were "fattened" by giving them plenty of water to drink!). Another example (*Partes animalium* 680a15), has to do with sea-urchins, their anatomical and reproductive features, and the "genera" (*géne*) according to which they are classified—here again, all with a view to the use of these tasty delicacies as food.

Such cases (and there are plenty of them) further emphasize how the "hard core" of Aristotelian zoology is based on a heritage of "technical" knowledge, the knowledge of the fishermen, hunters, breeders, farmers, but even of simple dealers and consumers of animal food. Such a knowledge is undeniably organized starting from the perception of natural (dimensional, morphological, ethological, etc.) differences, but—far from constituting an abstract "learning" serving no practical purpose—it functions in the light of the usage of the animal species considered, primarily as food, and it essentially remains the same in the majority of Aristotle's zoological treatises.

... and Abstract Taxonomies

Nonetheless, there is also a different "instrumentality" in Aristotelian taxonomies, which is specific to the philosopher's zoological work. Suffice it to mention two examples. The division of the classes of mammals into "animals with compact nails" (*mónycha*, more or less our single-digit perissodactyls), "animals with double nails" (*dichelá*, our fissipeds or artiodactyls), and animals "with many toes and no horns" (*polydáktyla kai akérata*), leads to an absolutely "irregular" and heterogeneous taxonomy. This is justified in relation to the stature and growth of the animals considered (*Partes animalium* 686b18).

The subsequent "solid nailed horned non-prolific" combination (*mónycha, keratophóra, oligotóka*), which also combines different classification criteria, is justified by the fact that it functions, instead, on the basis of the number and position of the mammary glands in the female. In fact, the group just mentioned, which has two such glands between the thighs, contrasts with the other group comprising "prolific fissipeds" (*polyschidê polytóka*) whose mammary glands can vary in number and position (*Partes animalium* 688a32).

In short, it can be said that in Aristotle's work "there is no apparent consistent taxonomic system, since the animals are grouped case by case according to the viewpoint of the treatise, which may be anatomical, physiological, ecological, ethological, or even pathological" (Vegetti 1976). The various and often contrasting classifications recurring in the zoological works do not appear to be an end in themselves—aimed at the creation of a general taxonomic scheme that Aristotle never traced, as we said earlier—but to provide the most useful reference grid within which to describe the various animal species' features. It is nonetheless a case of a different "instrumentality" from those identified before, and what could be called a "higher" level than the purely ethnozoological one, although the latter maintains nonetheless in Aristotle's works the place and relevance that we have already stated.

Generic Problems

The concept of the "genus" as the basic level of ethnonaturalistic classification has a key role in the works of Berlin, Lakoff and others. The genus, or "folk-generic level," would have various features, including that of more readily identifying and naming the objects, and with more straightforward names. Perception on a "generic" level would be holistic, gestalt, and also the easiest and most immediate; on a lower level, the situation is fully different, because "distinctive features have to be picked out" (Lakoff 1990).

A comparison with Aristotelian zoology appears distinctly more complex in this case, because it is difficult to isolate a level in the latter that equates to the "generic" one, and because of the polysemantic nature of the terms employed by Aristotle to designate "genus" (*génos*) and "species" (*eîdos*).

That the designation of animal types occurs, in Aristotle's works, at genus rather than species level (intending these two categories in relation to the Linnaean binomial nomenclature) is a hypothesis that remains to be verified (or disproved), although it is what we should expect if it is true (and it *is* true) that the majority of the animal names come from traditional (ethnozoological) Greek usage, with no effort on Aristotle's part to coin new names, especially at this level. And for Greek as for other folklore taxonomy we are authorized to surmise a general predominancy of the "generic" look.

The matter becomes more complex, however, as we have seen, because Aristotelian terminology often resorts without a neat distinction, and even with the same value, to the terms *génos* and *eîdos*, which for us have very different meanings, and ought to theoretically coincide with our "genus" and "species."

A few examples will suffice to illustrate this lexical ambiguity. Aristotle speaks of the cicadas, grouping them into a single *génos* ("the *génos* of the cicada," *Partes animalium* 682a18) and stating that there are several species of cicada (*pleío eíde*) that differ—once more !—mainly in size (*Historiae animalium* 532b11). But a little later on he identifies a particular *génos* of singing cicadas and, even further on (556a14), he declares that there are two *géne* of cicada, one smaller and the other larger. It therefore seems clear that the term *génos* is used differently at different times, either in its rightful sense or as an equivalent of *eîdos*. The same goes for the sponges, which should already constitute a "generic" group that is not suitable for placing on a higher level: Aristotle identifies three, or even four, different genera (*géne*), just when we would expect him to specify various "species."

More examples could be provided, but it may be better to put it briefly, as Vegetti said (1976): "For Aristotle, *génos* and *eîdos* are not genuine taxonomic keys, they simply represent levels of organization and groupings of animals that can be varied with great flexibility and that tend to be interchangeable." At most, the difference detectable between *génos* and *eîdos* is that *génos* always indicates a "group" (at various levels of genus, species, variety), while *eîdos* tends to maintain its original meaning of "form" in the sense of exterior physical appearance. The former disposes therefore of a higher degree of taxonomic potential.

As mentioned before, Aristotle's classification tools come from a totally different origin from their ethnozoological counterparts. The taxonomic categories he uses come from a completely different context, i.e., that of the logic of discourse and conceptual analysis, and they are superimposed on the zoological (descriptive) arrangement proper, without this combination succeeding in producing an entirely consistent unitary methodology.

Of course, it could also be said that a search for the "basic level" in Aristotle's zoology could, or should, be done without taking the lexical use of the two terms (*génos* and *eîdos*) into account, and conducting the analysis differently, by tangibly identifying those taxonomic entities that preserve their original, "generic" features, even before their inclusion in the Aristotelian scheme. But this would mean embarking on a study that, to the best of my knowledge, would have to start from square one.

References

Aristotle. 1976. *Opere biologiche.* a cura di D. Lanza e M. Vegetti. Torino: UTET.

Atran, S. 1985. Pre-theoretical aspects of Aristotelian definition and classification of animals: the case for common sense. *Studies in History and Philosophy of Science.* 16: 113–63.

Berlin, B. 1992. *Ethnobiological classification principles of categorization of plants and animals in traditional societies.* Princeton N.J.: Princeton University Press.

Lakoff, G. 1990. *Women, fire and dangerous things: what categories reveal about the mind.* Chicago: University of Chicago Press.

Manquat, A. 1932. *Aristote naturaliste.* Paris: Vrin.

Pellegrin, P. 1982. *La classification des animaux chez Aristote.* Paris: Vrin.

I wish to express my thanks to Professor Alessandro Minelli (Padua), to whom I am much indebted for kindly reading and emending this paper.

CHAPTER 4

Current and Historical Problems in Classification: Levels and Associated Themes, from the Linguistic Point of View

John B. Trumper

The Uniqueness and Claimed "Rigor" of the Scientific Paradigm and its Longterm Origin; Generalness and "Fuzziness" in Folk Taxonomy

M. Maddalon's part introduction opens up the question of the relationship between folk taxonomy and culture and states that opposing folk taxonomy to scientific classification oversimplifies this complex relation. Bulmer (1967) had argued that Karam taxonomy, though similar to scientific classification at the lowest levels (at its more or less gross morphological criteria, but not those based on habitat, ritual significance etc.), was at its upper levels less objective[1] and more general, perhaps fuzzier, so that, in an overall sense, "...the result shows little correspondence either with the taxonomy of the professional zoologist, which reflects the theory of evolution, or, for that matter, to our modern western European folk-taxonomies" (p. 6). Bulmer's 1970 paper went further, asserting that folk classification was more general and logical than scientific systematics. Ellen (1979b) contrasted Bulmer's position with that of Berlin, Breedlove, Raven (1968) who had argued that folk taxonomies were special-purpose, scientific ones, more general, tendentially universal, while more recent work by Berlin seems to acknowledge that both types of scheme are not so fundamentally different.[2] He criticizes the view expressed in Lévi-Strauss 1966 that so-called "primitive societies" are characterized by a dynamic, undifferentiated worldview, advanced societies, instead, by hyperdifferentiation and complex classification systems. To investigate the relationship between folk and modern scientific classification requires that we first investigate the historical determination of two apparently different and unrelated models of classification.

The Two "Models"

It can be reasonably shown, I feel, that two gross macro-models of classification have been in conflict in our particular culture over at least the last 2500 years, the first showing a unitary worldview in which there are precise symmetries between individual "souls," in which genus/ species etc. are analytical tools belonging to a particular model, individuals are characterized by "sums" of accidents or properties, fuzzy classes are either accommodated or relegated to the "wondrous," perhaps the older system. One might even identify it with the "ontological" model Aristotle proposed as the supreme analytical instrument, though it could be argued that what Aristotle did was formalize an already extant world-vision, making it a more rigorous, logical tool for coming to terms with both categories and real world entities.[3] The essentials of an ancient tradition refined by Aristotle[4] pass into Theophrastus in the successive generation,[5] through to Galen in ca. 160 BC,[6] thence to Dioscorides, and in the Roman world to Lucretius and Lucan[7] (ca. 60 BC) etc. The same basic set-up seems to hold good in ancient Chinese thought, at least in the period 300 BC–600 AD. Hsün Chhing (305–235 BC), an almost contemporary of Aristotle, using the basic water + fire *differentia* which is "chhi," constructs an organic life chain of the type (a) chhi + sêng ("life"?) = plants, (b) chhi + sêng + chih ("perception"?) = animals, (c) chhi + sêng + i ("justice" or "morality") = humans, an almost perfect parallelism with Greek (a) ψυχὴ θρεπτική (the "vegetative soul"), (b) ψυχὴ αἰσθητική (the "sentient soul"), (c) ψυχὴ αἰσθητικὴ διανοητική (the "thinking" or "rational soul").[8] If this be the case, where, then, does the insistence on a scaled hierarchy, in which *differentiæ* are all and work dichotomously in a quasi-binary manner, come from? It is this which, when perfected, creates the revolution that, proceeding from Descartes, Newton and Leibniz, moves to Linnæus, eventually to Darwin, creating a revolutionary rift between the world of the specialist and that of folk taxonomy and classification. I shall try to answer this question as briefly as possible, given the limits of the present paper.

The second model consists of a hierarchical worldview, in which there is a world-soul manifested in individuals; it contains tree hierarchies and is a strictly contrastive model, using features that are almost binary and are absolute defining values, so that fuzzy classes become embarrassments. This second "model" is usually called the "henological" approach and has long been associated with Plato. Though usually labelled in this way the model can also be traced in some sense to Homer's "golden cord," sometimes called the "golden chain" linking heaven = Jupiter = God with the created, already present in Iliad VIII, 19 ff. ("σειρὴν χρυ—σείην ἐξ οὐρανόθεν κρεμάσαντες" =" having hung a golden cord from heaven"). As such it is also present in the classical Latin model, the "descent" using the "golden bough" as entrance to the lower world (Æneid VI, 719–34), while more · explicit mention of the Homeric model is made, albeit in a negative fashion, in Lucretius.[9] In fact, Lucretius is explicitly in agreement with the older model in which there is substantial symmetry between the parts of a natural continuum that goes from plants to animals, animals to humans, humans to higher orders, and that even from the point of view of "parts" named, as in De Rerum Natura II. 669–671 "Furthermore bones, blood, veins, heat, damp, guts and nerves are the constituents of each and every creature that exhibits life, hence life emerges from all these constituents."[10] Within the Platonic model the One generates All-Soul which, in turn, generates and governs souls and created order (Phædrus 246B "All-Soul presides over the whole of the inanimate... when it is a perfect and winged being it flies high and governs the whole world, but once it has lost its wings it is borne down till it clutches something solid: once it has taken possession of this, it seems to move by the power conveyed it. The whole was called 'living,' this whole constituted of "soul" and "body," and took on the denomination 'mortal.'").[11] It is significant that for Plato the immortal descends to create the mortal, that the mortal ascends to the immortal: the relation of generation and creation implies a hierarchical descent and to some extent debasement of erstwhile perfection, cf. Phædrus 248A "But then follow the other souls, all desiring to attain to "ascent," though, unable to ascend, they are whirled down, submerged, trampling each other, bumping into each other, attempting to pass one in front of the other."[12] A similar sort of model approach is developed in his Timæus where the One generates All-Soul (ψυχή; πᾶσα) which mixed with other elements creates the Soul-of-All (ἡ τοῦ πάντος ψυχή), from which individual souls are planted in real-world bodies. All-Soul is then a dyadic demiourge or creator (Timæus 34a) who in a special crater mixes the elements to create (Timæus 41c: ὁ τόδε τὸ πᾶν γεννήσας) lesser beings (Timæus 41d: "…the crater in which I have tempered and composed, through mixing, the World-Soul",[13] where World-Soul = Soul-of-All and not All-Soul). Having generated World-Soul from All-Soul, from the former are created perfect souls (stars) and imperfect souls, so there is a descending hierarchy (Timæus 41e); finally souls are planted into less perfect bodies (Timæus 41e). Notwithstanding the common starting point between Phædrus and Timæus, there are certain contradictions between their points of view, the first expressing a substantially negative attitude towards nature, the latter a more positive one, though even in Timæus 42b-c women and animals are, in rebirth, lesser souls, and without «good living» the soul is necessarily downgraded. However, Plato on his own is insufficient to justify all aspects of the "henological" model, having too many internal contradictions between one work and another as his thought changes and progresses. Traces of the influence of this model are to be found in successive "philosophers," up to, say, Seneca (4 BC–65 AD), who seems to accept not only the Homeric equivalence Jupiter = heaven= God = World-Soul, in a tendentially monotheistic vision of things,[14] but the progression and equivalence World-Soul ≥ Soul, implying that for him World-Soul = All- Soul, as in *Naturales Quæstiones* III, 29, 2 "…siue anima est mundus, siue corpus natura gubernabile, ut arbores, ut sata, ab initio eius usque ad exitium quicquid facere quicquid pati debeat, inclusum est".[15]

The real development of this second "model" occurs much later with Plotinus (ca. 205–70 AD) who inherits in a sense Philo of Alexandria's attempt to combine the Old Testament (Genesis and Wisdom literature) with Plato's Timæus. In his famous *Enneads* he proceeds from the "One" (Τὸ῀Εν), which is above being (*Enneads* III,6, 6, 16–17) and which generates a series of dyadic principles or entities, in the first place the Nous dyad,[16] which in turn generates All-Soul (ἡ πᾶσα ψυχή in *Enneads* III, 9, 3, 1, mirroring Plato's ψυχὴ πᾶσα in the Phædrus, which is now a genus in Aristotelian terms); this, in turn, generates both World-Soul (ἡ τοῦ πάντος ψυχή, mirroring Plato's term in Timæus 41a etc., now another genus of which Nature is the lowest part) and individual souls which are, in Aristotelian language, species (εἴδη). Most commentators, however, agree that Plotinus often blurs the distinction between All-Soul and World-Soul. The intelligible or intelligence pre-exists with regard to the sensitive world, it orders and governs it, it permeates all life, so that life can only be classified as "intelligent",[17] but there is definitely a hierarchy in creation and creators that descend from the One in dyadic sets which anticipate descending binary steps (see Deck's [1967: 24] comments on *Ennead* V, 9, 9, 8–14). Extremely important is the emphasis on "descent" and degradation vs. "ascent" and progression from the One to Nature and vice versa, as in *Ennead* III, 8, 8, 1–6.[18] Concepts such as orders, and the related descent/ ascent from order to order, tend towards a hierarchizing of nature, though many commentators would leave the question open as far as Plotinus' original thought is concerned.[19] It is important that Plotinus not only takes up but develops Plato's descending/ ascending image (*Ennead* IV, 8, 4, 10–25) and is more negative than Plato as to the imperfection of the soul once "embodied," since he considers the body as being an "obstacle to the proper activity of soul" (see Deck's [1967: 36] comments on *Ennead* I, 8, 4, 1–4). Soul (ψυχή) that descends from All-Soul (ἡ πᾶσα ψυχή) is, on one hand, a "higher" entity, the very "soul quality" as superordinated genus, on the other, an extremely low-level species (*Ennead* IV, 8, 3, 10–12).[20] Nature is also a dyad, a superordinated principle underlying things and entities (*Ennead* III, 6, 4, 41–43; III, 6, 6, 33–34; III, 8, 4, 15–16), as well as the subordinated "vegetative soul" present in plants and earth (*Ennead* IV, 4, 27, 11–17). His system of intermediaries is infinitely more complex than any of Plato's original schemes, cf. Deck (1967: 113–114) "...it can be shown that, granted the outlines of Plotinus' view of the world, there must be intermediaries between the Nous and the sensible world... ."Some modern commentators try to attenuate the implications of Plotinus' language,[21] but it must be remembered that Nous, Soul and Nature are distinct, discrete ὑποστάσεις, therefore there are discrete superordinated entities like The One and Nous which descend into other subordinated ones, equally discrete (Nous > All-Soul ≥ World-Soul > Nature).

Later Developments of the "Two" Models

Were it solely for the ideas transmitted by Plotinus, perhaps the model would not have had the success it did. In early Christian authors reference is made to Homer's "cord" or chain or to some of Plato's assertions merely as a prophetic sign of extra-Jewish monotheism, but without any acceptance of the general model.[22] Whether as itself or as a "Jacob's ladder" revisited and reformulated, this type of interpretation, along with purely "spiritual ascent" defined in terms of Christian perfection, virtue or piety, is not uncommon in the early Fathers, as in Basil the Great's *Homily* on Ps.1, 4 (the ladder is correspondingly τὰ τέλεια, ἡ ἀρετή, ἡ εὐσεβεία), in Gregory of Nyssa's *De Anima et Resurrectione* in a more precise manner,[23] in the case of John Climacus (see comments below), or compare in the Latin West, say, Zeno of Verona or Chromatius of Aquileia.[24] However, two factors aided the transmission of Plotinus' thought and its mixing with Aristotle's sharpening and formalization of the earlier folk model. The first is that, in later Arab translations of the Syriac renderings of Plato's and Aristotle's works, much that belongs to their "schools" is labeled with their names, so that, for example, Nicolas Damascenus' Περὶ Φυτῶν is transmitted as one of Aristotle's minor works in the later Latin translations from the Arabic or Greek (Aristotle *De Plantis*). In the same way Plotinus' *Enneads* are transmitted as the work "Theologia Aristotelica," hence the Western Middle Ages' attribution and acceptance of Plotinus' ideas. In the second place, much of later acceptance of some of Plotinus' basic ideas is to be found in their transmission, suitably mixed with an Aristotelian analytical framework, in the Isagoge and Sentences of his pupil and spiritual heir Porphyry (ca. 233–305 AD). The One and the Many, descent/ascent, orders, intermediaries, the ambiguity of the soul, tempered, however, by Aristotelian categories, are all to be found here, see, for example, Isagoge 6, 14–25. This sort of reasoning is also present in his Sentences XI. Introducing Aristotelian analytical categories such as genus and species, it is obvious he is attempting some sort of synthesis, perhaps even a *captatio benevolentiæ* for Plotinus' model, though this is not given us to know, in a kind of new "onto-henological" model, where Aristotle is used to interpret and integrate, in cases such as τὸ Ἕν (Plotinus) = τὸ εἶναι (Aristotle), Πλῆθος (Plotinus) = τὸ μὴ εἶναι (Aristotle) etc. In this new model genera and species are no longer analytical tools but are used as effective intermediaries, binary steps in a descent among naturally ordered beings, real "substances" and no longer scientific properties, as is evident in Isagoge 17, 9–10.

In the Greek-speaking world this henological interpretative model, integrated with suitable Aristotelian analytical tools, belonging, however, to a purely ontological model, is carried on in the various manuals of Proclus (410–485 AD)[25] with his substantive categories Ἕν, Ἑνάδες and Noῦς in the works of Jamblichus and Pseudo-Dionysius Plotinus' categories are hierarchized and woven into a "chain-of-being" scheme.[26] In his *Commentarius in Platonis Parmenidem* Proclus even connects the Neoplatonic model explicitly with the "golden chain of being" of Homeric origin, cf. his comments on the One in the passage (Cousin 1961: 1100) "καὶ πάλιν ὅτι οὔτε ἀρχὴν ἔχει οὔτε τελευτὴν, διὰ τοῦ πρὸ αὐτοῦ καὶ ἑξῆς ἀεὶ κατὰ τὴν χρυσῆν ὄντως σειρὰν τῶν ὄντων ἐφ' ὧν πάντα μὲν ἐκ τοῦ ἑνὸς ἀλλὰ τὰ μὲν ἀμέσως τὰ δὲ διὰ μιας μεσότητος τὰ δὲ διὰ δυοῖν, τὰ δὲ διὰ πλειόνων, πάντα δὲ ἁπλῶς ἐκ τοῦ ἑνός".[27] The hierarchization of this chain is very largely the work of Pseudo-Dionysius at the end of the fifth century, as McGinn demonstrates (1972: 84) "...it is true to say that it is the Pseudo-Dionysius who makes the concept of hierarchy central to his system and who is the chief influence in most explicit medieval expressions..." etc. In the Latin-speaking world Porphyry's Isagoge is translated directly, first by Marius Victorinus, then by Severinus Boethius (480–526 AD). In the second case Boethius tries to further Porphyry's attempt at synthesis and mediation between Plato, Plotinus and Aristotle and his translation has an immediate and enduring success in the West. One might also note that, apart from his translation, Boethius hints at an ordered series of derivations, whose prime mover, however, is love, in a Christian sense, a "sacred bond" of ascension towards God, in his *Consolation of Philosophy*.[28] Already Augustine of

Hippo, more than fifty years before, had written in his *Contra Academicos* of attempts to mediate Plato and Aristotle, at least from the point of view of their philosophical premises rather than from purely religious ones, and he writes positively in this sense of Plotinus. More influenced by his reading of Cicero's *Academica Priora* than by direct acquaintance with Greek sources, therefore by Clytomachus' Academy's direct interpretation of Plato (Clytomachus > Philo of Larissa > Antiochus of Ascalona 130–67 BC who had Cicero as his direct pupil in ca. 79–78 BC), most of Augustine's references are to Plato so interpreted (see *Contra Academicos* III, XVII, 37–38), though towards the end of the third book he mentions Plotinus' work as being a genuine interpretation of and sequel to Plato's thought, as in *Contra Academicos* III, XVIII, 41, and to Plotinus' school he attributes the attempt to reconcile Plato with Aristotle, (see XIX, 42). The only serious moment in which a hierarchy is introduced into the reflections of Church Fathers like Ambrose, Augustine et al. is not in classification as such but in the naming process, owing to ambiguities suggested by Genesis (2.19), where all other creatures are brought before Adam for naming, while fish are not even mentioned, as explicated in Isidore of Seville's *Etymologiæ* XII.VI.4 "Pecoribus autem et bestiis, et uolatilibus ante homines nomina imposuerunt, quam piscibus, quia prius uisa sunt et cognita. Piscium uero postea paulatim, cognitis generibus, nomina instituta sunt, aut ex similitudine terrestrium animalium, aut ex specie propria, siue moribus, seu colore, uel figura, aut sexu.[29] This, on its own, certainly is not sufficient to justify a wholesale introduction of hierarchies into reasoning about classification. Probably as important as Augustine throughout the whole of the Middle Ages was A. Th. Macrobius' *Commentarii in somnium Scipionis*, with its continuous references to World-Soul = Jupiter (*Commentarii* I.17, 14),[30] quotations from Plotinus, from Plato's Timæus in Chalcidius' Latin versions, distinctive hierarchies, the progression of souls, descent and ascension, continuous use of Æneid VI, explicit mention of Homer's chain (*Commentarii* I.14, 15),[31] "man" the microcosm, in short the whole conceptual baggage of the Neoplatonic model and its expression.

In the 7th century AD, the "ladder" image was becoming commonplace, though at first not in any Neoplatonic, Gnostic sense, only that of "spiritual ascent," as in John Climacus' *Scala Paradisi*,[32]. The "ladder" is, of course, associated with Jacob's ladder and the cosmic tree, though no hierarchical order in creation is assumed, as is evident from the *Scala Perfectionis Step* XXX, 9–10, 12–13. Hierarchy, the cosmic tree and Jacob's ladder seem more associated in a Neoplatonic sense in Pseudo-John Chrysostom (VIth Sermon, Holy Week) in more or less the same period. Perhaps more importantly, however, a famous saint and influential Church Father, Maximus the Confessor in the sixth century, links the ascension theme to ordering principles and distinctions in a decidedly Neoplatonic fashion in his *Ambigua*, even though he makes no direct mention of the "golden chain" or ladder theme.[33]

In the West the ontological school that follows a more or less Aristotelian model appears to dominate, despite the success of Boethius' translation of Porphyry, as for example in the works of Anselm of Canterbury as late as the 11th century: an ontological model is constructed in his *Proslogion* proceeding from common sense criteria regarding the created up to the Creator. Ranking is an important criterion beginning with rocks, plants, animals, men etc. which are not, as he explains in *Monologion* I, defined in terms of themeselves but "per aliud,"[34] while, in ranking, relationships between elements are more important than order in itself, as Anselm says"…these very relations wherewith men are refered one to the other by no means exist by virtue of each other, because these same relations exist by virtue of the subjects they define…. Since then all things that exist do so by virtue of the One itself, doubtlesssly the One exists by virtue of itself. All else, therefore, exists by virtue of something else, it alone exists by virtue of itself."[35] Notwithstanding references to the One or *Summum Bonum* (= God), the hierarchical relation is relation rather than ordering between the One that is invariable and the many that present variability (see relativity in *Monologion* XV, invariable vs. variable in *Monologion* XXV). Even Thomas Aquinas in the thirteenth century subordinates order to relation (at the beginning of the *Summa Theologica*: "Hierarchia est gradus, id est relatio…"and later at Q. 108),[36] the ontological model is basic (*De ente et essentia*), while any reference to Jacob's ladder and ascension is appropriate to the contemplative life beyond and above an analytical model for generating creation and created order,[37] rather as in John Climacus, six centuries earlier. However, Pseudo-Dionysius and Macrobius have not gone unheeded, starting with faint echoes we may find at the beginning of the sixth century in Isidore of Seville or even Gregory the Great, as in the Man = Microcosm theme, with its Neoplatonic premises. The biggest contribution to the spread of Neoplatonism is in the works of the famous Irishman John Scot[t]us Eriugena towards the end of the 8th century: his translations of Pseudo-Dionysius, his mixing of Plotinus, Pseudo-Dionysius, Gregory of Nyssa, Maximus the Confessor and other Greek sources in his famous *Periphyseon* (= *De Diuisione Naturæ*) in ca. 850 AD are all amply dealt with in the literature (Edelstein [1953], McGinn [1972], Hintikka [1992] and many others), probably in a more correct way than in Lovejoy 1961.

Greatest acceptance of the disguised "henological" model seems to occur among a number of monks in the period between the eleventh and twelfth centuries, many of whom, surprisingly, were pupils of the French scholar William of Champeaux, some rebels with respect to him, like the famous Breton Pierre Abelard, others more securely chanelled into current Western monastic mysticism, such as the Augustinian Victorines, i.e. the Saxon (or Norman ?) Hugh, the Scot Richard, or the Frenchmen Absalon and Godfrey, or perhaps more importantly many Cistercian scholars such as the French brothers Thierry and Bernard of Chartres, with their pupils of the Chartres school, William of Conches, Bernard Sylvestre (of Tours), William of St. Thierry, and the Englishman Isaac, Abbot of Étoile, etc. The Noῦς = Mens = Christ, Anima Mundi (World-Soul) = All-Soul= Holy Spirit equivalences, with consequent properties, are commonplace in Abelard (e.g. *Introductio ad Theologiam* I. XVII) as are the hierarchical distinctions in ascending/ descending orders (e.g. *Theologia Summi Boni* III. 2. 54, 69 etc.). The ladder with descent as well as Man = Microcosm are common in Hugh of St. Victor,[38] with a long discourse dedicated to hierarchy following Pseudo-Dionysius (*Comment. in Hierarchiam Cælestem S. Dionysii Areopagitæ*). To the latter theme Godfrey of St. Victor offers a whole work (*Microcosmus*), to the former Absalon of

St. Victor's *Sermons In Adventu Domini* mix Jacob's ladder with Homer's chain and Virgil's bough, one side indicating this world's glory, the other otherworldliness, through both hierarchy and dependence.[39] Probably the most indicative works are William of Conches' *Glosæ super Platonem* and *Glosæ super Macrobium*, real compendia of Neoplatonism in its synthesis with Christianity, or Isaac of Étoile's *Epistola de Anima*,[40] as well as Bernard Sylvestre's *Megacosmus et microcosmus*, with its unfolding of hierarchies, orders and dependencies.[41]

A century later this new intellectual dominance seems to pass *via* the new Franciscans rather than Augustinian canons or Cistercian mystics, for example Bonaventura's *Itinerarium mentis in Deum*, especially in Prologue. 3, I. 2, I. 3 (an explicit reference to Genesis 28.12), where Christ is the ladder itself (I.3, IV.2) and where he uses the ladder, rungs/ steps, ascension/ climbing images in the sense of hierarchical ordering. See also "spiritus noster hierarchicus" in IV, 4 (our spirit ordered hierarchically) or "Et tandem manuducimur per hierarchias et hierarchicos ordines, qui in mente nostra disponi habent ad instar supernæ Ierusalem" in IV.7 (And finally we are lead by hierarchies and hierarchical ordering of our mental activities that are to be disposed towards the heavenly Jerusalem), where the Seraphim are levels of enlightenment on the ladder ("illuminationes scalares" in I.9). Here the ladder of ascent is definitely connected with the cosmic tree and strict ordering principles, since the rungs are *differentiæ* (II.10), Christ is the ladder iteself in a complex equivalence Christ = door = ladder = tree-of-life.[42] In more or less the same period another Franciscan, Raymond Llull, is taking the "tree of knowledge" image further, a tree which branches into distinctive, contrastive steps which are genera, species and individuals, as in *De Nova Logica* 4,6 "Cum arbor quidem ad primæ distinctionis significantiam ponitur? Ut per ipsam ars sit magis significatiua" and 204, 5 "Adhuc fit syllogismus, ponendo suas proprias dictiones in tribus gradibus, in arbore significatis, scilicet in genere, specie et indiuiduo... ."[43] The henological model and the tree-of-life image, later spread by Nicholas of Cusa at the Padua Studium through his appreciation and use both of Llull and the Chartres school, are here taking over the older ontological model.

The Two "Models" and Folk Classification

It can be convincingly argued that folk systems, *whether European or non-European*, are more in line with the first system than with the second, while the birth of modern scientific classification seems to be more the product of the second than the first, or perhaps of an overlay of the second on the rough patchwork of the first. Even if the second presents a "hierarchic conception of nature typified by scientific taxonomy" (Ellen 1996: 108–109), "hierarchy" in the sense of "ranking" and "relations" (see above observations) is also present in the first model with its capacity to reify nature, a reification around which I.K. can be instituted and built on (Descola 1992: 110 ff., Ellen 1996: 104 ff.). In the second model perhaps what is more important is the algebraic approach to hierarchy: in fact, scholars such as Leibniz and Descartes, the antecedents of Linnæus, Vallisnieri, Darwin and modern scientific models, ever more divorced from folk models, make explicit mention of Llull's particular algebra and his tree-of-knowledge distinctive model. Older Indo-European systems, the sources of European folk taxonomies, roughly follow the first general model, at the folk level, so that anthropologists' references to "modern western European folk-taxonomies" seem to be rather wide of the mark, hardly pertinent.[44] Indo-European specialists have realized for some time that there are precise symmetries at work between humans, other animals, birds, fish and even plants within naming processes which have nothing to do with similitude, metonymy or metaphor, but with an ancient worldview, even if much slighted by some anthropologists. That plants have "limbs," "sinews," "blood,""heads," "mouths" is already an intrinsic property and part of classification at the time of Theophrastus, so that similitudes, metonyms and metaphors are not involved. They may well be when "veins" are applied to streets or lanes (Lat. vena > OFr. venelle, Southern Italian vinedda), "wrinkles" to small backstreets (Lat. rūga > S. Italian ruga, French/ Portuguese rue/ rua, Albanian rrugë) etc. The mapping of territory in terms of human, animal or plant parts (e.g. "branches") is metonymic, sometimes metaphoric, as the relation in Greek between βρογχίη "system of water canals" vs. "bronchi" or "tracheal tubes," but not that between βράγχιον [= βρογχίη) "bronchi" (animals) and "gills" (fish). Hence, when Albert the Great in his *De vegetabilibus* V. 18 writes "sed [si] a radice, quæ est loco cordis in plantis, non videtur illi aliquid simile aliqua plantarum, nisi medulla...," he means "the root that occupies the position of the heart in plants," with a locative and functional equivalence, and is not employing metonymy. This latter is represented in processes such as IEW 474 *gwer-(gwer-) [< NIE *K'WER-[45]] "neck"; "throat" > Greek βρογχίη "canal," Albanian grykë "river mouth"; "strip of sea that wends inland," Celtic Irish brágh, bráighid, Gaelic brághad "neck"; "top, upper part" > "promontory," Lat. gurges, gurga (vs. gula "throat") "whirlpool"; "gorge" (in Toponymy we have Calabrian Vurga/ Vurganu "gorge" and/ or "whirlpool"), or Lat. sinus "bay," Alban. gji, gjiri "bay, inlet"<"breast" (Lat. sinus) etc. In the same way "breast" / "neck"> "hillock" is a common IE metaphor, cf. IEW 170.1 *bhreu-[s]- n- "swell" [< NIE *BREUS- < *BER-] > Celtic bruinne/ bron "breast" used as a metaphor for "hillock"/ "slope of a hill," derived *bhru-s-njo- > W. bryn "hill," IEW 694. 3 *mā- [NIE *MAH-]> *māmā, *mammā > Lat. mamma > mamĭlla "breast" (female), metaphorically for "hillock," also common in Romance toponymy (Màmmola in Calabria etc.), or Latin cĭrrus "cockscomb"> Span. cerro "nape of neck"> "hill," IEW 558.1 *ken- [NIE *KEN-]/ *k[e]n-okko- "neck" (as in Germanic) > Celtic cnoc/ cnwch "hillock," or Gr. λόφος "nape of neck"> "hillock," etc.

Natural Symmetries

Even in the last series of examples there is a great deal of doubt as to whether or not metaphorisation is at work or whether we are dealing with a more complex set of symmetries. One of the great Neoplatonic themes is that of "Man" constituting a microcosm of the macrocosm that is "World," in other words that man sums up within his being, within his defining and characteristic properties, all that is present in the cosmic chain of being. Such a position is more than evident in the second to third century AD, e.g. in Macrobius' works, in particular in the *Commentarium in somnium Scipionis* 2.12.11 ff. "Ideo physici mundum magnum hominem et hominem breuem mundum esse dixerunt, per similitudines igitur ceterarum prærogatiuarum, quibus deum anima uidetur imitari, animam deum et prisci philosophorum et Tullius dixit. Quod autem ait mundum quadam parte mortalem ad communem opinionem respicit, qua mori aliqua intra mundum uidentur, ut animal exanimatum uel ignis extinctus uel siccatus umor; hæc enim omnino interisse creduntur. Sed constat secundum ueræ rationis adsertionem, quam et ipse non nescit nec Vergilius ignorat... ."[46] In the centuries immediately following even a number of Church Fathers take up the theme, for example Gregory of Nyssa in his *De Anima et Resurrectione*,[47] Nemesius of Emesa in the Pseudo-Gregory *De Hominis Opificio*, even such unsuspectable authors as Gregory the Great, as in *Homeliæ in Euangelia* II, XXIX *In Ascensione Domini* (Migne P.L. 76 col. 1214B), or Isidore of Seville's *De Natura Rerum* IX (Migne P.L. 83 cols. 977C–978A). This is a claim that lies above and beyond any theory of metaphor or metonymy defined in contemporary authors of the sources mentioned, e.g. in Isidore of Seville's *Etymologiæ* IX. 2"…sicut os dici solet pro verbis, sicut manus pro litteris, genere locutionis illo quo is qui efficit per id quod efficitur nominatur …" etc. [just as "mouth" is used for speech style, "hand" for writing, following that rhetorical genre in which we use the actor's name for the action implied], it implies rather more than the essential unity of all living substance in a plant-animal-man continuum but is connected with the idea that earth itself, its rocks and soil, is a large animal,[49] so that eventually the whole of created, together with its Creator, forms a gigantic continuum earth, inanimates-plant- animal-man-planets, stars, divinities, as in Seneca's mega-scheme"…the idea appeals to me that the earth is governed by nature and is much like the system of our own bodies in which there are both veins (receptacles for blood) and arteries (receptacles for air). In the earth also there are some routes through which water runs, some through which air passes. And nature fashioned these routes so like human beings that our ancestors even called them "veins" of water…" etc.[50] This "chain-of-being," to keep to Lovejoy's expression, embracing the created and creators below and above the usual *plant-animal-man* continuum, with links between plant and mineral, between man, stars and divinities, in which Man is a compendium of the properties of the whole chain, the microrepresentative of World, its microcosm, spreads through Macrobius and Pseudo-Dionysius, via Eriugena's translations and compendia, into Western thought in the twelfth and thirteenth centuries: probably most representative in this sense are the statements in Isaac of Étoile's *Epistola de Anima*[51] and William of St. Thierry's *De Natura Corporis et Animæ*,[52] as well as being eminently explicit in the works of the Seraphic Doctor Bonaventura.[53]

The Projection of Parts

The fact that in the spreading of properties beyond the usual continuum we may well have to do with a more generally accepted penetration of the henological model into the historically older and more widespread model implies that the projection of human and animal properties and parts on to territory, rocks, geological formations etc. may not necessitate a particular cognitive model to explain metaphor, at least not in longterm, historical discussion, and how much or what precisely medieval, Renaissance and modern "metaphors," or even metonymy, owe to this past is not always clear in modern thinking. It might, then, be useful to look at the extended use of body parts, not just within the plant- animal- man continuum but beyond this continuum, in the works of classical authors. I propose Lucretius' *De Rerum Natura*, Seneca's *Naturales Quæstiones* and Macrobius' *Commentarium in somnium Scipionis, Somnium Scipionis* and *Saturnalia*: for two reasons, the first because they are important amongst the texts thusfar discussed, the second because they are historically pertinent to the Romance folk systems which I wish to deal with in this paper. They will be examined in detail in the extended essay based on this paper.

Obvious Symmetries

A symmetry principle involving the *plant-animal-man* continuum definitely antedates seventeenth-eighteenth century science, despite Vallisnieri's (1688: 426) assertion that "…queſta ſimetria regolatiſſima dell'Univerſo"[54] was uncovered by seventeenth century science. Parallels between plant and animal had already been succinctly dealt with in Albert the Great 500 years before, cf. *De vegetabilibus* I.136 "Obmissa autem comparatione partium plantæ ad partes animalium, loquamur nunc de generali diuersitate partium plantarum…" [But leaving aside the comparison of plant parts with animal parts, let us now discuss the general difference in plant parts], even if with due limitations and hedging.[55] As a clear demonstration of the parallelisms historically evinced between the relevant sections of this continuum we might take IE items for "pigs" and "fetus" (= suckling animal, < "womb"), which are definitely not confined to "pigs" or even to the animal world. We can set up a four-term IE subsystem as in figure 1. In the first case Greek and Latin do not have "porpoise" (sūs/ ὕαινα are other fish: according to Thompson [1947: 272] the reference is most probably to *Puntazzo puntazzo* [Cetti]), Celtic (morhwch/ mocc-sáil), and Germanic (marswīn) do; plants are usually spontaneous species of others or morphologically similar noxious plants (Gr. ὑοσέλινον, Celt. ffenigl-yr-hwch), though the Latin edible mushroom is suīllus < *su-in-lo-s (S. Italian sillu/ siddu *Boletus edulis* etc.). In the second one fish may well be salmon (Ir. orc, obviously related to orcán "pig"), but in Latin porcus marīnus, OFrench. porpeis > porpoise, pigs and porpoises/ dolphins are clearly associated. Insects are usually woodlice, plants species of cyclamin (French/ Italian dialects pa[i]n-porçin, Celtic bara'r hwch), though sometimes purslane is involved. "Pig" and "badger," "porpoise" and "salmon" are symmetrically arranged with each other and with determined plants. Both pig and porpoise seem to be culturally and cognitively central to symmetrical schemes, for often both represent life-forms in some European folk taxonomies, e.g. S. Italian fera"dolphin"(< Latin"wild animal"), opposed to mbestinu/ mmistinu, mmastinu"shark"< bestia"animal" as distinct from man, perhaps crossed with *mansuētīnus REW 5320 > mastino etc. > mastiff, since dogs and sharks are often symmetrically associated, N. Italian nima:l"pig" (Lombardy: "animal"), mas-cio"pig" (Veneto: "male," new fem. mas-cia, i.e. the "male" animal par excellence).[56]

Other well-known types are IEW 752-53 *MŪS- > (1) body part (muscle; arm), (2) animal (rodent: life-form ?), in almost all IE languages, (3) gasteropods etc., from Greek μῦς Cardium edule, μύα, μύαξ, μυίσκη, Lat. mūrex, myacēs,[57] mussel etc./ fish (Ital. pesce sorcio, SouthItal. pisci sùrici'i funnu *Chimæra* sp. etc.) (4) as plant derivatives of the sort μυοκτόνος, μυάγρος, μυός ὦτα = *μυωτίς etc.;[58] IEW 43-44 *ANGW-/ *EG(W)- [NIE *HANG-] and IEW 914. 3 *SEU(-LO-) > (1) body part (Latin lacertus"muscle"; Greek σαῦρος penis").

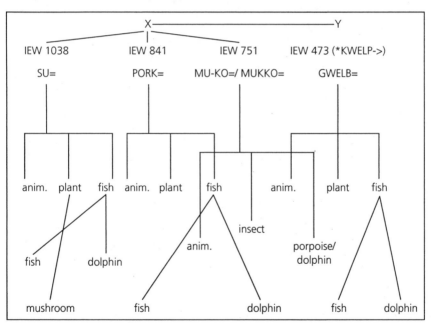

Figure 4.1 A Four-term IE Subsystem

Semantically akin is Albanian hardhith/ harith"uvula,"which is formally and historically a diminutive of hardh, hardhëlë, hardje, hardhucë "lizard," (2) animal/ reptile (Latin lacerta, Greek σαῦρος; cf. the Albanian example), (3) fish:[59] mackerel/ buck mackerel[60]/eel (Celt. eascung/ llysyw), (4) plants from Gr. σαύρα, σαυρίδιον, Latin saurion on (*Nasturtium; Sium; Arum*: cf. Lat. serpentaria for *Arum* sp., often associated with snakes in folk taxonomies as well as with lizards). For discussion see also Sanga (1992). These last examples are indicative from the point of view of a worldview, i.e. of how man related or thought he related to other animals and plants, in schemes that are definitely not metaphorical. This is rather different from the view expressed in Brown (1979: 266) who, commenting mūs 'mouse' > mūscŭlus > muscle, claims: "An absolute lexical change through metaphor results when the linguistic connection between a word and the descriptive label created through its modification is no longer recognized." In the cases discussed "shoulder muscle" (+ "muscle") and "thigh muscle" (+ muscle) are small animals that move under the skin of the man-animal as a composite higher animal who exists conjunctively with other animals in an animal continuum, therefore the starting point must be "mouse" + "shoulder muscle," "lizard" + "thigh muscle" + "penis," at least from the Indo-European point of view. Successive semantic movement is "mouse" + "shoulder muscle" > "shoulder muscle" + "muscle" > "muscle" by expansion, e.g. Italian muscolo, French muscle > English muscle, "mouse" + "shoulder muscle" > "shoulder muscle" + "muscle" > "shoulder" by restriction, e.g. Romance (Calabrian) dialect mušcu "shoulder." Birds are not excluded from this discussion but tend to be associated symmetrically with fish rather than with animals and plants[61] (long series such as κίχλη- tŭrdus, κόσ–συφος- mĕrŭlus, κίσσα etc.).

Examples are thick on the ground and show a systematic symmetrical patterning, except in cases where associations are formed with objects (rodents > crustaceans > boats etc.) or meteorological phenomena (dragons, snakes, reptiles in general, more than rodents),[62] the most obvious cases for postulating metaphorical processes. The thesis that common sense cognition does not involve symbolic thinking, as claimed in Atran (1996),[3] contrasts in too overt a manner the hypothesis, implicit in a certain linguistics, that common sense rationality underlying naming processes is directly associated with semantic relations and symbolic thought.[63] Symbols used to web over Cognitive Models are by no means irrational and seem to be a constant in Indo-European culture over large periods of time and over large spaces. To take the lizard-buck mackerel example and add to it a penis-salmon one,[64] we find Models continuously interacting at symbolic and Gestalt levels, as in figure 2. The continued association penis-stick-salmon produces in Celtic languages, when elements are combined, the symbol or image of the crooked stick = protruding jaw or snout of adult male salmons = penis = male salmon, as e.g. Welsh generic ehogyn > varietal camogyn (< cam "crooked" or "bent") "adult male (sea) salmon" (see relevant section in Part II for details).

Similar considerations may also underly Indo-European words for "large fish," i.e. IEW 958 *[s]kwalo-s or *kwalo-s, which Pokorny (1959: IEW) already felt to be a Hungro-Finnish (Uralic) borrowing into Indo-European. Alongside Latin squa#lus and pan-Germanic outcomes (Old Islandic hvalr, Anglo-Saxon hwæl > English whale, Old High German hwal > German Weller/ Waller), Bomhard (1981: 433; 1984: 241 item 165) also gives a Proto Indo-European form *KwÆL- [P-] "dog" or "young animal" (> Germanic, e.g. whelp, and Baltic), which he associates with a "Proto-Semitic" *KLB- (Hebrew keleb, Arabic kalb "dog" etc.) as one of his Nostratic items. Apart from such obvious parallelisms, Ruhlen (1994) associates IE *[S]KWAL-, not distinct from *KWEL-[P], with a Proto-Afro-asiatic form *KAL-

(Uralic *kala, Tungus *xol-sa, Gilyak *q'ol, Chukchi klxin, Eskimo iqaluk) which might be labelled"large fish"or"fish."Beyond parallel developments"fish">"small fish"vs."large fish"and associations"large fish" (voracious ?) and"dog,"for which see Italian"pesce cane"="shark,"a formal association might be set up between the *KwAL-/ *KAL- bases and the Proto-Dene-Caucasian root *K'VLČ'V- for "penis,"where V = vowel (see Ruhlen [1994: 26–7], table 4). Given such parallel series, it might not then seem strange to associate Proto IE *PÆS- "split" and "penis"(candidates according to Bomhard 1984: 194 item 17 for Nostratic) not just with the Proto-Celtic base for"salmon"but even with the Proto-IE "fish" root *PÆ[I]S-SK-. Symbols belong then to the semantic and cognitive "webs" with which we map our cognition of reality and are, if not universalistic (?), tendentially family-universals. I would also claim that they form, in some sense, the inborn "playful" tendencies alongside the"classificatory" urge that distinguish the *pars ludens* of *Homo sapiens ludens*, so I certainly would not exclude them from the model. Whether these are more than family-universals is another question, though they do not correspond to any homogeneity criterion of the ecological niche, since Tokharian B (China) had laks- for"fish,"like German Lachs, Anglo-Saxon leax and Russian lososvm"trout" (see Gamkrelidze-Ivanov [1990]). The British Isles, Central and Northern Europe, the Russian Steppes and part of Mongolia do not seem to have much in common ecologically. There is, however, I feel, insufficient data and analysis to push our theoretic arguments much further at the moment.

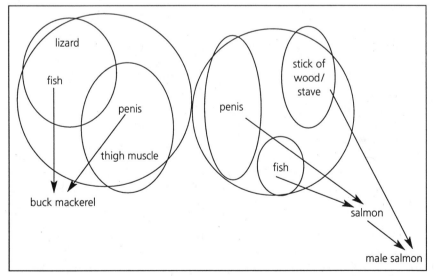

Figure 4.2 The Interaction of the Lizard—Buck Mackerel and Penis—Salmon Models at Symbolic and Gestalt levels

Folk Taxonomies vs. Scientific Taxonomy

To go back to Western European folk taxonomies, that they are distinct from scientific taxonomy becomes obvious when we look at real folk systems in Europe, not at imagined ones. The Apiaceæ in Trumper, Vigolo (1995, botanical folk system of the Veneto region, Italy) show a distinctive organisation along lines that I would suggest in figure 3.

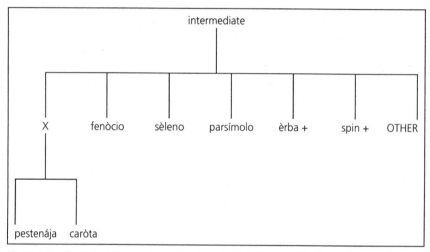

Figure 4.3 A Suggested Organization of the Botanical Folk System of the Veneto Region, Italy

Branches of X cover *Daucus* sp., *Pastinaca* sp., *Carum carvi* L., *Pimpinella major* (L.) Hudson (outside Apiaceæ also *Ranunculus ficaria* L.), *fenòcio* covers *Fœniculum* sp., *Anethum* sp., *Pimpinella anisum* L., *Ferula communis* L., *sèleno Apium graveolens* L., *parsímolo Petroselinum sativum* L. (with modifier it represents, outside this family, *Ranunculus acris* L., *Ranunculus repens* L.), "grass" + modifier *Heracleum sphondylium* L., *Peucedanum* sp., "thorn" + modifier *Eryngium* sp. (associated then with *Berberis vulgaris* L., *Paliurus* sp. and other "thorny" plants), while "other" covers rarer names referring to single, less common plants (e.g. s-ciavásene *Angelica sylvestris* L.). As in the case of Latin apium (> apiastrum = *Ranunculus sceleratus* L., perhaps also *Ballota nigra*(L.) Hayek: André 1985), the lexicalisation cuts across scientific genera, while some lexemes are raised to a level that is intermediate between the genus and Berlin's erstwhile intermediate level, inasmuch as they cover the *Ranunculus* genus. The situation is even more complex in the North Calabria Pollino mountains (Trumper, De Vita, Di Vasto [1997]), where there is even less correspondence with scientific genera (e.g. *funúcchjə* covers *Fœniculum* sp., *Tordylium* sp., *Carum carvi* L., though NOT *Ferula communis* L., differently from the first Romance situation, and is extended to *Anthemis* sp. but not to species of *Ranunculus*!) and some eleven productive lexical oppositions are involved. Many such cases might be discussed. Take another Indo-European example: in Celtic the Apiaceæ family involves eight productive lexico-semantic oppositions, together with eight unassociated, non-productive cases,[65] the productive cases being (1) ffenigl (*Fœniculum* sp., *Silaum* sp., *Meum* sp., *Crithmum* sp.), (2) pannas (*Heracleum* sp., some *Berula* sp., variably *Pastinaca* sp., *Sium* sp.), (3) persli (*Petroselinum* sp., *Aethusa* sp.), (4) perllys = mers (*Apium* sp., *Chærophyllum* sp.), (5) cegid (*Conium* sp., *Myrrhis* sp., *Oenanthe crocata* L., perhaps some *Anthriscus* sp.), (6) moron divided into moron (*Daucus* sp., *Pastinaca* sp.) and dyfrforon (*Berula* sp., some *Apium* sp.), (7) ceinioglys (lit."penny-plant": *Hydrocotyle* sp., *Sibthorpia europæa* L.), (8) derivatives of celyn (normally *Ruscus* sp., *Ilex aquifolium* L.) to cover *Eryngium* sp. There is a fuzzy ground between (2) and (6), indication of intergroup variability which would have to be investigated, a principle raised in Ellen (1979a), but which most investigators have not come to grips with, and a further problem of the supergeneric *celyn* which functions rather like derivatives of *spina* in Romance languages (plants with thorns and/ or thistles),[66] a sort of ill-defined "intermediate" to be commented on.

Some Tentative Conclusions

In conclusion, I would raise four points here: (A) "Western European" folk taxonomies are NOT directly related to MODERN scientific classification: relations that DO hold have to be investigated carefully;[67] (B) it is sometimes,though not always, unproductive in field situations to tie one's analysis too strictly to standard scientific classification, it may even be counterproductive in initial interviews;[68] (C) a great deal of apparent metonymy and metaphor is NOT such, as I have tried to show, in a historical (and perhaps longterm cognitive) perspective, so Lakoff's extremely productive Cognitive Model will have to take into account longstanding cognitive and semantic symmetries; (D) some fuzzy problems are decidedly not fuzzy in any historical sense and probably not in any cognitive sense, either: longterm overlapping between "seal" (*Phoca* sp.), "walrus" (*Trichecus* sp.), "porpoise" (*Phocæna* sp.) and "dolphin" (*Tursio* sp.) is historical-semantic—also etymological—in Indo- Europæan,[69] as well as being cognitive, for example small children seem only to have a "largeness" dimension operative here but no genus distinction between these natural types.

Notes

1. (1967: 6) "objective biological facts no longer dominate the scene."
2. Ellen (1979b: 18–19) "An example is the distinction made by Berlin and his associates (1968) between special-purpose and general-purpose schemes. Folk classifications are special, determined by specific subsistence and other culturally defined requirements; while scientific classifications are seen as general and topical... Bulmer has argued the opposite (1970), and now Berlin (1974 p. 267) appears to agree that in this sense folk and scientific taxa are fundamentally alike."
3. This, I feel, is the only correct way to interpret Atran's (1997: 21) rather exaggerated statement: "The boot-strapping enterprise in Western science began with Aristotle, or at least with the naturalistic tradition on Ancient Greece he represented. His task was to unite the various fundamental forms of

the world... into an overarching system of 'Nature'... It also implied combining the various life forms by 'analogy' into an integrated conception of life."

4. That these are refinements of a longstanding ancient tradition is also aknowledged in Atran (1996[3]: 30) "... one finds these groupings in Greece well before Plato and Aristotle proposed logical division... Granted, then, that such divisions were not Aristotelian in origin, but more primitive and popular, then they must be 'special-purpose' in the sense of utilitarian..." My contention is that (1) such divisions and structuring of reality are older Indo-European products and most probably cut across even large linguistic families like Indo-European, Afro-Asiatic, Sino-Tibetan etc., (2) such products are not necessarily utilitarian, rather have to do with Man as an innate classifier, though superimposed utilitarian considerations (hunting and gathering society > agricultural society ?) evidently highlight certain types of contrast, identification and lexicalisation.

5. For Theophrastus plants are ἔμψυχα like animals, have parts perfectly parallel with those of animals (*De Historia Plantarum* I, 1, 9), have veins, sinews, muscles, flesh (I, 2, 3), overall likenesses are stressed (I, 2, 5–6) etc. Plants, however, are not WHOLELY animals in ALL their respects (I, 1, 4).

6. In the Περὶ Φιλοσόφου ἱστορία" he accepts Aristotle's and the general opinion that plants are ἔμψυχα but animals have a different degree of "possession" of soul from plants, cfr. his discussion in chap. 38, which begins Ἔι ζῷα τὰ φυτά etc.

7. In the first case this is self-evident, in the second less so. However, Lucan equates *anima* and *sanguen*, both are "life-spirit" or "vital energy," blood spurting from a wound is the fleeing soul of animals and humans [3, 622–23 "volnere multo// effugientem animam lassos collegit in artos"], veins carry 'soul' [3, 640 "aera non passus vacuis discurrere venis// artavit clausitque animam"], with the usual Indo-European ambiguity between "breath" [> "air"], "soul" and "life." For a detailed comment on this author's transmission of traditional concepts see Migliorini (1997: 100–104).

8. For details see Needham 1956 vol. 2 § 9.5. Given the complex history of divers world-views, their reduction in the terms offered by Ellen (1996: 103) "... the opposition drawn between the holistic systemic vision of 'traditional,' 'tribal' or 'archaic' societies and the dualism of the modern scientific and dominant Judæo-Christian tradition." seems an unnecessary simplification and basically incorrect. The threefold Greek scheme is an opposition between vegetative «soul», sentient «soul» and sentient, rational «soul».

9. *De Rerum Natura* II, 1153–1157: "Haud, ut opinor, enim mortalia sæcla superne/ Aurea .de cælo demisit *funis* in arva/ Nec mare nec fluctus plangentes saxa crearunt,/ Sed genuit tellus eadem quæ nunc alit ex se." (my italics) [For, as I believe, there is no *golden cord*, that, from on high, let mortal beings and times descend on to nature, nor did the sea nor its sighing floods create rocks, but earth bore those very elements that it now nurtures from its very innards].

10. Hinc porro quamuis animantem ex omnibus unam/ ossa cruor uenæ calor umor uiscera nerui/ constituunt;...

11. "Ψυχὴ πᾶσα παντος ἐπιμελεῖται τοῦ ἀψύχου; τελέα μὲν οὖν οὖσα καὶ ἐπτερωμένη μετεωροπορεῖ τε καὶ πάντα τὸν κόσμον διοικεῖ, ἡ δὲ πτερορρυήσασα φέρεται ἕως ἂν στερεοῦ τινος ἀντιλάβηται, οὗ κατοικισθεῖσα, σῶμα γήϊον λαβοῦσα, αὐτὸ αὐτὸ δοκοῦν κινεῖν διὰ τὴν ἐκείνης δύναμιν, ζῷον τὸ σύμπαν ἐκλήθη, ψυχὴ καὶ σῶμα παγέν, θνητόν τ' ἔσχεν ἐπωνυμίαν."

12. "Αἱ δὲ δὴ ἄλλαι γλιχόμεναι μὲν ἄπασαι τοῦ ἄνω ἕπονται, ἀδυνατοῦσαι δέ, ὑποβρύχιαι συμπεριφέρονται, πατοῦσαι ἀλλήλας καὶ ἐπιβάλλουσαι, ἑτέρα πρὸ τῆς ἑτέρας περωμένη γενέσθαι."

13. "... κράτερα ἐν ᾧ τὴν τοῦ παντος ψυχὴν κεραννυς ἔμισγεν."

14. *Naturales Quæstiones* II, 45, 1–3: "Sed eundem quem nos Iouem intellegunt, rectorem custodemque uniuersi, animum ac spiritum mundi, operis huius dominum et artificem, cui nomen omne conuenit" (They recognized the same Jupiter we do, the controller and guardian of the universe, the mind [perhaps better 'soul,' given the general belief context] and spirit of the world, the lord and artificer of this creation. Any name for him is suitable). He is "destiny," "Nature" itself etc.: "hic est ex quo nata sunt omnia, cuius spiritu uiuimus. Vis illum uocare mundum, non falleris." (It is he from whom all things are naturally born, and we have life from his breath. You wish to call him the Universe? You will not be wrong).

15. Adjusted LOEB: "Whether the world is SOUL, or a BODY governed by nature, like trees and plants, there is incorporated in it from its beginning to its end everything it must do or undergo." I have

restored the manuscripts' reading ANIMA against the Loeb correction ANIMAL, since Seneca seems to be generally against the idea that the earth is ANIMAL but not that it has SOUL, cf. *Naturales Quæstiones* VI, 14, 2 "Nec ut illi paulo ante dicebant quibus animal placet esse terram. Nisi hoc est, quemadmodum animal, totum vexationem sentiet." (But not as they said a little above- those who are fond of the theory that the earth is a living creature [lit. 'animal']. Otherwise, the earth would feel the agitation all over, the way an animal does).

16. It is dyadic because Νοῦς "intelligence" implies νοητός "intelligible" (what can be "understoood" by "intelligence"), whereas the "One" has no necessary binary partner.

17. And every life is a kind of intelligence, even though one is fainter than another, just as life's very self (*Ennead* III, 8, 8, 16–17: Καὶ πᾶσα ζωὴ νόησις τὶς ἀλλὰ ἄλλη ἄλλης ἀμυδροτέρα, ὥσπερ καὶ ζωή.).

18. "Τῆς δὲ θεωρίας ἀναβαινούσης ἐκ τῆς φύσεως ἐπὶ ψυχὴν καὶ ἀπὸ ταύτης εἰς νοῦν καὶ ἀεὶ οἰκειοτέρων τῶν θεωριῶν γιγνομένων καὶ ἐνουμένων τοῖς θεωροῦσι καὶ ἐπὶ τῆς σπουδαίας ψυχῆς προς τὸ αὐτὸ τῷ ὑποκειμένῳ ἰόντων τῶν ἐγνωσμένων ἅτε εἰς νοῦν σπευδόντων..." (But, as contemplation ascends from nature to soul, and soul to intellect, and the contemplations become always more intimate and united to the contemplators, and in the soul of the good and wise man the objects known tend to become identical with the knowing subject, since they are pressing on towards intellect... (LOEB)).

19. See, for example, Deck (1967: 29, chap. 3) "... the notion of an ordering of lives, knowledge and theoriai according to clarity and obscurity... may be connected with a notion of continuity, but is not equivalent to it... The grades of life-knowledge could be more distinct than is envisaged in the continuity passage, but the mention of them has been prefaced by the allusion to the progress of the excellent Soul. It seems best, then, to suppose that the passage is not to be taken merely as an account of statically distinct grades of life-knowledge, but at least to imply their dynamic relations with one another." The comment is, of course, to *Ennead* III, 8.

20. Soul for Plotinus is not, then, a "universal, applying to the life principles of diverse beings," to use Deck's words (1967: 34), as it is in Aristotle's *De Anima*, or in works that depend on this, up to, say, Tertullian's *De Anima*, Basil of Cesarea's Ὁμιλίαι Θ᾽ εἰς τὴν ἐξαήμερον, Ambrose of Milan's *Exameron*, Mamertus' *De statu animæ*, Cassiodorus' 9th book of the *Variæ* entitled *De anima*, or, in the later Middle Ages, up to, say, Albert the Great's *De anima, De animalibus* or *De vegetabilibus*, Thomas Aquinas' commentary to the *De Anima* or his *De ente et essentia*.

21. Even the same Deck (1967: 114) when he states "The 'descent' of soul is a metaphor... if in truth there is no descent, ..., perhaps soul is merely Nous considered as related to the visible cosmos... nature would be Nous considered as related to plants, the earth, and the vegetative functions in animals."

22. An example would be Pseudo-Justin Martyr (perhaps 200–250 AD) in Cohortatio ad Græcos § 24 (Migne P.G. 6 col. 284) "Τῷ γὰρ ἑνὶ καὶ πρώτῳ Θεῷ τὴν ἐξουσίαν καὶ τὸ κράτος ἁπάντων Ὅμηρος διὰ τῆς χρυσῆς ἐκείνης ἀναφέρει σειρᾶς..." (= «Homer attributes authority and power over all to the One and Primary God by means of that golden cord image of his»).

23. See Migne P.G. 46 col. 89A.

24. In Zeno's case the twin sides of the ladder refer to the two Testaments (*Tractatus* XIII. *De somnio Jacob*), though for the first time we find the equivalence Ladder = Cross (*Tractatus* XIII, 4 Migne P.L. 11 col. 433A "Scala autem proprio nomine crux uocatur, quia per ipsam Dominus Iesus Christus mysteria uniuersa conficiens et concludens Patri et Ada reportauit..." (However, the Ladder is properly called the Cross, for by it our Lord Jesus Christ composing and finishing off the Universe's sacramental bonds brought Adam back to the Father). For Chromatius (*Sermo de Octo Beatitudinibus*) the Ladder has rungs and steps of beatitude and perfection.

25. There are already some ambiguous passages in Gregory of Nyssa's *De Anima et Resurrectione* and Nemesius of Emesa's *De Hominis Opificio*, once attributed to Gregory, though this is not the place for commenting them.

26. O'Meara (1975; 1987) is probably right in saying that Lovejoy's (1961) reading a rigorous theorising of hierarchies and "chain-of-being" model into Plato and Plotinus is *stricto sensu* incorrect, and that the model we know as the "henological" one is a later theorising on earlier notions which were not strictly a world-model. However, there is no doubt that this latter model stems from splinters of Platonic thought elaborated by Plotinus.

27. And again It possesses neither beginning nor completion by having something either preceding or following It, always in accordance with [the steps of] a genuinely *golden cord of beings* from which all things proceed, out of the One, but in an immediate procession, things produced by a single mediation, or even a dual or even multiple one, all things proceeding in a simple manner from the One. (My translation and italics).

28. *De Consolatione Philosophiæ* II, 8:"Hanc rerum *seriem* ligat/ Terras ac pelagum regens/ Et cælo imperitans amor." (And Love who commands from heaven, holding earth and sea in his sway, binds together this *concatenation* of things).

29. However, men gave names to tame animals, wild animals and birds before they gave them to fish, because the former were seen and perceived first. Shortly afterwards names were instituted for fish, once their genera were recognized, names either based on their similarity with earth animals, or by the recognition of particular species, or on the basis of their habits or colors or shape or sex. The statement is a partial, though more explicit, copy of parallel statements in Ambrose's *Exameron*.

30. "Ipsum denique Iouem ueteres uocauerunt, et apud theologos Iuppiter est mundi anima" (To sum up, our ancestors called It Jupiter, and in theologians'parlance Jupiter is World- Soul).

31. "...et hæc est Homeri catena aurea, quam pendere de cælo in terras deum iusisse commemoraret..." (...and this is Homer's golden chain, which he quoted the God as having ordered to be hung down from heaven to earth...).

32. The"royal way"(*Step* I, 29"...ἀλλ' ὁδῷ βασιλικῇ πορευθῃς") or"way of ways"(*Step* II, 14–15"...οἱ τὴν ὁδὸν τῶν προειρημένων ὁδῶν πορευομένοι,..."), is constituted by ASCENT (ἀνόδος in *Step* XXX, 9–10, ἀνάβασις in *Steps* II end/ VI, 7/ X, 4/ XII, 5/ XXIII, 19/ XXV, 26-27/ XXVI 2nd Part, 10/ XXVII, 11/ XXX, 9–10), with associated verb TO ASCEND (ἀναβαίνειν in *Steps* VI, 7/ XXIII, 19/ XXX, 9/ XXX, 12–13, XXX *Adhortatio*; ἐπιβαίνειν in *Steps* I, 30/ III, 21/ XII, 5; ἀνέρχομαι in *Steps* V, 31/ XXV, 26–27). The ascent to God is via the LADDER (κλῖμαξ in *Steps* IX, 1/ XXX, 9-10/ XXX, 12-13/ XXX, *Adhortatio*) which has spiritual RUNGS (βαθμοί in *Steps* IV, 100/ V, 31/ VII, 25/ VIII, 13/ IX, 7/ XI, 4/ XX, 10/ XXII, 27/ XXIII, 19/ XXIV, 1/ XXV, 26–27/ XXX, 9-10/ XXX, *Adhortatio*).

33. See *Ambigua* 222a in Migne P.G. 91 col. 1305C. Progressive distinctiveness as one descends the scale of beings, proceeding from the divine «forge» or perhaps «production line or belt» in modern parlance, is even more evident in *Ambigua* 221b [Migne P.G. 91 col. 1305A–B].

34. *Monologion* I (definition of summum bonum):"Illud igitur est bonum per seipsum, quoniam omne bonum est per ipsum. Ergo consequitur, ut omnia alia bona sint per aliud quam quod ipsa sunt, et ipsum solum per seipsum." [Thus it is good in itself, since all its good is self-defined. It follows hence that all other good things are defined in terms of something other than themselves, and it alone is self-defining].

35. *Monologion* III:"... ipsæ relationes quibus referuntur non omnino sunt per inuicem, quia eædem sunt per subiecta... Quoniam ergo cuncta quæ sunt, sunt per ipsum unum, proculdubio et ipsum unum est per seipsum. Quæcumque igitur alia sunt, sunt per aliud, et ipsum solum per seipsum."

36. Kuntz (1987: 4)"...Thomas is very modern to think of order as a relation, and to recognize the obvious asymmetry of such an ordering relation as above-below or before-after."He considers Thomas' ranking concept to be natural (8):"Hierarchy and the other modes of order are necessary because we find such structures in the cosmos and we live in societies based upon nature..."This seems less provable than his concept that ranking is a necessary part of all and every scientific model (10"Hierarchy is also found in science because nature requires a distinction between levels..."etc.).

37. *Summa Theologiæ* II–II Q. 181 art. 4:"Sed de his quæ pertinent ad dispensationem ministeriorum Dei, unus angelus docet alium, purgando, illuminando et perficiendo. Et secundum hoc, aliquid habent de vita actiua quandiu mundus durat, ex hoc quod administrationi inferioris creaturæ intendunt. Quod significatur per hoc quod Jacob uidit angelos in scala ascendentes, quod pertinet ad contemplationem, et descendentes, quod pertinet ad actionem."[But with regard to'stewards entrusted with the mysteries of God,'(I would say that) one Angel teaches the other, purging, enlightening and perfecting. According to this scheme of things, they (= Angels) partake in the active life as long as this world endures, because they have to do with the government of lower creatures. This is intended in the passage where Jacob saw Angels ascending a ladder, which is the pertinence of the contemplative life, and then descending it, pertinence of the active life].

38. Cf. *De Unione Corporis et Spiritus*, Migne P.L. 177 col. 285C.

39. *Sermo IV In Adventu Domini* (Migne P.L. 211 col. 36D–37A).
40. World-Soul, the Great Cosmos in a chain of derivation and hierarchy, Soul as a Trinity, the descending hierarchies of angels and beings, Man as Microcosm, are all present in his *De Anima*: the "golden rope" or "golden chain" is present not only in the *Epistola de Anima* (Migne P.L. 194 col. 1885C–D on the concatenation of the four elements: "Hac igitur quasi aurea catena poetæ, uel ima dependent a summis, uel erecta scala prophetæ ascenditur ad summa de imis" [Either by means of this the poet's golden chain, as it were, lower things, therefore, hang from [depend on] higher ones, or through the prophet's straightened ladder we ascend from lower orders to the highest ones]), but even in some of his sermons, cf. *Sermon In Nativitate Beatæ Mariæ* (Migne id. col. 1874D–1875A): "Et hic est primus naturalis mundus, aureum Saturni sæculum, aureaque catena poetæ" (And this is the first world, the natural one, the golden age of Saturn and the poet's golden chain) etc.
41. The most indicative passage is Book I (*Megacosmus*).4.68–80 which significantly concludes "In magno uero animali [sc. World, in its Platonic sense] cognitio uiget, uiget et sensus causarum præcedentium fomitibus enutritus. Ex mente [sc. Nous = Christ] enim cælum, de cælo sidera, de sideribus mundus unde uiueret, unde discerneret, linea continuationis excepit. Mundus enim quiddam continuum, et in ea catena nihil uel dissipabile uel abruptum." (In the GREAT ANIMAL flourishes cognisance, as well as sentience that feeds on the fuel of things gone before. For heaven proceeds from MIND, from heaven the stars, from the stars WORLD, by which it could live and sieve out things, there emerges a line that is a continuum. For WORLD is, as it were, continuous, and in that chain [of things] there is nothing discrete or abrupt). His "uel dissipabile uel abruptum" seems, however, to be antagonistic to pure Neoplatonism with its ordered hierarchies.
42. Similar troublesome images (if images) were already contained in the eleventh century in some exponents of the Chartres school, e.g. in Gilbert de la Porée and his pupils: there is a strange passage in the *Commentarium in Primam Epistolam ad Corinthios* (Gilbert, or his school ?) ch. X (Landgraf 1945: 162) where tree-of-life = cosmic order = cosmic serpent = Christ, cf. "Serpens figurat Christum, forma serpentis figurat similitudinem carnis peccati; palus in quo erigebatur serpens significat lignum crucis in quo suspensus est Christus" (The snake is a figure of Christ, the snake-form presents a figure of the likeness of fleshly sin; the pole on which the snake was raised on high signifies the tree of the cross whereon Christ was hung, etc.). We seem to have rather gone beyond redemptive equivalence, whether *ex simili* or *e contrario*, between the tree of life (Genesis) and the Cross, Aaron's snake-rod (Numbers 21) and the Cross, ever present to medieval man in the chants and lessons of the medieval solemnities of the Invention (May) and Exaltation (September) of Holy Cross, and we seem here to have finally reached cosmic figures.
43. Respectively 4,6: When is, indeed, the tree set to mean the first contrastive level? Its purpose is to make our model more meaningful, 204,5: Up to this point it becomes a syllogism that sets its categories in terms of three rungs (steps), indicated on the tree, steps which are genus, species and the individual.
44. They seem to be referring to a particular type of postagricultural, postindustrial urban individual who has little contact with the natural world except through hypermarket food chains, books, television and computer screens, not with European agricultural society and its millenary cultural schemes. This former is our immediate future, since in modern Europe less than 15% of our population have some vital contact with nature, bound up, that is, with agricultural subsistence models, above and beyond modern environmental movements.
45. NIE = New Indo-European is the "new-look" version of Indo-European incorporating both laryngeals and the new consonantism with *P, *P,' B = *p, *b, *bh etc., proposed in all studies since the pioneering work of Gamkrelidze and others. H represents any laryngeal, without further specification, since one cannot enter into the problem of the number and quality of reconstructed laryngeals in the present study. IE = Indo-European.
46. Naturalists thus claimed that World was a large version of Man, Man a limited version of World, and so according to a scheme of preferential likenesses, according to which Soul seems to imitate God, and earlier philosophers, as well as Cicero, claimed that God was Soul. But what Cicero says regarding a sort of mortality on the part of World is paying lipservice to common opinion, according to which some things [that exist] within World seem to die, just like an animal that loses its life force, a fire that has been put out, damp that has been dried out. For such things are in all respects believed

to have died. This is said to be rationally coherent in a truthful way, but he [sc. Cicero] was not unaware of what is rationally coherent, nor was Vergil really ignorant of what rationality entails...

47. *De Anima et Resurrectione* Migne P.G. 46 col. 28B–C, col. 28C–D.

49. Seneca, *Naturales Quæstiones* II.1.4"... ipsa animal sit an iners corpus, et sine sensu, plenum quidem spiritus sed alieni..." [whether [earth] itself is an animal or an inert substance, without sentience, to boot, certainly full of vital breath, but in the sense of "otherness"].

50. *Naturales Quæstiones* III.15.1–2"Placet natura regi terram, et quidem ad nostrorum corporum exemplar, in quibus est uenæ sunt et arteriæ, illæ snguinis, hæ spiritus receptacula. In terra quoque sunt alia itinera per quæ aqua, alia per quæ spiritus currit; adeoque ad similitudinem illa humanorum corporum natura formauit ut maiores quoque nosttri aquarum appellauerint uenas... ."Even Lucretius, in a previous period, certainly not the epitome of Platonic/ Neoplatonic thought, draws parallels that go beyond the usual plant-animal-man continuum to embrace in a substantial unity plants, rivers and rocks, as in *De Rerum Natura* II. 667–68"Tanta est in quouis genere herbæ materiai/ dissimilis ratio, tanta est in flumine quoque/..." [As great as are the material differences between plants, so are those between rivers, as well]: cf. *De Rerum Natura* II. 1013–1018 for further parallels.

51. Especially in the passage where man's connectedness with the whole of creaion, himself included, is expressed as a chain of properties hierarchically ordered (*Epistola de Anima* col. 1886A–B, Migne P.L. 194).

52. Book I (Migne P.L. 180 col. 698C).

53. See *Itinerarium mentis ad Deum* II.2"Notandum igitur, quod iste mundus, qui dicitur macrocosmus, intrat ad animam nostrram, quæ dicitur minor mundus, per portas quinque sensuum..."(Thus it must be noted that this world, called the"macrocosm,"enters into our soul, called the"microcosm," through the doorways of the five senses), II.3"Homo igitur, qui dicitur minor mundus, habet quinque sensus quasi quinque portas, per quas intrat cognitio omnium, quæ sunt in mundo sensibili, in animam ipsius." (Consequently man, who is called"microcosm,"has five senses which are rather like five doorways through which the cognition of everything in the sensitive world enters into his soul).

54. This extremely regulated symmetry of the Universe.

55. Cf. in this sense *De vegetabilibus* I.125:"Partes enim talium plantarum comparantur per imitationem quandam imperfectam membris animalium." [For the parts of such plants are thus compared by ⋅tue of a certain imperfect imitation with respect to animals' limbs]. He seems thus to be hedging rather in the fashion of Theophrastus I.1.4 (see note 5). Current Romance folk schemes still keep faithfully to such organic parallels, cf. "skin,""stalk,""pip" and similar concepts in Trumper,Vigolo (1995: 72–84), or the use of"eye,"'"mouth,"'"nail,"'"ear,"'"skin,"'"tongue,"'"tail,"'"foot"in plant specifications (Trumper, Vigolo 1995: 59–61), or the detailing of parts of Graminaceæ, in particular cereals, (140–144), as well as Scola, Trumper 1997.

56. It might be interesting to note that Latin and Romance outcomes of another of the"suckling"bases (IEW 241 *dhei-, NIE *DEI-) are referred to ovines rather than to swine, e.g. fœtus, fœtāre > Piedmontese féja, Central Veneto (VI shepherds' jargon, information M. T. Vigolo) féa [= còpana], N. Veneto (BL) féda, Carnian Friulian fède"sheep,"sometimes just"to bear"(of all animals) as in Veneto (North PD,VI) feáre; sometimes we find, with severe geographical restrictions, a shift towards chickens, as in Emilian [a]vdár/ [a]vdæ:r"to lay eggs" < fœtāre, while in more peripheral Romance we have reference to human progeny, i.e. Rouman. fắt"boy; princeling" (in Fairy Stories), fată «"girl"< older featặ etc.

57. For a discussion of Mediterranean loans of this type between Latin and Greek cf. Battisti (1960–61: 73 items 56, 57, 58, p. 86 item 152), as also μῦς > muraina"moray eel"= Lat. murēna subsumed under item 151, which Battisti tries to explain in terms of"bite"or"mouse-like teeth,"though this seems to me doubtful and will have to be rethought.

58. Boat names such as Greek μυοπάρων, Latin musculus = parua nauis, are similitudes or metaphors, probably based on the use of symmetrical terms for crustaceans, as in the case of caravelle < Gr. κάραβις derived from "crab"/ "crayfish" terms (ultimately from "horn" terminology o some other base?), which then become morphologically pertinent in the comparison.

59. The added association"fish":"penis,"accompanying"lizard":"penis,"is present in other Italian folk cultures, e.g. Salento (LE) zzobba = minchiale *Phycis* sp., which is connected with Sicilian zzubbu "penis"< Arab zubb id.(see Alessio [1960–61: 147] for observations); this is matched by a symmet-

rical association "jellyfish": "vagina" (NeoVenetian cón "jellyfish" < Lat. cunnus, Calabrian sticchj'i mari "jellyfish" lit. "sea-vagina").

60. This seems to be masked in Pliny (*Naturalis Historia* XXXII, 151) who calls the buck-mackerel sōrus. All we have to remember is that here Latin probably shows a folk Latial or Umbrian development with ō = au with respect to *saurus = σαῦρος which has normal Latin and Oscan -au-. N. Italian dialects show the -ō- development surèlo, suro (perhaps with back formation from the diminutive), S. Italian dialects the -au- development sáuru, sávuru etc. for this fish. See comments in Alessio (1966–67: 46). Latin in many cases shows central, standard -au- vs. provincial -ō-, a geographical and social problem in the Urbs: the Emperor Claudius was accused of calling himself in substandard fashion Clōdius, and the Roman upperclass was itself guilty of hypercorrection imposing -au- in plaudere and other cases where the onomatopœic base and form was always plōdere, based on plò-plò. It must be remembered that even as important a lexical item and concept as sōl 'the sun' shows a provincial, low-class -ō- instead of the expected -au- from IEW 881 *sau[e]l-, so social stratification in the Roman Empire had considerable weight. Even Pliny, then, with his proletarian, extra-urban form, is really attesting lizard = buck-mackerel.

61. A European exception is the cuckoo, associated with at least twenty plants in folk schemes.

62. Cf. the extension of "dragon" to whirlwinds, storms and heavy showers etc. in Romance, as in the excellent entry "dragunara" in Varvaro (1986: 294–95), with a diffusion from Sicily and Calabria to Lombardy, Romance Switzerland and France (FEW vol. 3, 150b), to which we might add snake derivatives for the same meteorological phenomena in NeoVenetian [b-, vissinèlo "whirlwind," < bissa *Natrix* sp. < bēstĭa], and in Celtic with extension to other meteorological phenomena (W. draig "dragon" = 1 mellt didaranau "lightning without thunder," 2. meteorite). For rodents we seem to have in Romance only Provençal and Tyrrhenian Calabrian [cud'i rattu] "rat-tail" items for "whirlwind." The "dragon" term is used classically in Latin authors applied, perhaps metaphorically, to coiled pipes for heating water, as in Seneca's *Naturales Quæstiones* III.24.2–3 "Facere solemus **dracones**... in quibus aere tenui fistulas struimus...ut sæpe eundem ignem ambiens aqua per tantum fluat spatii quantum efficiendo calori sat est; frigida itaque intrat, effluit calida. Idem sub terra Empedocles existimat fieri..." (We commonly construct **serpent-shaped containers**... in which we arrange thin copper pipes in descending spirals so that the water passes round the same fire over and over again, flowing through sufficient space to become hot. So the water enters cold, comes out hot. Empedocles conjectures that the same thing happens under the earth). It is a short step from this to whirlwinds.

63. Atran (1996[3]: 213) "Basic rationality, in other words, isn't a historical or cultural construct, but a cognitive universal. Symbolic speculations, though, are culturally idiosyncratic attempts to go beyond the immediate and manifest limits of common sense; that is, symbolic thought is post-rational, or at least posterior to common sense, which is rational."

64. In Part II I argue that the Celtic (eog/ éo) and Gallo-Latin (esox) generics for "SALMON" take their IE origin from the IEW 824.2 + IEW 824. 3 conjoint base sometimes defined "penis," which may be allied with the IEW 823.1 verbal root (a mooter point), i.e. *PES-OK-, sometimes with redetermination of the adjectival suffix *PES-A:K-.

65. These last are of the type buladd *Cicuta* sp., nodwydd-y-bugail *Scandix pecten-Veneris*, dulys *Smyrnium olusatrum*, paladr *Bupleurum* sp. etc. (Welsh examples).

66. Surprisingly Celtic languages do not have any intermediate covering both harmful Apiaceæ and Ranunculaceæ (Welsh indicates with crafanc "claw; talon" types of Ranunculus and Helleborus, cf. crafanc-yr-arth *Helleborus viridis*, crafanc-y-frân *Ranunculus repens, Ranunculus sardous, Ranunculus parviflorus*, crafanc-yr-yd *Ranunculus arvensis*, crafanc-yr-eryr *Ranunculus sceleratus*, more innocuous types of Ranunculus as blodau 'menyn lit. "butter flowers"), differently from Latin and some modern Romance languages and dialects. So much for Western European conceptual unity! The same goes for modern English folk taxonomy where the Ranunculaceæ are covered by "buttercup, crowfoot" and the originally learned "hellebore," harmful Apiaceæ by "hemlock" (also derived: "hemlock dropwort"), "cowbane" and "fool's watercress," without apparently any crossovers. Do Celtic and Germanic not, then, have the same lexical-semantic oppositions and perhaps cognitive categories as Latin and Romance?

67. Cf. Berlin (1992: 201) "One of the major tasks of ethnobiological description is to determine the several types of mapping relationships that hold between the two systems of classification, the folk and

the scientific."The debt of modern European folk taxonomies to medieval monasticism may be mentioned on this score (monks were largely responsible for the transmission of fruit-tree grafting and cultivation, systematic planting of particular cultivars, as also for knowledge of the plants involved in phytotherapy etc.). The details of this debt, responsible as well for folk translations of Greek and Afro-Asiatic originals, sometimes for their direct transmision in a garbled form with folk re-interpretations of no longer understood lexicon (examples in S. Italian dialects in Trumper, Scola [1997], Trumper,DiVasto, DeVita [1997]), often do not constitute a straightforward process. A simple N. Italian example: Piedmont *erba-dona* [= *erba d'i purìt*], Veneto *s-ciara—dòna = siligògna* [= *èrba da pòri*] for *Chelidonium majus* L., are folk re-interpretations of chelĭdŏnĭa < Gr. χελιδόνιον < χελιδών, i.e. swallow-wort, as a plant with skin-cleansing properties, cosmetically relevant for women, since in European folk tradition it was used to cure warts, hence the equally usual folk"wart-wort" names quoted. These last coexist alongside the deformed chelĭdŏnĭa products reinterpreted as "woman's wort" or "cleanse-woman." Other examples are more complex still.

68. I agree in part with Ellen's (1979b: 27) assessment"…it is not always helpful if we are tied to ideological distinctions a priori, such as those between'scientific'and'folk.' Inevitably, such an enterprise also involves an investigation into methods of demarcating boundaries of semantic fields, areas of overlap and fuzziness."

69. The story is too intricate to be gone into here but even English *walrus* < **hwæl-hros/* hros-hwæl < Anglo-Saxon hors-hwæl (cf. Celtic *morfarch* «sea-horse» = walrus) ties *Trichecus* in with whales and equines [in folk Romance systems"whales"are also associated with horses and donkeys, cf. S. Italian sceccu'i mari"sea-donkey"="whale"], whale was probably in origin connected with"shark"(IEW 958 [s]kwalo-: Latin *squalus*, Italian *squalo* etc."shark," etymological equivalents of"whale": these may all have to do with a"scale"root, as suggested by Alessio, though the problem is far from being resolved), the older Anglo-Saxon term for"whale" being *hran/ hron* (Bosworth, Northcote Toller, entry *hran*:"Her beop oft fangen seolas and hrones and mereswyn," glossing"Capiuntur sæpissime et vituli marini et delphinos necnon et ballenæ"), which is probably a Celtic borrowing from OWelsh *rhawn* [> mod. *moelrhon*] = OIrish *rón* [> mod. Gaelic *ròn*] usually"seal," sometimes"porpoise," as atttested in Canu Taliesin II (Trawsganu Kynan) lines 19–23 where the young princes sent as hostages to the Saxons ride their nags into the sea as if"seals"or"porpoises" ("kyfe[n]wynt y gynrein [donn] kygwystlon// gwanecawr [gwlychynt] rawn eu kaffon"], a text rather older than the Old English ones Bosworth, Northcte Toller quote [the battle reference is to ca. 580 AD, according to Nennius'*Historia Brittonum*, the language a little later, the extant copies much later, at least eleventh century or later]. In this case"whales,'"'porpoises" [including"dolphins"],"walruses"and"seals"are all interconnected in an intricate manner in the long, well-documented history of Indo-Europæan languages and dialects. The Celtic"seal"/"porpoise"items (Irish rón, Welsh rhawn > moelrhon) would seem outcomes of IEW *PER- "to bear"/ "young born," i.e. a derivative *per-a:n-a:, whence also Welsh erthyl"stillborn"; which Bomhard (1981: 405 IE *PER- = Proto-Afro-Asiatic *PRY- ; 1984: 194 item 16) considers a candidate for Nostratic.

References

AA. VV., 1985–1988, Thomas Aquinas: *Summa Theologiæ*, ESD critical edition of the O. P., 33 vols., Bologna: Ordo Prædicatorum.

Abelard, Pierre. *Introductio ad Theologiam* in Migne, Patrologia Latina vol. 178.

———— 1996. *Theologia Summi Boni*. Milan: Rusconi.

Absalon of St. Victor. *Sermones In Adventu Domini*, in Migne, Patrologia Latina vol. 211.

Albert the Great: see E. Meyer, C. Jessen, 1867.

Alessio, Giovanni. 1951. *The problem of balenare*. Word 7: 21–42.

———— 1960–1961. *Note etimologiche sulla terminologia marinaresca*, Bollettino dell'Atlante Linguistico Mediterraneo 2–3: 139–148.

———— 1966–1967. *Ichtyonymata*, Bollettino dell'Atlante Linguistico Mediterraneo 8–9: 43–58.

Ambrose of Milan. Exameron: see Banterle 1979.

André, Jacques. 1963. *Noms de plantes et noms d'animaux en latin*. Latomus XXII, Fac. 4: 649–63.

———— 1985[2]. *Les noms de plantes dans la Rome antique*. Paris: Société d'édition "*Les Belles Lettres*".

Angeli, P. F. and M. Pirotta. 1959. *Thomas Aquinas: In Aristotelis Librum De Anima Commentarium*. Rome: Vatican.

Anselm of Canterbury. 1995. *Monologion*, Milan: Rusconi [in the absence of a critical edition].

———— 1995. *Proslogion*, Milan: Rusconi [in the absence of a critical edition].

Aristotle, see Arthur L. Peck, David M. Balme and W. S. Hett.

Armstrong, A. H. 1966. *Plotinus: Enneads*. 7 vols, Cambridge Mass.: Harvard University Press, Loeb Classical Library.

Atran, Scott. 1996[3]. *Cognitive foundations of natural history. Towards an anthropology of science*. Cambridge: Cambridge University Press.

———— 1997. *Folk Biology and the Anthropology of Science: Cognitive Unversals and Cultural Particulars*. Preprint, Classification Session, Venice: Istituto Veneto di Scienze ed Arti.

Augustine of Hippona: *Contra Academicos*, in Migne, Patrologia Latina vol. 32.

Balme, David M. 1991. *Aristotle: History of Animals Books VII–X*. Cambridge, Mass: LOEB.

Banterle, Gabriele. 1979. *Sancti Ambrosii Episcopi Mediolanensis: Opera I, Exameron*. Milan: Biblioteca Ambrosiana and Rome: Città Nuova Editrice.

Barach, C. S. and J. Wrobel. 1964. *Bernard Sylvestre of Tours: De Mundi Uniuersitate I, siue Megacosmus et microcosmus*. Frankfurt on Main: Minerva.

Basil of Cesarea. *Homiliæ in Psalmos*, in Migne, Patrologia Græca vol. 29.

Battisti, Carlo. (a) 1960–1961. *Sui grecismi dell'ittionimia latina Parti I–III*, Bollettino dell'Atlante Linguistico Mediterraneo 2–3: 61–95, (b) 1962. *Sui grecismi dell'ittionimia latina Parte IV*, Bollettino dell'Atlante Linguistico Mediterraneo 4: 37–52.

Berlin, Brent, D. E. Breedlove and P. H. Raven. 1968. *Covert Categories and Folk Taxonomies*. American Anthropology 70: 290–99.

———— 1973. *General Principles of Classification and Nomenclature in Folk Biology*. American Anthropologist 75: 214–42.

———— 1974. *Principles of Tzeltal Plant Classification*. New York and London: Academic Press.

———— 1992. *Ethnobiological classification*. Princeton, New Jersey: Princeton University Press.

Bomhard, Alan R. 1981. *Indo-European and Afroasiatic: New Evidence for the Connection*, in Y. L. Arbeitman and A. R. Bomhard, *Bono Homini Donum: Essays in Historical Linguistics in Memory of J. Alexander Kerns*, Part 1: 351–475. Amsterdam: Benjamin.

———— 1984. *Toward Proto-Nostratic: A New Approach to the Comparison of Proto-Indo-European and Proto-Afroasiatic*. Amsterdam Studies in the Theory and History of Linguistic Science IV. Current Issues in Linguistic Theory vol. 27. Amsterdam: Benjamin.

Bosworth, J. and T. Northcote Toller. 1983[8]. *An Anglo-Saxon Dictionary*. Oxford: Oxford University Press.

Brown, C. H. 1974. *Unique Beginners and Covert Categories in Folk Biological Taxonomies*. American Anthropologist 76: 325–36.

———— 1979. *A theory of lexical change (with examples from folk biology, human anatomical partonomy and other domains)*. Anthropological Linguistics 21, no. 6: 257–76.

Bulmer, R. 1967. *Why is the Cassowary not a bird?* Man: 5–25.

———— 1970. *Which came first, the chicken or the egg-head?* in Pouillon, J. and P. Mananda, *Échanges de communications, Mélanges offerts à Claude Lévi Strauss*, 2 vols. The Hague and Paris: Mouton: 1069–91.

Bury, R. G. 1966. *Plato: Timaeus, Critias, Cleitophon, Menexenus, Epistles*. London, Heinemann, Loeb Classical Library.

Busse, A. 1887. *Porphyry: Porphyrii Isagoge et in Aristotelis Categorias Commentarium*. Berlin: Reimer (also includes Boethius' Latin version).

Chromatius of Aquileia. *Sermo de Octo Beatitudinibus*, in Migne, Patrologia Latina vol. 20.

Climacus, John Scholasticus. *Scala Paradisi*, in Migne, Patrologia Græca vol. 88.

Corcoran, T. H. 1971. *Seneca: Naturales Quæstiones*, 2 vols. Cambridge, Mass.: LOEB.

Cousin, V. 1961. *Proclus: Philosophi Platonici Opera Inedita 3, Procli Commentarium in Platonis Parmenidem*. Hildesheim: Olms.

Deck, J. N. 1967. *Nature, Contemplation and the One. A study in the philosophy of Plotinus*. Toronto: University of Toronto Press.

Descola, Ph. 1992. *Societies of Nature and the Nature of Society*, in A. Kuper, *Conceptualizing Society*. European Association of Social Anthropologists. London and New York: Routledge.

Duff, J. D. 1969. *Lucan: The civil war (Pharsalia)*. London: Heinemann, Loeb Classical Library.

Edelstein, L. 1953. *The Golden Chain of Homer*, in AA.VV. *Studies in Intellectual History dedicated to Arthur O. Lovejoy:* pp. 48–66.

Ellen, Roy F. 1979a. *Omniscience and ignorance: Variation in Nuaulu knowledge, identification and classification of animals*. Language in Society 8: 337–64.

_____ 1979b. *Introduction*, in R. F. Ellen and D. Reason, *Classifications in their social contexts*. London: Academic Press: 1–32.

_____ *The cognitive geometry of nature. A contextual approach*, in Ph. Descola and G. Pálsson, *Nature and Society. Anthropological Perspectives*. London and New York: Routledge: 103–23.

Englebrecht, A. 1885. *Claudius Mamertus: Opera, De statu animæ*. Corpus Scriptorum Ecclesiasticorum Latinorum XI. Vienna: Gerold.

Eriugena, John Scotus: *Versio S. Dionysii Areopagitæ; Periphyseon seu De Diuisione Naturæ*, in Migne. Patrologia Latina vol. 122.

FEW: see Von Wartburg, W.

Fowler, H. N., W. R. M. Lamb. 1971. *Plato: Euthyphro, Apology, Crito, Phaedo, Phaedrus*, London: Heinemann. Loeb Classical Library.

Fridh, Å. J. 1973. *Magnus Aurelius Cassiodorus: Variarum Libri XII*. J. W. Halporn *De Anima*, Corpus Christianorum XCVI. Louvain: Turnholt.

Gamkrelidze, T. V. and V. V. Ivanov. 1990. *The Ancient History of Indo-European Languages*. Scientific American, May Issue.

Granger, H. 1984. *Aristotle on Genus and Differentia*. Journal of the History of Philosophy XXII no. 1: 1–23.

Gregory of Nyssa: *De Anima et Resurrectione*, in Migne, Patrologia Græca vol. 46.

Gregory the Great: *XL Homeliæ in Euangelia*, in Migne, Patrologia Latina vol. 76.

Hamesse, J. 1972. *Bonaventura of Bagnorea: Thesaurus bonaventuranus I, Itinerarium mentis in Deum*. Louvain; new critical edition in J.G.Bougerol, C. del Zotto and L. Sileo, 1993. Rome: Nuova Collana Bonaventuriana, Città Nuova Editrice.

Heinsius, Nicolaus and Peter Burmannus. 1744. *P. Virgilius Maro: Æneid*, Amsterdam: Wetstenium.

Hett, W. S. 1980a. *Aristotle: Minor Works*. Cambridge, Mass.: LOEB (contains Damascenus' Περὶ Φυτῶν).

J. Hintikka, J. 1992. *La grande catena dell'essere e altri saggi di storia delle idee*. Ferrara: Artosi.

Homer, *The Iliad*. LOEB edition 1971.

Hort, Arthur. 1990[5]. *Theophrastus: Enquiry into plants*, 2 vols. Cambridge, Mass: LOEB.

Hugh of St. Victor: *De Unione Corporis et Spiritus*, in Migne, Patrologia Latina vol. 177; *Commentarium in Hierarchiam Cælestem S. Dionysii Areopagitæ*, in Migne, Patrologia Latina vol. 175.

IEW: see Pokorny, Julius.

Isaac of Étoile: *Sermones; Epistola de Anima*, in Migne, Patrologia Latina vol. 194.

Isidore of Seville: *De Natura Rerum*, in Migne, Patrologia Latina vol. 83.

Jeauneau, É. 1965. *William of Conches: Glosæ super Platonem*. Paris: Vrin.

Justin Martyr (Pseudo-): *Cohortatio ad Græcos*, in Migne, Patrologia Græca vol. 6.

Kühn, C. G. 1965. *Galen: Claudi Galeni Opera Omnia*, Vol. XIX. Hildesheim: Olms.

Kuntz, M. L. and P. G. Kuntz. 1987. *Jacob's Ladder and the Tree of Life. Concepts of Hierarchy and the Great Chain of Being*, New York. Bern, Frankfurt on Main and Paris: Lang.

Kuntz, P. G. 1987. *The necessity and universality of hierarchical thought*, in M. L. Kuntz and P. G. Kuntz 1987: 3–14.

Lakoff, George. 1990. *Women, Fire, and Dangerous Things*. Los Angeles: University of California Press.

Landgraf, A. M. 1945. *Gilbert de la Porée: Commentarium in Primam Epistolam ad Corinthios*. Rome: Biblioteca Apostolica Vaticana Studi e Testi 117.

Lévi Strauss, Claude. 1996. *The Savage Mind*. Oxford, New York: Oxford University Press.

Lindsay, W. L. 1966[3]. *Isidori Hispalensis Episcopi Etymologiæ siue Origines Libri XX*, 2 vols. Oxford: Clarendon.

Lohr, C. 1985. *Raymond Llull: De logica noua*. Hamburg: Meiner.

Lovejoy, Arthur O. 1961. *The Great Chain of Being*. William James Lectures 1933. Cambridge, Mass.: Harvard University Press.

Maximus the Confessor: *Ambigua*, in Migne, Patrologia Græca vol. 91.

McGinn, B. 1972. The Golden Chain, Appendix: Glosæ super Macrobium, *Cistercian Studies Series 15*. Washington D.C.: Cistercian Publications.

O'Meara, D. J. 1975. *Structures hiérarchiques dans la pensée de Plotin: étude historique et interpretative*. Leiden: Brill.

_____ 1987. *The chain of being in the light of recent work on Platonic hierarchies* in M. L. Kuntz and P. G. Kuntz, pp.15–30.

Menghi, M. and M. Vegetti. 1988. *Tertullian: De Anima*. Venice: Marsilio.

Meyer, Ernest and Carl Jessen. 1867. *Alberti Magni (Albert the Great): De Vegetabilibus Libri VII*, critical editino. Berlin: Unveränderter Nachdruck, Minerva reprint 1982: Frankfurt on Main.

Meyer-Lübke, W. 1992[6]. *Romanisches Etymologisches Wörterbuch*. Heidelberg: Winter.

Migliorini, P. 1997. *Scienza e terminologia medica nella letteratura latina di età neroniana (Seneca, Lucano, Persio, Peronio)*. Studien zur klassischen Philologie vol. 104. Frankfurt: Lang.

Murray, A. T. 1971. *Homer, The Iliad*. London: Heineman Loeb Classical Library.

Naldini, M. 1990. *Basil of Cesarea: JOmilivai Q jeij" th;n eJxahvmeron*. Milan: Valla Foundation.

Needham, J. 1956. *Science and Civilisation in China*. Vol. 2 *History of Scientific Thought*, Vol. 6 *Botany*. Cambridge: Cambridge University Press.

Needham, R. 1975. Polythetic Classification: Convergence and Consequences. *Man* (N.S.) 10: 349–69.

Nemesius of Emesa (Pseudo-Gregory of Nyssa). *De Hominis Opificis*, in Migne, Patrologia Græca vol. 44.

Paluello, L. Minio. 1966. *Severinus Boethius: Porphyrii Isagoge translatio Boethii et Anonymi fragmentum vulgo vocatum liber sex principium*. Paris: Verbecke.

Peck, Arthur L. 1993[3]. *Aristotle: History of Animals Books I–VI*. Cambridge, Mass: Loeb Classical Library.

_____ 1990[5]. *Aristotle: Generation of Animals*. Cambridge, Mass.: Loeb Classical Library.

_____ and E. S. Forster.1993[7]. *Aristotle: Parts of Animals, Movement of Animals, Progression of Animals*. Cambridge, Mass.: Loeb Classical Library.

Pokorny, Julius. 1959. *Indogermanisches Etymologisches Wörterbuch*, 2 vols. Bern and Munich: Francke.

Rackham, H. 1968[3]. *Pliny: Natural History*, 10 vols. Cambridge, Mass.: Loeb Classical Library.

REW: see W. Meyer Lübke.

Rouse, W. H. D., M. F. Smith. 1992. *Lucretius: De rerum natura*. Cambridge Mass.: Harvard University Press, Loeb Classical Library.

Ruhlen, Merritt. 1994. *On the Origin of Languages: Studies in Linguistic Taxonomies*. Stanford: Stanford University Press.

Sanga, Glauco. 1992. *L'ansia del serpente, Per un'etimologia etnolinguistica*. Proceedings Sodalizio Glottologico Milanese XXX: 74–79.

Stewart H. F., E. K. Rand and S. J. Tester. 1973. *A. M. T. Severinus Boethius, De theological tractates, The Consolation of philosophy*. London: Heinemann, Loeb Classical Library.

Thomas Aquinas. 1988. *De Ente et Essentia*. Hamburg: Meiner Verlag.

Thompson, D'Arcy W. 1947. *A Glossary of Greek Fishes*. London and Oxford: Oxford University Press.

Trumper, John B. and M. T. Vigolo. 1995. *Il Veneto Centrale. Problemi di classificazione dialettale e di fitonimia*. Padua and Rome: C. N. R.

_____ and A. Scola. 1997. *Nel regno delle graminacee: un excursus lessico-semantico e geolinguistico in Calabria*. Quaderni di Linguistica, Università della Calabria Serie Linguistica 6: 187–221.

_____ , Piero De Vita and Leonardo Di Vasto. 1997. *Classificazione botanica nella cultura popolare: le apiacee nella zona del Pollino*. Quaderni di Semantica a. XVIII, n. 2: 215–39.

Vallisnieri, Antonio. 1688. *Intorno all'ordine della progreſſione, e della conneſſione, che hanno inſieme tutte le coſe create*, in *Secunda Reſponſio Part III* chapt. IV. Venice [I wish to thank the Orto Botanico of Padua University for consultation of an original copy: the edition covers a large number of Vallisnieri's academic lessons at Padua in the late 17th century].

Varvaro, Alberto and R. Sornicola. 1986. *Vocabolario Etimologico Siciliano*, vol. 1 (A–L), Palermo: Centro di Studi Filologici e Linguistici Siciliani.

Von Wartburg, Walter. 1929–1970. *Französisches Etymologisches Wörterbuch*. vols. 25, Tübingen: Mohr, Bâle: Zbinden etc.

William of St. Thierry. *De Natura Corporis et Animæ*, in Migne, Patrologia Latina vol. 180.

Williams, Ifor. 1990[3]. *Canu Taliesin*, Caerdydd: Gwasg Prifysgol Cymru.

Willis, J. 1963. *Macrobius: Saturnalia, Commentarium in Somnium Scipionis, Somnium Scipionis*, 2 vols. Stuttgard and Leipzig: Teubner.

Zeno of Verona. *Tractatus*, in Migne, Patrologia Latina vol. 11.

Session I: Classification

Edited by Gabriele Iannàccaro

To prove how difficulties challenging the application of classificatory criteria some-times emerge, Maurizio GNERRE mentions what happened the first time he brought a young Shuar from the Upper Amazon to a village inhabited by white people. It was 1970 and the Shuar only had a single breed of dogs then, very small, ugly, lean and poorly kept ones, but excellent for hunting. In the village, seeing large and small quadrupeds of different colors, the Shuar started pointing to each of them, asking what it was and to the reply that it was a dog he commented that he could not believe it. The young Shuar thought the answer unlikely because, according to his own knowledge of dogs, he could not admit that any of the ani-mals he was seeing was a dog. Our word "dog" covers a wide range of shapes, col-ors, and sizes but the Shuar could not believe such a silly classification existed, a single word for a variety of animals obviously different from one another. Finally, when Gnerre pulled the tail of one of those sleepy dogs, the dog barked and the young Shuar, realizing that it manifested the transient attribute of barking, recog-nized it as a dog. Before that he had had no criteria to class all those perceptively different animals as similar. We must consider the importance of those attributes that appear momentarily and then disappear like barking; Kant mentioned them, as maybe did Aristotle. This side of classification is remarkable because it often becomes challenging and puzzling when people have to face totally new beings or creatures.

*

Oddone LONGO confirms that Aristotle mentions transient properties while con-sidering birds' voices, which are elements of distinction.

*

Alessandro MINELLI, as a zoologist (at Padua University), states by way of prem-ise that his observations differ from those of anthropologists and linguists. He wonders to what extent the language used by the scholars who discuss ethno-classification corresponds to the language used by systematicists-zoologists and botanists-today. He refers in particular to some interpretative semantic categories and to some themes that recur in the language of systematicists.

Today systematicists are apt to distinguish three or four operations: (1) group-ing objects-considering them a species approximately. (2) Establishing a rank for groups. (3) Naming, a separate operation possibly subsequent to the previous ones. Modern biological systematics questions important issues, the most per-plexing among them being the existence of many different definitions of species. We no longer know what a species is, and it would be interesting to know which of the current definitions in biology can correspond to those adopted by ethnologists. Moreover the ranks of Linnaean systematics are being questioned and modern systematicists, dissatisfied by them, almost seem to prefer systematics without any ranks.

*

Tim INGOLD raises three issues concerning, first, the status of prehuman universals, secondly the objectification of nature and thirdl, the meaning of classificatory acts.

As regards cognitive science and neo-Darwinian psychology, Ingold questions the supposition that human beings are pre-equipped with certain general capacities enabling the acquisition of culturally and environmentally specific contents. One cannot simply point to a stretch of DNA in the genome as if to say that there exists an innate device for the recognition of animacy or a built-in mechanism for the discrimination of colors. As there are no capacities, devices, or mechanisms that do not themselves arise within processes of ontogenetic development, any supposedly "innate" structure we may detect, at whatever stage, has already a history of development, in a certain environmental context, behind it. Therefore, no distinction can be made between the development of so-called "innate" structures and the acquisition of specific competence; it doesn't make sense to imagine a human mind prestructured by a Darwinian process of variation under natural selection, quite independently of the actual circumstances of human beings' life in the world.

Referring to Roy Ellen's observations about the "thinginess" of nature, Ingold maintains that explaining the universality of classification should not be a task for anthropology. In fact evidence from many regions of the world suggests that people do not immediately perceive their environments as discrete objects to be linked through the discovery of common attributes. On the contrary they perceive themselves as immersed in fields of relationships that are essentially continuous, so that the human task is not to discover parallels or similarities among diverse objects, but to discover or even construct discreteness out of a relational continuum. This suggests that classification is not a universal task for people.

Ingold concludes by wondering what is meant by "acts of classification." Every linguistic act arises from the social context from which it derives its force and meaning, and human beings, placed in the world, are involved in all kinds of linguistic acts all the time, constantly drawing each others' attention to certain aspects of reality. It would be wrong to consider these activities as acts of classification. In fact this would mean treating them as merely expressive of mental contents internal to the individual, ignoring their involvement in the social environment. A novice carpenter and a skilled carpenter do not differ because the master of the trade has a more detailed classificatory system in his head; rather, he is attuned through previous experience to notice qualities of the wood that he is working on which the novice simply fails to observe. Drawing the novice's attention to some aspect of the grain, the master is not performing an act of classification, nor is he transferring some portion of a classificatory scheme from his head into that of his pupil; in actuality, he is educating the novice's attention, so that the latter can feel and detect salient properties of the wood for himself.

Kenneth C. HILL (anthropologist of language) points out that by focusing our attention on the behaviour of non-humans, we can observe that some classifications are quite innate. A dog recognizes other dogs as dogs, and Gnerre's informant should have checked with some other dog he knew. In the same way human beings are defined by human beings as conspecific and, apart from any further consideration made by philosophers to define a human being, this classification is built into our nature. In fact, even if sometimes we try to rationalize certain groups as not worthy of being considered human because of their different shapes, sizes, colors, it is an act of the will to try to exclude, rather than to instinctively include. Both built-in classification and classification of another nature exist, and the latter is a matter of analysis for ethnography.

*

With reference to the form-function and perception-utility dichotomy, Glauco SANGA observes that Leroi-Gourhan's category of "functional aesthetics"provides a mean of interpretation. While a chopper, for instance, is not fully defined in its shape, a hand axe is symmetrical: now, this kind of symmetry is functional to the axe, but is also very closely concerned with the perception we have of the functionality of biological symmetry. In fact the kind of symmetry we see in a functional object is exactly the same as that which we see in the forms of animals, whose shapes evolved functionally with radial or bilateral symmetry—"functional shapes,"in fact. It is for this reason that form, selected by the biological and cultural evolution of the human race, should not be opposed to function in a form-function dichotomy. Neanderthal man collected polyhedra and odd-shaped shells, showing that he already had the perception of "deviant forms," of bizarre forms: it means that a theory of maintenance and recognition of shape was already available.

*

Ambrogio FASSINA, Professor of pathologic anatomy at Padua University, states that his work implies daily classification of what he analyses under the microscope. This classification demands truth criteria, above all in the interest of patients, and any suggestion about it is very welcome. Classification criteria derive from experience and from the subsequent results on-whether they live or die, for example. Unfortunately, some categories are statistically very difficult to identify, they can be either too similar or they can overlap; the generation of new categories is therefore a central problem. Pathologic anatomists are trying to use some systems of statistics that reduce parameters leading to diagnosis into simple terms and, through a machine, generate nodes that are very interesting from a statistical point of view, allowing the individuation of discrete categories.

*

In response to Minelli's comment, Brent BERLIN points out that in the folk systems that have been looked at, issues relating to classification would include the hierarchical organization of diversity, which is fairly typical of all systems of classification of plants and animals, scientific and prescientific. The one major difference that Minelli did not mention, is a concern for the basis of the activity of systematics in the modern sense: an understanding of the phylogeny or the historical relationships among the groupings.

Berlin invites Ingold to explain the results of an experiment he performs every year. After asking his students to classify various birds from the mountain region of the Jivaro area of Peru, he compares their classification, based on nothing but pure perception, with the classification established by ornithologists, based on scientific understanding, and with a further classification by the Jivaros themselves, based on real life experience. The groupings always turn out to be identical.

*

In response to Minelli's concern about the current ranking system in biological taxonomy, Roy ELLEN confirms that the matter demands attention and new reflection. New classifications, new ways of looking at biological diversity and particularly the impact of modern genetics, call for suitable new methodologies. Ellen suggests that Berlin should perform an experiment to discover what kinds of results students of ethnobiological classification would gather if they worked with a model for gathering field data, based on the kind of unranked systematics that Minelli talks about. The central problem about the current methodology used to generate data is that it is not known how independent the system of ranks that is discovered is from the kinds of concepts at the basis of classification.

Ellen underlines that his use of the term "act of classification" is purely rhetorical. Like Ingold he thinks that acts of classification are embedded in real situations, but that drawing somebody's attention to something in a real life situation, in his opinion may well involve the generation of some classification.

As regards the thinginess of nature, although Ellen considers the notion of innateness problematic, he recognizes the force of the experimental and ethnographic data. The universal human predisposition to turn the natural world into things should be regarded as the consequence of the interaction between normal developmental processes and environmental *stimuli*, since research is still far from identifying what its genetic determination might be. The ability to convert nature into discrete objects and to treat them as essence-based notions is one of those similarities that it is possible to detect in all systems of classification.

*

Oddone LONGO, taking up from Sanga, observes that while the separation of form from function may appear an aspect of homination, nonetheless forms maintain a certain functionality: for example the paintings of the Cro-Magnon of the Upper Paleolithic had their precise magical-ritual function, even though there is not only a correspondence of form-function, but already something else. In Plato the separation between form and function is taken much further, but not to the point of driving out form; with Aristotle there is a regression, and function once again penetrates form.

*

John TRUMPER claims that the scientific model cannot do without a ranking system or concept, because evidence based on informants attests that human psychology and cognition work in terms of order and kinship relations. However, neither Aristotle nor Theophrastus et al, in referring to *eidos* and *genos*, mean level x or y in a precise manner, and particular ranks could be redefined and reformulated. It would be advisable to reread Paul Kuntz's essay "The necessity and universality of hierarchical thought" (in M. L. Kuntz - P. G. Kuntz (eds.), *Jacob's Ladder and the Tree of Life* [Peter Lang, 1987]), and his reflections on the general validity of S. Thomas Aquinas'"Hierarchia est ordo, id est relatio, inter diuersos gradus,"in the first part of the *Summa*, where the emphasis is on *relatio* rather than on *gradus*.

The consistency of the basic concept of classification and the use of classificatory schemes are confirmed by what Lull wrote in the year 1200, in his Breviculum, Pars Dispositiva XVI: "Faber enim, quando facit clauum, priusquam ipsum faciat, figuram illius habet comceptam in imaginatione sua. Et illa figura, quæ sustentatur in clauo, est impressio figuræ, quæ est intra imaginationem" (When the smith creates a nail, before he actually makes it, he has its "figure" already conceptualized in his imagination. That very "figure," that is contained/maintained in the nail, is a print of the "figure" that exists within his imagination). Lull mentions a series of prototypes in human internal mental organization that are projected onto the real world and that guarantee the reproducibility of objects or the recognition of both natural phenomena and artifacts by a previous classificatory scheme that uses similar mechanisms. Leibnitz, Descartes, and other philosophers depended on Lull's concepts of rank, hierarchy, and classification; therefore it would be almost impossible to undo 700 years of Western European scholarship. It cannot be maintained that if there be no real classification outside the researcher's model, then there is nothing that might justify, in any formal manner, the continual development and repetition of classificatory images.

There must be similar prototypical and classificatory mechanisms behind some repetitions of images, configurations, and schemata that occur over long-term intervals and transculturally, and they do not seem to be random. A stick-penis-fish configuration underlying the history of the fish concept and its lexicalization in Indo-European provides an example: the same image, perhaps lexicalized in Indo-European more than 3000 years ago (the Indo-European base: *peisk-, *pisk- in Pokorny's *Indogermanisches Etymologisches Wörterbuch* [vol. 2, p. 796], which underlies the majority of lexicalizations of the "fish" concept in Indo-European, is perhaps *pes-"stick"="penis,"mixed with an elementary"water"base *pis-, modified with an adjectival formant in *-ko-) was lexically redeveloped in Celtic languages more than 2500 years ago (they redeveloped the image already present in their "salmon" word [Erse eó, Welsh eog etc.< Indo-European *pes- with -ŏkos or -ākos adjectival formants] but which because of the typical and continuous erosion operating in stress-timed languages cannot be retraced by the native speaker in the modern stage), and further recreated in Welsh (in Welsh a male spawning salmon is lexicalized now as *camogyn*, a reproduction of the "bent stick"image (*cam* = bent [stick]), no longer understood in *ehogyn*). This means that the speaker-user has some prototypes and classificatory mechanisms in his mind, that have validity outside any chosen scientific model.

*

Roy ELLEN maintains that the notion of rank is often confused with the idea of level in a taxonomy. Biological taxonomy appears to be facing the problem of level, wondering why a genus of a certain class of Invertebrates should be of the same significance as a genus of Mammalia.

*

Tim INGOLD responds to Berlin's challenge by way of an analogy. To illustrate the theoretical division between the models of perception of cognitive science and ecological psychology, he parallels the case of human walking to human perception. According to innatist theories, human beings are born with a built-in walking acquisition device, and then they acquire the cultural information that specifies how they should walk in the particular way they do in a particular culture. On the contrary, Ingold argues, the development of the devices is inseparable from the acquisition of culturally particular skills. Thus learning to walk is to learn to walk in one way rather than another. In the same way, cognitive scientists (Berlin included) regard perception as the operation of some sort of cognitive device located in the head, while ecological psychologists regard it as the functioning of a system constituted by the presence of the whole organism-person in an environment.

Of course, it is possible to separate out analytically all those features that are bound to be present in any environment from those present in some environments but not in others. For example, walkable surfaces are experienced everywhere, while the specific textures of asphalt, grass, earth, sand or snow would be encountered only in certain environments. By paring the environment down to its lowest common denominator, it would be possible to create the appearance of universality in the capacities that are supposed to develop therein. But the sifting of the particular from the universal is an artifact of analytic procedure, and does not reflect divisions in the world. From the start, the infant's experience of the ground surface is the experience of surface-textures specific to his or her environment; he or she does not experience walkable surfaces and specific textures separately. The perceptual experiments conducted and described by Berlin and the perceptual activity of ordinary life are different because the former are conditioned by experimental designs that structure environments in terms of artificial analytic divisions. These experiments tell us more about the mind-set of the experimenters than about the structure of the world or of the mind.

*

Brent BERLIN objects to Ingold's answer and claims that discontinuity is recognized by the Jivaro, while that of the students and the trained ornithologist are simply analytic distinctions, constrained by the nature of the experiment.

*

Oddone LONGO recalls Atran's demonstration that classifying is a human universal: for our first ancestors it was of vital importance to classify the surrounding world. Brent Berlin agrees.

*

Although Roy ELLEN finds it difficult to disagree with certain elements proposed by Atran, he objects to the notion of innate rank taxonomy and does not agree with those who maintain that human understanding of the natural world is based on some kind of innate module.

*

John TRUMPER agrees with Atran's position, adding that playfulness-playing with the things and beings of the environment, and the words and images to express them-seems as much a part of human nature as the very fact of classifying.

PART II

Naming

Glauco Sanga

The Ways of Naming Nature and Through Nature

The participants in this ethnolinguistics section are invited to think over ways in which "nature" is named and its names pass into other domains. We can roughly define "nature" as the physical and biological environment, not created but modified by human beings: animals, plants, minerals, territory, climate, atmospheric phenomena, body parts, colors, illnesses, etc. I should like to direct the discussion towards a few outstanding questions (of course other topics can also be brought into the debate).

1. Which linguistic mechanisms (phonetic, morphological, semantic, and rhetorical) are used for naming nature? Namely,
   (a) onomatopoea (phonetic mechanism), i.e., the imitation of external reality by linguistic means; or completely inside language mechanisms, where naming begins with the extant lexicon:
   (b) morphological derivation, i.e., the change in the shape of an extant name (e.g., through suffixation or composition or modification);
   (c) semantic extension, i.e., the addition of a new meaning to an extant name (also with semantic overlapping);
   (d) semantic specialization, i.e., selection of a meaning in a polysemic extant name;
   (e) metonymy (rhetorical mechanism), i.e., the use of an extant name with which it has connexions of contiguity or interdependence;
   (f) metaphor (rhetorical mechanism), i.e., the use of an extant name with which it has connexions of similarity.

Which mechanisms are outstanding or typical in naming nature? Suggestions could come from motivation (Alinei 1979, 1980, 1996a) and *appaesamento linguistico* "linguistic domestication" (Sanga 1997a, b).

2. Is nature lexicalized chaotically, through a progressive accumulation of names, or are there naming systems, i.e., lexical sets structured on the basis of some principle, whether physical, logical, utilitarian or social (Berlin 1978, 1992)? Many organizing principles are known, both of a natural or cultural type:
   (a) physical perception: e.g., brightness in colors, form in plants;
   (b) binary oppositions of the type big/little, heavy/light (Cardona 1985);
   (c) utilitarian considerations: e.g., cultivars vs. wild plants (Angioni 1976; Descola 1986);
   (d) social relations: e.g., animal names based on kinship terms (Alinei 1984);
   (e) belief systems: e.g., animal or plant names with magic-religious content, or magic-religious beings used for naming animals and plants (Alinei 1984).

3. I would like to draw your attention to a widespread phenomenon: the transfer of names from one domain to the other (Cardona 1981, 1985; Alinei 1984), for example,

   (a) animal names used for naming plants;

   (b) animal names used for diseases, illnesses, sensations (Sanga 1987, 1997c);

   (c) animal names used for body parts: e.g., Latin *musculus* "muscle" means "little mouse" (Cardona 1985);

   (d) body parts names used for naming territory: e.g., "head of a valley,'" foot of a hill," etc. (Cardona 1976, 1979, 1985; Cuturi 1981).

In the transfer of names from one semantic domain to another, are single names involved or entire naming systems?

4. There seems to be a relationship between "natural" naming and "social" naming, particulary evidenced in "totemic" naming. Is the relationship between nature and society of a realistic type, that is reflecting magico-religious belief (Alinei 1984), or of a linguistic and classificatory type (Lévi-Strauss 1962)?

5. Etymology shows that taboo names, i.e., the substitution of primary names (animals, body parts, diseases) with conventional names (e.g., "weasel" in Italian *donnola* means "miss," in French *belette* means "beautiful," Alinei 1986), are not sporadic or occasional but more widespread than believed.

   Which are the linguistic mechanisms underlying taboo names (derivation, metonymy, metaphor)? And are there cultural motives connected with symbolic, magico-religious, ritual systems or with economic ones (hunting)? Taboo names for animals might well derive from hunting language, i.e., from substitution names used by hunters-gatherers (Lot-Falck 1953; Sanga 1997a, b)?

6. The last question deals with one of the fundamentals points of twentieth century linguistics: linguistic arbitrariness as postulated by Saussure. Are names arbitrary, as some linguists suppose (Alinei 1996 a, b), or are they motivated, as some anthropologists and ethnolinguists believe (Berlin 1992)?

If names are motivated why do they seem arbitrary? Perhaps because they have to lose their motivations and become arbitrary in order to take place among the mechanical cultural behaviors upon which everyday life is based (Sanga 1997a, b)?

Are names motivated only at the linguistic level or at the extralinguistic level, as well, i.e., at the level of external reality?

References

Alinei, Mario. 1979. *The structure of meaning*. In *A semiotic landscape. Proceedings of the First Congress of the International Association for Semiotic Studies*. Berlin.

────── 1980. The structure of meaning revisited. *Quaderni di semantica* I, 2: 289–395.

────── 1984. *Dal totemismo al cristianesimo popolare. Sviluppi semantici nei dialetti italiani ed europei.* Alessandria: Edizioni dell'Orso.

────── 1986. "Belette". In *Atlas Linguarum Europae*, edited by Mario Alinei et al. Vol. I, 2, *Commentaire.* Assen / Maastricht: Van Gorcum.

────── 1996. Aspetti teorici della motivazione. *Quaderni di semantica* XVII, 1: 7–17.

────── 1996b. *Origini delle lingue d'Europa.* Vol. I, *La teoria della continuità.* Bologna: il Mulino.

Angioni, Giulio. 1976. *Sa laurera. Il lavoro contadino in Sardegna.* Cagliari: Edes.

Berlin, Brent. 1978. Ethnobiological Classification. In *Cognition and categorization*, edited by E. Rosch and B. Lloyd. Hillsdale: Erlbaum. Italian translation in Cardona (1981: 77–86).

────── 1992. *Ethnobiological classification. Principles of categorization of plants and animals in traditional societies.* Princeton: Princeton Univerity Press.

Cardona, Giorgio Raimondo. 1976. *Introduzione all'etnolinguistica.* Bologna: Il Mulino.

────── 1979. Categorie conoscitive e categorie linguistiche in huave. In *Gente di laguna. Ideologia e istituzioni dei Huave di San Mateo del Mar*, edited by Italo Signorini. Milano: Angeli.

────── , ed. 1981. Antropologia simbolica. Categorie culturali e segni linguistici. *La ricerca folklorica* 4: 3–98.

────── 1985. *La foresta di piume. Manuale di etnoscienza.* Roma-Bari: Laterza.

Cuturi, Flavia. 1981. Metafore, proiezioni e rideterminazione nella terminologia anatomica. In: Antropologia simbolica. Categorie culturali e segni linguistici, edited by Giorgio Raimondo Cardona. *La ricerca folklorica* 4: 25–32.

Descola, Philippe. 1986. *La nature doméstique: symbolisme et praxis dans l'écologie des Achuar.* Paris: Maison des Sciences de l'Homme.

Lévi-Strauss, Claude. 1962. *Le totémisme aujourd'hui.* Paris: Presses Universitaires de France.

Lot-Falck, Eveline. 1953. *Les rites de chasse chez les peuples sibériens.* Paris: Gallimard.

Sanga, Glauco. 1987. Affezioni animali. In *Aspects of Language. Studies in Honour of Mario Alinei.* Vol. II *Theoretical and Applied Semantics.* Amsterdam: Rodopi.

────── 1997a. L'appaesamento linguistico. Una teoria glottogonica. *Quaderni di semantica* XVIII, 1: 13–63.

────── 1997b. The Domestication of Speech: Towards a Theory of Glottogony. *Europaea* III, 1: 65–120.

────── 1997c. Passioni animali e vegetali. Per un'etnolinguistica delle sensazioni. In: Antropologia dell'interiorità, edited by Vincenzo Matera. *La ricerca folklorica* 35: 29–38.

CHAPTER 5

The Role of Motivation ("iconymy") in Naming: Six Responses to a List of Questions

Mario Alinei

Which Linguistic, Semantic and Rhetorical Mechanisms are used for Naming Nature?

The most general naming mechanism is undoubtedly the one that is traditionally called "motivation" (Alinei 1979, 1980, 1984, 1992, 1996a,b, 1997a, b, c, d). All the mechanisms listed by Sanga in the part introduction (onomatopoeia, derivation, extension, specialization, metonymy, metaphor and the like) are in fact motivational "subtypes," more or less productive within a natural lexicon.

Both the geographic scope and the theoretical importance of the notion of motivation have been proved by means of the geolinguistic pan-European "motivational maps" of the *Atlas Linguarum Europae* (Alinei 1983a,b, 1986, 1992, 1997d, Barros Ferreira and Alinei 1990).

Precisely because of the importance of the role of motivation in the genesis of words, I have recently proposed calling it *iconym* (from *icon* + 'name,' with the derivations "iconymy," "iconymic," and "iconomastic"), in order to avoid using the much too ambiguous and generic term "motivation" (Alinei 1997c). Only by a few linguists, unfortunately, have recently discussed some theoretical aspects of iconymy (e.g., Lakoff and Johnson (1980), Lakoff (1987) and Sanga (1997)).

I shall try to summarize the basic aspects of iconymy, as they have unfolded in the course of my research, by way of a simple example such as that of eyeglasses. At the time of their invention, in the late Middle Ages, different iconyms were chosen in different languages. Some were simple components of the conceptual definition of eyeglasses (which can be assumed to have been something like a "device to improve the eye-sight, consisting of glass or crystal lenses mounted on a frame which is hooked onto the person's ears"): "glass" in the English *glasses*, the crystal "beryl" in German *Brille* and Dutch *bril*; "glass for the eyes" in the full English form *eye-glasses*, Hungarian *szemüveg* and Finnish *silmäläsit*; "(something for the) eye" in Italian *occhiali*, Russian *očki* and Turkish *gözlük*; "lens" and "hook" in Spanish *lentes* and *gafas*; "glass-eye" in Swedish *glasögon*. In the French *lunettes*, however, the iconym "small moons" resulted from a free asssociation with the new, round lenses, and was thus metaphoric in character.

Notice, then, (A) that all the iconyms chosen from the components of a conceptual definition are, by definition, metonymic, since the relationship between a component of a conceptual definition and the definition itself will always be that of *pars pro toto*; and (B) that only one of the iconyms is metaphoric; (C) that most of the other categories listed in the questionnaire either are rare (onomatopoeia occurs in a limited set of natural objects), or are represented in the example: It. *occhiali* also represents derivation, Engl. *glasses* also represents specialization, and so on.

In fact, systematic research on iconymy and word origins (etymology) shows that there are basically four different ways to motivate words: (A) the most frequent is from the componential features of the concept that they aim at designating; (B) the second in frequency is through metaphor, in which case the reference to the new *designandum* is found in some similarity to the *designatum* of the old word; (C) the third is through the so called "expressive" words (a better term would be "phonosymbolic"), by which we usually mean words connected with the language of (or for) children (e.g. "papa," "mama," "popo," "pipi" and the like); (D) the fourth is through imitation in sound (onomatopoeic) (e.g. English and Dutch *chiffchaff, tjiftjaf,* a bird the song of which is based on a sequence of two repetitive notes; one rising, the other falling). Phonosymbolism, represented mostly by (C), has probably played a great role in the earliest layers of lexicon (see below).

From a theoretical point of view, it is important to observe that iconymy also fulfills the function of collapsing concepts into minimal lexical units. To understand this process one must recall what the Russian psychologist Vygotsky has taught us about words: they are "shortcuts to concepts" (Vygotsky 1962), by which we collapse a whole, articulated concept (the encyclopedic definition of a notion) into a single lexical unit. New words thus allow us to build manageable sentences without having to interconnect complex concepts, and thus overtax our memory and our syntactical capacity. This is exactly what iconyms do, by "representing," as it were, whole concepts. Any new concept that in the process of social communication has become standardized, can thus be collapsed, by means of iconyms, into a new word, allowing us to enrich our knowledge, without changing our abstract, mental categories.

It is, in fact, because of the discovery that iconymy fulfills the fundamental function of collapsing a concept into a word that one is justified in advancing the claim that every word is genetically motivated.

The question then arises: does the claim that every word is genetically motivated not contradict one of the founding principles of modern linguistics, namely Saussure's[1] *arbitrariété du signe*? To this question I will respond below, under point 6.

Is Nature Lexicalized Chaotically, through a Progressive Accumulation of Designations, or are there Designating Systems, i.e. Lexical Sets Structured on the Basis of some Principle, Whether it be Physical, Logical, Utilitarian or Social?

In my work on the structure of lexicon (Alinei 1974, 1979, 1980ab, 1984), I have shown how lexicon grows in many directions, by using pre-existing lexemes as features for new lexemes. For example, the word"computer"can be used as a feature for words designating new types of computer (e.g. "notebook"), or new notions related to it (e.g."surfing,""hacker,""virus"), all mapped onto the complex syntactical network that underlies lexicon, and which is the same that governs speech (Alinei 1974, 1980, 1984). All new lexemes that share"computer"as a feature thus form a semantic domain. Lexicalization of nature, i.e. the growth of semantic domains related to nature knowledge, obeys the same law. As such, it is cognition and culture bound. Consequently, its lexicalization does not take place chaotically but it is strictly determined by cognitive and cultural growth. In that sense, we can even predict, given a certain discovery (about nature or anything else), that a certain amount of new terms will be created to express the various aspects of that discovery, both in the scientific and in the standard vocabulary. We can also predict that within a new semantic domain determined by cognitive and cultural development, both hyponyms and superordinates will be created. For example when linguistics was hit by the Chomskian revolution, not only were a great number of new terms created, but also a structural reorganization of its semantic domain was set off, including new superordinates (psychology, cognitive science etc.). What is not predictable, at least in the present state of research, is to which extent lexicalization of a new semantic domain will be realized by means of verbs, or nouns, or adjectives, and in what mutual proportion. Needless to say, the conclusion that lexicalization is cognition and culture-bound implies that lexical sets (semantic domains and semantic fields or systems)[9] are structured on the basis of the same principles that underly cognitive and cultural development, and which are necessarily logical, physical, utilitarian, social etc.

Notice, however, that the structuring of lexicon, although quite similar in its formalism to that of iconymic recycling, has nothing to do with the iconymic process, as it operates at the level of already formed "words."The iconym "little moon"of the French *lunettes*, for example, has no function whatsoever in the structuring and/or in the growth of the semantic domain related to"eyesight/vision,"to which the word"eyeglasses"belongs.

In the Transfer of Names from One Semantic Domain to Another, are Single Designations Involved or Entire Designating Systems?

The transfer of designations from one domain to another is simply another basic aspect of the iconymic process. However, to understand it fully one must make use of the notion of "iconymic productivity" (Alinei 1997a, b, c), which in turn depends on the social prestige or popularity of the the semantic domain to which the iconym belongs (Alinei 1996a, b). Since iconymy is basically an extremely convenient means of obtaining the immediate socialization of new words (see below), it is obvious that the choice of an iconym, especially in the so-called "spontaneous semantic changes," will be dictated by its popularity or renown. This is the reason why so many Latin words evolved from designating typical traditional techniques, such as farming or shepherding or ceramics or weaving, to abstract notions; or why so many words and expressions appear in any modern European language showing the passage from horse-riding or from sailing to other, more general, aspects of life.

Here are some representative examples from Latin: the Lat. *puto* "to trim (a tree"), from which came the Latin word for "think," makes up many Latin derivations that still survive in the English notions of "deputy," "imputing," "computing," "disputing" etc.; the Lat. *grex* "herd," from which derives for example Lat. *egregius* "excellent," makes up many abstract Latin words that survive in English notions such as "congregate," "segregate," "aggregate," "disgregate" etc.; the Lat. *fingo* "mould an earthen vessel," forms many Latin abstract words that survive in English notions such as "fiction," "figure," "configure," "disfigure" etc.; the Lat. *texo* "weave" and its abstract derivations still exist in the English "textile," "text" and "context." Each of these semantic domains—respectively farming, shepherding, ceramics, weaving—is extremely well represented in Latin as the basis for semantic developments of this kind.

Developments from horse riding and sailing to other semantic domains can be found in any modern European language, as shown by Engl. "spur," It. *spronare*, It. *a spron battuto* "immediately," Engl. "to strain at the leash," It. *mordere il freno*, It. *perdere le staffe* (literally "to come out of the stirrup") "to fly off the handle," It. *prendere la mano* ("to lose control" said of the horse), Engl. "recalcitrant," It. *recalcitrante*, Engl. "to hold the reins,""to give free rein," cf. It. *tenere le redini, abbandonare le redini*, Engl. "to bridle," cf. It. *imbrigliare*, Engl. "unbridled," cf. It. *sbrigliato* etc.; Engl. "launching," cf. It. *varare*, Engl. "to sail against the wind," cf. It. *navigare contro vento*, Engl. "to sail through something,""to sail before the wind," cf. It. *col vento in poppa*, Engl. "to be at the helm," cf. It. *essere al timone* etc.

In both Latin and modern languages, the recurrence of the iconym lies in the popularity of its counterpart in reality: farming and shepherding had immense prestige in Latin archaic society, as did horse-riding and sailing in early modern European societies. For the same reason, many expressions taken from car-driving or from sport are nowadays acquiring new meanings in many European languages. For example It. *partire in quarta* (literally "to start in fourth gear") has become a very common expression meaning to start up an action with great thrust; *essere / andare su / giù di giri* "to increase / decrease the number of revs" (said of a car engine) (cf. German *auf Touren kommen*), is now said colloquially of a person who is respectively in high or low spirits.

Transferring the name of an animal to another semantic category is, then, the simple consequence of the enormous role that animals have played in human life.

To conclude, in answer to the question "are single names or entire systems of names transferred from one domain to another" I respond with: both, but the most frequent of the two possibilities is—by definition—the second, i.e. the most productive domains of iconyms.

In the So-called "Totemic Designations," is the Relationship Between Nature and Society of a Realistic Type, that is Reflecting Magico-religious Belief (Alinei 1984), or of a Linguistic and Classificatory Type (Lévi-Strauss 1962)?

As I have argued elsewhere (Alinei 1985, 1996c), the solution given by Lévi-Strauss (1962) to the problem of the nature of totemism is at the same time right and wrong. It is right because a totemic system is, indeed, a classificatory system, as Lévi-Strauss claimed. It is wrong, however, because totemism would not be a classificatory system if it had not been a religion as well. Any religion, whether historical or prehistoric, generates a highly hierarchized cognitive and classificatory system. The only reason that it is nowadays possible to talk about classificatory systems without involving religion is that science has gradually replaced religion in that function, and the role that mythical figures and mythical phenomena had in religious visions of the universe has been replaced by scientific laws and notions. Lévi-Strauss, uninterested as he was in any historical processes, and thus overlooking the historical development of classificatory systems, trivialized them, and thus trivialized totemism as well. In essence, the same criticism that Lévi-Strauss addressed to totemism could be used for Christendom or Islamism, or any other historical religion, the religious nature of which could be denied by claiming that they are only classificatory systems!

As to the question that arose during the discussion, whether "totemic" zoonyms basically reflect a hunting, male ideology, I would tend to give a cautious answer. Many totemic zoonyms apply to insects, very often to larvae, worms, and grubs, which are mostly the object of female gathering, and not of male hunting (there are of course exceptions that confirm the rule). Tchimanga Katangidiku's monumental dissertation on Zaire's zoonyms (Kutangidiku 1995) has confirmed that insects showing totemic or magico-religious names are those that are a part of the daily diet of the local populations, mostly larvae, worms, and grubs.

Which are the Linguistic Mechanisms Underlying Taboo Designations? And are their Cultural Motives Connected with Magico-religious Systems or with Economic Ones (Hunting)?

At first sight, one could expect the naming mechanisms determined by taboo to be a narrow subset of the iconymic ones. On closer analysis, however, noa names, i.e. those used to avoid the "real name," owing to taboo, appear to have the same typology as iconyms. This is because the prohibition of using the real name of an animal does not represent any limitation in the choice of a noa name, and thus of its iconym, which remains unlimited. On the other hand, the typology of noa names (see Alinei 1984, 1986, 1997a, b) is usually not based on the above cited general categories, but on more specific, idiosynchratic ones, such as lexical prohibition ("nameless," "unnamable"), names expressing contempt (e.g. "ugly," "monster"), endearing names ("darling," "little," "cute"), part for the whole ("horn," "tail," "foot"), absolute generalizations ("he," "she," "thing"), specialized generalizations ("animal"), characteristic aspects ("brown"), modified names, both motivated ((*porca*) *mattina* "damn morning" instead of (*porca*) *Madonna*), or immotivated ones ("gee," "gee whiz" instead of "Jesus," "gosh" instead of "God"), and so on. These specific iconymic categories, however, are nothing but the usual shortcut of the full conceptualization of tabooed realities.

In paleo-anthropological terms—which in my opinion are the most appropriate ones when studying language taboo—the distinction between magico-religious and practical-economical is hard to pursue. To an archaic society, natural and supernatural tend to coincide. Magic, ritual, and religion are in the first place means to obtain good results with hunting and farming, to chase away illness and death, to improve one's lot, and the like. Therefore, they are also practical-economical in essence.

As a consequence, the fact that hunting forms one of the main contexts for creating noa animal names (among which the "totemic" ones can also be included (Alinei 1997b)), does not contrast with their magico-religious and ritual origin. Even in modern times, many magico-religious relics are preserved in folk traditions related to taboo (Alinei 1984, Zelenin 1929–1930) (even in the aristocratic fox hunting, described by Leach 1962). Moreover, the interpretation of noa animal names should accord with the general interpretation of lexical taboo, connected with other aspects of life and nature, entirely independent of hunting, such as death, illness, sex, menstruation, magical and religious entities.

An interesting question that arose during the discussion is the possible dating of noa names of tabooed animals. In my opinion, noa names for animals imply the existence of a religion. According to most prehistorians, the existence of religion can only be proved on attestations of ritual burials. This ritual begins to appear with Neanderthal, in the Middle Paleolithic, but becomes a regular feature of the archaeological record only in Upper Paleolithic (cemeteries, however, appear only in Mesolithic). So I think it would be impossible to date noa tabooed names earlier than Neanderthal, and their likeliest dating would be Upper Paleolithic. For the same reason, the real name of animals—as reconstructed for example by Indo-European research—could be dated even to Lower Paleolithic (Alinei 1996).

Are Names Arbitrary or Motivated? And if they are Motivated why do they seem Arbitrary?

I have extensively responded this twofold question in the papers cited in point (1), and quite differently from Saussure. I will try to summarize my conclusions here.

It would be impossible, and perhaps pointless, to repeat here the many arguments that prove the basic arbitrariness of *signs* (and thus of words as linguistic signs), marshalled by the founders of semiotics Peirce and Saussure, further elaborated by subsequent generations of scholars, and today accepted as the founding principle of both semiotics and linguistics.

Even regardless of this fundamental conclusion of semiotics and modern linguistics, it is also our daily experience of language that proves that iconyms are not necessary. In fact, the very question I am answering explicitly states that words "seem" arbitrary to the speaker. For the words we use most frequently in our life are opaque. Many words that are motivated to linguists, in the sense that they might know their etymology, are opaque to the average speaker. Even transparent words lose their transparency after their original meaning has slightly changed, as for example Italian *penna* or French *plume*, which can no longer be associated to a feather (the original meaning of both words) by either Italian or French speakers. More generally, even unchanged and transparent iconyms might appear to the speaker as opaque, as shown, for example, by the word "computer," which even Anglophones use without ever thinking of "computing."

The question, then, could be put differently: if words are in essence—that is in their actual communicative function—arbitrary, how can they also be motivated? Or, more precisely, why are they almost always created with an iconym, and thus with a motivated word? In my own terms, having concluded that words are genetically, and thus necessarily, motivated, and at the same time that words seem to us, or actually become, arbitrary, the challenging question is: can these two conclusions be reconciled, and if so how?

Saussure, in a much debated passage of his *Cours* (Saussure 1989: 297–303), answered this question by claiming that iconymy is a device to diminish the inherent chaos of arbitrariness, to avoid the triumph of arbitrariness with the threat of total chaos: hence his notion of a "relative arbitrariness." The discussion that followed did not come to any substantial breakthrough. As I have argued extensively elsewhere (Alinei 1997c), in observing the coexistence of arbitrary and motivated words Saussure and semioticians have simply observed a basic aspect of linguistic reality, but have failed to explain it. In the same way, one can observe the apparent motion of the sun without discovering its real cause. To my knowledge, no attempt at solving the problem on a different basis has been made prior to the one I have presented in my own work (Alinei 1996b, 1997c). My own answer is based on some basic semiotic notions, which were known to Saussure and his followers, but the implications of which have been overlooked.

One of the conditions for the existence of a sign, of no matter which type, is its *socialization*. In simpler words, for a sign to exist, everybody who uses it must know it. The key of the sign code must be universally known. This principle is not in contradiction with its arbitrariness, but rather flows from it. Precisely because a sign is arbitrary, it requires adequate publicity to exist.

Starting from this universally accepted premise, however, semioticians have overlooked the fact that there are at least three ways to publicize or advertise a sign, prior to its usage, and that these three ways are quite different from one another: (1) On a public (state or institutional) level, for example for *monetary* and *metric* systems, a conventional, arbitrary sign is publicized by legal enforcement: metric standards are deposited in official institutions, and units of currency are issued solely by the state (Alinei 1979, 1980). (2) On a private level, the publicizing of a conventional, arbitrary sign (such as an artificial trademark) is made through costly campaigns of commercial advertising. (3) In language, and in a few other minor sign systems, the universal knowledge of the sign is obtained by means of a very ingenious, albeit extremely simple and economical, device, which consists of recycling pre-existing signs, adopted—and adapted—to designate new meanings/referents. Signs, in this way, need no longer be conventional, because they are born "already known." This procedure resorts neither to the state nor to the market, but rather to a sort of communal treasure—namely lexicon—which, in spite of the fact that it is available to everybody, is also inextinguishable.

Interestingly, this strategy, which can be defined as "reutilization of already existing means, modified in order to perform new functions," has a fundamental role in human evolution, the importance of which was underlined by Darwin, and in particular for language. He showed how breathing and eating organs were adapted to new linguistic function, and thus became linguistic organs. *Reutilization of existing lexemes in order to create new ones can then be seen as part of a single biological strategy, and as a fundamental component of glottogenesis* (Alinei 1996c).

We can then fully appreciate why linguistic signs can be in essence arbitrary, and at the same time motivated. In fact, *iconymy does not belong to the nature of the linguistic sign, but it is a mere device by which a new word can be immediately apprehended by everybody.*

In other words, when creating new words we usually do not invent names that give a totally new clothing to a meaning, as the "*arbitraireté du signe*" may induce us to think. We do this only with some trademark names, when we invent them and launch with gigantic publicity campaigns. But this is an exception that confirms the rule. In daily life, and even in commercial life, the usual procedure for naming is another: we simply recycle pre-existing words, as iconyms for new concepts. For the pre-existing words are already widely publicized, they belong to everybody, and their usage is entirely free. If one calls a new machine or a new product "car," or "dishwasher," or "answering machine," or "personal computer," one is using old words, easy to understand and to remember.[3] Then, once the new words, with their new meanings, have entered circulation, the original iconym can, and often does, fall into oblivion. Iconymy has the same function as the legal stamp in coined money. It represents official convention, it makes "currency" what it is, that is "current," valid everywhere. But once in circulation, the stamp can disappear, through usage or accident, without causing any loss of value. Also a word, like a coin, through constant usage and deterioration (sound change), can lose its official stamp.

As a consequence, on the theoretical level, we must distinguish between the *genetic* and the *functional* aspect of words, and see that iconymy, useless as it may be in the speech act, plays a fundamental role in the very naming process, that is in the moment we create the means to represent and classify reality and communicate about it. Genesis is indeed the "magic" moment, when things reveal their secrets. Obviously, in the arbitrary relationship between *signifiant* (word form) and *signifié* (meaning), there can be no place for iconymy. Even though the iconym of the so-called transparent words can be observed by speakers, we use words all the time without stopping to think about their iconym. Speakers need only to know what their words mean, and not how they were made to mean what they mean.

Words are genetically, and thus always, motivated, and the fact that many of them become opaque is the result of idiosyncratic sound changes, the occurrence of which, however, proves that word transparency is expedient and not related to their real function. Coexistence of opaque and transparent words is thus purely coincidental, and depends very much on what kind of developmental history each language has undergone. This is why I maintain that iconymy is a basic component of the *genesis* of the word, but not of its *function*. Moreover, also the *choice* of iconyms is arbitrary, since it is made from an infinite class of pre-existing words.

This is why I think that seeing arbitrariness as an "optical illusion," as claimed by Gnerre in his paper, is in itself an optical illusion. It would be closer to reality, even though I would consider it an oversimplification, to state that exactly the contrary is true, i.e. that word motivation is an "optical illusion," since we use it only to publicize our words, and we immediately dispose of it the moment we start talking or writing, and thus using our words for what they are, i.e. mere codes. The true nature of words (and signs) is demonstrated by those rare commercial words that have been created and launched with advertising campaigns.

True, the exciting discovery of iconymy and of its great role in language can lead us to think that a reversal of the current position is justified. However, the picture is far more complex. I would prefer to claim that both lexical arbitrariness and lexical iconymy can be demonstrated, at different levels, and that lexical iconymy, even though immediately disposable, and deprived of a really essential function in communication, is something that has also added complexity and depth to our linguistic code.

In fact, our theoretical efforts should address the problem by trying to understand how iconymy, even though not necessary in communication, once introduced in language as an original and ingenious expedient to publicize new words without social or institutionalized conventions, has become so interwoven in our linguistic systems as to become the basis for further developments, and thus acquire new functions. Perhaps the best way to appreciate this point is to recall what I have already said about the similarity between iconymic recycling and the re-use of non-linguistic organs to create language organs, as noted by Darwin. Originally, using breathing and eating organs to invent speech was also just an expedient. Eventually, however, what had been an expedient became the basis for an endless chain of developments, by which—to express it in an extremely concise way—*Homo erectus* became *Homo sapiens sapiens*. In this process, linguistic organs, despite their non-linguistic origin, acquired their own independence and became the motor of more and more activities, and the starting point of further developments. Just to give one example: human voice, which prior to speaking was not basically different from any other animal voice, first became the main speaking tool, and thereafter the tool of musical singing.

The same is true of iconymy. It is introduced into language only as an ingenious means to publicize new words. But once it is with us, it has not only become a stratigraphical archive of cultural developments, a sort of "living museum"—as has been said—available to scholars, but also, more importantly, a powerful means to give structure to grammar and lexicon, that is to map morphology onto language. Iconymy (whether transparent or opaque) is obviously the governing factor in declining nouns, conjugating verbs and to a certain extent also giving a shape to lexical domains, and thus classifying reality (Alinei 1997c).

Finally, another important aspect of the genetic role of iconymy can be seen in the research on phonosymbolism in the naming process, illustrated in the following chapter by Brent Berlin in his interesting comparative analysis of the names of "tapir" and "squirrel" in twenty-nine South American languages. For phonosymbolism, much more than onomatopoeia, provides an adequate answer to the intriguing question: "How could the first words uttered by *Homo loquens* be motivated, in the absence of pre-existing words?" Obviously, it can only be this kind of unconscious, and—as it were—highly "volatile," iconymy that can be attributed to the very first words uttered by *Homo loquens*. This conclusion is not contradicted by recent research on the origin of the internal structure of word forms, which has underlined the importance of biomechanical and self-organizational aspects in the formation of words in both infant speech and modern languages (MacNeilage and Davis 2000). In fact, these aspects, by strongly favouring a strict number of Consonant-Vowel and Consonant-(Vowel)-Consonant sequences, simply enhances the potential phonosymbolic load of each of them.

Significantly, thus, at the glottogonic level, iconymy and arbitrariness appear already inextricably interwoven, as they will continue to be, even though with more specialized functions, at the systemic level.

Notes

1. Saussure has had, of course, many predecessors, such as Whitney, but only with his contribution has the theory of the conventional character of language been embraced by modern linguistics.

2. On the difference between semantic domains and systems see Alinei (1974).

3. For brevity's sake I omit the illustration of the usage of words from dead languages (e.g. Latin and Greek) as iconyms, which is just a sociolinguistic and sociocultural variant of the more universal procedure.

References

Alinei, Mario. 1974. *La struttura del lessico*. Bologna: Il Mulino.

———— 1979. The structure of meaning. In *A semiotic landscape. Proceedings of the First Congress of the International Association for Semiotic Studies*, Milano 1974. The Hague, Paris, New York: Mouton.

———— 1980a. The structure of meaning revisited. *Quaderni di Semantica* I: 289–305.

———— 1980b. Lexical Grammar and Sentence Grammar: A Two-Cycle Model. *Quaderni di Semantica* I: 33–95.

———— 1983a. Introduction. In *Atlas Linguarum Europae I*, Vol. l. Assen: Van Gorcum.

———— 1983b. Arc-en-ciel. In *Atlas Linguarum Europae I*, Vol. 1: Cartes 6–9, 47–80. Assen: Van Gorcum.

———— 1984a. Lessico come romanzo, romanzo come lessico. *Lingua e stile* XIX: 135–55.

———— 1984b. *Dal totemismo al cristianesimo popolare. Sviluppi semantici nei dialetti italiani ed europei*. Alessandria: Edizioni dell'Orso.

———— 1985. Evidence for Totemism in European Dialects. *International Journal of American Linguistics* 51: 331–334.

———— 1986. Belette. In *Atlas Linguarum Europae I*, Vol. 2: Carte 28, 145–224. Assen/Maasstricht: Van Gorcum.

———— 1992. Lo studio culturolinguistico delle motivazioni. L'«Atlas Linguarum Europae» fra teoria e prassi. In *Atti del XXI Congresso Internazionale di studi della Società Linguistica Italiana. L'Europa linguistica: contatti, contrasti, affinità di lingua.*, Catania 10–12 settembre 1987. Roma: Bulzoni.

———— 1996a. Theoretical Aspects of Lexical Motivation. In *Mål i sikte*. Uppsala: Almquist & Wiksell Tryckeri.

———— 1996b. Aspetti teorici della motivazione. *Quaderni di Semantica* XVII: 7–17.

———— 1996c. *Origini delle lingue d'Europa, vol. I, La teoria della continuità*. Bologna: Società editrice Il Mulino.

———— 1997a. Magico-Religious Iconyms in European Dialects. A Contribution to Archaeolinguistics. *Dialectologia et Geolinguistica* 5: 5–30.

———— 1997b. L'Aspect magico-religieux dans la zoonymie populaire. In *Les zoonymes*. Nice: Publications de la Faculté de Lettres, Arts, et Sciences Humaines de Nice (Nouvelle Série, No. 38).

———— 1997c. Principi di teoria motivazionale (iconimia) e di lessicologia motivazionale (iconomastica). In *Lessicologia e lessicografia, Atti del XX Convegno della Società Italiana di Glottologia* (Chieti-Pescara, 12–14 Ottobre 1995). Roma: Il Calamo.

———— 1997d. Noël. In *Atlas Linguarum Europae* I. Vol. 5: Carte 59, 253–91. Roma: Istituto Poligrafico di Stato.

Barros Ferreira, Alinei, Mario. 1990. Coccinelle. In *ALE I*. Vol. 4: Cartes 42–44, 99–199. Assen: Van Gorcum.

Kutangidiku, Tchimanga. 1995. *La motivation dans la création lexicale des noms des petis animaux chez les Bantou du Zaire*, 3 vols. with 272 geolinguistic maps, and a *Questionnaire iconographique* of 272 items. Ph.D. Thesis Université Stendhal, Grenoble.

Lakoff, George. 1987. *Women, fire, and dangerous things (What categories reveal about the mind)*. Chicago and London: The University of Chicago Press.

———— and Mark Johnson. 1980. *Metaphors we live by*. Chicago and London: The University of Chicago Press.

Leach, Edmund. 1964. Anthropological Aspects of Language: Animal Categories and Verbal Abuse. In *New Directions in the Theory of Language*. Edited by E. H. Lenneberg. Cambridge, Mass.: The MIT Press.

Lévy-Strauss. 1962. *Le totémisme aujourd'hui*. Paris: Presses Universitaires de France.

MacNeilage, Peter F. and Barbara L. Davis. 2000. On the Origin of Internal Structure of Word Forms. *Science* 288: 527–31.

Sanga, Glauco. 1997. L'appaesamento linguistico. Una teoria glottogonica. *Quaderni di Semantica* XVIII: 13–63.

Vygotsky, L. S. 1962. *Thought and language*. Cambridge, Massachusetts: The MIT Press.

Zelenin, D. K. 1929–1930. Tabu slov u narodov vostocnoj Evropy i severnoj Asii. *Sbornik muzeia antropologii i ètnografii*: VIII: 1–151. IX: 1–166 (Italian transl.: Tabù linguistici nelle popolazioni dell'Europa orientale e dell'Asia settentrionale. *Quaderni di Semantica* IX (1988): 187–317. X (1989): 123–276).

CHAPTER 6

Tapir and Squirrel: Further Nomenclatural Meanderings Toward a Universal Sound-symbolic Bestiary

Brent Berlin

Sound symbolism is the technical term used to refer to the non-arbitrary association of the "acoustic and motor-acoustic" properties of sound and meaning (Ciccotosto 1991). Sound symbolism is at work when all of us are able to unambiguously associate the terms *maluma* and *takete* with the abstract figures reported in Wolfgang Köhler's famous treatise on gestalt psychology (Köhler 1933). Sound symbolism is at work when we effortlessly find it possible to assign the Tzeltal and Aguaruna names *p'ilich* and *bung* to the drawings of a toad and a grasshopper (Figure 6.1).

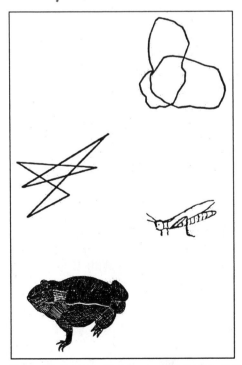

Figure 6.1 Drawings (Top) Representing the Non-sense Words *maluma* and *takete* (after Köhler 1933), and (Bottom) *bung* "Toad" [Aguaruna Jívaro] and *p'ilich* "Grasshopper" [Tzeltal Maya] (after Berlin 1992)

Sound symbolism is the process that leads us to intuitively appreciate the significance of developing words that capture the essence of the entities and events of the world in which we live. As St. Thomas Aquinas claimed centuries ago, "Names for things ought to reflect their [very] natures."

Such semantic non-arbitrariness, plausible as it may seem, was essentially abandoned in contemporary linguistics with the work of the founder of structural linguistics, Ferdinand de Saussure (1959), who popularized the theory that human languages function as referential systems precisely because there is an *arbitrary* link between the thing signified (the referent) and the signifier (the sounds used to indicate the referent). The empirical strength of such an observation (what is symbolically similar in the words "*dog,*" *ts'i,*' and *yawa,* all of which refer to *Canis domesticus* in English, Tzeltal, and Aguaruna), as well as the discrediting of a number of shoddy studies aimed to support speculative "bow-wow, sing-song, and ding-dong" theories of language origins, has led to the dismissal of much serious work on the nature sound symbolism. However, as we approach Clinton's bridge to the 21st century, I would not be far off the mark to say that the study of sound symbolism has again become an academically acceptable topic for exciting new scholarly investigation (see for example, Ciccotosto doctoral dissertation (1991), the collection of original papers found in Hinton, Nichols, and Ohala (1995), and the monographic treatment linking sound symbolism and grammatical processes in Quechua by Nuckolls (1996)).

In this short paper, I want to continue examining an aspect of sound symbolism as it relates to the names of living things, a subject that I take up in detail in Chapter 6 of *Ethnobiological Classification* (Berlin 1992). There, I presented evidence to support the view that words for "fish" and "birds" in Huambisa, a Jivaroan language of Amazonian Peru, were strongly marked by sound symbolic properties. Examination of the phonetic properties of the words for birds and fish in this language showed that the names for these creatures differ significantly in the distribution of phonetic segments of high and low acoustic frequency. "Bird names show a considerably larger number of segments ...which ...connote quick and rapid motion (i.e.,."birdness")...[in contrast] with the lower-frequency segments of fish names...[that connote] smooth, slow, continuous flow (i.e.,"fishness").

In addition, both fish and bird vocabularies demonstrate a prominent internal pattern of size-sound symbolism. Names of small birds and fish commonly show high frequency vowel [i] stems, while larger birds and fish are referred to by names made up of the low frequency vowels [a] and [u]. These vowels, in addition to connoting contrasting size, also connote quality of movement, as pointed out by Jespersen (1921) and many other linguists since his important work on sound symbolism. (e.g., "quickness" or "rapid movement"—Danish *kvik, livlig,* Swedish *pigg,* French *vite, vif, rapide,* Italian *vispo,* Japanese *kirikiri,* Tzeltal *chijchin*).

Building on these findings, I began to wonder what kinds of patterns, if any, would emerge if one were to examine the phonetic qualities of names of individual species (vs. groups of species based on their life forms). Focusing again on South America, where I already had a fairly complete inventory of terms for animals in at least a small sample of languages, I decided to look at names for "tapir" and "squirrel," two unmistakable mammal taxa for which I would likely have good data. I personally found the names of these two animals in Aguaruna Jívaro to be imminently appropriate to their nature (*pamau* and *wiching*, respectively). In addition, I had earlier predicted that these two creatures would likely receive linguistic recognition in South American Indian languages because of their taxonomic distinctiveness, tapir (*Tapirus*) occurring as the single member of the monotypic family Tapiridae, and the squirrel family represented by but two genera, *Sciurus* and *Microscirus*. My loosely-formed hypothesis was that if sound symbolism was at work, one should expect to find tapir marked by words connoting "largeness/slow movement" (hence sounds of low acoustic frequency, especially [a]) and words for squirrel marked by words connoting "smallness/rapid movement" (hence sounds of high acoustic frequency, especially [i]).

My sample of languages was drawn from the datafiles of the now defunct South American Indian Languages Documentation Project, a project that Terrence Kaufman and I worked on briefly in the early 1990s with the goal of pulling together reliable data on the major languages of South America with the goal of providing a database that could be used to test hypothesis of culture history and linguistic relationships (Berlin and Kaufman, n.d.). In addition, a colleague had recently called to my attention Conservation International's publication *Mamíferos Colobianos sus nombres comunes e indígenas* (Rodríguez-Mahecha et al. (1995) which included much usable data.

Combining these two sources of data, I was able to produce a sample of twenty-two languages from distinct linguistic South American Indian language families (Figure 6.2).

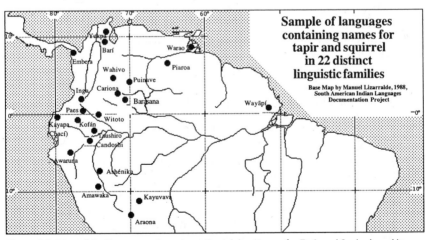

Figure 6.2 Map of the Twenty-two Languages Containing Names for Tapir and Squirrel used in Comparison (see text)

Each of these twenty-two language families, along with all other Indian languages except Eskimo-Aleut and Na-Dené, has been grouped into one super stock by Joseph H. Greenberg, a stock he calls "Amerindian" (Greenberg 1987, Ruhlen 1987). This remarkable proposal has not been widely accepted by linguists familiar with American Indian language classification and none of the twenty-two linguistic groups discussed here have been convincingly shown to be related above the level of family. Arguments that account for similarities in the words for tapir and squirrel among the languages in my sample as due to demonstrated genetic relationship would be considered by most linguists as implausible (see Campbell 1988, 1991; Chafe 1987, Darnell 1987, Goddard 1987, Golla 1987, Matisoff 1990, Rankin 1992, and the responses by Greenberg 1987a, 1987b, 1989 and Greenberg and Ruhlen 1992).

In any such comparison, however, one should attempt to control for borrowing. I tried to eliminate duplicate languages that showed possible influence of borrowing in one or both of the terms for tapir and squirrel. From cursory inspection, three pairs of languages looked suspicious, Aguaruna (Jivaroan) and Candoshi (Candoshan) of Amazonian Peru, Barí (Chibchan) and Yukpa (Carib) of Venezuela and Colombia, and Wayãpí (Tupian) and Wahivo (Wahivoan) of Brazil and Colombia. Borrowing is a plausible explanation of similarities for terms found in the pairs Aguaruna-Candoshi and Barí-Yukpa, as these groups occupy geographically adjacent zones of northwestern South America. Geography does not immediately suggest itself as among the causal factors involved in possible borrowing between Wahivo and Wayãpí. Nonetheless, one of the languages in each pair (Candoshi, Yukpa, and Wayãpí) was eliminated from the sample prior to testing for the possible effects of sound symbolism.

Using this resulting sample of nineteen languages, I then proceeded to test the sound symbolism hypothesis as follows. For every *pair* of words in each language, I predicted that,

> *if size-sound symbolism is at work, terms for* tapir *will show the central vowel* [a] *(low acoustic frequency, hence "large/slow") and terms for* squirrel *will exhibit the high front vowel* [i] *(high acoustic frequency, hence "small/quick").*

Table 6.1 shows that in fourteen of the nineteen pairs (74 percent), the native term for "tapir" exhibits [a] as one of its vowels, in contrast to a word for "squirrel" that includes a vowel [i]. These results are dramatic, and surely suggest that some kind of sound symbolism is at work.

The extent to which this contrastive [a]/[i] pattern might be general to all pairs of words in these languages can be determined by comparing the word of "tapir" with some other expression, e.g., "day," which might be used as a control (see Hays 1995, who, to my knowledge, was the first to use this procedure in sound symbolism experiments). I was able to find the word "day" in thirteen of the nineteen languages in my original sample. As can be seen in Table 2, comparable [a]/[i] contrasts found in "tapir"/"squirrel" are not observed for "tapir"/"day," supporting the view that the pattern observed for "tapir"/"squirrel" is not a general feature of the lexicons of the languages in the sample. Comparisons with other control terms will be needed to provide further confirmation to these findings.

Table 6.1 Terms for "Tapir" and "Squirrel" in Nineteen Indian Languages from Distinct South American Language Families

	Family	Language	"tapir"	"squirrel"	[a]/[i] contrast (+ yes, – no)
1	Panoan	Amawaka	ha'a	ka'iz	+
2	Takanan	Araona	awada	sipo	+
3	Maipurian	Ashénika	kemari	miyiri	+
4	Jivaroan	Awaruna	pama(u)	wiching	+
5	Tukanona	Barasana	veku	timoka	–
6	Chibchan	Barí	aerajbá	kariká	+
7	Cariban	Cariona	machihouuri	meeri	+
8	Chokoan	Embera	bi	Karkoromia	–
9	Kechuan	Inga	sacha wagra	waiwashi	+
10	Barbakoan	Kayapa	wagara	chijcun	+
11	Kayuvacan	Kayuvaca	baaetae	titi	+
12	Kofanan	Kofan	tsontinba	tiriri	+
13	Paesan	Paes	jimba chuch	shuma	–
14	Salivan	Piaroa	bwopoja	twarika	+
15	Puinavean	Puinave	yap	bik	+
16	Saparoan	Taushiro	he'hi	huven	–
17	Wahivoan	Wahivo	mezaha	kuzikuzi	+
18	Waraoan	Warao	naba	hura	–
19	Witotoan	Witoto	turuma	ii'kingo	+

14/19 "correct" (74 %), Chi-Square = 66.9, p <.0001

Table 6.2 Terms for "Tapir" and "Day" in Thirteen Indian Languages from Distinct South American Language Families

Family	Language	"tapir"	"day"	[a]/[i] contrast (+ yes, – no)
Panoan	Amawaka	ha'a	nutu'	–
Takanan	Araona	awada	tsein~e	+
Maipurian	Ashénika	kemari	Kitaiteri	–
Jivaroan	Awaruna	pama(u)	tsawa'n	–
Kechuan	Inga	sacha wagra	Puncha	–
Barbakoan	Kayapa	wagara	Malu	–
Kayuvacan	Kayuvaca	baaetae	ki'arama	+
Kofanan	Kofan	tsontinba	ja'n~o	–
Paesan	Paes	jimba chuch	En	–
Saparoan	Taushiro	he'hi	ichi'ku'ro	–
Wahivoan	Wahivo	mezaha	mata'kabi	+
Waraoan	Warao	naba	Yaja	–
Witotoan	Witoto	turuma	Mona	–

To what extent do native speakers of English share the intuition that the motor acoustic properties of these pairs of words for tapir and squirrel appropriately mark the "natural essence"? To find out, I ran an experiment with a class of eighty-two University of Georgia introductory anthropology students (thanks to Ben Blount for giving me permission to use his class as guinea pigs). The test was straightforward. Each student was given a test sheet onto which pairs of tapir/squirrel terms had been printed. Order of terms in each pair was determined by the toss of a coin. I assumed that most students would know what a 'squirrel' looked like, but at Professor Blount's advice, I asked how many students did not know what a tapir looked like. When I discovered that about a quarter of the class did not know the creature, I showed an overhead transparency of the animal, stating where it lived and how much it weighed. I then explained that the pairs of words on their test sheets were pairs of words for "tapir" and "squirrel" (in randomized order) in several South American Indian languages. Finally, I asked them to look at each pair of terms as I read them aloud (twice), and to circle the term that sounded most like the word for "squirrel."

Results of the test show that 79 percent of the students were able to correctly pick the term for "squirrel" in each of the nineteen pairs at a level greater than chance (Table 6.3).

Table 6.3 Pairs of Words for "Tapir" and "Squirrel" in Nineteen South American Indian Languages: Number and Proportion of Correct Responses in Experiment, [a]/[i] Contrast

Family	Language	"tapir"	"squirrel"	number correct	% correct and rank order	[a]/[i] contrast (+ yes, – no)
Panoan	Amawaka	ha'a	ka'iz	51	61 (12th)	+
Takanan	Araona	awada	sipo	52	62 (11th)	+
Maipurian	Ashénika	kemari	miyiri	49	58 (13th)	+
Jivaroan	Awaruna	pama(u)	wiching	61	73 (5th)	+
Tukanona	Barasana	veku	timoka	55	65 (9th)	–
Chibchan	Barí	aerajbá	kariká	54	64 (16th)	+
Cariban	Cariona	machihouuri	meeri	43	51 (15th)	+
Chokoan	Embera	bi	karkoromia	38	45 (18th)	–
Kechuan	Inga	sacha wagra	waiwashi	61	73 (6th)	+
Barbakoan	Kayapa	wagara	chijcun	65	77 (3rd)	+
Kayuvacan	Kayuvaca	baaetae	titi	69	82 (2nd)	+
Kofanan	Kofan	tsontinba	tiriri	62	74 (4th)	+
Paesan	Paes	jimba chuch	shuma	42	50 (16th)	–
Salivan	Piaroa	bwopoja	twarika	61	73 (7th)	+
Puinavean	Puinave	yap	bik	46	56 (14th)	+
Saparoan	Taushiro	he'hi	huven	23	27 (19th)	–
Wahivoan	Wahivo	mezaha	kuzikuzi	71	85 (1st)	+
Waraoan	Warao	naba	hura	41	49 (17th)	–
Witotoan	Witoto	turuma	ii'kingo	59	70 (8th)	+

Slightly less than half of the nineteen pairs showed correct response levels of 70 percent or greater, with the highest level of performance representing 85 percent in the pair *mezaha:kuzikuzi*. This proto-typical "pir/squirrel" pair suggests that the sound-symbolic effects of [a] "large/slow" and [i] "small/quick" may be combinatory. Polysyllabic terms for "tapir" with two or more [a] syllabics and terms for "squirrel" that contain two or more [i] syllabics, when they occur as contrasting pairs, show the highest levels of correct responses. Six of the 9 bisyllabic [a] tapir words show correct responses at the 70 percent level or higher. Six of the 7 bisyllabic [i] squirrel words show correct responses at the 70 percent level or higher (Table 6.4).

Table 6.4 Distribution of Bisyllabic [a] and [i] Terms for "Tapir" and "Squirrel" in Nineteen South American Indian Languages

"tapir"	+/–	"squirrel"	+/–
mezaha	+	kuzikuzi	+
baaetae	+	titi	+
wagara	+	chihkun	–
tsontinba	–	tiriri	+
pamau	+	wiching	+
sacha wagra	+	waiwashi	+
bwopoha	+	twarika	–
turuma	–	ii'kingo	+
veku	–	timoka	–
aerajbá	+	kariká	–
awada	+	sipo	–
ha'a	+	ka'its	–
kemari	–	miyiri	+
yap	–	biku	–
machihouuri	–	meeri	–
imba chuch	–	shuma	–
hura	–	naba	–
bi	–	karkoromia	–
je'hi	–	huven	–

Finally, an examination of those terms showed levels of correct responses at chance levels of less provide further evidence to confirm the sound symbolism hypothesis presented here (Table 6.5).

Table 6.5 Pairs of Words for "Tapir" and "Squirrel" Showing Levels of Correct Choices at Level of Chance or Less

first term	second term	number correct	% correct
machihouuri	Meeri	43	51
shuma	Imba chuch	42	50
naba	Hura	41	49
karkoromia	Bi	38	45
je'hi	Huven	23	27

In the first pair, subjects naturally (and in this single case, mistakenly) choose machihouuri as the proper word for squirrel in that two of its syllables contain the high-front vowel [i]. Subjects have no basis whatsoever in deciding on the word for squirrel in the naba-hura pair in that neither includes a high-front vowel. In the last two pairs, *bi* and *je'hi*, both sound like good squirrel names on the basis of their high-front vowels, and apparently students are making their choices (again, mistakenly in these cases) on this basis.

In conclusion, these data from languages of nineteen distinct linguistic families of South American Indian languages strongly support the claim that sound-symbolic processes are at work in the names these languages have selected as appropriate to the creatures "tapir" and "squirrel." Tapirs are natural kinds that are large in appearance and slow (relative to squirrels) in their behavior; their names show a preference for the vowel [a]. Squirrels are small and quick (relative to tapirs), and their names show, in contrast, a preference for the high front vowel [i].

Other factors than sound symbolism may, of course, be at work in the nomenclatural patterns described here and I invite alternative hypotheses that might account for these results in at least a parsimonious fashion as the one I have proposed here. Until other more convincing proposals are forthcoming, it would appear that sound-symbolic processes are at work in ethnobiological lexicons in ways that have only now begun to be explored. More generally, the study of ethnobiological sound symbolism may shed light on the evolution of humans representation of the perceived structure of the natural world in speech. St. Thomas was correct: "Names for things *ought* to reflect their [very] nature." And, if Neanderthals *could* pronounce a high front vowel, in their ancient words it's likely that they did not use them for "mammoth."

References

Berlin, B. 1992. *Ethnobiological Classification: Principles of categorization of plants and animals in traditional societies*. Princeton: Princeton University Press.

_____ and T. Kaufamn. n.d. *SAILDP: The South American Indian Languages Documentation Project*. Datafiles in the Laboratories of Ethnobiology, Department of Anthropology, University of Georgia.

Brown, R. 1958. *Words and things*. Glencoe, Illinois: Free Press.

_____ and R. Nuttall. 1959. Methods in phonetic symblism experiments. *Journal of Abnormal Social Psychology* 59: 441–44.

Campbell, L. 1988. Review of *Language in the Americas*, by J. H. Greenberg. *Language* 64: 591–615.

_____ 1991. On so-called pan-Americanisms. *International Journal of American Lingusitics* 57: 394–99.

Chafe, W. 1987. Review of *Language in the Americas*, by J. H. Greenberg. *Current Anthropology* 28: 652–53.

Ciccotosto, N. 1991. *Sound symbolism in natural language*. Ph.D. Dissertation, University of Florida, Gainesville, Florida.

Darnell, R. 1987. Review of *Language in the Americas*, by J. H. Greenberg. *Current Anthropology* 28: 653–56.

Fischer-Jørgensen, E. 1978. On the universal character of phonetic symbolism with special reference to vowels. *Studia Linguistica* 32: 80–90.

Goddard, I. 1987. Review of *Language in the Americas*, by J. H. Greenberg. *Current Anthropology* 28: 656–57.

Golla, V. 1987. Review of *Language in the Americas*, by J. H. Greenberg. *Current Anthropology* 28: 657–59.

Greenberg, J. H. 1987a. *Language in the Americas*. Stanford, California: Stanford University Press.

_____ 1987b. Author's précis (preceeding reviews of *Language in the Americas*). *Current Anthropology* 28: 647–52.

_____ 1989. Discussion Note: Classification of American Indian Languages: a reply to Campbell. *Language* 65: 107–14.

_____ and M. Ruhlen. 1992. Linguistic origins of native Americans. *Scientific American*: 94–99.

Jespersen, O. 1921. *Language: Its nature, development, and origin*. London: George Allen and Unwin.

Köhler, W. 1933. *Gestalt psychology*. New York: Liveright Publishing.

Hays, T. 1994. Sound symbolism in New Guinea names for frogs. *Journal of Linguistic Anthropology*.

Hinton, L., J. Nichols, and J. Ohala (eds) 1995. *Sound symbolism*. Berkeley: University of California Press.

Lapolla, R. 1995. An experimental investigation into phonetic symbolism as it relates to Mandarin Chinese. In *Sound Symbolis*, edited by Hinton, L., J. Nichols, and J. Ohala. Berkeley: University of California Press.

Matisoff, J. 1990. On megalo-comparison: a discussion note. *Language* 66: 106–20.

Nuckolls, J. B. 1996. *Sounds like life: Sound-symbolic grammar, performance, and cognition in Pastaza Quechua*. Oxford: Oxford University Press.

Rankin, R. 1992. Review of *Language in the Americas*, by J. H. Greenberg. *International Journal of American Linguistics* 58: 324–50.

Rodríguez-Mahecha, J. V., J. I. Hernández-Camacho, T. R. Defler, M. Alberico, R. B. Mast, R. A. Mittermeier and A. Cadena. 1995. *Mamíferos Colombianos: Sus nombres comunes e indígenas*. Occasional Paper No. s3 in Conservation Biology. Washington, D. C., Conservation International.

Ruhlen, M. 1987. *A guide to the world's languages*. Vol. 1: *Classification*. Stanford, California: Stanford University Press.

Sapir, E. 1929. A study in phonetic symbolisms. *Journal of Experimental Psychology*. 12: 225–39.

Swadesh, M. 1971. *Origin and diversification of language*. Chicago: Aldine/Atherton, Publishers.

CHAPTER 7

Jivaro Streams: from Named Places to Placed Names

Maurizio Gnerre

Introduction: Common and Proper Nouns, Places and Persons

In this paper I focus on Shuar (Jivaroan, Upper Amazon, Eastern Ecuador) place names, and, more specifically on stream and minor river names.[1]

In the study of names, place names have received little attention, compared to person names, as if the former lacked peculiarities needing to be specifically addressed. Person names have given rise to a long-standing philosophical debate, from the work of Stuart Mill to the present, on issues such as their denotation and connotation, sense and reference (Frege), "logically proper" names and definite descriptions (Russell, and more recently Searle´s "descriptive backing"), description and causal theories (Kripke).[2] Place names, for some reason, have been relatively neglected,[3] under the misleading assumption that the issues they give rise to are similar to those characterizing person names. Even in the anthropological linguistic debate, as Levinson has recently argued, the study and interpretation of place names "is in its infancy." (1996: 365). Place names have been mostly the object of etymological and historical research, often based on the hypothesis of an original transparency, tarnished to complete opacity by language drift and by sociocultural and population shifts.

The great amount of information contained in place names, their existence as such, their transparency or opacity, their motivation, are, among other issues, to be dealt with, and not simply data to be taken for granted. Thus, there is a need to rethink the problems posed by place names, in view of the limitations inherent in current philosophical approaches and the scarcity of research based on an anthropological linguistic approach[4] and cognitively-grounded[5] view of language.

The Shuar stream names that will be taken into account here are transparent. A structural-grammatical analysis (in Part I) leads to the identification of three groups. Investigation of their motivations, beyond their transparent reference, will show them to be either common nouns-names or names-names, carrying information differing in quantity and quality. While a small set of stream names is built on the reference to a feature of the stream, in most other names, a common noun apparently designating an animal or a plant or a person refers, in terms of the internal organization of the name, to a specifying "figure" (F), against the specified "ground" (G),[6] i.e. the generic spatial referent: the "river."

Most Shuar person names are connected, in turn, to an animal or a plant noun. It is evident that culturally-grounded metonymy and synecdoche are tropes extensively present in these names. Basic questions that will emerge in studying the nature of these tropes from a Jivaroan perspective are: what do Jivaroan proper names designate, what do they refer to? In other words, what does a person name refer to from a local perspective? Does it refer to a person, an actual individual, or rather to something else, to a more subtle and hidden aspect? In the same perspective, what does a place name refer to? Which evident or hidden feature of, say, a stream, does it designate or refer to? What does it express in itself? It is impossible to answer these questions without an anthropological understanding of local Jivaroan representations, which necessarily leads to a discussion that goes beyond a strictly linguistic-semantic one. Two conditions may be pointed out: only when a name is transparent and only when the representations from which those names derive are known is it possible to answer these questions. In other cases, reference is dimmed by opacity and by a referentiality which we may call "global" and aproblematic, i.e., the referentiality according to which a person name characterizing that individual refers to the person as a whole and *not* to a specific feature.

A necessary background to this discussion is represented by Jivaroan conceptualizations of humanity and their dynamic view of the natural environment, of relations between persons and places, but also, and perhaps mostly, of relations between living beings and the *wakan'* (to be understood as "self-consciousness, and by derivation as a representation of reflexivity"[7] and "image"). The interpretive mediation on the Shuar notion of personhood, of *wakan,'* and of the gendered interactions of humans with the natural world allows us to glimpse the relation, juxtaposition or overlapping between person and place names that circulate among them, on which humans exert a central regulating power (Part II). Stream names are ultimately to be interpreted as manifestations of the peculiar Jivaroan anthropocentrism.

Part I. Shuar Stream Names

Around 1955–56 an early survey of Shuar place names was carried out by Catholic missionaries (Barrueco 1959 and Centro Misional de Investigaciones Científicas 1961), using also pre-existing data.[8] In 1970 and 1971, I was able to check a small part of that data in a field survey of the (at that time) northern Shuar area, from north of Macas to Palora River (Upper Upano River, north of the Morona-Santiago province).

More than seventy river and stream names were either verified or gathered in the area. I found a high degree of both diachronic consistency between previously collected place names and those collected by myself, and synchronic consistency (i.e. social sharing) among the Shuar with whom I talked, who were mostly men. Almost all stream names (both those collected in 1955–56 and those found in the 1970 survey) were perfectly transparent: the Shuar were likely to have occupied a non-inhabited region and, from one generation to the other, of having deliberately "mapped" the area with operationally useful, and hence linguistically transparent, names.[9]

Considering the physical configuration of the territory, it seems logical that many Shuar place names refer to rivers and streams. On the other hand, a possible distortion introduced by the observers must be taken into account. Neither the missionaries who collected the early data, nor I focused on rivers or streams, taking into account the possibility that subsections of the same river would be named differently,[10] and the possibility that names might have been given to places that we, as Westerners, would not normally have named. In this sense, the data collected may have been influenced by the observers' perspectives. In fact, later evidence suggests that many names for places other than rivers are known and used.

Many things, however, have changed: not only have many place names changed (though this is not true of stream names), and many have been added as a consequence of demographic growth and increased pressure on the territory, but, what is more relevant, the focus on "places" has shifted, from "natural" references, such as streams or hills, to human-made places, such as villages and roads. In other words, the object itself of people's topographic "competence" has changed.

Territoriality is constructed as a magico-symbolic space, as well as a functional, physical knowledge of an area. Many Shuar men were used to traveling extensively in their once vast territory with their vast geographical knowledge; learning place-names was part of the traditional socialization of male children. They accompanied their fathers on hunting excursions, initially limited to a few hours distance from their residence. During these excursions, they were told the name of each stream they found on their way. As Bianchi (1981: 34) writes: "the boy will still accompany his father to learn from him, in the forest, the paths, the names of the streams, of the small rivers, of the lagoons, to know the plants and the habits of the animals. The most important thing is to learn to locate himself and to find his direction, because in the forest it is very easy to get lost or to walk around in circles."[11]

Seventy river or stream names, belonging to the six numbered sections of Map 1. (Number 2 to 7), are listed in Annex.[12] They are basically organized into two groups: a larger one (fifty-nine names, 1–54),[13] in which names refer to a living being, including animals, plants or a humans, and a smaller one (eleven names, 55–65) in which names refer to a quality of the river itself. This group of names could be called stream-centered, as opposed to the first group, hetero-centered, in which reference is to humans, animals and plants somehow connected to the stream. Each group raises different kinds of linguistic issues. In this paper, attention is focused primarily on the first group, and only a few mentions will be made to names belonging to the second group.

Internal Structures

In terms of their internal structure, Shuar river or stream names (Nm.) may be divided into two groups: (1) names consisting of two common nouns (Nn.), or in a person name and a common noun, and (2) names consisting of one common noun-name, using the expression to designate a common noun that has become a proper name.

In the first group, the second noun (N2) is usually (*é*)*ntsa* "river, stream, water of the land," with morpho-phonological alternating or deletion of the first vowel, i.e., a *sandhi* process in which the final vowel of the first noun (N1), when stressed, displaces the first vowel of (*é*)*ntsa*. In a few cases, the Quechua word *yaku*,[14] having the same meaning, is present. The first noun usually refers to an animal or to a plant (which may be connected to a person name), alternatively it consists of a person name (in many cases clearly deriving, in turn, from an animal or a plant noun).

In the second group, we find a single common noun-name that can (2a) end with -(*a*)*s*(*a*), derived from *éntsa*, or with -(*i*)*m*(*i*), (also -*m'*), probably derived from *yúmi* "water, water from the sky, rain."[15] Alternatively (2b), we find a common noun-name with none of these endings.[16] The phonological processes through which -(*a*)*s*(*a*) is connected to *éntsa*, and -(*i*)*m*(*i*), (or -*m'*) to *yúmi*, are not clear.[17] In the first subgroup of forms (2a) we may postulate either that **Vsa* was the old form (likely, with a nasalizaed V) and that an epenthesis of a nasal consonant (alternatively, a spread of nasality) took place, producing a further epenthesis of /t/ between the nasal and the sibilant, so that /s/ > /ts/. In the second subgroup (2b) a vowel shift /u/ > /i/, or the opposite, is involved. These names, however, are likely older than those listed in group 1.

While in groups 1 and 2a the first part of the name is its "distinctive" component, in group 2b the full common noun-name is "distinctive," and there is no component referring to the river or water. This is not a frequent pattern among transparent river names; it is found in only nine out of seventy river names and is characteristic of the few opaque river names used by the Shuar,[18] such as *Upano*, *Palora*, and *Sangay*.

The analysis outlined above may be represented through the following schema:

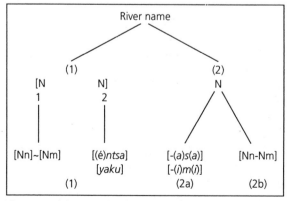

Figure 7.1 *Schema*, with Examples of River and Stream Names of the Two Main Groups

Following are a few examples of each of the three groups, along with an essential morphological analysis, when needed:

Group 1:

8   pankíntsa, *pánki*'anaconda,' generic, Eunectes murinus.

10   yantána éntsa,'black caiman,' Paleosuchus trigonatus.

33   chiánku éntsa, *chiánku*'a ginger.'

36   chíki yaku, Chíki'a female name.'

37   mamátentsa; Mamát(u), Mamátui'a female name.'

38   inchítu éntsa, Inchít(iu)'a female name.'

50   úshu yáku,'small plant, similar to taro, used to treat worms; alquitira, fam. Anturios; Araceae, Caladium bicolor.

51   mashiánta éntsa, Mashiánt'a male name.'

57   ayámtâi éntsa,'a hut where hallucinogens are taken,' from *aya-*'to dream,' *-tâi* 'verbal nominalizer, instrument, place where… .'

Group 2a:

1   namákim,' *namáku*'big fish, supra-generic;' related to a person name.

24   chiwiása, *chíwia*, probably Formicarius Colma, Reg. Sp'trompetero.'

41   paátmi, Paát'a male name,' from *paáti*'sugar cane, generic,' Saccharum officinarum, also graminacea ripicola ?, almost identical to Gynerium.

42   tsemántsmaim,' Tsemántsemai'a male name,' probably related to *tsentsém*, 'Hunting charm plant.'

53   nampírsa, Nampíri'a female name.'

56   wichími, *wíchi*'rotten logs.'

Group 2b:

16   warúsha, Wárusha'a male name,' with final vowel sonorization and stress displacement.

47   kukúshi, Kúkushi'a male name,' with final vowel sonorization and stress displacement.

56   pajának,'a slow stream in which the waters of a main river outflow when it is in spate.'

58   yutâi,'edible, food to eat, place where you can eat;' from *yu-*'to eat,' *-tâi*'verbal nominalizer, instrument, place where…'

62   púju,'white.'

65   yawá ujúke,'dog's tail;' *ujúk-*'tail,' *-e*'3rd person inalienable possession.'[19]

Inherently- and Hetero-centered Stream Names

As noted above, stream names may be grouped into two main categories: those in which the referential center is to be found in the river itself (we could call these inherently-centered names) and those with a displaced reference (we could call these hetero-centered names). Inherently-centered names are those in which a feature or a quality of the stream itself is predicated, without any reference to living beings; hetero-centered names are based on a reference to an "animated" being,—human, animal, or plant—or even a mythological being. Names of the first category are few. I will focus on the second category, represented in our corpus by fifty-four names; these may be further classified into three subgroups. In what follows I will discuss in some detail these subgroups, in two of which the role played by conceptual contiguity relations between human beings and other living beings is manifested through stream names.

Conceptual Contiguities

There are at least two levels of conceptual contiguity, and, as a consequence, two basic river name subtypes: those I call common noun-names (nn-nm), and names-names (nm-nm). As we have seen above, person names are present in each of the three subgroups of stream names taken into account here. Thus, by combining formal and semantic criteria each of these subgroups could be further divided into two types, one derived secondarily from a human-centered synecdoche, where a human name (in most cases derived from a common noun) plays the main role, and a second one based on a primary metonymic or synecdochic process centered on a common noun.

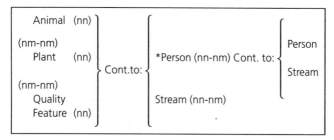

Figure 7.2 Contiguity Relations inside Person and Stream Names

This simple schema illustrates six types of contiguities. All of them are possible and altogether they cover a high percentage of all possible contiguity relations that can be established through naming.

The asterisk before the first occurrence of *Person* means that the condition of common noun-name for persons is postulated, but cannot be empirically observed: even if many person names may be interpreted as deriving from animal or plant nouns, or other features or quality nouns, *all* person names are actually based on a name "contiguity," or "chaining" with another person name. Stream names may be founded either directly on a common noun (common noun-names) or on a person name (name-name).

Both person and stream names imply a metonymy, but their nature is different: while person names imply an "homogeneous" metonymy (i.e., person-to-person, or their *wakan'*) metonymy, pointing to a contiguity with another (usually dead) person, in an endless diachronic (transgenerational) chain, river names are traditional (accepted, but not officially bestowed by anyone) "heterogeneous" metonymies pointing to contiguities between living beings (or their *wakan'*) and a river.

Things, however, are even more complicated, because, as we already stated, there is an "intermediate" group of common nouns highly suggestive of person names. Furthermore, the relation between place names and person names may be interpreted as hierarchical.

Place-to-place and Person-to-person Contiguities

All these relations may be interpreted as a figure (F) in contiguity with a ground (G). Shuar were very sparing in explicating these relations. The first relation indicates a certain material and (at least at times) observable feature of contiguity, while the second refers to a contiguity based on identity or kinship. In very few Shuar place and river names, explicit reference to another place name is provided.[20] A common noun, such as *tuná* "waterfall" may be used as a modifier for another place name: 24a *tuná chiwiás*, where reference is to 24 *chiwiás*. An adjective, such as *úchi* "small" may also play a similar role: 25a *úchi tayúntsa*, 42a *úchi tsemántsemaim,'* where reference is to 25 *tayúntsa* and 42 *tsemántsemaimi*.

As for the person-to-person relations, contiguity is explicitly provided in certain traditions, such as certain Semitic ones, where an explicit reference to a kinship relationship (mostly "father," "mother," "son," etc.) is part of the person's name. In the Western tradition, surnames play a similar, though not explicit role, providing an often misleading cue to contiguity. In the transparent Shuar person names I examined, when a conceptual contiguity was suggested, it was with a non-human being, as it will be shown below. No explicit synchronic clue to contiguity with other human beings was provided. There was, however, a "hidden" diachronic name chaining, as pointed out above, that may be traced back to Shuar understanding of personhood and humanity. We will discuss this dimension in Part II.

Person-to-place and Place-to-person Contiguities

We may notice that person-to-place relations were (and are) established only when it is necessary to refer to the area a person is from, e.g., Mashumar from Pajanak. In these cases, the person name is used as an F in contiguity with a place name, a G.

The second and opposite relation is the one that mostly characterizes, implicitly, Shuar place references. The number of "eminent" men,[21] and in a few cases well-known women, was limited, and their place of residence was identified through them.

Shuar local groups are part of a territorial network of intermarriages where the implicit reference is to the power of a few "eminent men." Certainly the names of these men are strong candidates to become names of the streams where they settled. In our reference corpus of stream names, we also find two women's names and several plant names that are connected to feminine names. In the past, in colonial and postcolonial accounts and maps, places in the Jivaro region were often identified under the name of an eminent man living in a determinate place. When the referent was a specific place and not a river, the person name was followed by Jivaroan locative suffixes (*-num, -nam*).

As far as I know, the Catholic Missionary Gioacchino Spinelli (1993, written in1926) is the only writer on Jivaroan culture and language who, while discussing names given to children, included"rivers"among the referents (together with"animals [...] fruits, plants") of those person names. He provided two examples of person names, which he interpreted as"river names": Unkuch and Tsamaraint (1993: 131). Father Spinelli maintained that certain person names had been derived from river names; it seems to me, however, that certain rivers have been named after persons.

Approximately half of the animal and plant nouns found in stream names, are also connected to person names. We may observe that many nouns referring to animals are connected to male names, while many nouns designating plants are connected to feminine names. When an animal or plant noun is also used as a river or a person name, this happens through the mediation of the latter, which became part of a river name at some point in history. Other animal or plant nouns are used only in river names, without being used for person names. Among the Shuar there seems to be a specific directionality, which allows only the transition from person name to place name, but not the opposite. In other words, if a common noun is found in both a river and person name, this implies that it became first a person name and as a consequence, mostly for focality reasons we will discuss below, a river name. In diachronic terms, I would suggest that person names are likely to have replaced animals and plants nouns in stream names.[22] Such a directionality from common noun to place name is highly iconic of the process of place individuation. Persons are already"given"as physical beings and necessarily named and culturally shaped, while places are"created"out of mere space through knowledge, reference and naming. Named persons exist in (come from, traverse and go to) places, and"places"arise because of human activities. Places are elaborations of human consciousness. That is, a place differs from mere space precisely because it is the object of consciousness.

Hetero-centered Stream Names

References to living beings, either animals or plants, are found in the fifty-four stream names listed in Annex 1. Four categories of names may be identified, in relation to person names: i) those based on a person name, ii) those for which it is not possible to decide whether reference is to a person name or to an animal or plant noun, iii) those in which there is a reference to both a common noun and a person name, and finally, iv) those in which no shared reference with a person name is found. Thirty-four names are classified under categories i) to iii). Most person names are male names.[23] All female names (1, 17, 18, 32, 33, 34, 35, 42, 36, 37, 38) and a few male names refer to'useful' (from a Shuar point of view) animals and plants (1, 3 (?), 9, 15, 24, 39, 40, 41, 46). Some person names either refer to 'useless'species, to features or, in two cases, are not interpreted at all.

In category i) only ten names (approximately one-fifth of our corpus of hetero-centered names), all based on person names, are included. As for their form, these names pertain to the three groups described above. Almost half of these names are female names (37, 38, 53, 54).

In twenty-three cases (category ii), the stream name corresponds to both a person name and the name of an animal or plant; in most cases it is impossible to decide whether the reference is to the person name or to the common noun from which the latter is derived, because the linguistic form of common noun and name is the same. In the other cases (category iii) person names are connected to common nouns. Finally in the remaining twenty-one stream names, listed below, all references are to useful animal or plant names, which, to the extent of my knowledge, do not correspond to any person name

Existence, Referential Status, Transparency, Focality, and Indexicality of Names

To understand certain cultural semantic dimensions of the Shuar stream names it is necessary to take into account some dimensions useful to our analysis; these are: existence, referential status, transparency, focality and indexicality. I will discuss them in a necessarily linear sequence, to show the chaining connections existing among them.

As a general statement, we may note that in the linguistic practice of small local groups, such as the Shuar and many other Amazonian societies, where an important role is played by diachronic (transgenerational) territorial mobility, transparent place names are likely to be used. In other words, places that have been known for a limited number of generations are likely to be referred to through clusters of common nouns (as Shuar *pamá éntsa* 'river of the tapirs'). Obviously, the same does not apply to person names: even for small, highly mobile groups, a "human continuity," a chain of tokens of the same type of person name, must necessarily exist, and a diachronic sedimentation of person names explains the loss of transparency (i.e., the full opacity) of at least part of them.

Here we find a central difference between place and person names, at least for the Shuar. Places are the result of the sedimentation of human usage. Only in few cases is there an individual choice and naming of them. Person names, on the other hand, are bestowed as part of a naming chain. They are usually grounded in a tradition of naming, and names are bestowed as such to an infant and are obviously unrelated to the child's future salient features or abilities. Person name formation and selection depends in different ways on local theories of personhood, and a significant role is played by metaphors and other tropes.

Existence and Referential Status of Stream Names

The existence of a proper name or, in other words, the relation established between a socially highlighted referent and a (common noun >) name, represents in itself an issue to be dealt with, rather than to be taken for granted. Proper names, with their particularly dense linguistic and referential status, function, much more than 'simple' common nouns, as meta-linguistic highlighters of socially perceptual saliency and/or physical discontinuity.

While all possible "natural" phenomena may be ideally organized, from a human perspective, in a continuum of perceptual saliency and discontinuity, only a limited subset of them is commonly "worth" being highlighted and referred to through a proper name, which signals by its very existence and creates through its usage, a specific referential status. Its existence not only signals some degree of perceptual discontinuity, but provides also the possibility of a relatively exact displaced reference.

Human beings are highly perceptible to other humans, and frequently refer to or address them in their verbal interactions. Individualizing reference to known human beings is certainly a basic need. As a direct consequence of these two conditions, proper names are assumed to be a necessary attribute for people. Although procedures and principles active in the naming of human beings vary greatly across societies, it is almost a truism to state that each person in the world has at least one name.[27]

Unlike people, a "place" is not taken for granted as a referent requiring a proper name. As travelers, we may be surprised to learn that a high mountain has no name. But here, nameless "places," are not our issue. On the other hand, we may be even more surprised to find out that a certain area in the midst of a desert has a name: "places" are such only through the active intervention of human consciousness. In other words, an undefined "space" may become a "place" only when it is socioculturally constituted as such: naming, and related displaced referentiality, are central, although not indispensable steps in this constitutive process. Once the need for individualizing displaced spatial reference emerges, it causes a place name to crystallize in local use. The question of how proper place names are formed is the question of which individualizing features are used to build the name out of one (or more) common noun(s).

Despite the role played by culturally perceptual saliency and the need for individualizing spatial displaced reference, we find that some perceptual discontinuities, such as that between land and water, are highly salient and we can predict that rivers, from major to minor ones, as well as lakes or lagoons, are likely to be named in most societies. Even in a region like the Amazon, where there are many rivers, these certainly hold a high position in the continuum mentioned above. As a consequence, they are very likely to "deserve" a name. Therefore, while human beings certainly occupy the highest position in the continuum of naming necessity, rivers (and lakes) hold a respectable position in the same continuum and, differently from many other potential "places" (for instance, a hillock in a flat land), are likely to be individually identified and referred to through a proper name. So far, so good. But, what is a "name" from a non-Western perspective? Or, more precisely, what is the referential status of a name?

Brown ascribes to the Jivaroan Aguaruna some kind of "nominal realism" (1985: 169), i.e., in very loose terms "the illusion that names have a special connection to the things they denote." More specifically, the author claims that the Aguaruna believe "that words can activate the things to which they refer." Both the general and the more specific versions of nominal realism are certainly shared by the Shuar and the Achuar (see below, Part II, "Restricted Choices") and I would spell out the second version in the following terms: if a proper name exists, then a referent (1) exists, and (2) (at some point in time and space) is identifiable by somebody. In other words, for a traditional Shuar a place name without a reference to a (somewhere or sometime) actually existing place (such as, for us, Camelot, Atlantis, Thule, Island of Utopia, etc.) is unthinkable. From a Westerner's point of view, the special feature is that this local theory of existential reality applies to proper names. We share a similar belief for common nouns.

Nominal realism was applied to both place and person names, a referential uniqueness constraint was limited to the former, at least for reference to space/place from the perspective of residential unit. A general statement including a person name, such as "Ankuash did not exist," or similar, would not have been possible in Jivaroan languages If a name exists, it must refer to a person who exists or existed in the past, and to deny this with an absolute statement is equivalent to denying the existence of the name just pronounced: it is a contradiction. Only a statement including place and time restrictions, such as "There was not a man called Ankuash (living) in Chiguaza in 1950" could have been possible.[28]

Still, the question of the referential status of a name persists. What is the referent of a name? Dimensions can vary very much from one cultural setting to another. In place names it is easy to recognize metonymic and synecdochic descriptive tropes. They evidence a specific (obvious or hidden) feature of a geographical area or event, and through it they refer to the entire area or event.

This is the basis of Shuar naming chains, active in at least two different dimensions. One is a "real" name chain, found both in the sparing use of person names, which leads to giving a new born child the name of a dead person, and, in the case of a stream, to name it after a person. The other dimension is that through which a common noun becomes a name, i.e., it combines with another common noun (e.g. "stream") and acquires a referential status of place individualizer; this is the case of stream names created by reference to plants or animals and not connected to any person name. From this perspective, inherently centered stream names, those containing an adjectival reference to special features or qualities of the stream itself, seem to constitute a group which, lacking any hetero-reference to beings, is positioned outside the naming chain. In this paper, I will not discuss this group in order to focus instead hetero-centered stream names.

Referential Status and Transparency

Further distinctions are needed, however, in relation to name existence. Two dimensions—both related to some degree of iconicity—will be taken into account here: name attribution and degrees of transparency. As for the first one, we can observe that bestowing a name onto a human being is a metalinguistic "right" explicitly exercised everywhere in the world. The more or less constrained or arbitrary selection of a name and the act of naming are procedures usually present in the naming of human beings, and not in place naming. This makes a big difference: humans individual names imply a specific salient act; place names only rarely imply such an act. In most societies, naming is treated as a special moment in social practice only when humans are named. As for places, in most cases individuals learn a name already sedimented and crystallized in social practice. A metalinguistic iconism or parallelism may be pointed out within this dimension: in many and perhaps in most cultures the highest degree of acknowledged discontinuity is the one between human beings, on the one hand, and all the other"natural" beings on the other; in a somehow parallel way, it is human naming which evidences again and again the human ability to name and human control over this practice.

In the second dimension, name transparency, i.e. a linguistic-referential identity with one or more common nouns, another aspect of linguistic-referential iconism is detected: in the case of humans, which are perceptually highly salient and discontinuous beings, strong signals of linguistic and referential saliency and discontinuity are to be expected. Opaque proper names fully realize this requirement: many of them are not intralinguistically connected to nouns and may form an independent lexical inventory which has to be learnt as such. The presence of opaque person names even in societies such as the Shuar one, which privilege transparency, is also connected to the"chaining effect", which, while playing an important role in many societies, in cases such as that of place names in Shuar society, applies only to a certain diachronic extent. One may speculate that where and when humans are not conceived of as beings highly discontinuous from other ("natural") beings, some degree of transparency in their names may be expected as a signal of this property. Following this assumption, transparency is conceptually much more challenging than opacity because, as in the case of Shuar stream names, it implies some degree of conceptual contiguity, which requires an interpretation.

Thus, an intrinsic iconism exists between humans, which rank high in the continuum of perceptual saliency and discontinuity, and the opacity of their proper names. This applies also to place names: when these are highly ranked in terms of saliency and discontinuity, as in the case of land/water discontinuity, we may expect a higher incidence of opaque proper names (for instance, main rivers names). In the Shuar case, only a few main river names are opaque, and this feature suggests that they have been, somehow,"inherited"by the Shuar (as well as by the white and mestizo settlers) from previous settlers of the area, before the Shuar were able, mentally and referentially, to "map" their new territory's hydrography in detail.

When the referent of a proper name is a minor stream, which in most cases is known only locally, or a certain"place"which has been socially built in local consciousness, without the support of any outstandingly salient feature, transparent names are more likely to be found. The above principle applies also to Shuar river and stream names, for which opaqueness is limited to main river names, i.e., the most highly perceptible rivers, which represent a high degree of physical discontinuity. Most minor streams, on the other hand, are referred to through transparent names. From the perspective presented here, degrees of name opacity and transparency may be both related to ways of signalling iconism. In more general terms, proper names transparency (or opacity) would have an iconic function in inverse relation to the degree of selfevidence, or perceptibility of the referent.

Transparency and Motivation

Despite the fact that some cultural patterns, such as with the Shuar and, in general, most Amazonian cultures, privilege name transparency, many Shuar person and main river names are opaque. Cultural resistance to name opacity triggers a search for intralanguage relations as well as for motivations.[29] A number of factors converge towards this rejection of opaqueness.

Deep (and not necessarily linguistic) cultural representations and attitudes underpin it. Among these, at least the belief in the presence of gendered *wakan´* should be mentioned. Not only among human beings, but also among many animals and plants, as well as from humans to non-humans, these *wakan´* are thought of as circulating "fragile images." They are conceived of as explicit intergenerational links, under the assumption of a humanity with a limited number of 'tokens' available. Most person and stream names, however, are either fully transparent, or connected to common nouns. Among river names, only a few are primarily opaque, while others are only secondarily opaque, as they are based on opaque person names; otherwise almost all rivers are known under perfectly transparent names (common noun-names).

At this point we should go back for a moment to the issue of place name existence. This issue is linked, as suggested above, to that of their transparency or opacity. These features are related on the one hand to local perspectives on "nature" and on human naming intervention (in the case of place names), on the other hand to local metalinguistic understanding and the intrinsic iconicity of opaque names, as opposed to transparent (nouns>) names. From this perspective we may not be satisfied with simply stating that in many traditions (such as the Shuar one) transparent names are "preferred" to opaque ones: this dimension is in fact central to our understanding of what a name is in conceptions that differ from those which have been assumed by several philosophers of language.

Transparency, as a general linguistic feature, is strongly related to the issue of intralanguage relations and of relative arbitrariness; as a characteristic of names it makes the lexicon "lighter" (as one does not need a long list of opaque names); many lexical entries, however, are "heavier" by the additional extensional load on them. Transparent names preserve within them a quantity of information that we can hardly disentangle; the quality of that information in many (if not most) cases may appear to be more accessible to a *prima facie* interpretation than it actually is.

Transparent place names, whether motivated or not, evoke common-sense knowledge. This is true both of common nouns and person names that are part of a place name. In the first case, when the synecdoche refers to an animal or a plant, reference is diffused, as in "tapir stream." In this case, tapirs are in a diffused, or scattered, relation to the stream: they can be found in different places along it. In the second case it is evident that reference is made either to a specific place where a person, usually an 'eminent' man with that name, used to live (e.g. Mashiánt, as in 51. *mashiánta éntsa*), or where some event in which such a person was involved happened.

When, on the other hand, a place name is an opaque lexical entry, no name internal"coordinate system"can be active. The experiential and encyclopedic echoing is erased and, along with it, all reference to, say, the relation between tapirs and the river.

To fine-tune our understanding, however, a distinction should be made between the concepts of transparency and motivation. The first term covers a much larger conceptual area than the second one: many names are transparent or subject to etymological reasoning (i.e., search for a motivation) by native speakers, as with Jimpikit and Antun,' but at the same time they are not clearly motivated.

No motivation precludes opacity: in fact this happens for many person names and, as mentioned above, for main rivers. For most fully transparent Shuar river names there is no obvious motivation, and we can only guess at one. When, as in most Shuar stream names, metonymy or synecdoche is present, motivation is only a sort of *ex-post* construction (produced either by a Shuar or by an outsider): a name such as"stream of the tapirs" (Sh. *Pamá éntsa*) is motivated on the ground that probably once there were tapirs along that stream. In a sense, N1, in synecdochic relation to N2, constructs the motivation of the name.

Although certain rivers or streams are actually named after some specific visually and/or acoustically salient feature, such as 24a *tuna chiwiasa*, where the reference to the waterfall (*tuna*) recalls a salient feature (and particularly so in the Shuar landscape and culture), most metonymic and synecdochic relations are intrinsically arbitrary. Such arbitrariness is mitigated only in part by cultural interpretation. When a stream name N1 corresponds to a person name, such as 51 *Mashiánta éntsa*, all we may safely assume is some kind of relation between that stream and a man called Mashiánt: for instance that sometime in the past he used to live along that stream, that he was killed there, that he killed some enemies there, and so on. Such guess work must be directed by knowledge of Jivaroan cultural and cognitive dimensions. When we compare the degree of arbitrariness in transparent place and person names we realize, however, that although in the former that degree is high, as we have stated above, in person names it is even higher.

Degrees of "relative semanticity" (the degree to which an expression is parseable in language) and of descriptive value of names are governed by these two continua: transparency and motivation.

Motivation and Focality

Fuzzy or diffused spatial reference is easily found in autobiographical oral accounts by Shuar men, as in the one by Carlos Angel Jimpikit Tiris,:[30] "According to him, his father came from a far away region, from the place named 'Patuka'" (16);"He undertook his third and last trip to the Wawaimi water-fall. This is found in the region of Kiim.'" (22);"When he was single, he used to go hunting with his brother, in the place of Yutui Naint'where he lives at present" (26);"For this reason [his father] moved him to the region of Chumpias, around the places [*Sic*, Sp.:"por los lugares…"] of Kunchaim…"(30). Most place names mentioned in this account are transparent and possibly also motivated. The last three names, for instance, mean, in their order,"hill of the large black ants [Cryptocercus atratus (?)],'''"stream of Chumpi (a male name),"and"stream of the copal [a Burseracea]."They are all, however, referentially fuzzy.

In many cases we can identify a (more or less extended) area that we can consider as *focal* in terms of the reference made by the place name; but, as in the case of marginal or 'border" areas, in many cases fuzziness prevails. For place names some degree of *focality* on some specific dimension is often present and even necessary. Two dimensions of focality may be identified: (1) a specific feature (a figure, F) pointed out by the name (an animal, a plant, a human being, and so on) vs. the type of geographical accident referred to (the ground, G), and (2) figure plurality vs. uniqueness.

Referential contents of transparent and somehow motivated names of both rivers and other types of places unavoidably include degrees of focality or diffuseness (or fuzziness?) in relation to the referent. Once some degree of motivation is assumed for a name, the problem of the quality of the relation between it and its referent is posed. Shuar (and, very likely other Amazonian societies) place naming strategies are intrinsically synecdochic and metonymic. The first trope is present in all cases in which a contiguity between an F and a G is present. The implementation of both tropes is grounded in most cases on a careful observation of the environment and consists in making explicit at least one relation of contiguity. What is the relation between tapirs and a specific stream named after them (e.g., *pamá éntsa*)? I would answer that a name like this suggests a scattered, or diffused, synecdochic relation: the environment surrounding that specific stream is particularly appropriate to tapirs' subsistence, so that they are or were usually found in different spots along the edges of that stream, or in some specific subsectors of it. So, tapirs are scattered Fs against a G (the river). It is from our external and generalizing perspective that we assume that the whole river or stream is a "tapir river." I would claim that a Shuar would not make the same assumption: his/her perception of the name is likely to be, in a sense, more focalized than ours. Quite differently, the relation between, say, Mashiánt and a stream named after him is a sharply focused one: whatever relation the name may suggest, everybody understands that it is a focused relation: Mashiánt was one single human being. This focality is, in Shuar understanding, even more than in ours, both spatial and temporal: Mashiánt is a single person, who lived at some time, and during a certain span of time, and the name refers to an individual being and not to a family or to a group.

It is very likely that in local subregional knowledge, a network of fine-grain place names, characterized by focused reference, was known and used. Such a network did not require extensive inferencing to identify relevant spatial referents. Names of rivers and stream subsections, in which reference was made to specific geographic accidents or locally known events (such as waterfalls, or a killed jaguar, or so) once represented a particularly focused or sharpened spatial reference.[31] During recent decades this kind of fine-grain spatial reference has gradually been disappearing, while other types of spatial references have been emerging.

Between 1950 and 1970, Shuar demographic growth and resettlement into small newly founded *centros* (i. e. villages) had triggered a general (and still ongoing) process of naming: the villages were "new" places that had to be named, while many previously "known" place names lost their referential use. The new names referring to "new" places offered the Shuar a richer and less fuzzy net of standard spatial reference, whose use requires, in a sense, less inferencing to identify each spatial referent. Even elders realize that their spatial reference is perceived by younger audiences as referentially fuzzy, if not empty. In the case of the life-story quoted above, the narrator either does not know the name of the place he refers to (in the case in which the place had a name at the time the account refers to), or (as he is tape-recorded by a young person) he does not provide an exact reference based on past knowledge; as a consequence his reference is based on present-day spatial re-organization: "a place presently named as Centro Shuar Chumpias and Napurak:[32]" Similar referential strategies are found in another contemporary autobiographical account, told by Tukup,' a Shuar "eminent" man. As the editor and translator of his account observes: "throughout the narrative [Tukup'] uses the names of Shuar and Achuar *centros* of the Shuar Federation to identify places where events took place [...] these places were not called by their contemporary names before the Federation was begun" (Hendricks (1993: 113).

At one extreme of a Shuar spatial focality continuum we find old-fashioned place names, such as Chumápnum "the place of Chumap" (i.e., his house), and present-day *centros* named after an "eminent" man, a *kakáram,* "a strong one." Even more so, as Hendricks observes, Tukup' "further identifies the location by mentioning that Nanchíram now lives in that *centro* [...] References to a contemporary resident of a place not only help identify the location, but also add to the story's believability" (1993: 114). In these cases, person (mostly male) names are used to sharpen reference focality.

At the other extreme of the continuum we can only assume the existence of the place being referred to, due to the very fact of it having a name, as in the spatial references quoted above from the autobiographical account by Carlos Angel Jimpikit Tiris.

Focality, Indexicality and Existence

Focality is a feature somehow related to indexicality. Reference in most Shuar place names is intrinsically diffused, or scattered, as I claim above, but even those place names can receive some degree of focusing through indexicality.

To clarify this point let us consider the dimension of fuzziness shared by two types of names: those which include a plant or animal noun, and those referring to a person, who may have died a long time ago. When we do not take into account this aspect, indexicality for the Shuar is quite similar to ours. While in our experience place names identify in many cases precisely defined areas (as in the case of an island or a technical-official definition of an administrative or political territorial unit), in many other cases (likely, in most) they refer to fuzzily defined spaces. Most of the time, we do not know exactly the spatial extension a name identifies, although some cases are less problematic than others. Shuar spatial reference corresponds to our second case; it is usually fuzzier than the one we are used to. Consequently, more inferencing is needed to identify the subsegment of the place a name refers to. Inferentiality, as is well known, is grounded on cultur-ally-shaped practices: if a Shuar says "I went fishing at the Stream of the Tapirs," the person he is talking to has to enact an inferential process, taking advantage of common-sense knowledge to pinpoint the spatial reference as much as possible, by inferencing, for instance, (1) that his interlocutor was possibly standing inside the water of the stream (and not fishing from the bank), or (2) that he went to a specific part of the stream particularly apt for Shuar fishing techniques. Quite dif-ferently, if the same Shuar says "I went to pay a visit to Chumap," until very recently very little inferencing was necessary, because of the high degree of indi-viduation intrinsic in Shuar person names.

Part II. Jivaroan "Vital Scarcity" and Name Referentiality

The above discussion of certain dimensions of Shuar place names in relation to person names was a preliminary step towards an interpretation of how Shuar relate to their environment. It provided some insights into more specific questions that can be answered, at least in part, in the context of Shuar ethnography. There is no point in trying to interpret the motivation of any set of transparent place names of a certain territory or even the acceptance by its inhabitants of opaque toponyms, without discussing, when possible, culture-specific representations. In this section I will move from "observable" ethnographic and linguistic dimensions to an interpretation of Shuar representations of the world.

The key challenge, and the main object, of this interpretation is represented by approximately thirty stream names, i.e., those under categories (i)–(iii) (excluding only a few under the first category: 2, 51, 52, 53 and 54) which either are person names, or share with a person name a common reference to an animal or to a plant. For all these place and person names I assume also a motivation in the con-text of Jivaroan conceptions of individuation, communication and naming, trace-able through various links.

Questions that must be posed are: what is (or was) a "place" for a Shuar, under the traditional way of territorial occupation? How are humans related to the world around them? And, ultimately, what is a person in this culture?

Starting from my contention that a place is a place of culture-specific sedi-mented consciousness, I will discuss, partly following Descola and Taylor's posi-tions, some dimensions of Jivaroan conceptions of personhood, according to which a person is a unique individual (gendered) realization of a *wakan,'* occupy-ing a place "vacated" by a dead person of the same gender. This discussion will lead to an interpretation of stream names as revealing a gendered attribution, related to the *wakan'* of the natural species referred to in names. Therefore, *wakan'* is the hidden link between persons and animals or plants, and it is the referent of many names.

The Observable

Bestowing a Name and Naming

In most societies, children naming is one of the few possibilities left to individuals to exercise some form of control on their language: while we are not allowed to choose common nouns, let alone our language as a whole, we do have the right (and often the duty) of selecting a name and bestowing it onto a child. The degree in which this right is awarded and exercised varies from one society to another.

While person names often imply some form of individual voluntary assumption of responsibility, place names, at least in premodern societies, are seldom bestowed. They are rather the product of a sedimentation of social consensus and of their own circulation.[33] Until recently, no individual responsibility was assumed among the Shuar, as well as among most Amazonian peoples, for any place name. In any case, no researcher has ever observed, as far as I know, an act of place naming. In fact, we know practically nothing of how places happened to be named.

Any question posed to the Shuar about the reason for a specific place name (though not in the case of an opaque toponym) is likely to receive a tautological, somehow functional, answer, a more or less made-up, ex-post explanation: the "stream of the tapirs" is so called because many tapirs are, or were, found along it. As for person names, however, things are quite different, due to the degree of individual and gendered responsibility in selecting and bestowing a person name.

In this sense, not only place names, but places themselves are more "legitimized" by social sedimentation than person names,[34] the latter having been selected by somebody (usually parents or relatives) at some point in time, to bestow a name to a baby. Behind each person name, probably in most societies, there is an act of individual and gendered assumption of responsibility. This does not imply, however, the absence of some form of social sedimentation. We do know something (though not much) about what the Shuar used to say about naming a baby. The only contemporary ethnographer of the Shuar, Harner, writing on data collected in 1956–57, states (1972: 84–5): "The child is given a name within a few days after being born. In some families, the father names both sons and daughters; but in others, the father may give the boys their names and the mother may bestow the names on the girls. If the father's father or the mother's mother are still alive, they may be asked to name the child of their own sex." The use of person names through the reference and the linearity of a parent naming the child are related to an important dimension of Jivaroan world, gender, which underlies many aspects of their everyday practice.

But, how do (or did) the Shuar select a specific name out of an apparently large gendered repertoire to bestow it onto a newly-born child? Some missionaries and ethnographers mention the habit of selecting the name of a deceased relative. Allioni,[35] a catholic missionary, devoted a section of his ethnography of the Jivaros (written in 1910 in southern Shuar territory) to person names, adding a list of male and female names. He stressed the fact that person names are always single, and stated also that male names were much more numerous than female names. As for the selection of the name, he wrote: "The name can be given in memory of a relative (in the case of a first-born) or of a friend or also for a fantasy of one of the parents. Often [the names] refer to birds, fishes, plants and colours," however, he added, "I do not exclude that there are names without meaning" (1993: 130). Another missionary, Spinelli, in a monograph written around 1925, stated that "[in imposing the name] the Jivaros do not have any fixed rule, but only that of their own opinion and whim. [...] They also put the name of some relative, both alive or dead ones. [...] there are names, transmitted only through tradition, which have a meaning and others which either do not have any meaning or it is unknown. Due to this, it is not unusual that, asking a Jívaro why he did put a certain name to one of his children, he answers promptly: 'así otro Jívaro llamando´ (i.e.,: this way another Shuar naming [in Spanish pidgin]" (1993: 167). Harner states: "In any event, ideally the child should be named for a deceased relative of the parental or grandparental generation on either side of the family who, if male, was respected for killing and working hard and, if female, respected for working hard" (1972: 85). Taylor writing on contemporary Achuar observes: "infants are never given the name of a known living person within the social universe of the parents. Children are given only one name [...] However, as soon as a person dies, his or her name can be given to a newborn. Ideally, children should be given the name of one of their grand- or great-grand-parents, but when this is not possible any ascending collateral relative's name is bestowed. No preference is given to paternal or maternal sides" (1993: 659).

Restricted Choices

It is evident from the above that choices in children naming are restricted. Both Taylor and Hendricks refer to a small stock of names, and the former writer mentions "scarcity," as one of the principles governing the "distribution and circulation of names" (1993: 659). Hendricks writes (1993: 114): "Shuar names are not a particularly reliable means of identifying people due to the relatively small number of Shuar names and the practice of repeating them in alternate generations." Contrary to the evidence provided by published repertoires, made possible by a recently-shaped super-local ethnic identity, we must observe that the Shuar had, and still have, limited local repertories of person names, and that as a consequence, the selection of a name is strongly constrained. Why is it so?

In Shuar culture, names are dealt with in a careful way. This is shown by their infrequent use in verbal interaction. For Shuar people (but particularly for men), it is extremely important to memorize a vast net of kinship relations, including the name of each person, but while kinship terms are used extensively, person names are not. Thus, in face-to-face interactions person names are used only to refer to absent persons. The name of the addressee (or of some other person present in the discourse setting) is rarely used, a kinship term being preferred.

When a Shuar meets an unknown person, he usually asks his name;[36] questions such as: *Yaitiam?* "Who are you?," or *Ame naaram yait?* "What is your name?" are common, and the usual answer is "I am X" e.g., *Chumapitjai* "I am Chumap'" After this initial question and answer and some additional reference by the newcomer to members of his family related to the interlocutor, through which he socially locates himself, the name is no longer used. When no close kinship relation between the newcomer and the interlocutor is discovered, a generic kinship term, such as *yatsúr* (*yatsúrua*, as address form) "my brother," or a more generic expression such as *amíkiur* (*amíkrua*) "my friend,"[37] or, among women, *impijiá* (address form), and a few others, are used. Otherwise, the appropriate kinship term is used in its address form.

The reasons for such a sparing attitude in relation to person names can be probably ascribed to different motivations. For the Aguaruna, Brown has remarked their avoidance of "too frequent use of names" of the magical stones (*yuka* and *nantag* in Aguaruna, *yúka* or *namúr*, and *nántar* in Shuar) which "may diminish the stones' power" (1985: 169).[38] But, if a similar motivation can be ascribed to the avoidance of person names, the question of their primary reference remains: do they refer to the physical human being, or to his/her *wakan*? I suggest that the latter is the right answer. The Finnish ethnographer Rafael Karsten (1935, Ch. VI) already suggested something similar: "It is believed that animals and plants are animated by human souls (*wakáni*), some of which are males and some females. Consequently, human beings are named as it corresponds." The fact that person names refer to a *wakan'* could account not only for Shuar avoidance of homonymy inside a local group, but also of the use of person names when these can be replaced by kinship terms. So, while the latter are associated to indexicality,[39] person names are associated to linguistic displacement. It is probably more than a pun to state that while many places are named after persons, person names are mostly used in dis-placed reference. The fact is that for the Shuar and the other Jivaroan peoples there is a tendency to safeguard limited resources. As Taylor writes: avoidance of homonymy is rooted in their "view of humanity as a finite collection of discrete potentialities of individuation" (1993).

Reference and Predation

Wakan' as Representation of Reflexivity

As nominal realism would suggest, to the scarcity of names, a scarcity of referents must correspond. If the set of nouns and names is finite, the set of their referents is also finite. From this perspective, however, a main difference between nouns and names would be that each type of the former may refer to many referent-tokens, while each type of the latter should identify, at least in the same social group and the same region, only one referent. The issue here is the referent. At least for person names, I suggest we should consider the referent to be the *wakan.'*

According to Taylor (whose data are from the Achuar) *Wakán'* "is by no means an exclusively human attribute: there are as many *wakan'* as there are things that may, contextually, be endowed with reflexivity" (Taylor 1993). Brown writing on the Aguaruna holds a different interpretation: "A *wakán* is the aspect of a human being that continues to exist after death. There is a limited number of *wakan'* which is endlessly recycled; birth is reappearance; when a person dies, the shadow soul becomes an anthropomorphic monster" (Brown 1985: 55). Even if the Shuar share several cultural features with the Achuar, it seems to me that their representation of the *wakan'* is closer, and very similar, to the Aguaruna representation.

Gendered *Wakan'* and Gendered Stream Names

Attributing a gendered nature to places—beyond that implicit in grammatical gender, for languages that have that category—is not uncommon. It applies to geographic accidents classified by Westerners under the same category, such as volcanoes, which, for instance, for the Nahuatl of Mexico are variously gendered (Popocatepetl, for example, is male, and Ixtacihuatl, female).

Linguistic and cultural-semantic evidence suggests that streams were perceived as gendered by the Shuar (and, likely, by other Jivaroan peoples). This is almost everywhere a lost dimension in the current conceptualization of water. The Amazonian distinction between "earth" and "sky" water is well known (Levi-Strauss 1964: 195) and made clear, as for the Achuar, by Descola (1986: 52–68). In group 2a of the set of stream names, I included those ending either in *-(a)s(a)*, or in *-(i)m(i)*. The first ending is derived from *éntsa* "water, from the earth" and the second from *yúmi* "water, from the sky."

Although in Jivaroan languages gender is not a grammatical category, it is, obviously, a cultural-semantic category; according to Guallart (1990), for the Aguaruna *yúmi* not only refers to the water from the sky, but also implies a masculine understanding of the water. Some Shuar male names, such as Yúma, Yumiák, Yumiákim, Yumíi, Yumínkias, Yumís are derived from this word, while some female names such as Entsa, Entsá, Entsámai, Entsánua are derived from *éntsa*.[40]

Only twelve of our stream names end in *-(i)m(i)*. These are likely to be old names in which reference is probably to a male *wakán.'* These are (listed in the category of overlapping with a person name): (i) 2 kumpáim, (ii) 1 namákim,' 41 paátmi, (iii) 34 wawáim,' (iv) 4 putuím,' 5 nayumpimi, 7 najempáimi, 11 úrikim, 30 shamkáim,' 43 kenkuími, and two in the group of "inherently-referring" names in which a feature of the stream is referred to: 56 wichími, from *wíchi* "rotten logs," listed by Pellizzaro (1976: 3) as one possible *arutam* vision, which is in turn an hypostasis of Shakáim, a male mythological being; and finally 64 ashkuáimi, probably from *asháku* "not inhabited by people" (or related to *ashkuai-*, possibly from Quechua, asharquiro"? Rubiacea, Hedyosmum racemosum.")

Most stream names either are formed with *(é)ntsa* (group 1) or end in *-(a)s(a)*. This last pattern, however, is extremely rare, with only three names found in our corpus: 24 chiwiása 45 pumpuís, and 53 nampírsa. Apparently, the first is associated to a male name, but the taboo involved in the noun *chíwia* is intriguing, the last of the three is based on a female name, Nampir.

So, we may conclude that in Shuar, as in Aguaruna, *yúmi* was (or still is) a 'male' water, while *éntsa* was a 'female' water. Current Shuar data, however, rather suggests that *éntsa* carries a genderless meaning. As a consequence, stream names of group 1, with the full form *(é)ntsa*, which I consider more recent than those of group 2a, include references to both male or female *wakán'*.

Vital Scarcity and Predatory Animism

There is a relation between Jivaroan patterns of territorial occupation and their ideology of *predation*, as Descola claims:"the Jivaro local groups seem obsessed by the idea of reproducing themselves *à l'identique*, and they poach from others the persons and identities they require. They deny compensation to game, through the subterfuge of a mock affinity, and also to the kin of their victims, who have to snatch it away from them by taking their lives. Their treatment of nature and treatment of others is governed by an ideology of *predation*, which is the negation of exchange"(1992: 120).

We should keep in mind that two main semantic types, inherently- and hetero-centered names, have been pointed out and that inside the second group a further distinction has been found, between names that share with person names a reference to an animal or to a plant and those that do not. These streams are incorporated into the *wakan'* network and circulation through their names referring to *wakan.'*

Following the same logic, I would claim that names as well are reproduced *à l'identique*. From this perspective, to associate game animal nouns and useful or domesticated plant nouns with rivers seems to extend"predatory animism"to the latter, through a sort of transitive property: human beings, animals and plants are united in a single social—more than"natural"—field. A congruity exists between human social relationships and man and nature relationships. In Jivaroan conception spaces become places through a conceptual contiguity to human beings. More in detail, a"predatory nominalism"is intrinsic to Shuar reference to many, but not all, streams: most of their names refer to entities to which a *wakan'* is ascribed. This is evident in the sets of stream names that reproduce person names, and also in those where reference to other *wakan'*—not the same as those referred to in person names—is found.

Part III. Conclusions

Person names constitute one dimension of language for which speakers have some control (posing, accepting, avoiding, innovating, etc.). Such control is deeply rooted in local conceptions of personhood and language. To understand local place names we need to interpret those conceptions, as well as those of space and place. We started our interpretation of Shuar stream names from two basic and connected statements: (1) we should not take for granted the referents, i.e., the spatial units that are the referents of place names, and (2) we should never assume *a priori* that in different cultures names refer to a distinct spatial unit (a 'place'), in the same way we assume it. This means that different ways of mediating referentiality should be assumed to exist.

As for the first point, we observed that rivers, and any place where a discontinuity between land and water is found, are highly ranked in the place naming hierarchy. In explaining names motivation, however, Shuar understanding of personhood, of"nature"and of language must play a central role. Shuar stream names have been grouped into three major sets: the first one echoing a person name, the second including an animal or plant noun which became a proper name, and the last (the inherently-referring) one, where a special feature of the stream becomes part of its name.Descriptivist (and eventually functionalist) explanations, which assume an image of local peoples projecting their observation and knowledge of the environment into place names, would be ruled out for the first two sets. It would hold, however, for the third (quantitatively and referentially minor) set. It is not surprising that names of the highest ranked beings, humans, are "placed" to name streams. Only as a second (although not in a diachronic perspective) choice, do nouns of culturally valued animals and plants become part of place names. Finally, only as a last choice are qualities or features evidenced in stream names. Any understanding of names as "meaningless" is ruled out by the analysis of most Shuar place (stream) names, which are linguistically transparent. Only a reduced set of opaque toponyms is widely accepted and used. The same is true for most person names.

Transparency should be interpreted as an attitude of continuity from reference to *wakan'* circulation among humans and other "animated" beings (plants and animals, and some minerals). Communication among them is the central issue and names transparency is an additional way to stress this: making explicit the reference, transparency is connected to an attitude open to communication, while name opaqueness somehow acts against this attitude.

From the second statement above, on different ways of mediating referentiality, some interpretive reflections derive, connected to what I have called name referential focality or diffuseness. Person names are likely to be preferred to animal and plant nouns to designate streams, due not only to their high status in the naming hierarchy, but, and arguably even more, to the chance of reducing the scattered or diffused reference, intrinsic in stream names based on animal and plant nouns. Focality is connected to the issue of Jivaroan person name uniqueness inside a local group and regional geographic reference.

Shuar stream names support a chain perspective on names. As already said, Shuar streams are, obviously, named places but they are also places for person names that thus became placed names. Thus, while most person and river names are based on plant and animal nouns, part of river names are primarily based on person names and only ultimately on plant and animal nouns. In Jivaroan languages no metalinguistic distinction between "common noun" and "proper name" is lexically or grammatically signaled. When we turn to proper names referentiality, however, we realize that the distinctiveness of most of them is primarily to refer to a *wakan.'* Human referential mediation plays a central role, parallel to the regulating role humans play in the dynamics of *wakan'* of plants and animals. A hierarchical and constrained order is symbolically established.

The discussion of certain dimensions of Shuar place names in relation to person names was a step towards an interpretation of how Shuar relate to their environment. It provided some insights into more specific questions that can be answered, at least in part, in the context of Shuar ethnography. There is no point in trying to interpret the motivation of any set of transparent place names of a certain territory or even the acceptance by its inhabitants of opaque toponyms, without discussing, when possible, culture-specific representations.

ANNEX 1

Reference List of Stream Names discussed in the paper. The reference CN, followed by a number will be found after each person name quoted below. CN refers to the appendix of *Chicham Nekatai* (1988), a Spanish-Shuar dictionary compiled by Shuar teachers after 1980. That appendix is the most complete repertoire of Shuar person names based on a list prepared by the Oficina de Registro Civil de la Federación de Centros Shuar de Sucúa (3. ix. 81) and enlarged in 1986 to include also Achuar names.

i)

2 (2a) kumpáim, kumpá; this stream name is related to the male name, Kumpanám. However this name can be interpreted as derived from the old Shuar term *kumpá*, from Sp. *compadre*, used in the past to refer to Christian Spanish and mestizo settlers from the Andean region.[24] Under this interpretation both could also be interpreted as 'River of the Andean settlers,' with generic reference to non-Shuar people;

12 (2b) wapúa, a male name [present also in 12a wapula (chico), a *mestizo* adaptation of the same Shuar name]; possibly related to *wapú* 'a large snake, not poisonous, 'Trachyboa boulengerii;

14 (2b) warúsha, Wárusha, a male name, according to Jimpikit and Antun' (1991: 65) from *wára*, a malodorous insect (unidentified), Reg. Sp. '*chinche;*'

37 (1) mamáténtsa; Mamat(u), Mamatui, a female name possibly related to *máma* 'manioc,' Dioscoreaceae, generic; Manihot esculenta (?); the most important cultivated plant; Jimpikit and Antun (1991: 76): "Name of a powerful (*waimiaku*) woman, who never lets the food to be short, a woman with beautiful hands, bound to work"; W;

38 (1) ínchitu éntsa, Inchít(iu), a female name possibly related to *ínchi* 'sweet potato,' Convulvacea, generic; Ipomoea batatas (?); Jimpikit and Antun (1991: 73): "It derives from *inchi*"; W;

42 (2a) tsemántsmaim, 'Tsemántsemai, a male name probably related to *tsentsém*, 'Hunting charm plant' (Brown 1985: 89); unidentifiied;

51 (1) mashiánta éntsa, Mashiánt, a male name, without a clear origin; Jimpikit and Antun' (1991: 51) provide a folk etymology: "Derived from the word *mashi*, 'all'";

52 (1) turák éntsa, possibly from Turík, at present a rare Achuar male name, not documented in Sh.; Turák is a possible form with wowel shift (frequent in Jivaroan);

53 (2a) nampírsa, Nampíri 'a female name,' possibly related to *nampér*, 'feast, party;' W;

54 (1) mirsíki éntsa (5), 'a female name;' W.

ii)

1 (2a) namákim, 'namáku 'big fish, supra-generic;' metonymic
term to refer to streams abounding in fish, e.g., *namák
kujáncham*, a variety of opossum (*kujáncham*) living close
to streams (see also Berlin 1992: 222, for Aguaruna data,[25] Namaku is a
female name, for CN (698) a male name;W;

3 (1) kumpántsa, kumpá, 'Rhamdia sp.;' for reference to person
name, CN (694), Kumpanám, see 2 (under i));

8 (1) pankíntsa, pánki 'anaconda,' Eunectes murinus, generic;
the anaconda is among the most frequently mentioned
arutam visions and is an hypostasis of Ayumpúm
(Pellizzaro 1976:3); in Pellizzaro (1973: 10) the facial
painting of a man of Unt Chiwias, Tiwi, symbolizing the
arutam power of the *pánki* is reproduced in a drawing;
CN (701);

9 (1) wampúchi éntsa, wámpu, Ficus insipida, a Moracea used by
the Shuar as an anthelmintic; -*chi* 'diminutive suffix;' the
same noun refers also to 'tadpole,' not edible, some
varieties of toads and frogs are eaten; ; *wampúsh* is a
Bombacacea; according to Jimpikit and Antun (1991: 65) male
names Wamputsar and Wampusrik are both related to wámpu
'Ficus insipida;' CN (713), also Wampachi;

13 (1) manchúntsa, manchu, 'mosquito,' Anopheles; CN (696);
according to Jimpikit and Antun' (1991: 51) there is a
connection between the fact that mosquitoes suck the
blood of all types of animals and the use of this name "for
a child to grow up in good health";

26 (1) jémpe éntsa, 'humming bird, generic; mythic being,'
trochilides; CN (692) Jempé, Jempékat; Jimpikit and Antun'
(1991: 46–7): "a child is bestowed with this name to be
nimble, quick and strong, and, more than anything else, to
be fast as a humming bird."

27 (1) máshu éntsa, 'razor-billed curassow, turkey cock, *paujíl*,
Mitu tormentosa; also a poisonous plant (?); CN (696);
Jimpikit and Antun' (1991: 52): Mashu, Mashumar and
Mashutak, all connected to máshu "a bird of the Kutukú
(mountains)";

29 (1) pítru éntsa, pítiur, a bird; unidentified; CN (701); Jimpikit and Antun
(1991: 55): "a long-legged bird [...] children are so called for them to have
long legs"; from Sp. Pedro 'Peter'?;

32 (1) tsetséma éntsa, 'a sweet-smelling herb,' tsentsem (W name)
a medicinal herb used to treat children's fevers. Its juice
is given to small children for them to grow strong;
unidentified. One of the plant names with a doubled pattern:
reference to other wild varieties with double-pattern names

mentioned in Nunkui myth (Pellizzaro 1978: 75); CN (709); W;

36 (1) chiki yaku, Chiki, a female name related to *chíki*
'arrow-root,' Maranta ruiziana; Callathea allouia ?; both
wild and cultivated; an edible tuber, in some cases it is taboo,
arrow-root is mentioned in Nunkui myth (Pellizzaro, 197.: 205, 23); the
same noun refers also to wild boar sperm; W;

39 (1) maikiúa éntsa, hallucinogen, Datura sp., Brugmansia sp.
(suaveolens ?), (three varieties are cultivated),
Solanaceae; CN (696); Jimpikit and Antun' (1991: 51):
Maitiunk"from *maikiua* [...] the strength for Shuar
survival. Children are called with this name to grow up
strong and powerful, the same as the plant."; Germani 1970:
Maík' (-kiua) and Maítiuk;

40 (2b) kukúshi, Kukúshi, a male name, possibly related to *kukúch*, an
edible Solanum,' Solanum coconilla (quitoense?), cultivated;

41 (2a) paátmi, Paát, a male name, related to *paáti* 'sugar cane,'
Saccharum officinarum, generic, and a graminacea ripicola, almost
identical to Gynerium saggitatum 'arrow cane' (Descola 1986: 101);

iii)
15 (1) tséma éntsa, 'a small monkey, mico,' Saimiri sciureus,
Callitrichidae (?); its meat is taboo for shamans and for adults when a
newly-born child is close; Tsemán CN (709); Tsemaik' is a female name;
CN (709) M., Tsemán; Tsenkush, Tsenkush, (tsenkush is the Achuar form
for Sh. tséma); according to Descola (1986: 107), the male of the
humming-bird Florisuga mellivora is called maikiua jempe (see 26 and
39), while the female is called tsemai jempe.

16 (1) tsénku éntsa, 'a mythic jaguar,' Hua. tsenkútsuk 'a variety of felin
living mostly on trees;' Tsenkuch, CN (709) Tsenkúsh;' Jimpikit and
Antun' (1991: 60): "a fierce animal *tsenkutsenku*, [which] ate a Shuar who
was hunting"; Tsenkúp (W.) is related to the name of a small monkey (the
same as 15 ?);

17 (1) tsuntsúntsa, tsúntsu 'a snail' (unidentified); CN (709); also
Tsuntsúmanch' (a female name); W;

18 (1) tsapúntsa, tsampuntsa??, tsampu, Calandra palmarum;
tsampú, Reg. Sp. *papaso*; an insect whose larva is the edible
suri; CN (709), Tsapáu; W;

24 (2a) chiwiása, chíwia 'trompetero,' probably Formicarius
Colma, its meat is taboo for a man, when his wife has
recently given birth; male name Chiwiant CN (707);

25 (1) tayúntsa, táyu 'a nightly bird, found in caves,'
Steatornis caripensis; CN (710) Táyu-jint, "path of the
nightly bird,' parallel to Kashi-jint, "nightly path;'

28 (1) ampúsha éntsa, 'owl, generic for Strigides;' not eaten,
taboo; Descola (1997: 367): "The wakan has no precise seat
in the human body, so the species whose form it takes

depends on the part of the body where it was residing immediately before its departure: it will be an owl if it was resident in the liver…"; CN (685); Jimpikit and Antun' (1991: 40):"the Shuar believed it to be […] the spirit of a dead *ancestor*";

33 (1) chiánku éntsa, chiánk*u*'a ginger; leaves are smoked to treat head ache;'used also to treat snake bites? leaves are used to wrap fish for baking it; Zingiberaceae, Renealmia hygrophila; grows close to rivers; CN (707); W;

35 (1) yuwíntsa, yuwí'various species of wild and cultivated squash,'used also to prepare infusions; Cucurbitaceae, Cucurbita maxima; CN (691); W;

46 (1) yunkuántsa, yúnkua'yellow sapán,'Lecythishians, Sapotacea (?). Its bark is used to prepare ropes to tie bundles or house parts. Heliocarpus popayensis, Sloanea fragrans, Elaeocarpaceae; also Yunkuankas;

iv)
Animals

4 (2a) putuím,'pútu'catfish;'(up to 70 cm. long), folk generic name for genus Hypostomus (at present also Cochliodon and Pseudancistrus; see Berlin 1992: 133), found in the hollows of river-beds;

5 (2a) nayumpimi, Hypostomus, a fish, Reg. Sp. *caracha*;

6 (2b) kamitán, probably'Serrasalmus sp;'

7 (2a) najempáimi, najémp*a*'a silvered fish of river head-waters;'

10 (1) yantána éntsa,'black caiman,'Paleosuchus trigonatus. The same noun can refer also to the ocelot, Felis pardalis; it is probably only as a name for the ocelot that it can be an *arutam* vision, functioning as an hypostasis of Etsa (Pellizzaro, 1976: 3);

11 (2a) úrikim,'an edible brown crab,'generic (?)'(unidentified);

19 (1) pamá éntsa,'tapir,'Tapirus terrestris. A reported *arutam* vision, an hypostasis of Ayumpúm (Pellizzaro, 1976: 3);

20 (1) káshai éntsa, Cuniculus paca, Reg. Sp.'lemucha, paca,'a rodent; its meat is taboo for the shamans;

21 (1) wishíwishí éntsa,'ant eater,'Myrmecophaga tridactyla; Also a fruit, Protium sp., Burseracee;

22 (1) újea éntsa, Bradypus sp.,Variegatus? Choloepus dudactylus; also an iwianch, usually said to live in the hilly regions, seen as a rushing horse, with firing eyes;

23 (1) sháiper éntsa, probably in Reg. Sp.'lomuchi;'

30 (2a) shamkáim,'shámak'a large green parrot with blue wings;'

31 (1) kuyúntsa, kúyu 'white-headed guan,' "turkey hen," "a lapwing' Aburria pipile, Pipile pipile (?), or low river water\dry [*kuyuktin*, 'season of dry rivers']·;

Plants

34 (2a) wawáim,' wáwa 'balsa,' Ochroma pyramidale (or lagopus or piscatoria ?); leaves are used to cover pots, it is planted around gardens or not too far from settlements; wáwa, wawách. Bombacacea gen.?;

43 (2a) kenkuími, kénku 'bamboo,' Guadua angustifolia, important in the myth of Nunkui (Pellizzaro, 1976: 21). Used in house-building;

44 (2b) yukípa, yukáip, yikíp 'resin used to glaze pottery;' the plant from which it is extracted;

45 (2a) pumpuís, pumpú, Heliconia Bijai, Musacea (?); Calathea altissima (winchu ?), its leaves are used to cover the top of the roof and to wrap raw food before cooking it and to block small rivers in order to fish ?. Cyclanthacea, Calathea altissima, Musacea; Heliconia Carludovica palmata; Rhopola complicata;

47 (1) káuchu éntsa, 'rubber tree;' Kuacha ??;

48 (1) káip éntsa, káipi, Mansoa aliacea, M. standleyi, Bignoniaceae, (?) onion-scented, '*sacha ajo*;' used to treat different diseases;

49 (1) mutíntsa, muút 'a small wild tuber, red fruit,' similar to sunkíp, sánku, 'pelma;' Aracea, Xanthosona;

50 (1) úshu yáku, 'small plant similar to taro, used against worms; alquitira, fam. anturios; Araceae Caladium bicolor, contains skin irritants such as calcium oxalate; grows close to rivers; Brown [on Aguaruna] (1985: 89): "hunting medicine or charm. Men apply the juice of ushu to their eyelids and the barrel of their shotguns before hunting. When game is near, the ushu alerts them to the animal's presence by making their eyelids itch. Different kinds of ushu have affinities with distinct animal species."

Notes

* This chapter has been read at different stages of its writing by Felice Cimatti, Antonino Colajanni, Flavia Cuturi, William Hanks, Jane Hill, Patrizia Violi. I thank them all for their insightful comments.

1. The Jivaroan peoples (Shuar, Achuar, Huambisa, Aguaruna and Shiwiar, altogether amounting to almost 90.000 people) live in southeastern Ecuador and northeastern Peru. Their respective ethno-linguistic dialects are very close, and as such it is preferable to refer to them not as a language family, but as a linguistic isolate with no close relation to any other Amazonian language (although a distant relation has been found to some Maipuran languages (Gnerre 1999)). "Jivaro" is a word of colonial origin (Gnerre 1973) which is not accepted, let alone used, by any of the local groups. Most of them are hunters and swidden-horticulturists. The Shuar and the Aguaruna occupy the western-most pre-Andean part of the Jivaroan area with some rainforest sub-areas (such as the one discussed in the present paper) at a relatively high altitude (between 800 and 1200 meters).

2. Many other philosophers should be mentioned even summarily to outline the debate on proper names; among them Peirce on semiosis and proper names, Wittgenstein on proper names and linguistic usages, and, among contemporary philosophers of language, Quine; I would also like to mention, as an anthropologically-grounded reflection, Lévi-Strauss' chapter (in *The Savage Mind*, 1964) on proper names, treated as forms of classification.

3. These are mentioned only as marginal examples of limited relevance, such as Stuart Mill's example on the fading motivation of the name of the town of Dartmouth, or Kripke's (1972: 305) discussion of an example provided by Quine of one mountain with two different names used by the inhabitants of two different regions.

4. This does not mean, of course, that in the anthropo-linguistic tradition there is no research at all on place names. At least since Boas's famous article on geographical names of the Kwakiutl (1934) a long set of contributions have been refining the reflection on this theme. A recent important contribution on North-American Indian (Western Apache) place names is provided by Basso (1996). The anthropological relevance of Lowland South-American Indian place names is discussed by Seeger (1981) and Viveiros de Castro (1993).

5. In a major collection of papers on language and space (Bloom 1996) the issue on place names is not considered in any of the fifteen contributions to the volume.

6. "Figure" and "ground" are terms first introduced by gestalt psychology and later introduced into linguistics by Talmy (1983): the former is the thing to be located, and the latter is the thing with respect to which the "figure" is to be located.

7. This is Taylor's (1993: 660) interpretation, which I share. *Wakán'* was translated in the past as "soul" (Karsten 1935, Harner 1962 and 1972). Some key connected terms that should be mentioned are: *uunt* "old, big" [man], and *kakaram* "strong, powerful" [man], the last two terms referring also to "eminent" men's hidden strength. *Arutam* is another key term in Jivaroan culture, referring to the protective entity met under special circumstances and under altered states of consciousness; the term could be probably glossed as "old thing." See also n. 23.

8. Shuar place names are still an almost entirely unexplored issue from a linguistic-anthropological perspective. Even if specific studies on the relation of Jivaroan peoples with their environment are available (Berlin 1976, 1978, Berlin and Berlin 1983, Brown 1985 for the Aguaruna, Descola 1988 for the Achuar, Bianchi 1981 for the Shuar), no reference to the process of place naming, of standardizing place-names and preserving toponyms is found. Men were used to traveling extensively, for the purpose of visiting, hunting and feuding, and were trained when little by their fathers and elder relatives to recognize places and learn their names.

9. On the basis of ethnohistorical evidence (Gnerre 1972; Taylor 1986) and recent archaeological findings (Rostain, 1997a and b, 1999) we may assume that the Shuar reached that area relatively recently, probably four centuries ago, arriving from the south.

10. In the northern Shuar area I have taken into account, the Shuar do not navigate on rivers. For this reason, fine-grained naming processes such as those observed by Seeger (1981) among the Suyá are not likely to exist; however, more detailed naming is likely to be used.

11. Bianchi goes on in his statement about boys' training on forest space; "Furthermore, he needs to know his position at each moment of the day, to know the time needed to go back, so that the night would catch him in a previously selected place, usually close to a stream or a spring, in a flat and dry place. They prefer to spend the night in places where somebody else has spent the night previously; in this way they can take advantage of an already built hut, where dry wood is stored..."

12. The westernmost part of the area represented in numbered sections 2 and 7 of the map is no longer inhabited by the Shuar. It is a region where the transition between the Andean cultural area and the lowlands has always been intense. There are approximately twenty-five (non-Shuar) toponyms, most of them known and used by the Shuar. These are: 2 Huamboya, 2,3,4 Palora, 3 Arapicos, 4 Pastaza, 6,7 Upano, 5,6 Makuma, 6 Domóno, 6 Kupainu, 6,7 Volcán, 7 Ramos yaku, 7 Sangay, 6 Maku tambillo; maku 'muchacha'?, 6 Sheláto, 6 Yulagia, 6 Lundigi, 6 Jurumbáinu, 6,7 Quebrada, 6 Oso-yaku, 7 Salado, 7 Arshí, 7 Abaníco, 7 Yana yaku, 7 Grande Yungalli, 7 Chico Yungalli, 7 Chuspiurcu. All these toponyms refer to major rivers; some of them are remnants of languages spoken by older settlers or of contacts with Andean peoples.

13. Five names are added to primary reference names under the same number.

14. We find this Quechua word only in Shuar areas where in the past some contact between Andean settlers and Shuar took place. This is the case of the area north of Macas.

15. On the Jivaroan lexical and conceptual distinction between two types of water, one from the sky and another from the earth, see in Part II, sections on"Gendered *Wakan'"* and "Gendered Stream Names."

16. It is likely that in some of the names listed here, the N2, *entsa*, was perfectly possible, but was not produced at the moment in which the name was written down, as it was obvious to both the inform-ant and myself that the referent was the river or stream we had ahead of us; in one of the names, 44. chikiáku, to be analyzed as chiki-iaku, (*iaku* means"river"in Quechua, so that the pattern N1 N2 is preserved.

17. Pellizzaro in a note to the"Nomenclatura generalizada"(in Centro Misional 1961: no page) suggests a derivation from -*imi*"far away"; it seems to me that such a derivation is based on a Shuar folk ety-mology, once the phonological and semantic links with *yúmi* were not perceived by young people and boys with whom Pellizzaro was in contact in the late 1950s. See Part II, sections on"Gendered *Wakan'"* and"Gendered Stream Names."

18. At present, taking into account the genetic isolation of Jivaroan languages, there is no definite evi-dence about phonological changes that took place in the past. Some reconstructions are hypothe-sized in Gnerre 1999.

19. Opacity is hardly admitted to exist by the Shuar who always try to explain even those names which are quite likely are remainders of a substratum, such as Upano, Makuma, Chankuap-Huasaga, and some other names of major rivers.

20. The complete list of stream names, with some comments for each one, is provided in Annex 1. Num-bers preceding each name refer to those of the list.

21. This name is the most diverging of all: it is a genitive construction, literally"of the dog's its tail."

22. Such as in *Stratford-on-Avon* (in relation to the *Avon* River), or *Anacapri* (in relation to *Capri; aná,* Greek'on, over'), or *Moji Mirim* (in relation to *Moji Guassu;* Tupi, *mirim*'small,'*guassú*'big'ptc).

23. The expression was introduced by Taylor (1993: 657), to provide some equivalence to Shuar expres-sions *uunt*"old (i.e., big) [man]"and/or *kakaram*"strong, powerful [man].'

24. During this century, the use of the locative suffixes disappeared from maps. Either the explicit refer-ence to the longhouse'*Jíbaria,*' followed by the name of its eminent man, or his name were used in local maps. Both practices are still in use in relatively recent Ecuadorian maps of the Jívaro area. For example, one well known place, Taisha—where a mission and a military base and, in more recent years, a growing colonist settlement have been established—was named after a famous'eminent man' who lived there from the'40s to the'60s. The same practice is still extant in recent naming of newly-founded Shuar'Centers' (i.e. villages resulting from self-resettlement of a group of families, most of them related), e.g. Tukup,' a Shuar Center named after a famous man (Hendricks 1993).

25. This historical interpretation derives from that of Shuar conception of names and personhood pro-vided below in the text.

26. In the complete list provided in Annex 1, female names are marked with W. There are some stream and place names related to beings referred to in mythological narrative. These names are present in the Southern Shuar areas and not in the one we focus on. For the analysis of some place mentioned in mythological narratives see Gnerre 1985.

27. Cases of no-name are known in some Australian Aboriginal societies, however.

28. It is likely that nowdays this is not the case anymore for young Shuar.

29. A tightly linked difficulty is represented by expressing a concept such as"to exist," different from"to live, to stay" (*puju-*).

30. This has been clearly shown by the recent archaeological work conducted by Rostain (1997a and b, and 1999) already mentioned, together with other references, in n. 9.

31. As this attitude is well rooted in contemporary bilingual young Shuar, we should ask ourselves whether bilingualism itself triggered, during the last few decades, such a meta-linguistic attitude Jimpikit and Antun' (1991), who do their best to provide a meaning, or at least a motivation, which we would classify as'folk etymology,' for each single name in their list of approximately 300 names recognized as Shuar, even when the names are not obviously transparent. Overall, the little book is an effort to reject opacity. We will try, in any case, to interpret this meta-linguistic attitude below.

32. During the last forty years the Shuar have adopted a person name system which usually includes: (1) one or two Christian-Spanish names, (2) in some cases a Shuar name, (3) the name of the father (or in recent times, that of the grand-father), treated as a surname and (4) in some cases, the name of the mother. In Shuar standard alphabet *e* represents a high back unrounded vowel; *A, E, I, U,* represent soundless vowels, occurring at the end of some word; *j* represents a laringeal fricative; *ts* represents a dental-alveolar affricate; apostroph ' is used to represent a slight palatalization of final consonants; other graphemes and digrams, such as *ch*, must be read according to Spanish convention. Representation of stress is still highly controversial among Shuar teachers and leaders. The text quoted here was translated into Spanish by Jimpikit and Antun (1991).

33. Things are, as obvious, much more complex than this, even for rivers in the Amazon: in many cases it is not clear why and where a river switches from one name to another. The best example of this is provided by the largest river of the region, the Amazon: upriver, before being known under this name, is known as Solimões and further upriver (in Peruvian territory) as Marañón. But, are we sure that it is "the same" river?

34. During the last forty years Shuar people reorganized their territory in a large number of previously not existing villages called *Centros*, which formed regional associations and a more general Shuar Federation.

35. The same statements can be made for most cases of place names in use in our modern societies, but, as should be clear, we also know the ceremonial context in which a place, e.g., a street, a square, or even a newly founded village or town is bestowed.

36. This is the function of nicknames or eponyms, acquired during one's existence or inherited from one's family. Nicknames, in societies where they occur, are more 'legitimized' than 'baptism' names (See, for instance, Zonabend 1977, for the case of a French village). Certainly, this is an aspect of their wide acceptance in many local groups where they are 'preferred' to bestowed names. Shuar do not use nicknames.

37. Michele Allioni wrote his ethnographic account in 1910, in Gualaquiza. Another missionary, Gioacchino Spinelli, wrote his text in 1926. In both ethnographies, a section is devoted to names and name-giving. Although their information is somewhat inaccurate, it is still evident that they were aware of the relevance of the issue. In the first account, a list of male and female names is included. The writer stresses the fact that the person name is always only one. He states also that male names are much more numerous than female names. Two typed copies of Allioni's work are known to exist: one in Spanish, privately-owned in Ecuador, published for the first time in 1978 by Mundo Shuar (Quito—Sucúa); the other one, in Italian, is in the New York Public Library collection. Both monographs (and others) were recently made available by Bottasso (org.) in 1993.

38. I refer only to verbal interactions among men. I have no data on women verbal exchanges in these contexts.

39. This is, obviously, a loan from Spanish *amigo*, as the form *kumpa*, found in a river name, derives from Sp. *compadre*.

40. Magical stones, called *yuka, namur* and *nantar*, enhance their owners' good performances: women in their gardens and men in game hunting.

41. There is only one male name that could be connected to *éntsa*: Entsakua, that Jimpikit and Antún (1991: 44) interpret as "boiling water."

42. An old name to refer to the Christian God was *Kumpanama*; this name could be interpreted as 'among the Christian people,' a reference to some special protecting spirit (*arutam*?) the *compadres* (see N. 39) were believed to have.

43. Berlin's Jivaroan data are from Huambisa and Aguaruna of Northern Peru. Huambisa language and culture is very close to Shuar. Berlin (1992) refers to Huambisa common noun-names of individual species that are the same for Shuar.

References

Allioni, Michele. 1993 [1910]. El pueblo Shuar. In *Los Salesianos y la Amazonia.* Tomo II, *Relaciones Etnográficas y Geográficas,* edited by J. Bottasso. Quito: Abya-Yala.

Barrueco, Domingo. 1959. *História de Macas.* Quito: El Girón. Publicaciones del Centro Misional de Investigaciones Científicas.

Basso, Keith H. 1996 *Wisdom sits in places. Landscape and language among the western Apache.* Albuquerque: University of New Mexico Press.

Beckerman, S. J. no date. Swidden in Amazonia and the Amazon River. In: *Comparative Farming Systems* edited by B. L. Turner and S. B. Brush. New York: The Guilford Press.

Berlin, Brent. 1978. Bases empíricas de la cosmología Aguaruna. In *Etnicitad y Ecología,* edited by A. Chirif. Lima: Centro de Investigación y Promoción Amazónica.

_____ 1992. *Ethnobiological classification. Principles of categorization of plants and animals in traditional societies.* Princeton: Princeton University Press.

_____ and Elois-Ann Berlin. 1983. Adaptation and ethnozoological classification: Theoretical implications of animal resources and diet of the Aguaruna and Huambisa. In *Adaptive responses of native Amazonians,* edited by R.B. Hames and W. T. Vickers. New York: Academic Press.

Bianchi, Cesare. 1981. *El shuar y el ambiente. Conocimiento del medio y caceria no destructiva.* Quito-Sucúa: Mundo Shuar.

Bloom, Paul, et al. eds. 1996. *Language and Space.* Cambridge. Mass: The MIT Press.

Boas, Franz. 1934. Geographical Names of the Kwakiutl Indians. *Columbia University Contributions to Anthropology, No. 20.* New York: Columbia University Press. Partial reprint in: *Language in Culture and Society. A Reader in Linguistics and Anthropology,* edited by D. Hymes. New York, Evanston, and London: Harper & Row, Publishers.

Brown, Michael. F. 1985. *Tsewa's gift. Magic and meaning in an Amazonian society.* Washington and London: Smithsonian Institution Press. Smithsonian Series in Ethnographic Inquiry.

Centro Misional de Investigaciones Científicas. Sección de Antropologia. 1961. *Mapa antropogeografico. Ficha de investigación para la Zona "A" Correspondiente al Vicariato Apostólico de Mendez—Ecuador,* edited by S. Pellizzaro. Quito: Cuadernos de investigaciones científicas, mimeo.

Chicham Nekatai. 1988. *Chicham nekatai. Apach chicham—shuar chicham. Diccionario comprensivo castellano-shuar.* Sucúa: Sistema de educación radiofonica bicultural Shuar.

Descola, Philippe. 1992. Societies of nature and the nature of society. In *Conceptualizing society,* edited by A. Kuper. London and New York: Routledge.

_____ 1994. *In the society of nature: A native ecology in Amazonia.* Translated by Nora Scott. Cambridge: Cambridge University Press [Original title: *La nature domestique: Symbolisme et praxis dans l'écologie des Achuar.* Paris: Fondation Singer-Polignac/Editions de la Maison des sciences de l'homme. 1986].

_____ 1996. Constructing natures: symbolic ecology and social practice. In *Nature and society. anthropological perspectives,* edited by P. Descola and G. Pálsson. London and New York: Routledge.

Fausto, Carlos. 1997. *A dialética da predação e familiarização entre os Parakanã da Amazônia oriental: por uma teoria da guerra ameríndia.* Doctoral thesis. Rio de Janeiro: PPGAS/UFRJ/MN.

Gnerre, Maurizio. 1972. L'utilizzazione delle fonti documentarie dei secoli XVI e XVII per la storia linguistica jívaro. In *Atti del XL Congresso internazionale degli americanisti,* edited by E. Cerulli. Roma-Genova: Tilgher.

_____ 1973. Sources of Spanish Jívaro. *Romance Philology* 27: 203–204.

_____ 1985. Lo spazio del mito. *La Ricerca Folklorica* 11: 29–34.

_____ 1999. *Profilo descrittivo e storico-comparativo di una lingua amazzonica: lo shuar (jívaro).* Napoli: Istituto Universitario Orientale. Quaderni di AION.

Guallart José. M. 1990. *Entre Pongo y cordillera. Historia de la etnia Aguaruna-Huambisa.* Lima: Centro Amazonico de Antropologia y Aplicacion Practica.

Hallpike, Charles. R. 1979. *The foundations of primitive thought.* New York: Oxford University Press.

Harner, Michael. 1962. Jívaro souls. *American Anthropologist* 64: 258–72.

_____ 1972. *The Jívaro. People of the sacred waterfalls.* New York: Natural History Press. [sp. translation: *Shuar: Pueblo de las cascadas sagradas.* Quito-Sucúa: Mundo Shuar.]

Hendricks, Janet. W. 1993. *To drink of death. The narrative of a Shuar warrior.* Tucson: The University of Arizona Press.

Hockett, Charles F. 1958. *A Course in modern linguistics.* Toronto: The Macmillan Company.

Hornborg, A. 1996. Ecology as semiotics: outlines of a contextualist paradigm for human ecology. In: *Nature and society. anthropological perspectives,* edited by P. Descola and G. Pálsson. London and New York: Routledge.

Hunn, Eugene. 1993. *Columbia Plateau Indian place names: What can they teach us.* Seattle: University of Washington, unpublished paper.

Jimpikit, C and Antun,' G. 1991. *Los nombres shuar. Significado y conservación.* Quito-Sucúa: Abya-yala-INBISH.

Karsten, Rafael. 1935. *The head-hunters of western Amazonas. The life and culture of the Jibaro indians of eastern Ecuador and Peru.* Helsinki: Societas Scientiarum Fennica, Commentationes Humanarum Litterarum, Vol. 19, No. 5.

Kripke, Saul. A. 1972. Naming and Necessity. In *The semantics of natural Language,* edited by D. Davidson and G. Harman G. Dordrecht: Reidel.

Lévi-Strauss, Claude. 1962. *Le Totémisme Aujourd'hui.* Paris: Presses Universitaires de France.

_____ 1964. *La pensée sauvage.* Paris: Libraire Plon.

Levinson, Stephen. C. 1996. Language and Space. *Annual Review of Anthropology* 25: 353–82.

Pellizzaro, Siro. 1973. *Técnicas y estructuras familiares de los Shuar.* Sucúa: Federación de Centros Shuar.

_____ 1976. *Arutam. Mitos de los espiritus y ritos para propiciarlos.* Quito-Sucúa: Mundo Shuar.

_____ 1982. *Etsa, el modelo del hombre shuar.* Quito-Sucúa: Mundo Shuar.

Rappaport, Roy A. 1993. Humanity's evolution and anthropology's future. In *Assessing cultural anthropology,* edited by R. Borofsky. New York: McGraw-Hill.

Rostain, Serge. 1997a. Arqueologia del Rio Upano. Amazonia Ecuatoriana. Quito: Instituto Francés de Estudios Andinos.

_____ 1997b. Nuevas perspectivas sobre la cultura Upano del Amazonas. Paper presented at the Symposium Arqueologia 13, 49th International Congress of Americanists. Quito, July 7–11, 1997.

_____ 1999. Occupations humaines et fonction domestique de monticules préhistoriques de l'Amazonie équatorienne, *Bulletin de la Société suisse des Américanistes* 63: 71–95.

Seeger, Antony. 1981. *Nature and society in central Brazil. The Suya indians of Mato Grosso.* Cambridge: Harvard University Press.

Spinelli, Joaquin. 1993 [1926]. Etnografía de los Jívaros. In *Los Salesianos y la Amazonia.* Tomo II, *Relaciones Etnográficas y Geográficas,* edited by J. Bottasso. Quito: Abya-Yala.

Talmy, Leonard. 1983. How language structures space. In *Spatial orientation: Theory, research and application,* edited by H. Pick and L. Acredolo. New York: Plenum.

Taylor, Anne.-Christine. 1993. Des Fantômes stupéfiants. Langage et croyance dans la pensée achuar. *L'Homme,* XXXIII (Avril-Décembre): 429–48.

Viveiros de Castro, Eduardo. 1986. *Arawa, Os Deuses Canibais.* Rio de Janeiro: Jorge Zahar/ANPOCS.

Zonabend, Françoise. 1977. Pourquoi nommer? Les noms de personnes dans un village français: Minot-en-Châtillonais). In *L'identité,* [Séminaire interdisciplinaire dirigé par Claude Lévi-Strauss], edited by J. M. Benoist and others. Paris: Bernard Grasset.

CHAPTER 8

What is Lost When Names are Forgotten?

Jane H. Hill

Introduction to the Problem of Lexical Loss

In indigenous communities in the Americas, where the replacement of Native American languages by world languages like Spanish and English is a matter of acute concern, community members often seem to think of their language exclusively as an inventory of words—especially, the"old words"that are seen to capture in a special way the privileged essence of local tradition. A good speaker is"a person who knows a lot of old words."Language loss is characterized as"nobody knows those old words any more." Code switching by bilinguals is explained as "something people do when they can't remember the words in our language." Attempts at language revival often focus on collecting and memorizing individual words, and the walls of schoolrooms in communities where language revival movements are under way are papered with number words, color terms, animal names, and names of artifacts associated with traditional culture.

Such an understanding of language and its obsolescence reflects a vernacular "metapragmatic awareness"(Silverstein 1984) rather than a scholarly comprehension of the complexities of language structure, and linguists correctly try to guide language planners in such sites to attend to morphology and phonology, syntax and discourse, pragmatics and poetics, as well as to lexicography.[1] Yet it may be that linguists have been guilty of the opposite ideological bias. Lexical semantics has not been very fashionable in recent years, even in language and culture studies, and students of language obsolescence have paid little systematic attention to the loss of lexicon. Linguists have often assumed that the loss of lexicon is an automatic result of culture change, comparable, perhaps, to the loss of words for the horse complex among English speakers[2]—a loss that I, at least, experience as trivial. However, the loss of lexicon should receive more attention. Since word—centered linguistic ideologies make lexical loss a focus of speakers' insecurity (with consequent reluctance to speak the local language), vocabulary enhancement programs can have an important function in boosting their confidence. Furthermore, the idea that certain"old words"are important to local heritage deserves respect; such words can have the same function in the humanistic life of a local language community as the knowledge of archaic and rare words has for educated speakers of regional and world languages.[3] Apart from the impact on communities of speakers, lexical attrition is very problematic for historical and comparative linguistics. For instance, it may be impossible to find a speaker who can confirm the identification of a plant or animal mentioned in an old text, so that even if the item was recorded at some point, it must be used in comparative work with great caution. Attention to the systematic dimensions of lexical loss in the study of language obsolescence can contribute to our understanding of the structure of lexicon. But my focus here is less on these problems than on the way that words seem to be inextricably tied to local knowledge: both discursive knowledge that exhibits an almost "lexicographic" quality, and the complexly embodied and affect-laden domains of "practical" knowledge that are foundational for performance as a skilled bearer of culture. This is the case for the lexicon of biosystematics (the lexicon for indigenous flora and fauna) among the Tohono O'odham.[4] For Tohono O'odham speakers, the phonological forms of names seem to be very important mediators for the recall and use of forms of local knowledge of the natural environment. Loss of words may not be epiphenomenal to the loss of this knowledge. Instead, as words are lost, knowledge fades as well, even when there is no concomitant cultural or environmental change.

Such an association between words and knowledge was proposed by Benveniste (1971) in his famous challenge to Saussure's concept of the arbitrary relationship of word form to word meaning. Benveniste argued that Saussure's account of arbitrariness in semiosis depended on an ontological error, the implicit assumption of a third term—qualities that resided in the world—in addition to the signifier and the signified. Benveniste himself took the phenomenological position (and a position that is, of course, more profoundly structuralist) that the relationship between signifier and signified was not arbitrary, but necessary: meanings reside only in the language and signifiers project their signifieds rather than somehow matching up to categories in the world.

Many students of language and cognition, stimulated by findings such as that by Berlin and Kay (1969) on color terminologies, have lately returned precisely to Saussure's implied position: that meaningful categories—at least for natural kinds of the type that concern us here—do indeed reside in the world, and that possible human languages will include only lexical items whose meanings match those categories. That is, we now believe that human languages will have words for tigers, and words for grasses, but that no human language will have a lexical item that has as its referent something like "the back left leg of a tiger and those grasses against which the leg brushes as the tiger walks." The task for the child is then to overcome arbitrariness: to learn the conventions that match particular meaningful forms to particular learnable categories. Our understanding today of the child's task is not, however, precisely Saussurean. Berlin argues (in this volume and elsewhere, e.g., 1992) that the scope of arbitrariness is limited, since for many natural kind terms the child gets a good deal of help because the convention that he or she must acquire is considerably constrained by iconic principles of sound symbolism that reflect properties of the referent such as size and quality of movement.

I wish to suggest today a different sort of move, a return to something more like Benveniste's position. I want to argue that Benveniste was correct, at least at the level of the "practical," with respect to the sort of human occupation of a language, and through it a world, that Gnerre (in this volume) refers to as an "interactive relation," or what Slobin (1996) has called "thinking for speaking." I present evidence from the Tohono O'odham language, that access to names mediates access to practical understandings of three types. The first two types are discursive, including abstract, encyclopedic knowledge (shared broadly among the O'odham and constituting part of D'Andrade's (1995) "cultural knowledge"), and personal knowledge, including details of life history that may figure importantly in self-construction. The last type is not strictly discursive, but involves affective relationships to nature that may be expressed in postures and spontaneous gestures and "cries" (Goffman 1978), and in recollections of tastes and smells and sounds that can be only vaguely articulated. The types of sound symbolism that Berlin has identified may be involved at this level of understanding.

These systems of practical knowledge, while they must include universal components, are in their totality profoundly local and specific. They must constitute an important component of what it means to "know nature" as a cultural being. The data that I present here suggest that names are particularly efficient routes of access to these systems, such that when names are lost, practical knowledge begins to erode. A strong sense of this power of names to mediate a way of living in the world, and not linguistic naiveté, may lie behind the intense concern for "words" among indigenous peoples that I have noted above.

Elicitation of Names in the Dialect Survey of Tohono O'odham (Papago)

The Tohono O'odham (formerly "Papago")[5] live in towns, villages, and isolated ranches spread across a unique environment, the Sonoran Desert of south-central Arizona and the northern border of Sonora, Mexico. Approximately 50 percent of the population on the main reservation in Arizona are bilingual in Tohono O'odham, a language of the Uto-Aztecan family, and English.[6] I discuss materials collected by myself and Ofelia Zepeda[7] during a study of dialect variation in Tohono O'odham conducted from 1986 to 1990. The study, which included ninety one speakers (all but one over fifty years of age) focused neither on the lexicon nor on people's knowledge of nature. Our goal was to identify the linguistic variables— phonological, lexical, morphological, and syntactic—that would permit an analysis of the system of regional dialects in the language. As a part of the study we elicited from speakers a large number of names of plants, animals, and cultural items. These had been identified as possible points for phonological and/or lexical variability. For instance, the word *jiawul*[8] "barrel cactus" (*Ferocactus* spp, *Echinocactus* spp.), a loanword from Spanish *diablo* "devil," has several different pronunciations, including ['dʒiawur] and ['dʒiawur]. The difference between the presence of [i] or [ˀ] in this and many other words with stressed vowel clusters signals a major dialect boundary that separates the southern and western parts of the Tohono O'odham Reservation from the eastern and northern regions.

Words were elicited by showing speakers pictures of the named item. These were clipped from magazines and natural history field guides and mounted in clear plastic holders in a three-ring notebook. Elicitation took place in Tohono O'odham. The interviewer, a native speaker, pointed to the picture and said, *"Ṣa: 'o 'e-'a'aga 'i:da?"* (What is this called?). After giving the name, the speaker was asked, *"Mat o ṣa mu'ijk?"* (If there are many?), in order to elicit a plural form (these are also variable in interesting ways). If necessary, speakers were prompted in Tohono O'odham with folk definitions constructed on the spot by the interviewers (I discuss these in more detail below), by English words, and, as a last resort, by the Tohono O'odham word itself as pronounced in the interviewer's dialect. The pictures included twenty different wild plants and twenty seven different wild animals (this was only a small part of an interview instrument that included terminology in many other domains, as well as elicitation of language attitudes and life-history materials). Because some of the pictures included items other than the main stimulus element, and because some speakers proposed the "wrong" name, over seventy names for wild animals and plants appear in our data. All elicitation sessions were tape recorded, so we have a complete record of interviewer prompts, speakers experiencing a "tip of the tongue" phenomenon, discussions between interviewers and speakers about the item, emotional reactions by speakers to the items (such as a sound of disgust when giving the name for "rattlesnake"), and remarks about the items ranging from very brief modification of the main word— for instance, *s-cuk judum* 'black bear' uttered immediately after *judum* 'bear, *Ursus americanus*' or *si s-ke:g* 'that's real pretty' after *taṣ mahag* (*Lupinus* sp.)—to quite extended narratives about uses of or experiences with the item in question.

These materials suggest that "names" in this domain are an inextricable part of encyclopedic "nature knowledge" among the Tohono O'odham. The more names a speaker knew, the more likely he or she was to elaborate and amplify beyond merely saying the name. Even the spontaneous emotional reactions seemed to be tied to being able to say the name, and affective "cries" often began immediately preceding the utterance, with the special affective vocal quality spreading into the name itself. Brief remarks and stories also seemed closely linked to the name, following immediately upon its utterance. Speakers rarely offered such elaborations unless they knew and could say the name, even in cases where the interviewer had given fairly elaborate information about it through prompting, or where the speaker knew the English name. Their behavior in these interviews suggested that their "nature knowledge" was closely tied to, and perhaps most easily accessed through, mention or use of names.

Loss of Biosystematic Lexicon in Tohono O'odham

First I review briefly the data on loss of this lexicon for the non-domesticated (or semi-domesticated, as in the case of some basketry plants, e.g., Nabhan and Rea 1987) flora and fauna of the Sonoran Desert. The first published work on this question, by Nabhan and St. Antoine (1993), determined to no one's surprise that children hardly know this lexicon at all (very few children are learning Tohono O'odham at home or in school). Nabhan and St. Antoine (1993: 244) report that in a study of twelve O'odham and Yaqui children living in southern Arizona, they found that the children recognized on the average only 4.6 names out of seventeen native species shown to them in pictures. In contrast, their grandparents averaged 15.1 names for the seventeen species.

Zepeda and I found that lexical loss was even more advanced than the Nabhan and St. Antoine (1993) study had revealed. Substantial attrition in the biosystematic lexicon was obvious even among people in their sixties. In order to design a maximally efficient instrument for our dialect survey, we presented about 100 pictures of local flora and fauna to five elderly speakers in a pilot study. Based on the results of this study, we selected only the forty seven pictures that were very easy for all pilot subjects to identify and name. This was important, because we did not want subjects to have to pause and think about the names; instead, we wanted to collect as many lexical items as quickly as we could in order to develop a statistical base for a study of language variation.

We recruited speakers through personal networks and through the officers of the Gerontology Division of the Tohono O'odham Health Department. We always asked before contacting a speaker if that person could see and hear well, or if he or she had any mental problems that might make participation in the study difficult, and excluded from participation speakers with known impairments. When we invited speakers to participate, we asked them again whether they felt they would be able to see pictures and hear our questions. Where there was any doubt, we asked them again after the free conversation part of the interview and before starting the picture section. Thus all speakers were screened for this kind of problem.

In spite of this screening of both the elicitation instrument, for ease of recognition and high saliency of vocabulary, and the speaker sample, for adequate vision, hearing, and mental capacity, in the main sample of eighty six speakers we found that a substantial minority had trouble with this biosystematic lexicon.

In order to explore the impact of lexical loss, I selected a subsample of thirty three speakers for whom I had no reason to believe, based on having been present myself at the interview or having heard the tape, that the speaker had trouble seeing or hearing. These elders split into three large groups, as shown in table 8.1. One group not only easily named all of the items, but were reminded of other related plants and animals and gave their names too, and named small plants or animals that were in the background of pictures intended to elicit the names of more prominently focused species. A second group (the largest) knew nearly all the names, requiring extensive prompting on one or two items and confessing not to know perhaps one other. These speakers were probably like the "grandparents" in the Nabhan and St. Antoine (1993) sample. A third group, however, had great difficulty with the list, being unable to provide Tohono O'odham names for up to almost half of the pictures. This group included some of the best educated people in the sample (many elderly Tohono O'odham have had only a few years of primary school), and two speakers who had been high tribal officers. Four of them are quite fluent and use O'odham daily in their communities; all of them speak the language at least some of the time. Three had lived off-reservation for long periods, although they were living on the Reservation at the time of the study. The other three lived at San Xavier, technically a district of the Reservation but for practical purposes a suburb of Tucson. Interestingly, none of these speakers tried to satisfy the interviewer with hypernyms, even when these are available (biosystematic hypernyms in Tohono O'odham include *ṣa'i* "weed, brush," *hoi* "cactus" [literally "thorn"], *kui* "tree," *'u'uwhig* "bird," and *ha'icu dodakam* "animal"), although they might say the hypernym in trying to remember the word. Instead, they observed that they could not remember the correct names for the species. The unknown items included even "charismatic megafauna" like bighorn sheep and bobcats, saliently dangerous small animals like scorpions and gila monsters (*Heladerma suspectrum*) that are common in Tohono O'odham communities, and plants used in basketry, the main craft form for Tohono O'odham women. Probably 25 percent of O'odham women make baskets for sale, and no one would not have at least one close friend or relative involved in this craft. Thus the loss of this lexicon cannot be attributed entirely to culture change, but seems instead to represent the advance of a language shift.

Table 8.1 Attrition in the Biosystematic Lexicon in Tohono O'odham

8 speakers in Group I, with no items missing and many additional terms volunteered. 19 speakers in Group II, averaging 1.89 missing (range 1–3) 6 speakers in Group III, averaging 13.5 missing (range 4–21):			
Speaker	**Age**	**Words Missing**	**Sex/Biography**
LF	51	21	(female, schoolteacher, college degree)
EF	52	20	(male, high tribal officer with some college)
MP	67	15	(female, high school degree, San Xavier)
EF	61	13	(female, high school degree, San Xavier)
CL	73	8	(female, high school degree)
PF	65	4	(male, tribal officer, San Xavier)[9]

This sort of lexical attrition has profound consequences. Nabhan and St. Antoine (1993) note that it cuts speakers off from their environment; they may be better able to name African megafauna on a television nature program than animals that are common in their own backyard. Furthermore, it makes speakers globally more ignorant. As I will show below, the expertise about the flora and fauna of the Sonoran desert that is encoded in these names does not transfer to English, in two ways. First, with the exception of a few desert buffs and biologists, English speakers in southern Arizona make far fewer distinctions among the various plants and animals of the desert than do traditional Tohono O'odham, even those who are not considered particularly expert.[10] Second, while certain kinds of discursive knowledge about plants and animals seems to transfer, many elements of practical knowledge, especially including the very interesting spontaneous emotional reactions, do not.

How People Behaved in the Interviews

The Interviewers: Definitional Prompts

Two interviewers participated in the study, both fluent and literate native speakers of Tohono O'odham. Zepeda was at the time Assistant Professor of Linguistics at the University of Arizona and held a doctoral degree. She had published poetry and linguistic studies in English and O'odham. Belin was a teacher's aide, a high-school graduate with some college credits, who was well known as one of the best readers and writers of the first generation of people to be literate in O'odham.[11] The two interviewers were thoroughly familiar with the elicitation instrument. Zepeda had designed the interview and done the pilot study; Belin accompanied her on the first four interviews in the main study to learn how to administer it. The interviewers were free to design their questions and to prompt speakers in any way they saw fit. We had agreed that it was better not to prompt with the Tohono O'odham word for fear of influencing the subject's pronunciation. However, when they were fairly sure that the speaker's problem was not ignorance but simply confusion about the picture, both interviewers did occasionally prompt with some version of the desired word. This turned out not to be a problem. Subjects often rejected the word or pronunciation given by Zepeda or Belin (who speak two different dialects—Zepeda often converged strongly on the speech of the subject, but Belin was much less likely to modify her usual pronunciation); they would repeat the word in their own form, saying "I/we say it this way." If a speaker did not recognize the word, they would often say, "No, I never heard that." Where a "tip of the tongue" problem was resolved by the Tohono O'odham prompt, this was usually very obvious: subjects would immediately say something like "'*Ah! He'u! Taiwig!*" "Oh! Yes! Firefly!" and often continue with an elaborating remark.

Zepeda and Belin started by gesturing to the picture (the notebook with pictures was on the subject's lap, or in front of them on a table) and uttering the stock question, "*Ṣa: 'o 'e-'a'aga 'i:da?*" (What is this called?). After a number of items, when the task was clear, they would sometimes just gesture at the picture and say "'*I:da*" (This one). Sometimes, if there were several elements in a picture, they would put their finger on the one to be named and say something like "These things sitting/standing/hanging/going around here?" If the subject hesitated, they might say, a little louder, "*Has ce:gig? Has 'ap ha-ce:ceg 'a:pi?*" (What is its name? What do you name them?).

When a speaker could not immediately produce a name, both interviewers would often prompt with a brief description of the item in Tohono O'odham. Casagrande and Hale (1967) enumerate thirteen types of folk definitions produced by a speaker of Tohono O'odham, and Zepeda and Belin used all the types that were appropriate for prompting plant and animal names. These include the following:

1. "Attributive" definitions of several types, including attribution of "intrinsic stimulus properties," e.g., for blue palo verde, *Circidium floridum,* "This is big, with yellow flowers.

It's green." Or, for cottonwood, *Populus deltoides*, "This one has leaves that are real big. They are soft." An example of the attributive type "distinctive marker," for skunk, *Mephitis* sp., is "Its tail is black and white." Exemplifying "behavior or action" attributives, Pinacate beetle, stinkbug was often defined as "When you step near it, it makes a smell." Vultures, *Cathartes aura*, were defined as "the ones you see on the road, eating dead things." Exemplifying "habitat," for cattails, *Typha* sp., they would say "This one likes water", or, for foothill palo verde, *Circidium microphyllum*, "There are a lot of those on the mountains."

2. "Functional" definitions were preferred in several prompts. For creosote bush, *Larrea divaricata*, Zepeda and Belin would usually mention "that which is used for curing." Basketry plants were defined as "This one is for making baskets." To elicit the term for *Agave deserti*, Zepeda often said, "They pound this and use it to make rope." Mesquite beans (the fruit of *Prosopis juliflora*) were defined as "O'odham people eat them."

3. "Comparative" definitions were used for some items. Thus, organ pipe cactus, *Cereus thurberi*, was often defined as "not saguaro (*Cereus giganteus*), but different." Cottontail (*Silvilagus audoboni*), was often defined as "smaller than jackrabbit (*Lepus americanus*)". Bobcat (*Lynx rufus*) was said to be "like a mountain lion, but short" (*ṣopolk*).

4. "Class inclusion" prompts included "It's a kind of bird (*'u'uwhig*)" or "a kind of animal (*ha'icu dodakam*)" or "It's an animal that walks around on the ground (*'am jewed da:m 'oimmed*)", a subclass of animals including insects, turtles, snakes, and small rodents noted by Mathiot (1962). Occasionally "Pinacate beetle" was prompted by an English class name: "It's a *bug*."

5. "Synonymy" for Casagrande and Hale (1967) included only the proposal of an "exact equivalent" in a different O'odham dialect. However, Zepeda and Belin often gave English words, if they thought the subject would know them, e.g., "The White people call this 'greasewood,'" or "The White people call this 'stinkbug.'" As far as I can tell, they never intentionally gave a different dialect form although sometimes their own O'odham prompts turned out to be different from the forms the speaker preferred. While the pilot-study subjects all easily recognized our picture of cottonwood, it turned out that many of the subjects had never seen these trees.[12] In order to elicit the word, the interviewers would say "Do you know the name for Florence Village?" This place is called *S-'Auppag* "Many Cottonwoods" in Tohono O'odham. This may be a case of "synonomy."

Once or twice, Zepeda and Belin were able to use an "ostensive" strategy, not noted by Casagrande and Hale (1967). For instance, in trying to elicit "blue palo verde," Zepeda once said "Like that one, over there by the car."

Many of their most frequently used prompts were quite elaborate, including several of Casagrande and Hale's (1967) definitional types in a single utterance. For instance, for firefly, one prompt offered was "You see those flying at night, especially around water. On their tail it looks like a little fire is burning." An elaborate prompt for creosote bush (*Larrea divaricata*) was "These have yellow flowers, and they're kind of greasy. They smell good when it rains. They're used for curing." Or, for Pinacate beetle: "This is a *bug* [English]. It's real little and it's black. It walks around on the ground and when you step near it it makes a smell."

These prompts closely resemble some of the kinds of elaborating remarks on names offered by interview subjects. However, unlike interview subjects, Zepeda and Belin did not draw on stories of personal experience or on their own affective reactions in developing their prompts (with the possible exception of Zepeda's occasional prompt, for *Larrea*, that"it smells good when it rains"—but this is a universally appreciated property of this plant). Instead, these definitional prompts drew exclusively on a local system of discursive encyclopedic knowledge, a "vernacular lexicography" (to use the term proposed by Casagrande and Hale 1967) that apparently is widely understood. While subjects were often confused by the pictures, they never seemed to be confused by these prompts (although the prompts did not always elicit the desired term, of course!). These prompts are part of a fairly old and widespread cultural system. They were often nearly identical not only in function but in form to the"folk definitions"collected in 1961 from an elderly Tohono O'odham man, Luke Preston, by Kenneth Hale (Casagrande and Hale 1967). Preston came from the village of Sikol Himadk, which is about thirty km from Belin's home village of Jewak (Lower Covered Wells), and in the same dialect area. However, it is about eighty km, and in a different dialect area, from Zepeda's home town of Stanfield, Arizona or from Zepeda's grandmother's home in Quitovac, Sonora, where Zepeda spent time as a child. For instance, Preston said of the Pinacate beetle or stinkbug:"[that] which is black; and it goes around; and it sticks its rear end up; and it releases a scent."Zepeda and Belin both consistently mentioned"when you step near it"as a condition under which"it releases a scent,"but otherwise their prompts were identical. Preston said of vulture,"and that also goes around which eats dead things."Zepeda and Belin often added that"you see them on the road,"but otherwise almost always said that"these eat dead things." For hummingbird (various genera of the *Trochilidae*), Preston said,"... and those birds which are small." Zepeda and Belin often used the prompt, "It's a bird, but it's small."Preston used exactly the same"comparative"definition for jackrabbits and cottontails that was usually provided by Zepeda and Belin:"This one is big, and this one is small." In addition, very similar definitions can be found in example sentences in Mathiot's (1973) *Dictionary of Papago Usage*.

In summary, the interviewer prompts reveal a quite solid and consistent local "natural history"expressed through definitions and descriptions of plants and animals.

The Behavior of Interview Subjects

In preparing the present study, I listened to the tape-recorded naming sessions of twenty eight speakers, in alphabetical order by last name starting with the letters A through J (except for CL, a speaker with many words missing, included to further explore this effect).[13] I did not include any of the five pilot study speakers (although we have used their responses in the dialect survey maps), because these speakers were asked to name about twice as many items as the speakers in the main study. Furthermore, I did not include any speakers who were interviewed together with others—for instance, we interviewed several husband-and-wife couples, a mother and a daughter, and a pair of sisters. These materials were useful for the dialect survey, but not for the present study, since people often prompt one another, or do not always both say the word even when they both know it, so I cannot evaluate the number of words that they know. Furthermore, the kind of conversation about items that ensued when two or more subjects were participating in the interview may be somewhat different from the kind of narratives and remarks that ensue when only the interviewer and the subject (and the nearly monolingual English-speaking linguist, silently tending to the tape recorder or keeping intrusive grandchildren entertained) were present.

In this analysis of the taped interviews, I attended to the following questions:

1. Whether the subject gave the correct name. By "correct," I mean the name of the indicated species, regardless of lexical form (most items in the study have several variants). For instance, an example of an "incorrect" form was where speakers gave *'umug* "sotol" (*Dasylirion wheeleri*) for the picture of *'a'oḍ* (*Agave deserti*), and stated that they did not know the word *'a'oḍ* when this was given as a prompt. Sometimes speakers gave no name at all, usually remarking that they had never seen the plant or animal in question, or that they did not remember the name.

2. Whether the subject gave a paralinguistic affective response. Two of these were quite stereotyped: One, given only by women, was a sound of disgust for "rattlesnake": "[ʔɨ::] *Ko'owĭ!*" Yuccch! Rattlesnake!" (Our stimulus picture was of western diamondback (*Crotalus atrox*). The second was a chuckle at a picture of a white-winged dove, *'okokoi* (*Zenaida asiatica*) sitting atop a saguaro cactus and eating the ripe red fruit. This picture was actually intended to elicit the word for the fruit, *bahidaj*, since *'okokoi* is a word with no variant pronunciations nor does the bird have more than one name. This would often be pronounced something like "@@@@! *'O@'@ok@oi!*"[14] (Heh heh heh heh! whitewinged dove!), with the chuckling continuing into the pronunciation of the word. The white-winged dove eating *bahidaj* is a very meaningful image. The O'odham greatly value the saguaro fruits, which are collected and boiled down to make a thick, very sweet syrup used to make candy and jams, but especially saved and fermented into *nawait* "saguaro-fruit wine," for the Wine Festival. The Wine Festival, today celebrated in only one or two villages and not, sadly, accomplished in every year even there, was traditionally celebrated in July to "call down the rain" of the summer thunderstorm season on which O'odham maize and tepary bean agriculture depended. People collected around the great pots of *nawait* and drank and sang and danced and prayed all night, drinking until they vomited. The color of the vomit was said to resemble the color of the sunset clouds (the favorite color of the O'odham, translated as "crimson" by Ruth Underhill [1951]),[15] and people were said to be "vomiting clouds." The white-winged doves are thought of as a sort of charmingly greedy competition for the ripe saguaro fruit; several subjects said something like "The O'odham had better hurry or the *bahidaj* will all be gone." Of course the appearance of the white-winged doves also serves as an important signal that the fruit, which is often six to eight meters above the ground, is ripe and ready to be pulled down with long poles. Zepeda sometimes prompted this name with "The one that eats the *bahidaj* first."

3. Whether the subject amplified the name with any remarks about the item, and if so, whether the remarks were very brief or more extended.

4. When the subject did not immediately come up with the correct name, but instead volunteered others, how these were related to the "correct" name.

5. Finally, I attended to the content of remarks, since these are an important indication of the range of encyclopedic knowledge that seems to be accessed by the name.

Briefly, the Findings are These:

1. Most people, for most pictures, simply give the name and the plural or multiple, and move immediately on to the next item. If they make any remarks, these are metacomments on the task, like "What is its name? I don't remember." Or "That's an X, or something like that."

2. The more Tohono O'odham names people provide, the more likely they are to make amplifying remarks, and the more likely they are to give paralinguistic affective responses.

3. People do not make amplifying remarks or give paralinguistic affective responses unless they also say the Tohono O'odham name. That is, women do not see the picture of the rattlesnake and say "Yucch!" and then give the name after hesitation or intervening remarks. The paralinguistic element virtually coincides with the name and the voice quality often continues into the name. Similarly, people do not make amplifying remarks just based on the picture when they do not know the name. If any amplifying remarks precede the name, they are extremely brief, like "Oh, you mean the little ones! Those are called *ho:hi* "mourning doves" (*Zenaidura macroura*). In almost every case, the amplifying remarks immediately follow the name.

4. In contrast with the "prompts" given by the interviewers, which apparently realize a widely known system of discursive encyclopedic knowledge that we can think of as a vernacular lexicography, the amplifying remarks frequently deal with some personal experience or connection that the subject associates with the item at hand. Occasionally people will do some "lexicography" (for instance, one subject gave the word *'o:bgam*, a local word for Sonoran palo verde (*Parkinsonia aculeata*), and continued *"mo ce'ecem g ha:hag"* (the one with little leaves). However, the "personal associations" are very common. Sometimes these are very simple. For instance, in giving the expression for "many organ pipe cacti," one woman said, *"B 'o ce:ceg g ñ-'o:gbad `Mu'ij g cucuis'"* (My late father used to say, "Many organ pipe cacti."). Or someone might say of cholla buds, "Those are very good to eat" (in contrast to "People eat them" in the prompts given by the interviewer).

5. Where people give a variety of alternative proposals before coming up with the right name, these are always names in their own right, and are usually somehow related to the item at hand even if this relationship is mistaken. For instance, many people found it difficult to figure out the picture of the Pinacate beetle. This is a fairly large black beetle, but the picture, about 8×13cm square with the beetle on the ground in the middle with its abdomen raised in a characteristic posture that means the beetle is about to spray an odorous liquid to deter a predator, was difficult for many people to interpret. However, it was very clear that they classed it correctly with "animals that go around on the ground." For instance, one man suggested first two other insect names, *makkumĭ* "tomato horn-worm" and *ṣo:'o* "grasshopper." In a particularly remarkable case, a woman rapidly offered for the Pinacate beetle picture six terms for kinds of mammals, using a lexicographical discourse following several of the terms, in the following order: *jewho* "gopher" (*Thomomys* sp.), *ka:w* "badger" (*Taxidea taxus*), *koson* "packrat" (*Neotoma lepida*), *ṣelikĭ* "ground squirrel" (*Citellus* sp.) or "prairie dog" (*Cynomys* sp.), *dahiwua* "kangaroo rat" (*Dipodomys* sp.), and *nahagio* "mouse." For *koson*, she said "the ones that sleep under cactus thorns," virtually identical to Luke Preston's definition from 1961, "... they gather them [the "balls" of *Opuntia bigelovii*] and live underneath them" (Casagrande and Hale 1967: 171).

I turn now to a brief analysis of the behavior of the twenty eight respondents whose tapes I have reviewed in detail. Their behavior is summarized in table 8.2, where they are ranked from most to least comments offered on names during the elicitation of the forty seven wild plant/animal words.

Table 8.2 Interview Behavior for 28 Dialect Survey Respondents, Ranked by Number of Comments on Names.

Speaker	Sex	Number of Comments	Words Missing
1 FJ	F	34	1
2 AC	F	26 *1	0
3 SJ	M	26	1
4 MG	M	26	1
5 SA	F	23	0
6 SF	F	21 *2	0
7 MD	F	18 *1	2
8 MC	F	17	2
9 TA	F	15	2
10 CHM	F	13	0
11 DF	F	13	0
12 EH	F	13	0
13 JAh	F	11	1
14 CAn	M	11 *1	1
15 EF	F	11 *2	13
16 JA	M	10	5
17 PF	M	10 *1	4
18 BA	M	9	2
19 TG	M	8	1
20 VA	M	7	1
21 AJ	F	5	2
22 VJ	F	5	1
23 MA	F	4	1
24 EA	F	3	0
25 LA	F	3	5
26 EnF	M	1	20
27 CL	F	1	11
28 LF	F	0	21

*Number of comments in total made about item although name not known

The table makes clear the general trend, that people who know more words make more comments, and that people who do not know many words make very few. Second, these results show that people make elaborations only when they know the names, with very few exceptions. The major exception to the first point is EF, who ranks in the middle of the list (15) on number of comments, but, with EnF and LF, was one of three speakers who were unable to name a very large number of stimulus items. EF, a woman from San Xavier who was fifty two years old at the time of her interview, code-switched into English at a very high frequency during her interview and had rather long sentences in English. Some of her comments (one is given below) were in English. She is the only exception to the evidence from these data that the characteristic discourses of commentary do not seem to transfer into English, at least in the interview context. EF is very interested in improving her O'odham competence and several times mentioned that she consulted with older speakers about the language. Once she interrupted the interview so that she could write down the Tohono O'odham name for an item that she did not know.

In only eight cases (the starred examples in table 2) did speakers offer amplifying remarks without saying the name. Seven of these eight cases involved two words: *gewiol* "Thistle" (*Cirsium* sp.), and *taiwig* "firefly." These were the two most frequently "missed" names in the survey; they are conspicuous but relatively rare, and while the pilot study consultants all recalled the names without difficulty, only about half of the participants in the main study remembered them. In the main study, speakers remembered having seen the item and had rather elaborate discourses about how pretty the *Cirsium* are when many are in bloom, or about how they had seen fireflies at night near the water when visiting the Mexican river town of Magdalena. However, when prompted with the Tohono O'odham word, they said they did not remember it or had never heard it. PF said, "I saw those with my parents but I don't think they had a word."

Some of these comments included paralinguistic affective responses occurring with the names for rattlesnake (a sound of disgust), scorpion (only one case of a disgust sound). White-winged dove, Pinacate beetle, and desert tortoise (*Gopherus agassizi*) all elicited chuckling, as "funny".

About half the remarks are folk definitions drawn from the repertoire of local natural history also seen in the interviewer prompts. Examples include the following, summarized from the O'odham remarks:

"Those are real thorny" (of *ceolim* [*Opuntia acanthocarpa*], a cactus with edible fruit).
"The ones with little leaves" (of foothill palo verde [*Cercidium micophyllum*]).
"Like in Florence" (giving the word for "cottonwood;" the name of Florence Village means "Many Cottonwoods")
"That's lupine" (giving English name in addition to *ta:ṣ ma:hag* [*Lupinus* sp.])
"Long ago, they used to eat a lot of that" (of *'a'od*, [*Agave deserti*])
"It's like a cat, but it's wild, so they call it *miscin mistol* ("wild cat," instead of *gewho* "bobcat").
"It's a kind of cactus" (of *ceolim* [*Opuntia acanthocarpa*], a cactus with edible fruit)
"You usually see them around water" (of *taiwig* "fireflies").
"A *ho:hoi* ("mourning dove," (*Zenaida macroura*)) is not an *'okokoi* ("white-winged dove" [*Z. asiatica*])
"They get yellow like that when they're ripe" (of *wihog*, the fruit of mesquite [*Prosopis juliflora*])

Examples of commentary involving personal experiences include the following:

"I like this kind, it has a lot of different kinds of flowers" (of *kokaw* [*Opuntia versicolor*]).

"Yeah, it really stinks" (of Pinacate beetle).

"You don't see them around here" (of *ku:sjim* "caracara" [*Caracara plancus*])

"That one really likes the ripe saguaro fruit!" (of the white-winged dove, *'okokoi* (*Zenaida asiatica*) in the picture).

"That one really hurts!" (of *to:ta hanam* "teddy-bear cholla" [*Opuntia bigelovii*].

"One of those ate a colt here not too long ago" (of *mawid* "mountain lion" [*Felis concolor*]).

"My mother used to call that kind the *je'ekam*" (saguaro with no arms) (elaboration on *ha:san* "saguaro cactus" [*Cereus giganteus*])

Regarding *takwui* [*Yucca elata*] an important basketry plant used for a greenish color, her grandmother could make it come out a really beautiful light color by soaking it a long time.

The name *jiawul* "barrel cactus" [*Ferocactus* sp.], reminds her of the name of her town, *Jiawul Da:k* 'Devil Sitting,' and she tells the story of the sighting of the devil that gave the place its name.

In English, speaking of *'ihug* "devil's claw" [*Martynia parviflora*], a basketry plant often cultivated in home gardens: "One [presumably White—JHH] lady came over and said, "Oh, I didn't know you could plant okra!" I said, "Those aren't okra! I wouldn't plant okra!"

Note that some of these examples are, in a sense, "mixed," including widely shared folk-definitional material as well as reference to personal experience or affective posture. This is the case of the first three, for instance. But in each case, the speaker added a personalized and practical twist to the folk lexicography.

In some cases, these amplifying remarks are extremely elaborated. Zepeda and I have used some of these texts for teaching discourse analysis and Tohono O'odham translation skills, so they have been carefully prepared and translated. I give below translations of three examples that show the complex mix of discursive and practical encyclopedic knowledge, and the intricate intertwining of affect and cognition, that is accessed by these speakers immediately upon the utterance of a name.

Examples of Extended Commentaries Blending Different Types of Knowledge and Affect:

Marie Mandre, Bobcat Village

Melhog [ocotillo (*Fouquieria splendens*)] is what this is called. And before, the old-time people, especially the children, the children never had things to eat that were sweets, like candy, oranges, or soda. So they used this *melhog*. When it bloomed, they wrapped it and wrapped it and set it somewhere under a piece of wood, and then they'd just take off those wrappings and suck it. Because it was sweet, they would suck it. What they really enjoyed was also to make a little pile of dirt. They would stand up a stick there, and just take the flowers off and press them on where they were sticking out along that stick and it [the nectar from the flowers] would run and settle there [in the pile of dirt]. It was clean dirt. They would pour it there and there was really a lot of juice, heh heh heh, talking about dirty things! There would really be a lot of juice there, and they'd take it to eat. You know, it's said, that, how to also take one, then they would really take it off here, and they'd also take it off on the side, they would be taking it to press it, and they'd pour a little bit there, and they would drink that liquid that spilled out. They certainly did eat it, the children long ago certainly ate it. Now they don't even like it. One takes a little and tries to give it to them, but they just handle it and just drop it and run off somewhere.

In this interview response, the speaker moves immediately from the name *melhog,* for ocotillo (*Fouquieria splendens*), a plant with bright red flowers with a very sweet nectar, to a very stereotyped Tohono O'odham discourse that I have heard several times from elders, often in public meetings. It always contains the element: "In the old days, we did not have candy, or soda," and proceeds to discuss the use of some desert product that they see as a rough equivalent. This discourse has developed in the context of serious concerns about diabetes, which reaches world-record levels among the O'odham; elders are often told that they can help fight diabetes by encouraging young people to return to nutritious traditional foods. Marie Mandre's complaint about today's children, who won't eat the traditional treats, is an obvious extension of this discourse. However, in this case it is clear that she is talking about real experiences with her own grandchildren and their friends. She also chuckles about the old-time children (no doubt including herself) eating dirt, an affective element that is not a stereotyped component of the "No candy, no soda" discourse of health.

Marie Mandre, Bobcat Village

Jiawul (*Ferocactus* sp.) is what they would call that. And I don't know much about it, about what kinds of properties this *jiawul* has. Also, for us, when we try to touch them, I mean for everybody here that plant is really frightening, because if the thorns go in, it's hard to take them out, so it is frightening, because of their very big thorns. That is about all I know. These foreigners, the Mexicans, when they would get together for a fiesta, well, there was one old lady who would take that cactus when it was really small, she would take that and when she had grated it she would put it in the oven and put sugar in it and boil it. Oh, she made it sweet, sweet. Yes, one of our neighbors at Cobabi used to prepare it that way.

This remark is especially interesting because in it Marie Mandre makes a metareference to the existence of some traditional encyclopedic discourse about *jiawul* that she should know, but does not (except for the fact that it has especially formidable thorns). As a substitute, she mentions that she knows about an old Mexican lady who would make cactus candy from the plant, which then reminds her that her own neighbors in a nearby village used to make it. Again, the text mixes encyclopedic discursive knowledge with personal affect and experience.

Marie Domingo, Cu:lig Village

"Yuucch! RATTLESNAKE![16] Yuucch! *ko'oi! Ko'oi.* [When there are many?] *Ko:koi.* One almost bit me; these are real bad some of them. Here's right where it fell, right here (gesturing to her shin), and tore MY DRESS. It even got all the way to my SLIP. I thought perhaps, if my DRESS had been short, it would have fallen against my leg and bitten me, and yet I always wear a long dress; so it happened to fall right here. It didn't bite me. IT WAS A BIG RATTLESNAKE. It crawled back there, UNDER A BENCH IN THERE which is there in that little shed from where I was trying to take some PAINT … it almost bit me. I just took my DRESS straight off and threw it right away. I GUESS IT'S THE POISON, a sort of, a sort of GREENish-YELLOW,[17] just ran down there, where it presumably squirted out from where it fell and struck. Where it fell, presumably thinking that it fell against my leg, and squirted out all that poison. And when I had thrown it away, I said to MY AUNT, "Bring me something so I can put MY DRESS ON!"

Here, Marie Domingo begins with a unique case of the association of [ʔɨ::] "Yuucch," the Tohono O'odham sound for disgust about rattlesnakes, with the English name. She has code-switched throughout the interview (and continues to do so during the story). She immediately repeats both the disgust sound and the Tohono O'odham name, gives the plural, and continues with a narrative of personal experience that entirely lacks elements from the discursive system of vernacular lexicography. It is heavily laden with affect, especially in the gruesome detail of the exact color of the venom. One of its rather interesting points is that the story is partly about Marie Domingo's appropriate O'odham female modesty; she always wears a long dress, and so is saved from being bitten, and even knowing that the snake is still nearby under a bench, her first thought is for a clean dress to cover herself.

Discussion

The data presented above suggest that knowledge of a Tohono O'odham name for a plant or animal is extremely important in permitting speakers to access not only "traditional" encyclopedic knowledge, but anecdotes and affective elements of personal experience as well. We must make some qualifications of this generalization, of course. First, this effect almost certainly involves "thinking for speaking" (Slobin 1996), and not some absolute cognitive capability. It is possible that, were the interview in English, speakers might say more about the plants and animals even if they did not know the O'odham names. For instance, EF, who used the most English of any speaker except for one other woman from San Xavier in her interview, made eleven comments, an average number, even though she could not remember names for thirteen of the forty seven stimulus items. Second, it may be that some speakers experience a collapse of confidence once they cannot remember a few names, and refrain from commenting because they want the embarrassing interview to proceed as quickly as possible. This is possible even though we assured people that the interview was not some kind of test, and that we simply wanted to know how they said the names that they remembered (which was in fact the case—since each dialectological variable was represented by several stimulus items, it did not matter very much if a speaker did not remember one or two of them). However, usually when people could not remember names they joked about it, saying something like"I must be getting old!"or"I don't know where my mind is today!" Finally, it may be that the mental effort involved in trying to remember a name that one has forgotten simply blocks other kinds of contemplation.

One question, raised in the discussion at the"Nature Knowledge"conference, was whether some speakers might have difficulty with identifying plants and animals from pictures. Some speakers did indeed comment that it would have been easier to identify animals from life, since often the way an animal moves, or some feature that is hidden in some postures (such as the black bars under the wing of a species of hawk), is not apparent from a photograph. However, the speakers who had the most problems with giving names were in fact among the most educated people in the sample, and thus are presumably thoroughly accustomed to looking at color photographs. In any case, regardless of why speakers might have behaved in the ways shown above, their behavior suggests strongly that names mediate rapid access to both cultural and personal knowledge as well as to the affective "feel"of the interactive relationship with particular elements of nature.

What I am proposing is more than the usual understanding, that when knowledge is lost, names are lost with it (as in the case of the English-language horse terminology mentioned above). Instead, I am suggesting that the reverse is also true: when names are lost, knowledge is lost with them. Some of this loss of knowledge and names may partially involve cultural change. For instance, many O'odham basket makers no longer go into the desert to collect their own materials; instead, they buy them from vendors who drive right up to their front doors with a car trunk full of dried plant materials. We must also note some degree of "cultural loss." Traditional experts had access to vast systems of metaphors accessed through animals that are simply not used today. For instance, traditional O'odham were divided into two patrilineal moieties, the"Coyotes"and the"Vultures."Since the word"vulture"apeared in our list, we occasionally asked people if they knew to which moiety they belonged. Nobody did; several volunteered that they had heard about the old moieties but did not think people used them any longer. Some native plants are today little used. However, I have never met an O'odham person who has not had some experience, even if only on a school field trip, with the traditional desert products such as saguaro fruit or cholla buds (the flower buds of *Opuntia acanthocarpa*). Furthermore, these plants all grow everywhere in the O'odham environment and are common even in heavily settled areas. The animals and birds represented in our sample are very common, and it is important to teach children to avoid some of the dangerous animals such as scorpions, gila monsters (*Heladerma suspectrum*), and rattlesnakes. Furthermore, most O'odham believe that offenses to animal spirits can cause disease; several of the most important animals associated with *ka:cim mumkidag*"staying sickness," animal spirit diseases (Bahr, Gregorio, and Lopez 1974), are represented in our name sample. So the loss of names and knowledge for them is not like the loss of horse terminology in the generation between my grandfather and myself. Instead, names and knowledge that a person alert to the local environment can use in many contexts are simply disappearing from the lexical and associated practical-knowledge repertoire of some speakers of Tohono O'odham.

This knowledge, the yield of thousands of years of human adaptation to the Sonoran Desert environment, is not only locally useful and an important source of pride for the Tohono O'odham, but is valuable more broadly. We are learning that the products of the Sonoran Desert can play a role in recovering the traditional good health of the O'odham themselves. In addition, they may be useful on a global scale. To mention only one case, flour ground from *kui wihog*, the dried beans of mesquite (*Prosopis juliflora*), one of the commonest trees in the southwest and now used mainly for making charcoal, is extremely nutritious. It is an excellent source of roughage, and tastes sweet yet does not induce or aggravate diabetes, because the sugars in the flour are released very slowly into the body (Nabhan 1997). Knowledge of the habits of animals may be very important; a recent case in point comes from the 1993 epidemic of the deadly hantavirus, when Navajo medicine men alerted investigators from the U.S. Center for Disease Control, who were seeking reasons for a cluster of deaths from mysterious influenza-like symptoms, that a good crop of pine nuts had been followed by a population explosion of wild mice. The mice turned out to be the main vector for the virus. These are only two examples where the "indigenous knowledge" of Native American elders has proven to be important in the use and manage of local resources.

In addition to the loss of practical knowledge in the usual sense, when names are lost people may experience weakened access to elements of their very own life stories and even their emotional life that are tied into experiences with plants and animals. Thus the loss is not only a matter of cultural knowledge, but a matter of an attenuation of the very selfhood of these speakers.

In summary, these data suggest that names themselves, often considered a relatively trivial dimension of language, must be added to the inventory of linguistic materials in which culture is very deeply embedded. These data contribute new examples to the argument developed eloquently by Woodbury (1994), that much more than mere linguistic material is at stake when a language passes out of use.

Notes

1. See Hinton and Hale 2001 for recent practical models of language revitalization.
2. In this connection I recently had an entertaining e-mail correspondence with William Bright as we tried to understand the meaning of words for horse colors that have been borrowed into Southwestern Indian languages from Spanish. We could translate Tohono O'odham *s-'alṣa:g* (< Sp. *alazán*) as "sorrel" and *s-wa:yodag* (< Sp. *bayo*) as "bay," but neither of us knew exactly what "sorrel" and "bay" meant!
3. Wolfram and Schilling-Estes (1995: 718) have pointed out the fascination that "unique relic lexical items" have for the people of Ocracoke, an island off the coast of North Carolina, where an obsolescent dialect of English has come to seem worthy of preservation.
4. The distinction between "discursive" and "practical" knowledge comes from Giddens 1979.

5. The Tohono O'odham are the "Desert People": *tohono* means "desert;" *'o'odham* means "person, people." "Tohono O'odham" and simply "O'odham" are used interchangeably in this paper, both with reference to the people and to their language.

6. A small number of people are monolingual in Tohono O'odham. In addition, some people speak Spanish instead of, or in addition to English, and/or other indigenous languages, especially Yaqui (and of course there are Tohono O'odham who know languages like French and German as well).

7. We were assisted in the study by Mary Bernice Belin and Molly DuFort. Ms. Belin did about half the interviewing and nearly all the initial transcription of the main interview text (not including the word lists that are the focus of the present study). The dialect survey was funded by the National Science Foundation (BNS 80608009) and grants from the Institute for Social and Behavioral Sciences of the University of Arizona.

8. Tohono O'odham words, unless marked by phonetic brackets, are given in the official orthography adopted by the Tohono O'odham Nation (the so-called Alvarez-Hale orthography, cf. Alvarez and Hale 1970). Vowels are *i* [i], *e* [ɨ], *a* [a], *o* [ɔ], and *u* [u]. Consonants are *p, t, c* [tʃ], *k* (these four stops are voiceless, and are pre-aspirated in syllable-final position); *b, d, ḍ* [ɖ], *j* [dʒ], *g* (these stops are voiced in initial and intervocalic position and pre-glottalized and voiceless in syllable final position); *s, ṣ* [ʂ], *l* (normally identical to the tap *r* [ɾ] of Spanish, but sometimes more lateral, especially in syllable-final position, where it is devoiced), *w* [w ~ β], *y* [j], *m, n, ñ* [ŋ].

9. This speaker was included in this group because on the basis of the criteria I used at the time, he was missing seven lexical items. In preparing this paper I determined that if a speaker was able to give a name after extensive prompting, I would count that name. In the present study even lexical items that were quite difficult for speakers to remember are counted as "known."

10. To give an example, for the members of the genus *Opuntia*, most English speakers have only two names, "*prickly pear*" and "*cholla.*" In Tohono O'odham there are at least twenty words for the various kinds of *Opuntia*. There are close to a dozen mononomials for economically important species like *ceolim* (*O. acanthocarpa*). The economically insignificant chollas are all usually called *hanam*, but a number of speakers know mononomials such as *hadṣadkam* (*O. bigelovii*) (this is a dialect variant found mainly on the eastern side of the reservation; most speakers use the binomial *to:ta hanam* "white cholla" for this variety. English speakers know it as "teddy bear cholla" or "jumping cholla" [for its pernicious ability to attach to shoes, pants legs, the lips of cattle, etc. at even the slightest contact]). Even relatively expert English speakers have only binomials for the various species of *Opuntia*.

11. Literacy in O'odham is known from an earlier period, a famous case being that of Juan Dolores, consultant for Alfred Kroeber and a scholar in his own right who published several articles on his language. However, only within the last twenty years have more than a handful of people learned to read and write the language.

12. Cottonwood trees were once common around O'odham villages, but the drilling of deep wells and the consequent relatively careless use of water and the drawing down of the water table has led to the death of most cottonwoods; their roots can no longer reach water even in riparian zones.

13. MP, a woman from San Xavier who switched to English often during the interview and was counted in the earlier study (shown in table 8.1) with fifteen missing words, is not included in this sample. Her husband, who had declined to be himself an interview subject, was present during the interview and prompted her constantly, so that it is quite difficult to tell exactly how many words she really does not know. The number is clearly large; based on her own comments such as "Oh, I never heard that one" or on her volunteering clearly erroneous forms, she is missing at least eleven words. Nonetheless, I decided not to include her materials since I have not included other speakers where two people were interviewed together.

14. "@" notates a bit of laughter.

15. Many O'odham have told me that they particularly like Underhill's book because they find the title, *People of the Crimson Evening,* to be very beautiful.

16. Code switching into English is marked with SMALL CAPS.

17. *green–*ma* yellow, using English words with the Tohono O'odham derivational element *–ma* "somewhat, resembling."

References

Alvarez, Albert and Kenneth L. Hale. 1970. Toward a manual of Papago grammar: some phonological terms. *International Journal of American Linguistics* 36: 83–97.

Bahr, Donald, Juan Gregorio and David Lopez. 1974. *Piman shamanism and staying sickness (ka:cim mumkidag)*. Tucson: University of Arizona Press.

Benveniste, Emile. 1971.The nature of the linguistic sign. In *Problems in general linguistics*. Translated by Mary Elizabeth Meek. Coral Gables, FL: University of Miami Press.

Berlin, Brent. 1992. *Ethnobiological classification: Principles of categorization of plants and animals in traditional societies*. Princeton, NJ: Princeton University Press.

_____ and Paul Kay. 1969. *Basic color terms; Their universality and evolution*. Berkeley: University of California Press.

Casagrande, Joseph B. and Kenneth L. Hale. 1967. Semantic relationships in Papago folk definitions. *In Studies in southwestern ethnolinguistics*, edited by Dell H. Hymes and William E. Bittle. The Hague: Mouton.

D'Andrade, Roy. 1995. *The development of cognitive anthropology*. Cambridge: Cambridge University Press.

Giddens, Anthony. 1979. *Central problems in social theory: Action, structure, and contradiction in social analysis*. London: Macmillan.

Goffman, Erving. 1978. Response cries. *Language* 54: 787–815.

Hinton, Leanne and Kenneth L. Hale. 2001. *The green book of language revitalization in practice*. New York: Academic Press.

Mathiot, Madeleine. 1962. Noun classes and folk taxonomy in Papago. *American Anthropologist* 64: 340–50.

_____ 1973. *A dictionary of Papago usage*. Indiana University Language Science Monographs 8–1 and 8–2. Bloomington: Indiana University Publications.

Nabhan, Gary Paul. 1997. Diabetes, diet, and Native American foraging traditions. In *Cultures of habitat*. Washington, DC: Counterpoint.

_____ and Amadeo Rea. 1987. Plant domestication and folk-biological change: The Upper Piman/devil's claw example. *American Anthropologist* 89: 57–73.

_____ and Sara St. Antoine. 1993. The loss of floral and faunal story: The extinction of experience. In, *The biophilia hypothesis*, edited by Stephen R. Kellert and Edward O. Wilson. Washington, D.C. and Covelo, CA: Island Press/Shearwater Books.

Saxton, Dean, Lucille Saxton, and Suzanne Enos. 1983. *Papago/Pima-English, English-Papago/Pima Dictionary/O'othham-Mil-gahn, Mil-gahn-O'othham*. Second Edition.Tucson: University of Arizona Press.

Silverstein, Michael. 1984 The limits of awareness. *Southwest Laboratory for Sociolinguistics Papers*. Austin, TX: University of Texas.

Slobin, Dan. 1996. From "thought and language" to "thinking for speaking." In *Rethinking Linguistic Relativity*, edited by John Gumperz and Stephen Levinson. Cambridge: Cambridge University Press.

Underhill, Ruth. 1951. *People of the Crimson Evening*. Washington, DC: United States Indian Service.

Wolfram, Walt and Natalie Schilling-Estes. 1995. Moribund dialects and the endangerment canon: The case of the Ocracoke brogue. *Language* 71: 696–721.

Woodbury, Anthony C. 1993. A defense of the proposition, "When a language dies, a culture dies." In *Proceedings of the First Annual Symposium about Language and Society*, edited by Robin Queen and Rusty Barrett. (*Texas Linguistic Forum* 33: 102–30). Austin: University of Texas.

Addendum

Responses to Questionnaire with reference to Tohono O'odham (Papago), a Uto-Aztecan language of southern Arizona.

1. A wide range of linguistic mechanisms of used for naming nature in Tohono O'odham. No single strategy is dominant, and many plants and animals are not relatable to any other meaning.

A. *Onomatopoeia.*
This process is especially noticeable in bird names, e.g.
   ho:hoi "mourning dove" (*Zenaida macroura*)
   kokoho "burrowing owl" (*Speotyto anicularia*)
   kolo'ogam "whipporwill" (*Phalaenoptolus nuttali?*) The name means "the one who says/does
    kolo'o."
   kuhigam "*Phaenopepla* sp." The name means "the one who says/does *kuhi*" (this is the verb root
    for speaking of the calls of animals)
   ku:ku'ul/ku:kwul "elf owl" (*Micrathene whitneyi*)
   'okokoi "white-winged dove" (*Zenaida asiatica*)
Note that the O'odham word for "dog," *gogs*, comes from Pre-Tepiman *wowosi*, almost certainly sound-imitative, and so provides an example where regular sound change (Uto-Aztecan *w>Tepi man *g*) and phonological innovation (deletion of vowels in unstressed syllables) has over-ridden sound symbolism. Another example of this type is *ba'ag* "eagle," from Pre-Tepiman **kwa'aw*), also quite possibly sound-imitative in its original form (many Uto-Aztecan words for large birds include this **kwa-* root).

B. *Morphological derivation.*
These examples involve change in the shape of an extant name through suffixation, composition, and modification.
There are many binomials, where the modifying element is usually a color word or other adjective such as "soft" or "hard."
For *hu:ñ* "maize" we find binomials with several color terms. In addition, we encounter the following:
   s-hiosigam hu:ñ "flowery corn" (popcorn)
   cecpa'awĭ hu:ñ "whore's corn" (corn with many colors on a single cob, also known as *totonto hu:ñ*
    "crazy corn," from Spanish *tonto* "crazy")
For *hanam* "cholla" (various species of *Opuntia*), we find
   to:ta hanam "white *hanam*" (*O. bigelovii*)
   cehedagĭ hanam "green/blue *hanam*" ("nipple cholla")
For *winoi*, usually called by the reduplicated form *wipnoi*, we find the following:
   ce'ecem wipnoi "little *wipnoi*" ("christmas cactus," *Cylindropuntia leptocaulis*)
   ge'egeḍ wipnoi "big *wipnoi*" ("pencil cactus," *Cylindropuntia ramosissima).*
Many plants exist in both cultivated and "coyote" (*Canis latrans*) forms.
   bawĭ "cultivated tepary bean" (*Phaseolus acutifolius*)
   ban bawĭ "coyote's tepary bean" (of uncertain meaning)
   'ihug "devil's claw" (*Martynia parviflora*, cultivated for fiber length)
   ban 'ihug "coyote's devil's claw," a wild variety with short fibers
   toki "cultivated cotton"
   ban tokiga "coyote's cotton" ("Milkweed")
Other examples of binomials include
   ko'owĭ "rattlesnake" (*Crotalus atrox*)
   a'agam ko'owĭ "horned rattlesnake" ("sidewinder")
A series of characteristic final syllables appear with high frequency in plant and animal names, and may be a residue of old classifying suffixes. These include *-li* (sometimes without final /i/, and some times appearing as *-ḍ), -wĭ, -wigĭ, -poḍ, -oi.* These elements invite etymological research.

C. *Semantic extension.*
A notable example of semantic extension, with marking reversal and apparent loss of the indigenous name, is the following series:
   ko:ji "peccary, javelina, domestic pig" (from Spanish *cochino* "pig")
   misciñ ko:ji "wild pig" (Peccary)
   do'ag c-eḍ ko:ji "mountain pig" (Peccary)
Similarly, we find *ca:ngo* "coatimundi" from Spanish *chango* "monkey," with no trace of the indigenous name for the animal.

A number of names are organized around prototypes. For instance, *wihog* is prototypically the fruit of mesquite (*Prosopis juliflora*), which is dried and ground into flour, but the term is also used for the fruits of other Leguminaceae.

D. *Semantic specialization*
An example of this type is the specialization of *'ihug*"Devil's Claw" to refer only to the cultivated variety of this basketry plant, with *ban 'ihug*"Coyote *'ihug*"always used to refer to the wild variety of *Martynia parviflora*.

E. *Metonymy*
I suspect that the words *bitotoi/bitokoi/bicitoi*"Pinacate Beetle"(the words are dialect variants), derive from *bi:t* "dung, excrement." People often laugh at this word, perhaps evidence in favor of this hypothesis. Furthermore, long-vowel shortening in derivation is fairly common in Uto-Aztecan languages, although it is not predictable in this stressed environment (Tohono O'odham has initial stress).

> *Wa:mog*"mosquito" contains a root for water, *wa:-*, found only in derived forms; cf. *wa:mul* "swamp."
>
> *koson*"pack rat"(*Neotoma* sp.) probably contains the element *koṣ-*"nest."*koson* is an unusual case of a sequence /so/; we would expect /ṣo/ (/s/ normally appears only before /i/). These animals make very large nests, often made out of hundreds of"cholla balls," the spiny ends of branches of *Opuntia bigelovii*.

An especially interesting example of metonymy is *ba:p ce:po'ogam*"the one who says/does'Where is your bedrock mortar?,'"the name for the Preying Mantis. This is said to be the name because the position of the arms of the insect look like the arms of a woman grinding seeds.

F. *Metaphor.*
Examples of metaphor include *ko:magĭ hidoḍakuḍ*"grey cooking pot,"the name of a lizard, and *jiawul* "devil"(from Spanish *diablo*), the term for several species of *Ferocactus* and *Echinocactus* with clusters of very long, hard, curved thorns.

2. The organizing principles for naming nature in Tohono O'odham are of limited scope. Many names bear no clear relationship to one another or to other words in the language. There are many relatively onomasticized mononomials, with limited zones of"system"in the form of binomials. While cover terms are relatively few, it is clear from people's behavior that there are covert categories. For instance, a cover term meaning"things that go around on the ground" includes insects and many small animals such as rodents. However, it is clear that a covert"rodent" and"insect" category are both available to speakers.

A. Some names seem to be related to the perceptual effect of the entity from the O'odham point of view. For example, *ce:mĭ*"senita cactus"(*Cereus schottii*) is a cactus that exhibits clusters of tall stems that seem to grow directly from the ground. The name seems to be related to the verb *ce:m*"gather into a group, of humans." (Note that the tall cacti are all considered to be human beings). *Cuck kaikam*"the one that has black seeds"is the term for a kind of melon with such seeds. *'U:w ha:l*"fragrant melon" is the term for"canteloupe."*Hiwcu wegĭ*"red groin"is the name for the black widow spider, which has a conpicuous hour-glass shaped red spot on its black abdomen. *Hoho'i*"multiple thorns"is the name for"porcupine."*Hugṣum wamad*"rake-marked snake"is used for the king snake and coral snake, which have colored stripes around their bodies. Interestingly, this name does not distinguish the harmless king snake from the very dangerous venomous coral snake. *Bi:bhiag*"many things twisting, wrapping"is the term for"morning glory," a twining vine. *Keli hu:c*"old man's toenail"is the name for the curve-billed thrasher (*Toxostoma curvirostre*), and apparently refers to the long curved beak of this large bird. *Taiwigĭ*"firefly"includes the element *tai*"wild fire, spark."One of my favorite names with a perceptual basis is *cukuḍ ṣoṣa*,"owl's snot," the word for dates and date palms.

B. "Utilitarian" considerations appear mainly in the series of distinctions noted in (1B) above, where plants exist in unmarked and "coyote" varieties. Coyote is said to "spoil" plants, making them inedible or otherwise undesirable.

C. In the area of social relations as an organizing principle, we can identify a number of kin terms. These include *wihogmaḍ*, a small weevil that eats mesquite beans, from *wihog* "mesquite bean" and *maḍ* "woman's child." *hewel mo:s* "western kingbird" (*Tyrannus verticalis*) means "wind's daughter's child" (or "wind's mother's mother"). *'O:bĭ maḍ* "gecko" means "Apache's child, enemy's child" (using the term for "woman's child"). A series of plants are known as "mothers." *Howij je'e* "banana yucca" (*Yucca batata*) includes the word for "mother," *je'e*, and the word *howij*, the term for the banana-shaped fruit. *Şu:na je'e* "elephant tree" (*Bursera microphylla*) includes the word *şu:na* "fig." *'Utko je:j* "Yucca schottii," includes *'utko* "yucca stalk," used for fencing and ceilings. *Je:kam* "armless saguaro" (*Cereus giganteus*) means "the one that is a mother." When the arms of the saguaro appear, they are called *mamadhog* "various woman's-child elements."

A second kind of social relation that appears in names is "ownership." This is dealt with in more detail in (3) below.

D. A variety of names appear to reflect systems of myth and belief. For instance, a large ant is called *kuaḍagĭ*, where *-dagĭ* derives abstract nouns. *Kuaḍk* is the term for a shaman's trance song, where *-k* is the suffix for participial elements. Several flowers include the element *gi:kho*, the name for a woven basketry crown used in ceremonials. The night-blooming cereus (*Cereus greggii*), which blooms only once a year and produces very fragrant flowers that last only through the night, is called *Ho'ok wa:'o* "Ho'ok's tongs." The *Ho'ok* is a female cannibal monster. *wa:'o* are paired sticks used to pluck thorny cactus fruits. The stems of the night-blooming cereus are covered with short, fine, nasty thorns. *Naw* "prickly pear," for several species of flat-stemmed, edible-fruit-bearing members of the genus *Opuntia*, seems to contain the same element that appears in *nawmk* "drunk," and *nawait* "saguaro-fruit wine." This may be a reference to the pleasures of the land of the dead, where people drink, dance, and enjoy themselves, because the fruits of *naw* are said to be hearts or souls. The main source for *nawait* is not *naw*, but *bahidaj* "saguaro fruit" (literally, "its ripeness," perhaps a euphemism, since such fruits are usually called *'i:bhai* "breath, heart"). The fruits of *naw* ripen in the fall, while *bahidaj* is available in the end of June and early July, just before the coming of the summer thunderstorms. *Naw* is a reflex of a Uto-Aztecan etymon (it is related to the Nahuatl word *nopalli*, source of Spanish *nopal*); the possible relationship of this element to drunkenness and the land of the dead needs further exploration. Certainly the idea that the fruit of these cacti is a metaphorical "heart" or "soul" is very widely-attested among Uto-Aztecan peoples. Another plant name that may be associated with understandings of the land of the dead is *noḍ* "prickly poppy" (*Argemone platyceras*), an exceptionally large and beautiful spring flower. The O'odham name means literally "crazy, spinning." The land of the dead is said to be filled with flowers (Hill 1992).

3. Animal names seem to transfer to plant names, but plant names do not seem to appear in animal names. Certain animals are thought to own or cultivate certain plants. I have mentioned above the series of "Coyote" plants. Other examples include *babad 'i:wagĭ* "frog's greens," *kukuwid ha-ha:ḍ* "antelopes' lily" (*Rudbeckia* sp.), *ko:koḍ 'oipij* or *ko:koḍ ho'ibaḍ*) "crane's awl" or "crane's needle" (*Erodium cicutarium*), and *ho:hoi 'e'es* "mourning dove's crop" (*Escholtzia* spp.). Nabhan (2001) has discussed the way in which indigenous knowledge of plant-animal interactions is often incorporated in such names.

Some plants are named for animal parts. Examples include *cu:wĭ taḍpo* "cotton-tail rabbit's foot-hair" (*Orthocarpus purpurescens*) and *ko'owĭ ta:tamĭ* "rattlesnake's teeth" (*Cassia* sp.).

Animal names are common in disease terminology. *Ka:cim mumkidag* "staying sicknesses," so-called because they "stay with" the O'odham and do not affect Whites) are said to be caused by offenses, witting and unwitting, against the spirit natures of animals. Bahr, Gregorio, and Lopez (1974) give an extended treatment of these.

While animal names do not, as far as I know, appear in body-part terminology in O'odham (there are no examples of the *musculus* "muscle" from *mus* "mouse" type), body parts are used for naming features of the landscape. For instance, the world-navel mountain, home of the Creator, called *Waw Giwulk* (Baboquivari) is said to have a forehead, nose, shoulder, and waist. The earth, *jeweḍ*, can be shown to be a metaphorical body, in that elements that protrude from and intrude into the earth take long-vowel reduplicated plurals, a rare pattern shared with protruding and intruding body parts such as noses and mouths (Hill and Zepeda 1998).

Animal names appear in O'odham social organization primarily in the names for the two patri-lineal moieties, the "Coyotes" and "Buzzards." These moieties no longer organize social life, and most people do not know today which one they belong to. The symbols of the two animals persist as important symbols of "O'odham culture," and appear in contemporary murals on the reservation.

4. Some O'odham animal names seem to be euphemisms. For instance, *ko'owĭ* "rattlesnake," contains the element *ko-* meaning "to cause pain." The same element appears in *ko'okol* "chili pepper." This euphemism is, however, very ancient; the rattlesnake word is a reflex of a Uto-Aztecan eytmon. *Ban*, the word for "coyote," comes from Pre-Tepiman *kwa-nɨ*, and probably means "the one who barks." The O'odham word for "to bark" is today *hi:nk*, but the stem *kwa-* appears in related languages. In many Uto-Aztecan languages the coyote is called "liar." The word for "bear," *juḍum*, may mean "jumper." The word for "big-horn sheep," *ceṣoñ*, probably means "climber." The resemblance to the verbs is not exact, however.

5. A crucial question for the study of nature knowledge is whether names are "arbitrary" or "motivated" in the sense of Saussure. To me, it seems that the very arbitrariness of names is itself motivated—by local functions and ideologies. The tendency of names to "onomasticize": to become name-like by dropping out of regular sound change and bleaching semantically, is apparently not a universal phe-nomenon. Many of the plant and animal names discussed above, such as those noted in (4), exhibit these properties. While there are a few clearly descriptive plant and animal names of the "Owl Snot" and "Old Man's Toenail" type, most names in Tohono O'odham cannot be linked to other words in the language except by the professional etymologist.

Interestingly, Tohono O'odham place names do not seem to onomasticize. Most place names have very obvious and transparent meanings, such as "Where the dog burned up," "Where the sad-dle was hanging," "Where something iron is standing," "Going around in circles," etc. Some names clearly express knowledge of natural properties of a site. For instance, "field names" describe the properties of fields that are used for late-summer rainfall agriculture: "Cold fields," "Bitter fields," "Big fields,", etc. The important properties of water sources are also clearly spelled out: "Sweet water," "Hot water," "Big pond." This transparency persists even though there is evidence that O'odham has been spoken in its present location for at least 2,000 years. We can show this through studies of con-tact with Yuman languages (Shaul and Hill 1998) and the reconstruction of hot-desert flora and fauna names to Proto-Tepiman.

This contrast suggests that place names may have a rather different function from plant and ani-mal names (see Basso (1996) for the functions of place names among the Western Apache). Thus it may be difficult to draw generalizations about the linguistic properties of the encoding of the knowl-edge of nature.

References

Bahr, Donald, Juan Gregorio and David Lopez. 1974. *Piman shamanism and staying sickness (ka:cim mumkidag)*. Tucson: University of Arizona Press.

Hill, Jane H. 1992. The flower world of Old Uto-Aztecan. *Journal of Anthropological Research* 48: 117–44.

———— and Ofelia Zepeda. 1998. Tohono O'odham (Papago) plurals (with Ofelia Zepeda). *Anthropolog-ical Linguistics* 40: 1–42.

Nabhan, Gary. 2001. Cultural perceptions of ecological interactions: An "endangered people's" contribu-tion to the conservation of biological and linguistic diversity. In *On biocultural diversity: Linking lan-guage, knowledge, and the environment*, edited by Luisa Maffi. Washington, DC: Smithsonian Institution Press.

Shaul, David Leedom and Jane H. Hill. 1998. Tepimans, Yumans, and other Hohokam. *American Antiq-uity* 63: 375–96.

CHAPTER 9

Examples of Metaphors from Fauna and Flora

Giovan Battista Pellegrini

When prehistoric people discovered or invented the "phoneme," they made enormous progress and gradually began to abandon signals. By using an elaborate language they acquired the ability to apprehend and distinguish the world around them. They thus came to describe the surrounding *realia* initially vaguely and then with greater accuracy. Whether this grew out of the attempts to imitate the sounds of nature is an unresolvable question, which we cannot explore here. Detailed efforts have nonetheless been made on this question (e.g. Bruni 1958). What is certain is that our very distant forefathers continued to enrich and hone their expressions gradually, but for various reasons and on the grounds of simplicity they soon resorted to various, metonymic and metaphorical expressions. This involved attributing identical names to very different phenomena, but with a number of definite similarities, evidently perceived as such.

I will not dwell on a few cases in which very different referents with some similar aspects have the same name and the designation of one is transferred often through proximity to another object in the same semantic category. Such procedures go back a very long way. They may also be found in parallel cases and may be useful in providing etymological explanations for otherwise difficult and incomprehensible phenomena. I will thus illustrate a case that helped me to understand unknown etymons, based on what are to my mind well established parallel processes.

While studying various sections of the Friulian and Alpine vocabulary, I came across various names for "cuts of hay." In Friulian they may be distinguished between four or five mowings and with the same number of denominations (although not all areas and their related dialects have such a wealth of distinctions). Thus in some Alpine and Cisalpine areas for the second cut, the Latin term *(re)cordum*, the technical term for haymaking, is used. The metathetic form *dork*, on the other hand, survives in Valcellina and also at Erto, Cimolais and Claut, due to the Bellunese influence, since the word *dòrk* means *guaime* "evening/late hay" (Nazari 1884: 84); the word **recordum* is mainly found in the Alpine foothills and the Po Valley plain in the Veneto and in Piedmont. The nearest form to *dork* is found in the Valsugana (*cordo*) and the Bergamo valleys (*cort*). On consulting a good Latin dictionary, we find that in Latin the mainly agricultural adjective *cordus* is extended to the fauna (see Varro, RR 2, 1, 19 "dicuntur agni qui post tempus nascuntur" and P. Festo 57, 13 "corda frumenta quae sero maturescunt, ut fenum cordum..."). The basic meaning must have been "late" or "evening." But in the same semantic area of haymaking we find a parallel example in Friuli for third-cut hay (possibly inverted, i.e., borrowed from fauna and applied to flora). This word (I am possibly the first to analyze it) is the Friulian *mujàrt* of which Pirona, Carletti and Corgnali (1992: 629) gives the following definition:"*mujàrt* adj. said of the last thin cut of hay... said of the first soft fine body hair (also *pel mat*), a variant of *mugnart.*"Here we note that in Friulian there is a phonetic alternation between [j] and [ñ] found in many cases studied by Carla Marcato (1983–84). In the ASLEF (Pellegrini 1972–86) *mujàrt* is much more frequent than *muñàrt*, which only appears in three places—Prato Carnico, Luincis and Ilegio; whereas *mujàrt* is found in thirty-seven places. We may surmise, therefore, that the primitive form (rarely changed into *muñ-*) was *mujart*. As regards the etymological interpretation we must first clarify the ending in -*art* (from -*ardo*), given that it is extended by a word that has the same meaning and stands for the"third-cut hay" called *bastardo*, a word also found in Friuli, for example in Cimolais, Barcis and Poffabro. According to Pirona, Carletti and Corgnali (1992: 42),"*Bastàrt* is said of an animal and also of a plant or fruit (it is the equivalent of *mat* 'not authentic,' 'false;' also *mul*)."Identifying the origin of the base *mui-* is more difficult. From the phonetic aspect it may derive from *mui-*, *mugj-* or *mudj-* or even the palatalized *muga-*. A comparison with *cordus* suggests a parallel meaning to the lexical theme through an exchange between fauna and flora. In reality there are probably plausible comparisons in the realm of fauna, which I have already identified (Pellegrini 1972: 394–97). I believe we must begin from a pre-Roman and Alpine base of *MUGIO-* "steer" (cattle), cf. Upper Engadina. *muj-a*, Sent *muoi* "two-year-old steer" or, more vaguely, "not mature, "not genuine" or so on. The clue comes form a series of terms cited by me and investigated by some scholars. We may use as a base an article published by Jud (1911: 17), also on the subject of *mugio* "steer," cf. alto engad. *muj-a*, Valtellina di Livigno *motc* "one-year-old calf,"Valtellina *močč* "young bull," Bormino *mùghera* "heifer,"Soprasilv. *mutg*, *muje* -*a*"two-year-old steer,"Bresciano *muǧ* "young steer," Piedmontese *mogia*"heifer," etc.There is some useful information in Stampa (1937: 47–48), and especially the article by Hubschmid (1966: 187–89). Here I would add from the recent *Dizionario etimologico grosino* by Antonioli and Bracchi (1995: 545) *moc'*"castrated bull." The derivation of *moč* "truncated" in the specific meaning of castrated seems unlikely, because of the vast family of words round the meaning of"young animal"from a pre-Latin base *mugio* thus see also *mùgh-era* (Tognina 1967: 210; Lurati 1968: 169; Quaresima 1964: 268; Battisti 1922: 45). Analogous denominations or those of similar meaning may be found in German-speaking areas (von Wartburg 1922ff.; cp. Bavarian *motschen*"kuh") and in Provençal French, many denominations refer to young animals. Rohlfs (1940: 365) explains the whole vast family as"lautmalenden Ursprungs" (without advancing any etymology) and Hubschmid (1966) thinks it may be a very old pre-Roman word, possibly a primitive *Lockruf* (call). We, too, see it as a pre-Roman word possibly even pre-Indo-European, which cannot be assigned to a specific language (Pellegrini 1972: 383–405).

We have thus apparently come a long way from our *mujàrt*, but we believe it very likely derives from a *moio-* plus suffix in a way near *chordus* (*fenum chordum*) "Spätgeboren"(Meyer-Lübke 1935: n. 1383).[1]

To stay with the conference theme I will try to bring together a number of cases of metaphors (or the like) that mainly refer to the flora. I gathered them by looking through books of botanical illustrations and in the case of Friuli, by consulting Pellegrini and Zamboni (1982), and two excellent volumes for the Biella area by Alfonso Sella (1992, 1994), who often includes denominations of plants derived from fauna, but also from other sections of the vocabulary.

We can easily observe how terms for the vast lexical field of botany were often created by beginning from the animal world. And the generator of so many metaphors was often chosen at popular level to indicate various plant species. We will leave aside the well-known theory that many names were attributed to plants by Medieval monks (Bertoldi 1927), but because of the choice of metaphors, many plants have the same or similar names thus often creating confusion. There may be various reasons for this lack of distinction, but I would suggest that it is mainly due to the real or imaginary form or similar functions. There are also cases in which the real reasons for the uniformity are not known.

Here I offer a list that could be greatly lengthened. The *campana* (bell) and *campanella* (little bell) are often alluded to: in Friuli we have *campana* for *Iris germanica* (Pellegrini and Zamboni 1982: 107) and also the narcissus: *campanelas, -es -èi* and the *Galantus nivalis* or "snowdrop"; at the same time we have derivatives from **clocca* "bell," e.g. *ciocchìt* (Canton Ticino) and l'*Aquilegia vulgaris aquilina/* columbine and the *Leucojum verbum* called *campanellino* (Pellegrini and Zamboni 1982: 142) or *campanelutas* for *Convolvulus arvensis* "field bindweed" (Pellegrini and Zamboni 1982: 103).

In the Biella area (Sella 1994: 126) we often find *galineta* (from *gallo/gallina* "cock/hen") e.g., *gallinaccia* (fungus), *gallinetta* for "clover," and *gallinetta* for "honey mushroom." *Gambe di gallina* is the fungus *Lepiota procera*.

Gatto (cat) and *cane* (dog) also lend their names to plants. Thus in Friuli (Pellegrini and Zamboni 1982: 97) we find *giate* as *Carduus nutans*, for the prickly species of thistle reminiscent of the cats claws, and *Carlina acaulis, Eryngus campestre*; these two animals are also combined with colors (*blancie, nere, turchine, celestis, zalis*) and many other plants.

In the Biella area *gatto* (cat) is often found, e.g. for *Aruncus deoicus gata* (*zal* species) and the willow *gate* (Sella 1992: 46); *gati* refers to the catkins while in the *Creus flagelliformis* the *gati* are caterpillars.

Cane (dog) is commonly found in compounds. In the Biella area we have *denc' de can* for "dandelion" (*Taraxacum officinale*) (Sella 1994: 17); *lengua de can* for the fungus *Anchusa*; *picci de can* for a fungus; *piotte de can* for "cock's foot" (*Dactylis glomerata*). In Friuli we have *cian* for *Trifolium repens* "white clover" (Pellegrini and Zamboni 1982: 215).

It has been pointed out by A. Zamboni that the presence of the element *can* in many compounds may have come from the adjective *canus* "white-haired, white."

In the Biella area (Sella 1994: 132) the *peir de la gaža* or *Pyrus communis* (a variety of small sweet summer pears) comes from *ghiandaia* (jay). *Oca* (goose) lends its name to *pe d'oca* (crowfoot) and *pista d'oca* (goutweed), *Aegopodium podagraria* (Sella 1994: 137). Many names originate in *lupo* or "wolf" (Sella 1994: 52), e.g., *ai dal lü* is "ransom" (*Allium ursinum*), or *buca de lü* (dandelion), and *cua dal lüu* (lit. "wolf's tail ") is *Conyza canadensis; pëtta du lü* (lit. "wolf's fart") is "puffball" (fungi in the *Lycoperdon* genus), etc. Many names are also derived from *pidocchio*—"louse" (Sella 1994: 204): *erba de piöchj* ("louse grasss") is *Helleborus viridis; piöch dal lüf* is "burdock" (*Aretium lappa*); and *masa-pjöch* ("louse-killer") is "bladder campion" (*Silene vulgaris*).

From asino (ass) we get *bjava di aʃu* for "barley" (*Avena sativa*) (Sella 1994: 8), *salata d'aʃu* for "thistle," and *ureggi d'aʃu* for "mullein" (and leaving aside flora, we also have *urgjun* for the illness "mumps"; and similarly from *gatto* we have *gattoni* for "parotitis" and also *mal del moltone*); *pasta de eʃu* is *Berberis vulgaris*. From camoscio (chamois) we have *erba camussa* for *Bupleurum stellatum* (Sella 1994: 10).

Capra (goat) gives *crava* for "box" (*Buxus sempervirens*), *crava, craveta russa*, for the fungus *Boletus rufus* or *barba di crava* (goat's beard) for "wisteria." *Cavallo* (horse) gives *cavalini* for "butterbur" (*Petasites albus*), *cua cavallina, de caval* "box with which children play"; *erba cavalinna* is "grass" *Nardus stricta*. From *lepre* (hare) we have *flor dla levra* for "dog's tooth" (*Erythronium dens-canis*). *Porco* (pig) gives *erba del purchët* (pig's grass) for *Portalaria oloracea*; and *erba purcalina* for *Polygonum arviculare*. *Orso* (bear) lends its name to *braje de ors* for lycopodium or clubmoss. *Topo* (mouse or rat) provides *bërle d rat* (mouse snots) and *cue d rait* for "lycopodium"; *ureggi di rat* is "hoary plantain" (*Plantago media*). From *vacca* (Sella 1994: 92) or "cow" we get *vacca d'aularo*, for the fruit of meadow saffron (*Colchicum autunnale*). *Bue* (ox) is widely found as *lenga de bu* or "cuckoo-pint."

Further examples may be found in Pellegrini and Zamboni (1982), i.e. the Friuli dialects: from *pidocchio* (louse) we have *erba da peduois* for common "mullein" (*Verbascum thapsus*) (222) and cf. *peduč* for "goosegrass" (*Galium aparine*) (124); *erba del lof* for *Arum italicum* (60); there is also *Convallaria maialis* and many compounds with *jarbe* or "grass" (727); *jarbe love* for *Cuscuta europaea* (110); *jarbe dai pedoi* for *Pedicularia palustris* (385); *jarbe dai pulz* for *Galium aparine* (124); *boča de čan* for *Delphinium consolida* (116); *code di čan* for *Cynosuruta Cristatus* (307) (the ear is like a dog's tail); *code di volp* for *Alopecus pratensis* (this is an old metaphor); *kode di mus* for *Equisetum Arvense* (119). As we have already seen the "tongue" (*lingua*) is often used for plant names see Pellegrini and Zamboni (1982: 734): *lengua cervina* for *Scolopendrum vulgare* (197); *lenga de vača* for *Arum italicum* (60); *lenga de bo* for *Anchusa officinalis*. Note also *merda de ǧat'* (cat's shit) for *Ilex aquifolium* (135) and *Ruscus aculeatus* (191); *-de kuk* "fir sap" (with analagous forms in the Belluno area).

Among birds *si÷ila* (swallow) found in botanic names, e.g., *voi de* (eyes of) *si÷ila* for *Myosotis silvestris* (45) and also *Pulmonaria officinalis; voi de siʃile* for *Dianthus carthusianorum* (117), etc.

One category of animals is used in devices for carrying, bearing or sustaining. Such animals are usually beasts of burden or work. From *capra* (goat) we have the derivative *capriata* (truss). The meaning is of a large support; but even without the suffix *capra* can mean "trestle" or a tool used for support, etc. (in the Belluno area it is a device used for support while working on horseshoes). Similarly *asino* (ass), especially in the diminutive *asinello*, is the beam across the top of the trestle; cf. the Spanish *asne* "trestle," the French *poutre*, etc and *putrella* (beam) derive from **pullitra* (filly). Similar meanings are given to *mulo* (mule) and *bardotto* (hinny) as well as *cavallo* (horse) and *cavalletto* meaning support, and similarly *cantherius* (castrated horse), used for a "rafter" and other related meanings; thus too *bordone* in the Latin meaning of *burdo -onis* meaning "mule" has become a common metaphor for "beam" (already in Medieval Latin) and later a "pilgrim's staff."

We might also cite the various other metaphoric uses of *cane* (dog) (Sella 1994: 10–17), from *cane del fucile* (cock of a gun) (also in the Biellese *cagnét* and *cagnò*; in the Agordo dialects *cagnòl* is a special kind of hook in the sawmill) to *cagna*, "a wooden device with iron teeth used by carters," and an "iron nail."

I will not dwell on the many cases of metonymy. There are some very well known parts of the body used for neighboring parts in late and vulgar Latin. Thus the case of *bucca* "cheek" came to substitute *os -oris* (because of its conflation with *osso*—"bone" it was almost always abandoned); or the *gula* became *gurā* "mouth" in Romanian; we can also cite *maxilla* which in Spanish has come to mean "cheek" (*mejilla*). Similarly *perna* "thigh" in Romance Iberian, Spanish *pierna*, Portuguese *perna* came to mean "leg" like *coxa* (thigh) found in the Neapolitan *cossa* (leg), etc.

There are many interesting cases of an ancient name surviving but with a complete different meaning for various reasons, but especially through changes in techniques. I could thus take some of the examples I quoted in Pellegrini (1975: 347–348). Thus for example the French word *ruche* meaning "beehive" continues the Gallic word *ruska* meaning "bark of a tree" (Meyer-Lübke 1935: n. 7456). The earliest beehives in some peripheral regions were built with bark, as were some other recipients. The English "taper" (a long thin candle) and the Irish *tapar*, go back through *tapurus* to *papyrus*, i.e., the paper used as an elementary means of lighting. In the Veneto the word *pavér* or *pavier* is similarly derived from **papyreus* (Meyer-Lübke 1935: n. 6217) and means "a small light." But the German *Kerze* "candle" also goes back through the old German *Charza* to the Latin *charta*. The Italian *penna* (for "fountain pen") goes back to the Latin *penna*, which although had a similar function had a completely different meaning.

One last example from our dialect area concerns *versorium*. In the Veneto the form *varsor*, means "plough" (there is a famous *Indovinello veronese* "Verona riddle" *...et albo versorio teneba*). But before it came to mean plough it signified "the ploughshare." From this meaning, surely because of the shape, we arrive at the derivation of "snowplough" throughout the region. In the Vicenza area, however, there is now a distinction due to the Cimbro influence changing *v* into *f* between the more archaic *farsór*, meaning "plough" and *varsor* "snowplough."

We could go on at length exploring the interesting theme proposed by the conference. Indeed on close examination we would find very many examples.

Notes

1. For our purposes there is a very interesting article by Gardette (1966). But see Salvioni's (1902–05) attempt at etymology i.e. deriving *muje-* from the ancient French *mude* from *mutāre* (this explanation then passed to Meyer-Lübke 1935: n. 5785) and see also Vidossi(ch) (1905). Mention must also be made of the work by André (1963), which I will not use here.

References

André, Jacques. 1963. Noms de plantes et noms d'animaux en latin. *Latomus* 22, 4: 649–63.

Antonioli, Gabriele and Remo Bracchi. 1995. *Dizionario etimologico grosino*. Grosio.

Battisti, Carlo. 1922. *Studi di storia linguistica e nazionale del Trentino*. Firenze: Le Monnier

Bertoldi, Vittorio. 1927. Monaci e popolo, "calques linguistiques" e etimologie popolari. *Revue de Linguistique Romane* 3: 137–64.

Bruni, Lorenzo. 1958. *L'origine del linguaggio*. Milano, Studium.

Gardette, Pierre. 1966. Ancienne lyonnais *cuer* 'qui reste en dernier.' *Revue de Linguistique Romane* 30: 71–87.

Hubschmid, Johannes. 1966. In von Wartburg (1922 ff.), vol. VI, 3.

Jud, Jakob. 1911. Della storia delle parole lombardo-ladine. *Bulletin de Dialectologie Romane* 3: 1–18.

Lurati, Ottavio. 1968. *Terminologia e usi pastorizi in Val Bedretto*. Lugano: Società svizzera per le tradizioni popolari.

Marcato, Carla. 1983–84. Osservazioni sul passaggio da *ja_ń* esempi friulani. *Quaderni Patavini di Linguistica* 4: 193–212.

Meyer-Lübke, Wilhelm. 1935. *Romanisches etymologisches Wörterbuch*. 5th ed. Heidelberg: Winter.

Nazari, Giulio. 1884. *Dizionario bellunese-italiano*. Oderzo: Tipografia Bianchi.

Pellegrini, Giovanni Battista and Alberto Zamboni. 1982. *Flora popolare friulana*. 2 vols. Udine: Casamassima.

———— 1972. *Saggi sul ladino dolomitico e sul friulano*. Bari: Adriatica.

———— 1972–86. *Atlante storico-linguistico-etnografico friulano (ASLEF)*. 6 vols. Padova—Udine: Istituto di Glottologia dell'Università di Padova—Istituto di Filologia Romanza della Facoltà di Lingue e Letterature Straniere di Trieste con sede in Udine.

———— 1975. *Saggi di linguistica italiana*. Torino: Boringhieri.

Pirona, Giulio Andrea, Ercole Carletti and Giovanni Battista Corgnali. 1992. *Il Nuovo Pirona, Vocabolario friulano*. 2th ed. by Giovanni Frau. Udine: Società filologica Friulana.

Quaresima, Enrico. 1964. *Vocabolario anaunico e solandro*. Venzia—Roma: Istituto per la collaborazione culturale.

Rohlfs, Gerhard. 1940. Zu einigen Etymologien Alessios. *Zeitschrift für romanische Philologie* 60: 362–70.

Salvioni, Carlo. 1902–1905. Appunti sull'antico e moderno lucchese. *Archivio Glottologico Italiano* 16: 395–478.

Sella, Alfonso. 1992. *Flora popolare biellese*. Alessandria: Edizioni dell'Orso.

———— 1994. *Bestiario popolare biellese: nomi dialettali, tradizioni e usi locali*. Alessandria: Edizioni dell'Orso.

Stampa, Renato Agostino. 1937. *Contributo al lessico preromanzo dei dialetti lombardo-alpini e romanci*. Zürich-Leipzig: Niehans.

Tognina, Riccardo. 1967. *Lingua e cultura della valle di Poschiavo*. Basel: Società svizzera per le tradizioni popolari.

Vidossi(ch), Giuseppe. 1905. *Archeografo Triestino* 30: 416–17.

von Wartburg, Walther. 1922 ff. *Französisches etymologisches Wörterbuch*. Bonn: Schröder.

CHAPTER 10

Lexicalization of Natural Objects in Palawan

Nicole Revel

Thought works with symbols that constitute the knowledge of a given human community. Starting from the notion of "sign" as defined by F. de Saussure, C. S. Peirce, R. Jakobson and E. Benveniste, one can attempt to grapple with the problem of words related to things (signs as symbols of the external world), of words and notions linked to life; of words and concepts (signs as symbols of concepts); in other terms, one can attempt to deal with the problem of signs and referents in the world or in thought, the problem of sign and meaning.

Hence we slip from words as symbols of things, to words as symbols of mental categories that order a perceived reality by gathering things into sets and sub-sets, which have to be brought to surface, to be circumscribed.

In the real world a discontinuity brings out unique objects. It is necessary to identify the perception and the denomination elaborated by the people of a particular culture, in this case the Palawan, through their language and oral tradition. In order to carry out properly this lexicographic and semantic undertaking, we must look for semantic features of distinctive, hence relevant values. This allows the contrasts between words to operate hence to distinguish them. Thus we conducted a systematic structural and componential analysis of the lexemes designating the geographic, mineral, celestial, animal and plant worlds.

Cognition of the phenomenal world is filtered through the five senses: sight, smell, hearing, touch, and taste.

As every language has its specific lexical, morphological, and grammatical processes with which it refers analyses and categorizes human experience, it is necessary to understand how the Palawan language organizes, through its semantic system, the world of experience and phenomena in which its speakers evolve, with which they live.

Within the extensive Austronesian family, the languages of the Philippines group is part of the Western branch. Palawan language is part of the subgroup of the Visayas. It is an agglutinous language characterized by a simple phonology (four vowels, sixteen consonants and no relevant stress), a relatively complex morphology, then the subtle topicalization process and verb concord so manifest in the Philippine languages' syntax and morpho-syntax.

Starting from a section of vocabulary limited to the designation of a part of the real world, we shall attempt to capture the morphological and semantic features uniting lexemes in a particular realm. Hence, step by step, we shall attempt to bring to surface the rules of naming as manifest in the various nomenclatures of

the natural world in Palawan. We shall illustrate two major processes at work in nouns formation, namely:
— derivation and composition processes and
— borrowing of foreign languages, but also methaphorical borrowing.

Lexicalization of the Plant World

The nomenclature of plants in Palawan reveals a peculiar lexicografical treatment distinct from the one prevailing in the nomenclature of birds and insects, as we shall see later. I shall list the inventory of the various lexical processes of the "basic plant names" and "plant types," according to H. C. Conklin's terminology (in French: "termes de base" des différents "types de plantes").

I shall not focus upon the terms related to the taxonomy and the intermediary levels of a tree structure that goes from the particular to the general, instauring sets and sub-sets between general categories, (catégories englobantes) and segregates, (catégories englobées or sous-catégories); I shall limit my presentation to the basic plant names (taxa terminaux) in a taxonomic perspective, knowing that at the superior level of the hierarchy, few covert categories, (catégories latentes) are also present, they are unlabelled but semantically present and perceived by the speakers. Hence covert categories and segregates will not be referred to in this paper. However, polysemic categories will appear.

The description of the formation of names of plant types, or the lexicographic treatment of naming plants, will drive us to dissociate grammatical categories from semantic categories. As a matter of fact, if a lexeme is "a minimal entry" of the dictionary (J. Lyons, 1977) or an "elementary lexical unit" (M. Swadesh 1946 and W. Goodenough, 1957), the meaning of a lexeme cannot be always deduced from its apparent grammatical structure as F. Lounsbury has expressed it in 1956 and H. C. Conklin has illustrated in 1962: "The relation between formal linguistic structure and semantic structure need not be isomorphic."

Today, we shall focus upon the morphological study of plant names, and this will allow us to distinguish between: "specific words," "non motivated words" and "motivated words," according to the terminology proposed by A.-G. Haudricourt (1963) for New Caledonia and the Malay world. The first are "bare root-words" and manifest the old strata of a given language, while the "motivated" terms, which look like derivates and compound words, are based on a composition morphological process and are descriptive words, probably less basic to the given culture. Eleven sub-types appearing, within the six main lexemic types (A to F), namely:

Table 10.1 Nomenclature of Plant Types

A. Lexemes of plant types non motivated and unreductible:
Type 1   Non motivated lexemes, monolexeme or bare root-word
   ex: *ubi.* "yam"

B. Lexemes of plant types non motivated and complex:
Type 2   Non motivated complex lexems: bisyllabic root-word fully reduplicated.
   ex: *ilang-ilang.*
Type 3 | Root-word with the frozen verbal infix *-in* |
   ex: *b-in-ulungan* < *bulungan.*

C. Lexemes of plant types motivated and plain:
Type 4   Diminutive Derivation: "little, tiny…"
   | Reduplication first syllale CV+ Final cons. of root-word+ Root |
   ex: *lig-lisag* < *lisag* "little hour."
Type 5   Derivation marking an analogy in the relationship:"looking alike," "related with."
   | Prefix mäg- + root word |
   ex: *mäg-mamaqan* "like areca nut"
Type 6   Derivation by suffixation: | root-word + *-an* |
   ex: *ipus-an* < *ipus* "in the shape of a tail."

D. Lexemes of plant types motivated and complex:
Type 7   Composition: | Noun + Adjective |
   ex: *imlung mäbanglu* "perfumed cucumber."
Type 8   Composition: | Noun + Noun |
   ex: *ipus amuq* "monkey's tail."
Type 9   Noun phrase: | Noun + determination particle + Noun |
   ex: *babaq ät tjäw* "mayna bird beack."

E. Covert categories lexemes
Type 10 Specific lexeme covert by a more general category

F. Lexemes based on borrowing from:
Type 11 Spanish and languages from México; English;
   Metaphorical borrowing to the animal realm, parts of the body,
   manufactured objects and geographical spaces.

There are in Palawan some ambivalent affixs. In general, the nominal system of affixs is distinct from the verbal system of affixs. However two exceptions are noticeable:

— The frozen infix *-in-* is the mark of accomplish aspect in the verbs and used here to characterize grain varieties and cuttings of root crops.

— The prefix *mäg-*, works usually as the mark of the active voice in verbal morphology. It expresses also a diadic relationship in nominal morphology It is used in the vocabulary of wild plants to designate an analogy, a similitude.

We notice a clear tendency to use monolexemes:

There are 167 wild edible plants, 150 fruits trees and wines bearing edible fruits, and seventy seven cultivated plants, hence a total of 317 edible wild plants, and 394 edible plants.

540 non motivated lexemes designate edible, technological and some medicinal plants types.

But composition is also frequent to designate varieties of vegetables and fruits, while derivation with frozen infix -*in*- to designate fifty nine varieties of grains on one hand, and eighty eight clons of tubers on the other hand is also attested. In the various realms of lexicon a sentential, or quasi sentential compound word like English "forget-me-not," is not attested.

The same types can be observed in the nomenclature of mushrooms and shells.

Table 10.2 Nomenclature of Shells

A.	*The diminutive relates the shell to a tool, a celestial body, a quality, a type of habitat:*
kay-kay	(Mesodesmatidae)
kadkad	(Psammobiidae)
pulid-pulid	(Natricidae)
pala-pala	(Muricidae)
sikad-sikad	(Strombidae)
lätag-lätag	(Turbonidae)
ulas-ulas	(Pectinidae)
umang-umang	Tiny hermit crab
nipaq-nipaq	Shell living in Nipaq fructicans
layug-layug	(Cerithridae) "the little far away"
bawing	"Tiny bawing" with a scent of acanthacae
bantal-bantal	(Olividae) "tiny pillow"
bungkäl-bugkäl	(Cerithridae)
bituwän-bituwän	(Naticidae) "litte star"
danglas-danglas	(Melampidae)
gängas-gängas	(Muricidae)
wasäy-wasäy	"Small adze" in the shape of...
bägat-bägat	(Veneridae) "small heavy"
indang-indang	"Joint together"
lambi-lambi	
ranga-ranga	(Strombidae)
sigäy-sigäy	(Cypraeidae)
saduk-saduk	(Acmeidae)
sapiq-sapiq	(Cypraeidae) "smallcow"
B.	Compound words based on borrowing parts of the body of human beings and animals by visual analogy.
duduq umawäy	(Trochidae) "Maiden's breast"
duduq bängkan	"Saw's utter"
tänäb uyäw	"Monkey's jawls"
lalak uyäw	(Conidae)
tigäm uyäw	(Spondilidae) "Monkey's fang"
kima ägdanan	(Tridachna squamosa) "Notched bivalve"
susu duyung	(Terrebridae) "duyong's snail"
susu bwaya	(Nassariidae) "crocodile's snail"
C.	Onomatopeic name
singär	Sounding shell when sucked

The principle of an analogical visual perception underlies the composition processes at work in naming the plant world. It will be overun by hearing perception in the building of the nomenclatures of birds and singing insects, but analogy is the basic principle.

We are confronted here with ideophones—making: "A picturesque and vivid way of expressing perceptibles (visual, auditory, tactile, etc.) properties of living beings and objects, mental characteristics or various types of motions" (C.Hagège 1993: 23).

Language builders' creativity exists only to a certain extent...: "Whether conscious or not, it is goal oriented. LBs make the words that fulfil definite needs, even when the formation process seems to be complicated, if looked at from the outside. They seldom build words for purely aesthetical purposes. LBs do not start from meanings, they meet their communicative needs... Through the compound words they build, LBs strive to find, for a given meaning, the most convenient expression for the speaker as well as for the hearer" (C.Hagège 1993: 173–74).

Lexicalization of Birds and Insects

In Palawan, as well as in many languages of the Philippines and South-East Asia, ideophones do have a highly productive lexemic potentiality. This mimesis process extends from birds songs to all the sounds of nature and its various beings. A very extensive set of words attracted my attention and moulds children' speaking capacity as they acquire the knowledge of their native language. Within this culture, the relationship between people and birds and the actual dialog developing among them is quite exemplary. By listening to birds, copying the sounds they produce through syllables, mimicking their rhythms and melodies, creating brief phrases and sentences, adults and children speak to the animals and initiate a tiny dialog with birds, as they are very keen of them. Again the abstract principle at work is a principle of analogy. An ideophone is governed by a relation of analogy between the signifying and the signified. It is precisely an analogy between the two components of the linguistic sign that makes up an onomatopoeia.

One should make a distinction between acoustic onomatopoeia, whose number is rather restricted (interjections, screams and clicks), and articulatory onomatopoeia, whose number is open. This relationship of analogy is not an absolute one, but it retains a certain value of reality and plausibility by locutors and listeners. According to the various phonological systems and syllabic structures of the given languages, building up lexemes shall be different. As a matter of fact the link between sound and meaning is not an absolute one, it is rather to be considered within the respective constraints of the language where it appears. As R. Jakobson as analyzed, "symbolism, although conditioned by the neuropsychological laws of synesthesia—and according to these very laws—is not the same for all" (6th lesson, 1946).

These are "sound icons" and "visual icons" according to the definition provided by C. S. Peirce of an icon. Through the perception of the phenomenal world, language builders make up lexemes. There simple and basic qualities of soundscape and perception can stimulate a reorganization of ***conduite motrice***. The mimetic behavior of the child is genuinely capturing the behaviour of the bird, it is a transfer of the vocal behavior of the bird to the child. A process of shaping this sensorial experience of the world is present here, for speaking is tightly linked to listening and hearing.

Ideophone-making

In this realm the root-word is made from an onomatopoeia, or an ideophone based on auditory perception, in two ways:

A. The root-word is totally moulded on imitation of the actual scream, or the song of the bird, according to a bisyllabic or a polysyllabic pattern, as the following musical and phonetic and syllabic transcription for each bird shows:

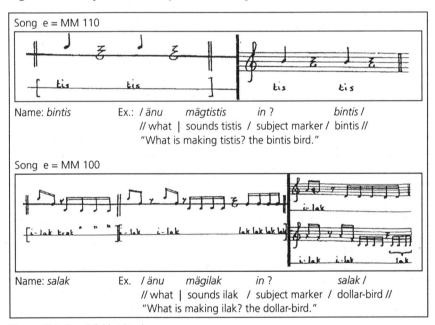

Song e = MM 110

Name: *bintis*   Ex.: / *änu*   *mägtistis*   *in* ?   *bintis* /
             // what | sounds tistis / subject marker / bintis //
             "What is making tistis? the bintis bird."

Song e = MM 100

Name: *salak*   Ex. / *änu*   *mägilak*   *in* ?   *salak* /
             // what | sounds ilak / subject marker / dollar-bird //
             "What is making ilak? the dollar-bird."

Figure 10.1 Two Syllable Ideophones

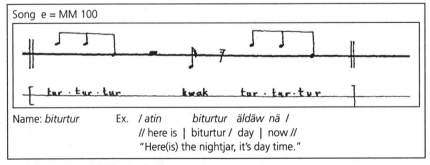

Song e = MM 100

Name: *biturtur*   Ex. / *atin*   *biturtur* *äldäw nä* /
             // here is | biturtur / day | now //
             "Here(is) the nightjar, it's day time."

Figure 10.2 Three Syllable Ideophones

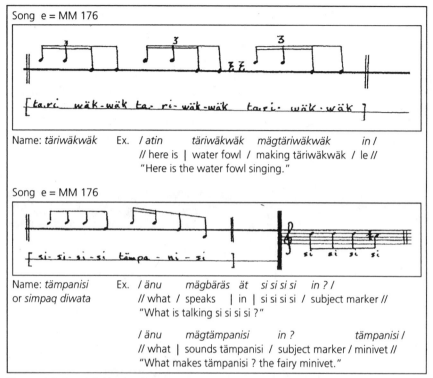

Name: *täriwäkwäk*   Ex. / *atin*   *täriwäkwäk*   *mägtäriwäkwäk*   *in* /
                         // here is | water fowl / making täriwäkwäk / le //
                         "Here is the water fowl singing."

Name: *tämpanisi*   Ex. / *änu*   *mägbäräs*   *ät*   *si si si si*   *in* ? /
or *simpaq diwata*   // what / speaks | in | si si si si / subject marker //
                           "What is talking si si si si ?"

                         / *änu*   *mägtämpanisi*   *in* ?   *tämpanisi* /
                         // what | sounds tämpanisi / subject marker / minivet //
                         "What makes tämpanisi ? the fairy minivet."

Figure 10.3 Four Syllable Ideophones

Consequently the ideophone is respectively bisyllabic or polysyllabic. Bisyllabic root-words are the most frequent: of 105 birds names, twenty two are bisyllabic; twenty four are trisyllabic; eight are quadrisyllabic, as the following table shows:

Table 10.3 Nomenclature of Singing Birds

Two-syllable ideophones	Three-syllable ideophones	Four-syllable ideophones
tibyäw	*täläläw*	*tampaliklik*
tigtig	*tärtägar*	*tämpanisi*
tuking	*tälili*	*täriwäkwäk*
wahwah	*kuryasäw (kurisäw)*	*rumaruku*
kwahaw	*bäkbäräk*	
widwid		*tämpayuhyuh*
gukguk	*karuwang*	*bäransisiq*
täkräw	*kulingsyan*	*wirik-wirik*
tuklug	*säritsit*	*bäräk-bäräk*
	tärikwäd	
kälit		
butbut		
tanguk		
puwäk		

B. A monosyllabic ideophone root-word is kept and incorporated as a final syllable in its plain or reduplicated form. While in A. it is made of the whole of the root-word, here in B. it is made of only part of the root-word.

In Palawan most nominals are bisyllabic or polysyllabic, although there are some exceptions. Monosyllabic forms are used for personal pronouns, articles, markers and other particles. The language seems to intent to pressure the syllabic structure and makes a polysyllabic root-word out of a monosyllabic song. On 16 birds names there are 8 bisyllables and 8 trisyllables. Palawan expresses the act of singing by a verb specific for each bird. Then, signified and signifier are signs and there is no way to dissociate them.

Table 10.4 Nomenclature of Singing Birds

Two-syllable ideophones		Three-syllable ideophones		
		without reduplication	with reduplication	
bintis	< [tis-tis]	*sasawih* < [wih-wih]	*bälilit*	< [lit-lit]
salak	< [lak-lak]	*sungkäling* < [kling-kling]	*täringting*	< [tring-tring]
päras	< [ras-ras]		*bikurkur*	< [kur-kur]
tämsik	< [sik-sik]		*biturtur*	< [tur-tur]
pikpyak	< [pyak-pyk]		*sungsuring*	< [suring-suring]
pätjyag	< [tyag-tyag]		*käkasa*	< [kasa-kasa]
tuklis	< [lis-lis]			
mäntud	< [tud-tud]			

Table 5 Nomenclature of Stridulant Insects

Extension to stridulant insects		
ritrigät	< *rigät* (onomatopée)	"rigat"
rikrik	< *rik-rik-rik*	"rikrik"
lilyah	< *liyah-liyah-liyah*	"cicada"
gigyah	< *gigjyah-gigyah-gigyah*	"cicada"
gigi	< *gi-gi*	
sisih	< *sih-sih-sih-sih*	
andurayan	< *andura-yan*	
yäyagaq	< *yaga-yaga-yaga*	
kutkulit	< *kulit-kulit-kulit*	
mänggaras	< *mänggaras-mänggaras*	
ämantik	< *pitik-pitik*	

In addition to the 105 bird names in Palawan, and the transcription of fifty distinct human voice imitations, in the context of the Highland, and the eleven sounds of stridulant insects, I have further identified fifteen ideophones for wind sounds, sixteen for the various kind of rains; twenty eight for the sounds of human steps, beats and tremors on the ground; twenty eight lexemes related to sounds of works in the upland field, from trees slashing to rice harvesting; twenty two lexemes expressing the sound of falling bodies and objects; twenty seven expressing the sounds of people presence at home; eighteen for sounding activities at home; twenty two for the sounds of musical instruments… The semantic nuances in relation to the experience of the "Sensible" and the sound associations upon which they rely, are very subtle. However the phonic constituents set into motion are certainly not new, nor marginal, they are in total adequacy with the Palawan phonological system. Hence expressiveness is not free, it is filtered by a precise framework (N. Revel, 1992: 41–66).

Compound Words for Non Singing Birds and Non Stridulant Insects

As there are in nature non singing birds and non stridulent insects, ideophone-making in these cases turns to visual perception, as we observed in the naming of plants, mushrooms and shells.

The root-word is made up of a morphological feature of the bird, or a habit of life. Three formation processes appear and characterize these motivated lexemes based on visual perception.

A. By metonymy, a word suggestive of the shape of the bird is borrowed from the general lexicon and comes to signify the very bird.

ex: *tulabung*: a) "long curved neck"; > "cattle egret."

A word can also be borrowed and slightly altered, but very suggestive of the bird's manner stepping on the ground:

ex: *tustwis*: a) " rump"; > a bird moving its rump as it steps: "wagtail."

B. By derivation, either reduplication of the whole root, or by reduplication of the first syllable, following the derivation pattern for diminutive, a bird is named by a distinctive feature: for instance, the colors of its feathers;

ex: *kutkunit* < *kunit* "tumeric":"tiny tumeric" (a yellow flycatcher). (Timaliidae).
ägsäm-ägsäm < *ägsäm* "bitter, astringent":"the tiny bitter" (a kind of wasp).

This naming process is also very productive for insect nomenclature:

ex: *lilibu* < *libu* "making circles":"the tiny turning one." Hydrophilidae.
yangyagäw < *yangäw* "the small gripping walker." Nepidae.

C. The bird or the insect name is sometimes a compound word. Although less frequent two types come up: An isomorph compound-word made of the sequence Nominal+ Nominal and a relationship of determination:

bägit-Buntäl   "bird of the Globe-fish constellation" (the Southern Cross).
bägit-Nyug   "bird of the Coconut constellation" (Scorpio).
A determinative construction underlies each one of these sequence, with the succession:

Determinated + Determinating.

This naming process is particularly relevant to designate various migrating bird, which function in Palawan culture as almanach, they signals times in the year cycle.

We have focused our attention upon the relationship between perception and action and the correlated behavior in language building (N. Revel, 1997 pp. 18–28).

Soundscape has certainly an incedence upon sensitivity and the process of lexemic building is transposed at the level of musical creation too. The Palawan Highlanders do have an elaborated pentatonic anhemitonic musical scale named *läpläp bägit*, "bird touch" or "bird scale," to imitate the sounds of Nature and of People. In music creation this process becomes less iconic and more transcendental.

References

Barrau, J. 1967, *Introduction à l'enquête ethnohistorique dans le monde melanésien*, Ploycopié du Museum d'Historie Naturelle, Paris.

Benveniste, E. 1996, *Problèmes de linguistique générale*, Vol. I, Ch. 4. Nature du signe linguistique. Gallimard (NRF), Paris.

Conklin, H. C. 1962, Lexicographical treatment of folk taxonomies, *International Journal of American Linguistics* 28 (2 IV),: 119–41.

Goodenough, W. H. 1957, *Cultural Anthropology and Linguistics*, Monograph Series on Language and Linguistics, Washington.

Hagège, C. 1985. *L'homme de paroles. Contribution linguistique aux sciences humaines*. Ch. 5, Le territoire du signe. Fayard, Paris.

———— 1993, *The Language Builder. An essay on the human signature in linguistic morphogenesis*, Chap. 6 Lexicalization. J. Benjamins Publishing Co., Amsterdam / Philadelphia.

Haudricourt A.-G. 1963. Vernacular Plant Names in Melanesia. Some examples from Northern New Caledonia. In *Plants and the Migration of Pacific People. A Symposium*, edited by J. Barrau. Honolulu: Bishop Museum Press.

Jakobson, R. 1976. *Six leçons sur le son et le sens* (6ème leçon, 1942–43). Les Éditions de Minuit, Paris.

Lounsbury, F. 1956. Semantic Analysis of the Pawnee Kinship Usage. *Language* 32: 158–94.

Lyons, J. 1977. *Semantics*, Cambridge University Press, 2 vol.

Merleau-Ponty, M. 1945. *Phénoménologie de la perception*, Part II: Le monde perçu,. Ch.1: Le sentir, Gallimard, Paris.

———— 1997. *Parcours 1935–1951* Ch.XVIII: Les relations avec autrui chez l'enfant, Verdier, Paris.

Revel-Macdonald, N. 1979. *Le Palawan. Phonologie. Catégories. Morphologie*. SELAF, Paris. (ASEMI 4), 280 p., 2 cartes, 10 ph.

———— 1990. *Fleurs de Paroles. Histoire naturelle palawan* Vol. I: *Les dons de Nägsalad*, Peeters-SELAF, Paris (Coll. "Ethnosciences" 5), 374 p. 80 fig., 16 pl. ph.

———— 1991. *Fleurs de Paroles. Histoire naturelle palawan*, Vol. II: *La maîtrise d'un savoir et l'art d'une relation*, Peeters-SELAF, Paris (Coll. "Ethnosciences" 6), 322 p., 50 fig., 30 pl. ph.

———— 1992. *Fleurs de Paroles. Histoire naturelle palawan*, Vol. III: *Chants d'amour/ chants d'oiseaux*, Peeters-SELAF, Paris (Coll. "Ethnosciences" 7), 208 p., fig., 16 ph.

———— 1998. "Morphogénèse des langues: les idéophones," workshop "Embodied cognition or situated cognition" organized by M.-J. Fernandez-Vest. CD-Rom des conférences, *Congrès International des Linguistes*, Paris 25–30 juillet 1997, B. Caron éd. 1999

———— N. and Maceda J. 1991. *Philippines.Musique des Hautes-Terres palawan. Palawan Highland Music*, CD, Harmonia Mundi (Coll. CNRS/Musée de l'Homme, Paris), [1ère éd. 1987, disque 33 t.].

Sapir, E. 1949, *Selected Writings of Edward Sapir in Language Culture and Personality*, University Press of California.

Zimmermann, F. 1994. "Chants d'oiseaux palawan," *L'Homme* 132, XXXIV/4, oct.–déc.,:151–58.

CHAPTER 11

Levels and Mechanisms of Naming

John B. Trumper

The Pertinent Linguistic Mechanisms Whereby Nature is 'Named': Which is the Pertinent Level at Which Naming Processes Begin? A Discussion of Plant and Seed, Pip and Stone

One of the problems in a discussion of naming is to decide which is the pertinent level at which naming begins as a linguistic process, whether at the taxonomic and perceptual generic, intermediate or life-form level, and whether from a particular level lexical spreading is then directionally bottom-up or top-down as a productive process. Berlin and his collaborators have always insisted on the pertinence of the generic at the relevant taxonomic, linguistic, cognitive and psychological levels. Although initially the basic opposition between a monotypic generic and polytypic higher and lower levels was emphasized, Berlin (1992: 118) has since declared that the mononomial representation of generic or higher groups vis-à-vis the binomialization of the subgeneric is a *tendency* and not an absolute rule, while he also underscores (1992: 140) the mid or intermediate level between generic and life-form as a supra-generic grouping creating in some definite instances "a chain of polytypicality over … ethnobiological ranks." Atran 1996[3], in his discussion of the search for a "basic level" of organization, with a precise reference to Owen 1866, links the basic-level taxon concept, in this case the species, to the lexical level in an explicit manner, (26) "The various exemplars thus become token 'namesakes' of the specific type." He equates this level first with Aristotle's ἄτομον εἶδος (*Historia Animalium*), then with the generic,[1] arguing in a fuller manner that this is the pertinent level in all approaches, whether taxonomic, linguistic, cognitive, or psychological, see, e.g., Atran 1996[3]: 27 "This historically and conceptually primary stage of scientific classification, then, would appear to correspond to the basic operation in folk biological classification the world over. According to Bartlett (1940: 351), the concept of the genus is the most natural and useful level of classification, inasmuch as it is 'the smallest group that everyone might be expected to have the name for in his vocabulary.'" Berlin (1972: 55) notes that basic-level folk taxa fit this requirement because they are, in fact, "the first to be encoded in the ethnobiological lexicons of all languages. The same appears to be true of folkzoological taxonomies." The basic-level is thus firmly latched on to the lexical-semantic contrastive level, though whether it be at a possible generic or specific level may be a less vital matter.[2] Although everybody perceives "gaps" between groups of organisms, an effective distinction between generic and specific is often unimportant because (1) many natural genera may well be monospecific, without other known subgenerics, (2) at a local level the species observed in the native surrounds may well be considered locally a genus, since extant related subgenerics are not known in that particular environment (Atran 1996[3]: 29–30). Atran, as a result, develops a particular generic-specieme level. This, however, is still formally "taxonomic" and "hierarchical," so that one does not meet Hamill's 1979 particular criticism that, though the "generic" level in a hierarchical chain of the type unique beginner (covert, not necessarily labelled)-life-form (often not labelled)-generic-specific ("labelled by secondary lexemes" (Hamill 1979: 147)-varietal (polynomial) seems to give lexical structures that are "the building blocks of folk classification systems." There is overinsistence, without adequate formal proof, that the lexicon is organized taxonomically in this way (Hamill 1979: 152"...we seem to insist that lexical structures can only be of the taxonomic or paradigmatic variety"), and that one needs, instead, to look for "more powerful formal devices."

Two questions are raised; the first regards the history of classification and possible repercussions at the naming level; the second involves the linguistic model as such. Aristotle, for example, seems to be saying, at the beginning of the *Historia Animalium* (487a), that important differences occur at a typological and generic level, but slightly later he is already stating that differences are established in accordance with species and grade (*Historia Animalium* 488b), and one can compare the relevant quotations adduced by Atran. The *differentiæ* used at the level of similarity/ dissimilarity between parts and organs establish *genera*, while the different ways in which "common parts" actually feed create the final *species* of that progression according to which individuals are recognized. *Differentiæ*, however, for Aristotle are not strictly hierarchical, binary options that define, but sums of characteristics that do establish classes. At the same time, *genera* may be *species* with respect to higher levels, so that the actual use of "species" may seem ambiguous, as in some of the examples quoted. It is interesting, for example, that Theophrastus (*Historia Plantarum* 1.III.1) says that his basic-level plant classification in δένδρον ("tree") vs. θάμνος ("bush") vs. φρυγανόν (shrub or smallish "woody" plant) vs. πόα (a "non-woody" grass-like plant)[3] is established κατ᾽εἴδη, apparently "according to species," but here the term seems to represent "species" with respect to some unique beginner. To clarify this school of thought one needs to take up Aristotle's point in his *De Anima* that genus is not "determinate," in that it does not specify "individuals," who are identified as such by form and "species," since genus is "potentiality," species the realization of this classificatory potential (*De Anima*, 412 a, 6–10). Specific categories are thus established cognitively (ὡς ἐπιστήμη) and visibly (distinctions visible to the naked eye: ὡς τὸ θεωρεῖν), and have lexical reflexes. Obviously, it must be noted that Theophrastus brings in the concept of *similarity within the genus* based on a twofold criterion, the first using appearance (*sameness judgements operated by the naked eye*) but also utilitarian (*Historia Plantarum* VII. VII. 1–2), whereas in Aristotle *the species level* is important for *establishing dissimilarities* and individuals with respect to other individuals (the determinate, τόδε τι, in which the genus' potential is realized).

At this point both genus (abstract) and species (concrete) have their specific content, both have their relative importance in the model, but we may ask whether dissimilarity or similarity be the key point and what status the *differentiæ* (διαφοραί) have. In classical authors these last are not in their own right hierarchical in a strictly binary fashion, but are binary and may be sums of apparently binary features (e. g., X = living + feeding + growing = plant/ animal/ human, X + Y (= voluntary movement + particular types of sentience) = animal, X +Y + rationality + moral sense = human animal etc.). They are also of two types, i.e. definitive and characteristic. This is the traditional theoretical position up to S. Thomas Aquinas: (1) likeness belongs to the genus and does not constitute identity, which is established at the specific level (*Summa Theologiæ* Q. 74 Sol. 2), (2) the "characteristic" seems to belong to the genus and also holds "within" the species for varietals, the "defining" to the specific,[4] so that "corruptio" (the process of being ruined, of going off, as in Cæsar's *hordeum corruptum* "barley that has gone off," also used of a sick limb, an unhealthy thought etc.) changes the species but not the genus, as in S. Thomas' example of the unnamed genus (covert ?) that contains the species *vinum* and *acetum* (= *vinum corruptum*). *Vinum* contains the subspecifics *agresta* (= *vinum in via generationis*) and *mustum* (= *vinum impurum*). It is important that "in necessitate" (= from the standpoint of definition) *agresta* and *mustum* can substitute *vinum* as belonging to the same species, *acetum* can not substitute it, even though *vinum* and *acetum* belong to the same genus, they are "alike." Evidently [± *corruptio*] is a defining feature, [± *in via generationis*, ± *impuritas*] are "characteristic" *differentiæ*. A similar line of argument is construed to define the relationship between *frumentum, hordeum, triticum, far, spelta* and *siligo*. However, concentrating on a hierarchical tree-structure whose branches are determined in some way by binary features would prevent us from extracting from classical Greek (Theophrastus) and Latin (Seneca, Pliny) basic plant distinctions, in fact any generalisations at all. Concentrating on the generic vis-à-vis the specific, mid (intermediate) and life-form levels, would inhibit any serious comparative semantics or investigation of the "building blocks" of lexical creation and spreading. The two classical "hierarchies" are given in figure 1. Although we know that πόα may be characterized as [- woody], Theophrastus leaves us with the basic four-term opposition κατ᾽ εἴδη, whereas we know from Latin authors that the category *sata* which is "reproductive part" + "product" for plants, animals and humans, is further subdivided into the binomial *sata et arbusta*, the second term a collective that includes groupings identifiable as *arbor* and *frutex*. More is not given us to know. In the Greek scheme φυτόν "plant" (- φύω "to grow; to exist" < IEW 146 *bheu-, NIE *BAW-) is **unrelated** to σπορά σπόρος/ σπέρμα "seed" (< σπείρω "to sow" < IEW 993. 2 *sper-, NIE *SP'ER-): they must be related through separate Cognitive Models,[5] whereas the Latin contrasts belong to a single Cognitive Model as in figure 2. The difference between these two related Indo-European cultures and languages is a property of the Model and the network of lexical-semantic relations established within each separate language.

   More than a thousand years later Albertus Magnus in his *De Vegetabilibus* reorganized the Latin scheme according to universals, introducing "per diuisionem corporum plantarum" (I.110: "by a separation involving plant forms") two hierarchical contrasts, one explicitly formulated in a binary fashion as [±ligneitas] = [±

woodiness], the other opposing the possession of a "stipes fortis" or "caudex" to a vine-structure ("per modum palmitum").[6] The oppositions might be formalized as *planta* ⊂ X [+woody] + *herba* [-woody], X ⊂ *arbor* [+large, +trunk] + *arbustum* = *ambragyon* [-large, +vine-structure], where we now have a covert category at the mid-level and where the relationship between Lat. *arbustum* and *arbor* is now inverted with respect to Seneca and Pliny. The resulting scheme is nearer the Greek separation than the original, complex Latin Model. Three hundred years later Andrea Cesalpino from Arezzo, in his *De Plantis*, modified this set-up into a first binary opposition [+woody] vs. [-woody] with two covert categories. These are then divided into [+woody] [± trunk-structure (= caudex)], [-woody] [± short-living]. We might formalize as planta ⊂ X [+woody] + Y [-woody], X ⊂ arbor [+trunk-structure] + frutex [- trunk-structure], Y ⊂ suffrutex [- short-living] + herba [+ short-living].[7] Cesalpino divided linguistically derived structures into separate branches of expansion (frutex, suffrutex) and expanded his life-form into two covert categories rather than one as in Albert the Great. It is interesting to note that Latin semantic expansion and inversion neither begins nor ends with a "large plant" vs. "small plant" opposition, contrary to the universal hypothesized in Brown 1979.[8]

One of the basic problems of the original Latin scheme (figures 11.1, 11.2) is that "plant" seems to be a covert category, and such it is for many classical writers: in fact *sata*, formally the neuter plural of the past participle of *serere* "to sow," apart from its poetical meaning of "progeny," "son" / "daughter," may mean "cultivated fields" or "fields sown with cultivated crops." This is true, say, in Ovid's *Tristium* IV.I.79–80, but more often than not it means "crops," as in Varro's *De Re Rustica* I.IX ("...ut radices **satorum** comburat": so as to burn the roots of **crops**), I.XXXI ("sata admixta": mixed crops), Virgil's *Georgics* I.443–44 or IV.330–32, Pseudo-Ovid's *Nux* verse 61, Seneca's *De Beneficiis* VII.XXXI.3, in Germanicus, *Fragment* IV.8–9, and even Pliny in his *Naturalis Historia* (for example in XVII.III.36, XVIII.XLIV.154, XVIII.LXIX.281) or Seneca (*Naturales Quæstiones* V.18.13 "ad alendos **satorum** atque arborum fructus" = LOEB: to nurture the **crops** of the fields and the trees). Sometimes there is ambiguity between the two senses (e.g. Cæsar's *De Bello Civili* III.XLIV). However, the general meaning of "plants," "vegetation," as well as "seedlings," is attested from a fairly early date, e.g. in Varro's *De Re Rustica* I.II ("Eæ enim omnia nouella **sata** carpendo corrumpunt, non minimum uites atque oleas"),[9] I.VII, Virgil's *Georgics* I.443–44 ("...namque urguet ab alto/ Arboribusque **satisque** Notus pecorique sinister," where the sense may well be "foe to trees and **plants** and animals"), II.349–350 ("Inter enim labentur aquæ, tenuisque subibit/ halitus, atque animos tollent **sata**..."),[10] Ovid's *Fasti* I.351–52 etc. This meaning seems best attested in the writings of Seneca (*Epistolæ Morales* VII.LXVI.XI, XX.CXXIV.XVIII, *Naturales Quæstiones* III.XXIX.2) and Pliny (*Naturalis Historia* XI. VIII, XVII. II. 18, XVII.XVIII. 90–1, XIX. XXXV. 117, XIX. LVII. 176, and other well-known passages), though it is not absent from authors such as Quintilian and Columella[11] or even minor authors such as Germanicus (*Aratea* 336–37) or Cornelius Fronto (*De Feriis Alsiensibus* II). In later Christian texts the "plants" / "vegetation" general meaning becomes considerably rarer, e.g. in Lactantius' *De Ira Dei* 13.7.38

("...ut **sata** imbribus inrigentur, ut uitis fetibus, arbusta pomis exuberent..."),[12] Hilary's *De Trinitate* 9.4 line 9 ("Cerne arbores, **sata**, pecudes" [Choose well your trees, your **plants**, your animals]), 12.53 line 24 ("omnia sata" [all vegetation]), Marius Victorinus' *Ars Grammatica* 3. 109 ("... omnia **sata** arboresque et herbæ parturiant..." [that all **plants**, both trees and grasses, bring forth fruit...]), while the absolute majority of authors, from Jerome and Ambrose, through Augustine, Paulinus of Nola and Sulpicius Severus, up to Cassiodorus,[13] Isidore, even to Bede in the sixth—seventh centuries.,[14] use *sata* with the restricted sense of "crops"/ "cultivated plants," even "cultivated fields." Ambiguity also seems to be present in Gregory the Great's use, cf. *Dialogues* III.9 line 6 "quæque sata ac plantata," III.9 line 19 "sata uel plantata," where the opposition appears to be "plants" (generic) vs. "shoots" (partonymic), though it may be "crops" (sown) vs. "trees" (grown from shoots).

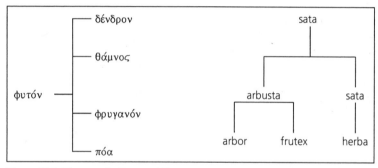

Figure 11.1 Diagrams for Trees and Plants

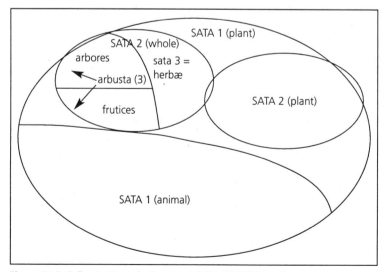

Figure 11.2 Collective Reproductive Agent: SATA (IEW 889.2 se-)

From early literature on *planta*, instead, and its neuter derivative *plantare* pl. *plantaria*, mean either "sucker" or "slip"/"shoot"/"young plant or shoot used for grafting," while the collective *plantarium* corresponds to *puluinus* "nursery" (where slips and shoots are kept for grafting). The first we find in Virgil's *Georgics* II.23–24, II.65, II.79–80, in Pliny's *Naturalis Historia* XIX.41 ("...cuius **planta** extremo uere plantatur...," and the **seedling** is bedded out at the end of spring, LOEB), XIX.XLIII.153 ("[carduus] autumno **planta**...," from a **slip** planted in autumn, LOEB), so in *Naturalis Historia*. XIX.XLVII ("**planta** uel, si nondum germinat, spongea," using a **shoot**, or if it is not yet making bud, a matted tuft, LOEB), XIX.LIV.170 ("melius tamen **planta** tralata," though it [sc. the **shoot**] is improved by being transplanted, LOEB), in Columella's *De Re Rustica* X.147–148, X.153, etc. This seems to be true up to Isidore shortly after the sixth century, cf. *Etymologiæ* XVII.6.12 "**Plantæ** sunt de arboribus; **plantaria** uero, quæ ex seminibus nata sunt cum radicibus et a terra propria transferuntur."[15] The same reference to a plant part, usually the young slip or shoot ready for grafting or planting, is present in *plantare*, usually attested in the plural, as in Virgil's *Georgics* II.26–27, in Calpurnius Siculus' *Bucolica* II ("...irriguo perfunditur area fronte/ Et satiatur aquâ, sucos ne forte priores/ Languida nutata quærant **plantaria** terrâ." [the plot is bathed, its brow all wet, and sated with water, lest the languid, hesitant **shoots** seek former moisture deep in the earth]), in Pliny's *Naturalis Historia* XVI.LX ("...uulgoque dotem filiæ antiqui **plantaria** ea appellabant" [the ancients used to call these slips "daughter's dowery"]), XVII.XXXIV.149–150, XXI.XXXIV (à propos of Abrotanus: "**plantaria** transferuntur" [**slips** are transplanted]), though more often than not we are dealing with the plural of *plantarium* as above, cf. *Naturalis Historia* XIII.VIII.37 ("Ergo **plantaria** instituunt anniculasque transferunt et iterum bimas" [So they set up **nurseries** and transplanted annuals and biennials]), XVII.XII.65. In later Christian authors such as Rufinus, Ambrose, Augustine, Jerome, Ennodius, Leo the Great etc., the meaning of *planta/ plantæ* and *plantaria* is still restricted to this particular set ("new shoot"),[16] though more often than not the "new life" content dominates in the metaphorical extension usually present in homilies. This dominance, together with the new restriction added to *sata*, which confines it to its original meaning of "crops," with new metonymic possibilities (*pars pro toto*) that produce lexicalisation of the yet again covert category "plant," obviously takes *planta* from "shoot"/"slip" to "plant" in an undefined period between 800 and 1200 AD. It might, however, be remembered that in Ambrose *plantaria* are not only the shoots and slips of woody plants or vine-structures but the term is applied to the "shoots" or "seedlings" of greens or herbs that do not enter into the classical scheme, as in Ambrose's *Exhortatio Virginitatis* 5.29 line 7 "...uelut olerum **plantaria** sunt, in quibus frequens gelu est; et ideo sicut holera herbarum cito cadunt atque marcescunt,"[17] so that evidently we have here a first generalisation of the term. One might even say that later developments of planta "slip," "shoot,""young plant," even "seedling," are adumbrated in some relatively ambiguous passages of Pliny's *Naturalis Historia*, as in XVII.XII.66 ("discedentes virus relincunt dum radici auellitur **planta**..."),[18] XVII.XXVIII.124 ("...ut inlabito cortice atque ut sectura inferior ponatur semper et quod fuerit ab radice, adcumuleturque germinatio terra donec robur **planta** capiat...”),[19] or XIX.XL.133. The translator evidently senses a certain amount of ambiguity, even if one feels that the balance is weighted in favour of "shoot" or "slip." The later history of the lexeme and its position in the general "plant" Metonymic Model, however, is rather obscure until about 1100–1200 when **planta** began unambiguously to mean, in all late medieval authors, the life-form "plant."

What seems to be happening in Latin is originally a bottom-up movement, involving the naming of a life-form built from lexical substance existing at the intermediate level, i.e. *quercus* (generic) ⊂ *arbor* (intermediate) ⊂ *arbusta* (life-form: a collective neuter pl., derived originally from *arbor*) ⊂ *sata* (unique-beginner),[20] whereas in medieval Latin we have the converse, top- down expansion in which the life-form becomes a new intermediate, aligned with *arbor*, leaving a covert life-form: *quercus* (generic) ⊂ *arbor* vs. *arbustum* (intermediate) ⊂ X (life-form) ⊂ *planta* (unique-beginner). Thus we note that partonymic *planta* (originally tree-cutting, vine-cutting, slip, shoot) has taken the place of partonymic *sata*. We have movement between levels but naming processes are not tied to the generic level, indeed here they seem to involve the intermediate. Perhaps it is not so much a question of what level is pertinent in naming, but the fact that we have a network of relations in continuous movement within a Model. The movement may begin at any given level, the only condition being that it must be above the specific.

With M. Maddalon we have been working on a model for "seed" which may be represented as in figure 3.[21] What is specified [+fruit seed], [-woody pith] may occasionally offer the further size opposition [± small]. The model has been tested with numerous speakers from two Romance groups (NeoVenetian; North Calabrian) and the resulting configurations seem relatively similar, as shown in figure 11.4. Within "seed" NeoVenetian distinguishes the kernel (*lóla* for cereals, sometimes all Graminaceæ, otherwise *mególa*) from the husk (*bula* or *spigaróla* for Graminaceæ, otherwise *sgussa*), as does North Calabrian (kernel: *civu*; husk: *jusca* for Graminaceæ, otherwise *corchja, còrchjula* or *scorza*). Lexical creation and movement appear to begin at the unique-beginner stage, in some cases, with top-down movement partonymic > unique-beginner > life-form > generic (derivatives of Latin. sēmen < IEW *sē- 889.2), at others with a bottom-up movement partonymic> generic (NeoVenetian *mególa*) > life-form (Calabrian *civu* < *cibus* «food"). In the first case the movement is from "sowing," a typically agricultural concept, in the second, from "seed" = "food," which would seem still characteristic of a plant- or seed-gathering society, a strange departure point for any Indo-European group, let alone a Romance one, and requires reflection. Germanic and Celtic schemes, still only partially tested, seem to show lexical spreading from the highest level down: e. g. English has unique-beginner [*seed*] ⊂ life-form [*seed*] ⊂ generics:(cereal [*grain*] + other [*seed*]) + (woody-pith [*stone*] + non woody-pith [*pip*])."Pip" vs."stone" may also involve a largeness contrast. Welsh, on the other hand, would seem to have unique-beginner [*had*] ⊂ life-forms [*had*, singulative *hedyn*] + [*carreg*, which can be glossed *hedyn ffrwyth*], life-form [*had*] ⊂ generics (cereal [*grawn*] + other [*hedyn*]), life-form *carreg* ⊂ generics (woody-pith [*carreg*] + non-woody pith [*graean*]), i.e. both top-down (*had* > *had, hedyn*) and bottom-up (*carreg* > *carreg*) developments. Albanian[22] seems to have simplified in a spreading unique-beginner [*farë*] ⊂ life-form [*farë*] ⊂ generics (woody-pith [*bërthamë*] + non-woody-pith [*farë*]), where historically ⊂ = >. Interestingly, the kernel of a fruit-stone used for planting is also called *farë*, so that *bërthamë*[23] is both contained in and contains *farë*. In conclusion, lexical productivity and semantic spread may be bi- or multidirectional and the starting point for naming processes may be any of the levels proposed in the classical works of Berlin and collaborators (e.g. Berlin 1992), providing it be above the specific. Naming processes connected with cognitive, taxonomic and psychological categories begin with the most inclusive member of Metonymic Models like those of figures 11.2 and 11.3, though the problem remains open!

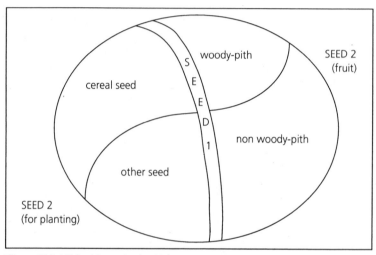

Figure 11.3 Minimal Reproduction Unit

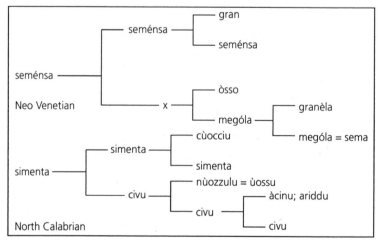

Figure 11.4 The Resulting Configurations Found after the Model for "Seed" was Tested with Numerous Speakers from Romance Groups

Part II. Onomatopœa as a Cultural Phenomenon

There is no intention to go back to the Greek discussion on language development κατὰ φύσιν, *ergo* onomatopœic and spontaneous, vs. that which is κατὰ φύσεως, conventional and a social product, since I believe the question not to be an aut-aut but an et-et. Nor should we attempt a serious discussion of P. Lieberman's 1975 model of the physical articulatory possibilities of *Australopithecus africanus* vis-à-vis *Homo sapiens neanderthalensis* vis-à-vis *Homo sapiens sapiens* "newborn" and "adult" to gain maximum acustic and perceptive effect, which might give us a limited number of onomatopœic bases. From these Indo-European and other language families, perhaps even Nostratic, would have built other bases. Nor do I wish to touch on the problem of Nostratic as the sum of those basic elements that Indo-European, Afro-Asiatic, Sino-Tibetan and yet other language groups might or might not have in common.[24] If we have to start from somewhere it must be from Whitney's 1867 dismissal of a three-pronged alternative that moves towards the universal, i.e. bow-wow vs. pooh pooh vs. ding dong (lecture XI). Obviously the third must be dismissed as "idealistic-positivistic," while the other two are not real alternatives but coexist and are short-term solutions on the road towards "…attempts to find an intelligible sign for a conception which the mind has formed and desires to communicate" (Whitney 1867: 428). As far as animals and birds are concerned, there is "an evident disposition to give an imitative complexion to words which denote matters cognizable by the ear" (429), but such an imitative principle must perforce be shortlived, since it "…on the one hand, has its natural limits, and, on the other hand, would soon begin to admit the concurrence of a new principle of word-making: namely the differentiation and various adaptations of the signs already established in use." (434). If we admit there ever was an imitative stage in human proto-linguistic development, then it must be a short-term phase in evolution: "…the onomapoetic stage was only a stepping-stone to something higher and better." (Whitney 1867: 433).

The ever-present discussion of the first chapters of Genesis these days pinpoints attention not on the perhaps incorrect idea of the "perfect" tongue, lost, that humanity must strive to recover, but on two basic problems: (1) when in Genesis (2, 19 ff.) God brings animals etc. before man to be named, does Jerome's translation "nominibus suis" entail that other-than-man has a "natural" name or a name given by man's later socially-imposed usage here sanctioned,[25] (2) when S. Augustine discusses the interpretation of Genesis is he correct to think in terms of a pre-linguistic basic language present in thought and in the newborn child's mind which deal directly with the "things"[26] that are later given linguistic signs rather than any "perfect" proto-language? As Eco (1996: 22) says "Ma sant'Agostino non manifesta alcuna nostalgia per una lingua verbale che qualcuno possa o debba di nuovo parlare." [But S. Augustine shows no nostalgia for a verbal language that someone may or again may have to speak]. This last point was dealt with in a more circumstantiated fashion in Hagège (1978: 470) where in a certain sense the passage from a unique language to a plurality of languages is considered as a first approach to evolution: "… cette unité de la langue n'était autre que l'absence de toute langue" [This language unity was none other than the absence of any language]. In other words, evolution takes humanity from pre-linguistic signs (pre-Babel unity) to organized languages (post-Babel disunity), and this process is enshrined in a mythical, religious framework in Genesis. Pre-linguistic signs must in some way be mental"imagines mundi,"which would explain the greater preoccupation with drawings and representations of reality from this period on, since they are pre-linguistic and nearer to both the human fœtus' and a newborn child's possibilities than other systems, but this is not our direct concern here. Prelanguage, with particular acoustic manifestations, would, however, only give us a series of howls, cries and screams in common with other animals, whereas an Australopithecus'oral cavity, like that of a newborn child, would give us maximum consonant closure [b] and aperture [g], maximum vowel closure [z] and aperture [ʌ], with an almost impossibility to combine oral closure with lip rounding, at the most with enormous difficulty owing to the type of jaw and mouth protruberances, to produce a kind of [ʌ] or very closed [o]. These possibilities would be connected with an intrinsic hyponasality, i.e. would produce tinny vowel sounds and an absence of nasal sonorants. We would have, then, optimal structures of the type maximum closure [bé], maximum aperture [gʌ], intermediate, non-optimal [bʌ], [gé].

As far as onomatopœa is concerned, I would underline that, as Whitney says, such a stage in human evolution is short-lived, once humanoids move from the prelinguistic state (from the raw possibilities of *Australopithecus* or *Homo neanderthalensis* or *Homo habilis*). One might add that this short-lived stage is not a socially or anthropologically significant state, but determined by what humans biologically have in common: it has absolutely nothing to say about the monogenesis of language.[27] Much seems to depend, as well, on language typology. An agglutinative or isolative language may—because of its very morphology and biunivocal relationship between morph and abstract morpheme, in the second case even morph = morpheme = word—keep associations with onomatopœic bases longer, while flexional languages, more especially if stress-timed, may destroy initial onomatopœic symmetries in their normal development. The wi-wi- or iw-iw repetition that gave Indo-European squirrel names, starting from Gr. σκίουρος and Lat. uiuerra, is partially kept in Celtic (Welsh gwiwer), partially destroyed (Irish feoróg /'foro:g/ with two back vowels,< *we-wer-āko-, with adjectival suffixation and the suffix that later takes stress, the erosion of the initial syllable, the development e > eo > jo in certain circumstances, loss of palatalization in the labials etc.). In Romance it is destroyed by the addition of suffixes, e.g. NeoVenetian. schirát[o] < *sciur+attus, like Italian scoiattolo, sometimes kept despite lexical decay (Lombard nisciurín), sometimes destroyed by lexical substitution that chooses a different onomatopœa, unlike the Lombard example (Lucanian, Calabrian zzaccanella, zzaccaredda). Sometimes borrowings change type, cf. Italo-Albanian a-a in cakarele < Romance [Calabro-Lucanian] vs. Albanian original e-e in ketër (< i-i ?). There is more motivation to keep to such schemes in the case of birds and a few particular animals, for obvious reasons. For Italy Arrigoni degli Oddi (1929) gives a wide panorama of relevant bird calls for "warblers" and the like, which are extendible to the same European types found in other countries. The ones that are of interest here are, in his very rough phonetic dress, TRR-TRR for *Troglodytes* sp., TSI-TSI, TSAK-TSAK/ TSIK-TSIK, TUK-TUK/ TAK-TAK-/ TUT-TIT for *Turdus* sp. (for details see Arrigoni degli Oddi 1929: 277). Sylvidæ and warblers are all said to have "un cicaleccio basso, dolce, continuato e piacevole," precise types are CHERR=CHERR/ CHIRR-CHIRR/ RERR-RERR/ ERR-ERR (256). These classes of small birds, apart from some members of the Turdus genus, have similar calls and similar European names, some being generically named on the basis of their singing or warbling prowess, e.g. English "warbler," German "-Sänger," Greek ajhdwvn, Albanian "(tushë) këngëtare," which are taxonomically generic or even intermediate. A basic type recognized in Indo-European handbooks is present in Pokorny as IEW 255.3 *dher=, IEW 203.2 *der-, *der-der-, IEW 1079 *tet[e]r-, and IEW 1096 *trozdo-; these, I feel, can be reduced to a very fundamental type *DÆR=, reduplicated *DÆR=DÆR=, derived *[S=]D[E]R=SDO=, onomatopœa DRR=DRR or TRR=TRR with vowel expansion in accordance with first and second formant values of an [r] type, i.e. /DR:/. From such a complex base stem the Celtic names for *Troglodytes* sp., *Acrocephalus* sp., *Hippolais* sp., *Locustella* sp., *Phylloscopus* sp., even *Sturnus* sp. and *Pastor roseus* (Welsh dryw ± modifiers; Irish dreán, Gaelic dreathan ± modifiers; for starlings Welsh drudwy, Irish druid, as a bird that imitates other birds' calls),[28] as well as Latin turdus (> Italian tordo, NeoVenetian dórdo, Calabrian turdulice, Roumanian sturz etc.), in Germanic English "thrush," German "Drossel," probably Albanian tushë, and marginally Celtic (Welsh tresglen *Turdus viscivorus*), for many Turdus species, to which we might add Albanian tetedyzë for warblers and some Sylvidæ, which may have influenced tushë. In some way Greek τροχίλος *Troglodytes* sp. and warblers, Latin trŏchĭlus and verbs trŭcŭlāre, trŭcĭlāre, are also related, via *DOR= reduplicated *DOR=DOR,[29] as o-grade of the first group, which in all likelihood explains the IEW *trozdo- > turd-forms, which may turn out to be S=DOR=D[OR]=.

The story of the onomatopœic base for the names of Troglodytes, warblers, Turdus and Sylvidæ in a cross-section of Indo-European languages does not, however, finish here. Albanian for *Sylvia borin* (= simplex) has beng, -u,[30] which may be the "beak"="breaker" base IEW 114 *bhe[n]g= [NIE *BAK'-/ *B´K'-/ *B´NK'-"to break, shatter," denoting the action of the bird's beak][31]- crossed with the onomtopœic base *pink=.[32] This last is renowned as the PINK=PINK call of the Fringilla genus (see also observations in Arrigoni degli Oddi [1929: 120]), and explains names for *Fringilla* sp. in Celtic (Welsh pincyn, ysbincyn) and Germanic (English finch, chaffinch, German Fink etc.). The reason it has passed to the Sylvia genus in Albanian is perhaps the contamination with the "beak" base. An apparent onomatopœa for *Troglodytes troglodytes troglodytes* L. in Albanian, ČERR= > thërrak = fërrak, is, however, derived from the Albanian name for the plant *Rubus* sp. ferrë (i.e. brambles-bird).[33] Italian dialects also present a plethora of other onomatopœic formations: ČERR=ČERR= > NeoVenetian cercèr *Troglodytes troglodytes* (VI, VR), ČUWÍ =ČUWÍ > NeoVenetian ciuín *Phylloscopus* sp. (PD, VI), BUWÍ=BUWÍ > buín id. (VI, VR), though this last may be an ellipsis of òcio buín "ox-eye," as in Roumanian ochiu-boului *Troglodytes troglodytes*, TSÍ=TSÍ > Calabrian zívulu, zijúlu, zijúolu, zizí, zijú:ghə *Phylloscopus* sp., *Acrocephalus schœnobœnus, Locustella nævia,* Albanian and Italo-Albanian cirlë for warblers [connected with çirrem "cry; shout" etc.?], PLI-PLÍ > Calabro-Lucanian pli-plíjə for some *Phylloscopus* sp. and so on. There is a continuous remotivation of sound representation in these and similar cases, though immediately one goes outside this restricted field of name production, onomatopœa is decidedly culture-specific, even sometimes in the case of domestic fowls, as the cock-a-doodle-do vs. chichirichì contrast teaches, and is often destroyed by suffixation and other morphological devices in highly flexional languages, even more so if such languages are stress-timed rather than syllable-timed, with considerable syllable erosion.

Part III. Trade Jargon and Specialization: Complex Metonymy and Metaphor as Productive Process

I have argued so far that in the case of natural phenomena (animals, birds, fish, plants) naming processes involve either straightforward similitude or the application of particular symmetrical schemes between animals and plants, animals and fish, fish and birds, etc. in terms of a particular, historically valid, cosmic vision. Many of those between animals and plants are illustrated for antiquity in André 1963. In table 11.1 are listed some correspondences between the animal"wolf"and plants characterized as "wolf-bane,""wolf-wort" or "wolf-mouth" etc. from Latin into three Italian dialect groups (Piedmontese, NeoVenetian and Calabro-Lucanian). Three constants are observed in the symmetry, viz. (1) poisonous, whether to humans and animals or to other plants (*Aconitum, Arum, Cuscuta, Orobanche, Ranunculus, Rhinanthus* sspp., though only *Aconitum* sp. is constant over 2000 years), (2) bitter fruits (*Lupinus albus,* so called, says André 1985, for"l'amertume de ses graines," etc.), (3) "wide-mouthed" = "wolf-mouth" (*Antirrhinum* sp. constant throughout Romance).[34] The symmetry is likewise extended to fish in Italy, probably based on voraciousness alone, therefore part of case (1), i.e. Latin lupus *Dicentrarchus labrax, Dicentrarchus punctatus,*[35] which in NeoVenetian marine coastal dialects is also"lóvo," the feminine form"lóva" being applied to the young of this fish or to adult *Merluccius merluccius* of the Gadidæ family. In the extreme South "wolf-fish"seems confined to the Gadidæ, since, for example, we only find in Calabrian coastal dialects *Phycis blennoides:* "lup'i rina" or merely "lupu," generic extended to specific, *Phycis phycis:* "lupu jancu"/"lupu faròtu"/"lup'i fundali"/ "lup'i scògghju" (rarely lupu on its own), *Onos tricirratus:*"lupu nìvuru, lupu niru, lupu nigru"/"lupu šcuru"/"lup'i rrocca"/"lupu masculinu"/ (rare) "lupu 'nzitatu" (never lupu on its own). On the Ionian Calabrian coast (northern section)"vucch'i lupu" ["wolf-mouth"] applied to anchovies and their young is obviously a calque on the shape of the mouth, though the reference is not at all modern, even present in classical Greek, cf.Thompson 1947: 152 λυκόστομος already used in Ælian. The data can be presented in taxonomical form as in figure 11.5.

None of these cases represent instances of classical metonymy or metaphor but the way in which particular Cognitive Models are interrelated with each other in a particular vision of the natural world. On the other hand, trade jargon, which represents a necessity for naming objects, artifacts, proto-industrial or even industrial methods, metals and similar substances which are elaborated from raw materials (ore) etc., has nothing particularly natural about its objects, their elaboration and working methods. From the last century up to the period between the two world wars such jargons were said to be fundamentally cryptolalic,[36] and even some of its users seem to have accepted the hypothesis of the "secret" nature of their code, rope-makers in Sanga (1979: 207), tinkers in Trumper (19962: 28). On the other hand, it has been justly said that the only "secretive" part of such jargon

Table 11.1 Correspondences Between the Animal "Wolf" and Plants Characterized as "Wolf-bane," "Wolf-wort" or "Wolf-mouth" etc. From Latin into Three Italian Dialect Groups

family: animal	plant: Latin	plant: Piedmont (Sella 1992)	plant: Veneto (Trumper-Vigolo 1995)	plant: Calabro-Lucanian (Pollino) field work	bird (if any associated)
CANIDÆ: wolf	Aconitum; Humulus; Lupinus; Ranunculus; Verbascum; Lycopodium; Arum; Ricinus; Verbena; Echium.	Lupinus; Allium ursinum; Leopoldia comosa; Lycopodium; Antirrhinum; Conyza canadensis; Duchesnea.	Lupinus; Antirrhinum; Orobanche; Cuscuta epithymium; Rhinanthus major.	Lupinus; Antirrhinum; Aconitum; Lycopus; Veronica; Melittis.	Lat. Verbena/ Lycopus: dove; cockrel. Piedmont Ranunculus: goose. Veneto Ranunculus: crow. Calabro-Lucanian Arum: crow; Rhinanthus: cockrel.

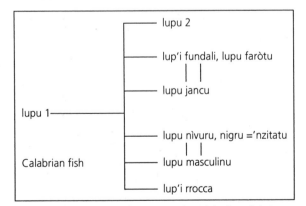

Figure 11.5 A Taxonomical Representation of the Data Relating to the use of "Wolf-fish" in Calabrian Costal Dialects

is that which covers erstwhile trade secrets on the part of people who consider themselves a sort of élite group vis-à-vis others, as in Guiraud (19736: 98) "...la pègre se considère comme une aristocratie... superstitieusement jalouse de ses codes, de ses rites, de ses hiérarchies...". [Cant-users consider themselves an aristocracy ... superstitiously jealous of their codes, rites and hierarchies...] In the 1960's some investigators went further, completely denying the particularly "secret"nature of trade jargon, distinguishing it completely from criminal jargon,[37] the only cryptolalic part of such codes being that minimal nucleus that regards trade secrets, or matters considered to be such, as stated in Stein (1974: 82).[38] In the tinker jargon of Southern Italy (Calabria: Trumper 1996[2]) I noted a terminology for metals and metal parts that was far richer than that of any dialect, richer even that of some standard languages, in which the primary element in naming seemed to be synecdoche and other mechanisms of classical metonymy, e.g. campanaru"lead"is"object made of lead"(bell) >"substance lead,"this is probably also the shift observed in camággiu"zinc"< Byzantine Greek καμάκι(ov) "fish-gaff" (see Alessio 1960, 1961: 140 item 6, Rohlfs 1964 καμάκιον), so"object made of zinc"(fish-gaff) >"substance zinc,"while for the shift"property">"substance possessing that property"we have mprácchja"function: for sticking on">"sticker-on" >"tin-foil," šcarafílice "repair-fern" (in dialect one of the meanings of šcarari is "repair") >"repairer" >"tin-foil," again function > object or substance, in some cases color X > metal color X, X = nívuru"black"> culu-nívuru"iron" (lit."black-bottom").

One of the best examples is a double synecdoche Greek χρυσός "gold" > χρυσοῦλον"gold container/ vase">grisciólu >"precious metal,"so that these tinkers have"precious metal" + "white" grisciólu bbianchèparu "silver,""precious metal"wihout modifier grisciólu"gold."The most complex term seems to be gritta "copper"connected with the"paint"verb ngrittari: we suggest in Trumper-Straface (1998, item gritta) that the metonymic development is"sticky" > (property of) "paint" >"painted" >"pellucid" > (property of being) "pellucid,""bright" metal > "bright metal"+"copper"(expansion) > (member for class)"copper,"thus the most likely starting point is Middle Greek (Byzantine) γλιττος"sticky" connected with terms for"stickiness" and"glue"like Middle Greek γλίδα, γλία, γλιά < classical γλοιόν. This basic"copper"name carries with it specialized terminology such as "copper ore" = bbrušijána, "unworked copper" = scorza, "copper-ingot" = masséllu,[39]"roll of copper"= 1 kg. of copper = mbrógliu,"strip of copper"/"strip of metal" = priciána, "residual powder left over from the working of a metal" = fuscáglia, etc. The first term is probably connected with Central Italian terms for "crumb"(brùciolo = brìciolo), though a connection with the name of the Belgian town of Bruges cannot be excluded, the second is a plant term ("bark"), the third connected with"mass,"the last a derived color name (< fuscu"dark"), etc., obvious instances of classical metonymy. Specialisation, whether in agriculture (say, Graminaceæ and cereals as food plants, Apiaceæ as culinary and therapeutic plants, and so on), or in industry or rather proto-industry, as in the last examples, creates an overall situation that brings utilitarian criteria to the fore and imposes them on all other possible criteria. The pure classificatory instinct in specialized societies or groups seems to give way to a series of utilitarian criteria which then determine intricate metonymic chains, thus the creation of metaphors. My hypothesis is, then, that specialisation provokes true metaphor, even exacerbating its effects on naming processes. In general, apart from few instances, we might claim that agricultural society inhibits pure metaphorical development, except for precise culturally determined cases, e.g. Latin plant and agricultural terminology > the thinking process (cernere > discernere, putare etc.), Greek plant and agriculture > sex (including σαῦρος for penis, σύκη for cunnus) etc., themes amply dealt with in the literature, while proto-industry and industry encourage, instead, metaphorical development based on Gestalt in the passage nature > metal > artifact,[40] an argument which I feel has not yet been fully explored.

Notes

1. Atran 1996[3]: 26 "In many cases, it seems that generic, rather than specific, groupings are easiest to sort."
2. Atran 1996[3]: 27 "The apparent dilemma is resolved once it is realized that the distinction between species and genus is largely irrelevant to a basic appreciation of the flora and fauna of a local environment."
3. The glosses are only indicative at this first level of discussion.
4. *Summa Theologiæ* vol. 28 Q. 74 Art. 4. 4, Resp.: "Respondeo dicendum quod circa materiam huius sacramenti duo possunt considerari: scilicet quid sit *necessarium*, et quid sit *conueniens*." (italics mine. We might translate "I answer saying that two considerations may be brought out with regard to the matter of this sacrament: viz. its defining features ['necessary'] and its characteristic ones ['appropriate'].").
5. Lakoff 19902: for similar diagrammatic formalizations see Berlin (1992: 36–51), Italian discussion and application Trumper- Maddalon 1995.
6. *De Vegetabilibus* I.153, 154, 155: "Arbor enim est, quæ habet ex sua radice stipitem fortem, super quem nascuntur rami plures... Plantæ autem, quæ sunt mediæ inter arbores et herbas, quæ dicuntur ambragyon græace, et latine arbusta, habent emissos a radicibus suis multos ramos per modum palmitum. Et sunt trium generum... et tertium est sicut frutex. Vocamus enim frutices ea(s), quæ multos ex una radice ramos emittunt, cum tamen sint lignei rami... [herba]... parum aut nihil habet ligneitatis." [For a tree has (growing) from its roots a strong trunk-structure, on which grow many branches... But there are plants which are intermediate between trees and grasses which in Greek are called "ambragyon" and in Latin shrubs: they have many branches springing directly from their roots in the manner of vine-structures. And they belong to three types, ...and the third is a sort of bush, for we call bushes those plants which grow many branches out of a root, though they are woody branches......(grasses) have little or no woodiness].
7. Cesalpino, *De Plantis* I. XII: "Sunt quidem omnibus recepta quatuor genera plantarum ex totius habitu & vita distincta: Arbor, Frutex, Suffrutex, Herba; ex totius igitur habitu Arbor & Frutex à reliquis differunt: nam corpulentia illorū longe durior grauiorq; ſpectatur, quod lignum vocatur. Reliquorum autem mollior ſubſtantia, & laxior eſt, ideo & diutius viuunt Arbores, Fruticeſque, & vt plurimū magnitudine quoque vincunt. Arbor autem à Frutice differt ſimplicitate; vnico enim caudice apta nata eſt arbor aſſurgere, niſi impediatur, eoque habitiori, quàm Frutex; ... Suffrutex autem & herba inuicem ex viuendi tempore diſtinguuntur. Suffrutex enim perplures annos viuit, aut radice tantum... aut etiam caule..." etc. (Certainly four plant types are accepted by all, holistically distinct by form and manner of living: Tree, Shrub, Bush and Grass. Tree and Shrub are formally distinct from the rest, for their sturdiness gives, as expected, a by far a harder and hardier exterior called "woodiness." In the others this is a softer substance and more flexible. Thus Tree and Shrub live longer and are larger in size than the others. The Tree, however, differs from the Shrub in simplicity, for while a

Tree develops as a single stem (="trunk"), this is not the case of the Shrub which is multiform. Bush and Grass, on the other hand, are separated by the periods they live, for the Bush lives for many years, even if only supported by its root or even by its stem…). Note that here in Cesalpino Frutex is more or less Shrub, differently from S. Albert or even earlier Latin sources.

8. Brown 1979: 258: "…initial suprageneric classification of plants universally entails a 'large plant'/ 'small plant' distinction."
9. For by plucking them they ruin the new **suckers of plants**, not least in the case of vines and olives.
10. LOEB: "For the water will glide between, the air's searching breath will steal in, and the **plants** will take heart."
11. The most general meaning is recorded in Quintilian's *Institutiones Oratoriæ* XII.X.19 "uelut **sata** quæ-dam cælo terraque degenerant…" [LOEB: just as certain **plants** degenerate as a result of change of soil and climate…], while we seem to have "seedlings" = "young plants" in VIII. Preface "propter id quod sensus obumbrant[ur] et uelut læto gramine **sata** strangulant[ur]" [LOEB: because they darken the sense and choke the good **seeds** [better "seedlings"] by their own luxuriant growth], Columella's *De Re Rustica* X.320 "Exurat **sata** ne resoluti pulueris æstus" [LOEB: Lest the burning heat of earth dissolved in dust the **seedlings** scorch] etc.
12. "…that **plants** be watered by the showers, that vines be rich in produce, trees with fruit… ."
13. Cf. *Sententiæ* III col. 677 line 7 "…in quibus quanto tardius **sata** semina exeunt, tanto ad frugem cumulatius crescunt" […in which the later seeds come out as **crops**, the more abundant is growth and the crop harvest].
14. Venerable Bede's *In Marci Euangelium Expositio* I.4 line 1972 "…uel cum in toto mundo **sata** fuerit non exsurgit in holera sed crescit in arborem ut in aliis euangelistis apertissime dicitur" (…or just as such **seedlings** throughout the world do not grow into vegetables but into trees, as is most clearly stated in the other Gospel writers…). In this case **sata** seems to be substituting **plantæ** as "seedlings" or "shoots," so that in ca. 800 AD a certain interchangeability seemed to be functioning. See further.
15. The **slips** come from trees: indeed they are the **shoots** which spring from seeds and are transplanted with their roots and with their own earth.
16. Cassiodorus as late as the sixth century presents "shoots" as the usual meaning, cf. *Variæ* VI.14.6 "Agricola diligens præueniendo adiuuat imbrem cælestem et ante rigat **plantaria**, quam pluuias mereantur optatas" (The careful farmer anticipates and encourages the sky's showers and waters the **shoots** before they need the long desired rain.)
17. …they are like the "shoots" of greens when there is frequent frost; like the green shoots of grasses they quickly fall and rot.
18. The LOEB translation is actually "Whether on going away they leave their venom behind when the **plant** is torn up from the root," where what is rendered "plant" could equally be "shoot," "slip" or "young plant" from the context.
19. Again the LOEB has "that they are planted without injury to the bark and always with the cut end and the part that was nearest the root downward, and during the process of budding the **plant** is kept heaped over with earth until it attains strength." Once more the translator has used "plant" instead of a contextually more suitable "young plant" or "shoot."
20. arbūstum < *arbos-to- < arbos= classical arbor < accus. arbōrem by intervocalic rhotacism in Latin extended to the nominative, perhaps originally from the "stave; pole" base IEW 63 *ardh-.
21. This represents a revision with respect to a previous position in Trumper Vigolo (1995: 88–89).
22. I wish to thank for a detailed discussion Sh. Rrokaj (Tirana University) and G.M. Belluscio (Calabria University).
23. If Meyer (1891: 34) is correct, then bërthamë < *petr-amen < Lat. petra indicates the same choice of lexical substance as regional Ital. "òsso," English "stone" and Welsh "carreg," though a Latin borrow-ing. I reject his explanation of farë (a Germanic borrowing for such a basic concept?!) and, in absence of other cogent proposals, would accept Huld's 1984 proposal < *spor-eH o-grade of *sper- IEW 993.2 which alignes Albanian with Greek and South Slav (Bulgarian, Macedonian).
24. See Bomhard (1981, 1984), Dogopolsky (1986) and Shevoroskin (1991) for what might be achieved in this sphere. I still feel that a large number of claims in Ruhlen (1994) are exaggerated, though this

does not imply that I do not consider it possible to construct some sort of Nostratic. However, this is not the appropriate place for dealing with this thorny problem.

25. This takes us back to the two horns of the Greeks' dilemma. An important observation is made by Hagège (1978: 475) that the Hebrew pronoun LO here (="for himself") may well indicate the necessity of expressing man's relating himself to natural things by means of words, cf." …comme pour rappeler que l'homme n'augure sa relation au monde extérieur qu'à travers le language par lequel il le désigne…" [as if to remind one that man only wants his relationship with the external world to be mediated by language with which he is designing it [sc. this relation]…"]. One might object, using S. Augustine's observations, that man may relate himself to the outside world pre-linguistically. It remains, however, an interesting and fecund comment on the Genesis passage.

26. Eco 1996: 21"…egli pensa ad una lingua perfetta, comune a tutte le genti, i cui segni non sono parole ma le cose stesse." [="… he is thinking of a perfect language, common to all, whose signs are not words but the things themselves"].

27. Hagège (1978: 47) "Aussi, il n'y aurait pas monogénétisme des langues, alors que pour le genre humain, les données de l'anthropologie, aussi que des considérations biologiques générales, suggèrent une unité d'origine sur un territoire restreint." [Thus there would be no monogenetic origin of language, whereas for mankind anthropological data as well as general biological considerations suggest unity of origin over a restricted territory].

28. Older Gaulish *driwo=, *dreno= [> W. dryw, Ir. dreán] may well underlie French draine *Turdus viscivorus*, grive *Turdus* sp.

29. Probably relating English dor-hawk = night-jar, dorr as flying beetle or cockchafer, also bumble-bee, as well as the word drone. Similar names in other languages will have to be re-examined in the light of this.

30. Italo-Albanian presents sirkofán,-i for the same bird, probably < Greek συκοφάγος 'fig-eater,' a common appellative of this bird in Europe.

31. This is also claimed in Bomhard (1984: 202 item 37) to be a Nostratic theme and he compares his *BAK'- with Proto-Semitic *BAK'-Aʔ- / *BAK'-AR- and Proto-Afroasiatic *BʔK'-"to cleave, break."

32. Meyer (1892) gives this last as the only explanation for the term, p. 32 item "benk" Goldamsel.

33. It gives one of the very few Albanian borrowings in Italian dialects, Calabro-Lucanian Pollino range pulləcîll'i ferachə *Troglodytes troglodytes*.

34. This last is even present in Fr. "gueule de loup," French not being the most conservative of Romance languages.

35. Labrax already means 'voracious' and the same fish is similarly called λύκος in Greek. As usual Latin denomination of the Dicentrarchus labrax "sea-basse" cf. Ovid, Halieutica vv. 23 ("lupus… inmitis et acer"="the sea-basse cruel and impetuous"), 39 ("lupus acri concitus ira…"="the sea-basse spurred by impetuous anger"), 112 ("rapidique lupi"="and the swift sea-basse").

36. For France cf. Dauzat (1917: 9) "…le but cryptologique de ces langages … un de moyens de défense collective du groupe" right up to Guiraud (1973⁶: 5) "Jusqu'à Vidocq, tous les témoignages attestent le caractère cryptologique du jobelin…,"(22)…ce caractère cryptologique de l'argot, tant ancient que moderne…," for Italy Rovinelli (1919: 7)"… il parlare artificioso e incomprensibile ai non iniziati" up to Pellis 1930 who insists on the *animus occultandi* of argot and trade jargons, etc.

37. Esnault (1965: V) "…le mot argotique n'est ni conventionnel, ni artificiel, ni secret." [… the slang word is neither conventional nor artificial nor secret].

38. "Ainsi proposons-nous notre théorie, aussi surprenante qu'elle paraisse, selon laquelle l'argot des malfaiteurs n'est un lexique secret qu'en ce qui concerne son corpus de termes techniques. Or, l'existence des termes techniques n'est pas un monopole argotique. Nous en concluons donc que ce qui est intentionnellment secret dans l'argot des criminls, ce n'est pas l'élément caractéristiquement argotique, mais c'est l'élément professionel:" [We thus propose our theory, as surprising as it seems, according to which criminals' slang is not secret vocabulary as far as its corpus of technical terms is concerned. Indeed the existence of technical terms is not a monopoly of slang. So we conclude that what is intentionally secret in criminals' slang is not the characteristically slang element but the professional one].

39. Meanings such as "block of metal," "iron-bar" (Seneca, Columella), "block / ingot of gold" (Pliny, Ovid), "block / ingot of copper" (Seneca, Vergil), as well as "block of marble" (Pliny), attributed to

massa are well known in antiquity. Latin massa also passes into early Celtic with the meaning "metal"(mâs). This sort of transfer is then perfectly normal.

40. The development in an agricultural society of this type of metaphor connects natural entities with non-natural ones, i.e. animals with mounds of collected plants or fertilizer, with outbuildings and building partonomy etc., so that the dominant element is always the non-natural final one, be it arti-fact or whatever. Preindustrial or proto-industrial development, including the non-natural aspects of agriculture, involve the use of a binary opposition "natural" vs. "artifact," which seems extremely important and is already well-known in the development of colour adjectives first in the medieval cloth and dying trade, later during the Industrial Revolution (e.g. Japanese is a well-known exam-ple). The Roman introduction of vegetable paint and dying techniques probably influenced the development of Celtic colour naming contrasts, e. g. Welsh glas (bright grey + green: nature, blue: artifact) vs. gwyrdd (< Lat vir[i]dis, green: artifact) etc., a problem not understood by Hjelmslev, whose first impressions heavily biased all consequent discussion of colour oppositions in Celtic lan-guages! Glas as"bright grey"was perhaps originally opposed to llwyd"dull grey,"as in the Black Book of Carmarthen "Nud **glas** minit tenev/ vy llen imi nyd llonit./ **lluid** yv vy blëit" (The bright-grey hillock, my threadbare clothes and my greying hair are no happiness to me!). The poet-monk is referring to a dull, wintery or fall colour, which is less dull than the grey of his hair. However, we also have the old adjective cann = canwelw"whitish,""white-to-grey,"that considerably complicates any analysis. Ancient colour oppositions are, on the whole, rather difficult to reconstruct, even in well-documented cultures.

References

AA. VV., 1985–88. *Thomas Aquinas: Summa Theologiæ*, ESD critical edition of the O. P., 33 vols., Bologna: Ordo Prædicatorum, particularly QQ. 73–83, vol. 28 *De sacramento Eucharistiæ*, 1970, Bologna Ordo Prædicatorum.

Alessio, Giovanni. 1960–61. *Note etimologiche sulla terminologia marinaresca*, Bollettino dell'Atlante Lin-guistico Mediterraneo 2–3: 139–48.

André, Jacques. 1963. *Noms de plantes et noms d'animaux en latin*. Latomus XX11, Fac. 4: 649–63.

_____ 1985. *Les noms de plantes dans la Rome antique*. Paris: Société d'édition "Les Belles Lettres".

Arrigoni degli Oddi, E. 1929. *Ornithologia italiana*. Milano: Hoepli.

Atran, Scott. 1996³. *Cognitive foundations of natural history. Towards an anthropology of science*. Cambridge: Cambridge University Press.

Augustine of Hippona: *De Genesi contra Manichæos; De Genesi, Liber Imperfectus; De Genesi ad Litteram libri XII*, in Migne, Patrologia Latina vol. 34.

Banterle, Gabriele. 1979. *Sancti Ambrosii Episcopi Mediolanensis: Opera I, Exameron*, Milan: Biblioteca Ambrosiana and Rome: Città Nuova Editrice.

Berlin, Brent. 1992. *Ethnobiological classification*, New Jersey: Princeton University Press.

Bomhard, Alan R. 1981. *Indo-European and Afroasiatic: New Evidence for the Connection*, in Y. L. Arbeitman and A. R. Bomhard, *Bono Homini Donum: Essays in Historical Linguistics in Memory of J. Alexander Kerns* Part 1: 351–475. Amsterdam: Benjamin.

_____ 1984. *Toward Proto-Nostratic: A New Approach to the Comparison of Proto-Indo-European and Proto-Afroasiatic*. Amsterdam Studies in the Theory and History of Linguistic Science IV, Current Issues in Linguistic Theory vol. 27. Amsterdam: Benjamin.

Brown, C. H. 1979. A theory of lexical change (with examples from folk biology, human anatomical partonomy and other domains) *Anthropological Linguistics* 21, no. 6: 257–76.

Cesalpino, Andrea. 1583. *De Plantis lib. XVI*, Florence [I wish to thank the Orto Botanico of Padua Uni-versity for allowing me to consult an original copy].

Dauzat, A. 1976. *Les argots de métiers franco-provençaux*. (Paris: Champion 1917) Geneva: Slatkine reprint.

De Saint-Denis, É. 1975. *Ovid: Halieuticon*. Paris: "Les Belles Lettres."

Dogopolsky, A. in V. Shevoroshkin, T.L. Markey. 1986: 27–50.

Eco, Umberto. 1996. *La ricerca della lingua perfetta nella cultura europea*. Bari: Laterza.

G. Esnault. 1965. *Dictionnaire historique des argots français*. Paris: Larousse.

Fridh, Å. J. 1973. *Magnus Aurelius Cassiodorus: Variarum Libri XII*. J. W. Halporn *De Anima*, Corpus Christianorum XCVI. Louvain: Turnholt.

Gigon, O. ex rec. I. Bekker. 1960. *Aristotelis Opera: vol. 1* Berlin [containing the Organon, Physica, De Cælo, De Generatione et Corruptione, Meteorologica, De Mundo ad Alexandrum, De Anima, Parva Naturalia, De spiritu, Historia Animalium, De Partibus Animalium, De Motu Animalium, De Animalium Incessu, De Generatione Animalium].

Guiraud, Pierre. 1973⁶. *L'Argot*. Paris: Presses Universitaires de France.

Hagège, Claude. 1978. *Babel: du temps mythique au temps du langage*. Revue philosophique de la France et de l'étranger CLXVIII: 465–79.

Hamill, J. F. 1979. General Principles of Classification and Nomenclature in Folk Biology: Two Problems. *Anthropological Linguistics* 21 no. 3: 147–153.

Hercher, R. 1864–1866. reprint 1971. *Claudius Ælianus: De Animalium Natura Libri XVII, Varia Historia Epistolæ Fragmenta*, 2 vols. Leipzig (Graz): Teubner.

Hort, A. 1968. *Theophrastus:* Περὶ φυτῶν ἱστορίας *(Enquiry into Plants)*, 2 vols. Cambridge, Mass.: LOEB classical library.

Huld, Martin E. 1984. *Basic Albanian Etymologies*. Columbus Ohio: Slavica Inc.

Lakoff, George. 1990². *Women, Fire, and Dangerous Things*. Los Angeles: University of California Press.

Lieberman, Philip. 1975. *On the Origins of Language*. New York: Macmillan.

Lindsay, W. L. 1966³. *Isidori Hispalensis Episcopi Etymologiæ siue Origines Libri XX*, 2 vols. Oxford: Clarendon.

Meyer, Ernest and Carl Jessen. 1867. *Alberti Magni (Albert the Great): De Vegetabilibus Libri VII*. critical edition, Berlin: Unveränderter Nachdruck, Minerva reprint 1982: Frankfurt on Main.

Meyer, Gustav. 1881. *Etymologisches Wörterbuch der Albanesischen Sprache*. Strassburg: Trübner.

Naldini, M. 1990. *Basil of Cesarea:* Ομιλίαι Θ' εις τὴν ἑξαήμερον *(Basilio di Cesarea, Sulla Genesi)*. Milan: Valla Foundation.

NDDC: see Rohlfs, Gerhardt, 19902.

Owen, R. 1866. *The anatomy of vertebrates, vol. 1: Fishes and reptiles*. London 1866 (quotations from Atran).

Pellis, Ugo. 1930. *Coi furbi*. Udine: Tipografia Del Bianco & Figlio.

Pokorny, Julius. 1959. *Indogermanisches Etymologisches Wörterbuch*, 2 vols. Munich: Francke.

Rohlfs, Gerhardt. 1990². *Nuovo Dizionario Dialettale della Calabria*. Ravenna: Longo (ref. NDDC).

_____ 1964². *Lexicon Græcanicum Italiæ Inferioris*. Tübingen: Niemeyer.

Rovinelli, A. 1919. *Il gergo nella società, nella storia, nella letteratura con alcuni saggi di vocabolario di vari gerghi*. Milan: Sonzogno.

Ruhlen, Merritt. 1994. *On the Origin of Languages: Studies in Linguistic Taxonomy*. Stanford: Stanford University Press.

Sanga, Glauco. 1979. *I cordai di Castelponzone. Da "dritti" a proletari*, in Leydi, Roberto and Guido Bertolotti. *Cremona e il suo territorio*. Mondo popolare in Lombardia 7: 199–221. Milan: Silvana Editoriale.

Sella, Alfonso. 1992. *Flora popolare biellese*. Alessandria: Edizioni dell'Orso.

Shevoroshkin, Vitaly and T. L. Markey. 1986. *Typology, Relationship and Time*, Ann Arbor, Michigan: Karoma Press.

_____ 1991. *Dene-Sino-Caucasian Languages*. Bochum: Universitätverlag, Brockmeyer.

Stein, André L. 1974. *L'Écologie de l'Argot Ancien*. Paris: Nizet.

Thompson, D'Arcy W. 1947. *A Glossary of Greek Fishes*. London and Oxford: Oxford University Press.

Trumper, John B. and M. T. Vigolo. 1995. *Il Veneto Centrale. Problemi di classificazione dialettale e di fitonimia*. Padua and Rome: C. N. R.

_____ and Marta Maddalon. 1995. *Sesso femminile, genere maschile*, in G. Marcato, *Donna & Linguaggio*. Padua: CLEUP, pp. 459–74.

_____ 1996². *Una Lingua Nascosta. Sulle orme degli ultimi quadarari calabresi*. Soveria Mannelli: Rubbettino.

_____ and Ermanno Straface. 1998. *Varia etymologica*, in Alberto M. Mioni et al. Quarta Raccolta di Studi Dialettologici in Area Italo-Romanza. Per G. B. Pellegrini. Padua and Rome: C. N. R.

Whitney, W. W. (1867) 1997. *Language and the Study of Language, Twelve Lectures* (London 1867). London and New York: Routledge.

Data regarding plant, animal, bird and fish systems in the Italian regions of the Veneto, Lucania and Calabria are from my or my collaborators' published works or unpublished field notes, some of which are being prepared for publication in book form (sea-fish, crustaceans and boat-building in the whole of Calabria, birds, fresh-water fish and insects in the Veneto, plants in the Pollino range and elsewhere). NIE means New Indo-European according to more modern reconstruction models.

Session II: Naming

Edited by Gabriele Iannàccaro

Marta MADDALON points out that there is a tendency to underestimate the relationship between Western industrialized humans and classification, since, in discussion, the analysis of indigenous cultures or the historical-philosophical approach is usually privileged. This emerged clearly when Gnerre stated that there is an evident difference in toponomastics between more recent types of place names and the older types. For example, the biologist's observations about the move towards a model of classification is no longer based on morphological consideration, but mainly on criteria of a genetic type that cannot be shared with most native speakers. As they can only see those differences that are observable to the eye, their limited level of observation and of classificatory judgment ends up sharing in the vision of nature held by systematicists and botanists. Plant names can illustrate Western humans' relationship with classificatory levels: there is a tendency towards a polarization between few higher-level categories, usually life forms, sometimes generics, and a greater number of lower levels. Hyperdifferentiation is usually linked to experiences of a sectorial nature, i.e., food specialization, or esthetic hyperdifferentiation in the case of ornamental plants.

*

Referring chiefly to Gnerre's paper, and to a lesser degree to Pellegrini's, Rita CAPRINI (Professor of linguistics at Genoa University) suggests an index of arguments worth meditating upon.

Gnerre's investigation into Shuar (Upper Amazon) place names, especially river names, attests that the Shuar have changed them three times in a relatively short space of time. In Europe, on the contrary, the names of courses of water are the part of a toponymy least exposed to change and European river names are perhaps the most ancient linguistic material we have. Considering that European linguistic records also cover a much longer period of time than those available for the Amazon area, the difference is striking.

Gnerre's information about the repetition of the same river name (the "Anaconda river") in a small Shuar territory raises a fundamental point in the general problem of name-giving: for most of humankind, throughout most of history, proper names are not intended to identify, or rather they are not only and principally intended to identify. Françoise Zonabend, a French anthropologist who worked in a small conservative French village called Minot, told the story of a husband who had suggested that his wife should give their newborn second daughter the same name they had given to the first-born. Calling the daughter by a name that represented a meaning with respect to their beliefs and their social environment was plainly more important than "distinguishing" her even from her sister.

Linguists studying onomastics have the bad habit of quoting the name of some philosopher or logician before entering the discourse. Recently they have felt obliged to refer to Saul Kriepke's work. In the past decades Alan Gardiner's *Theory of proper names* dominated (based on the false belief that the only aim of proper names is "identifying"), and it fell to Lévi-Strauss to rebut Gardiner's assertion that good proper names are those that "have no meaning," as they are more apt to "identify." Moreover, affirming that Vercingetorix is a better name than Mont Blanc ("White Mountain"), Gardiner made a mistake in historical linguistics, since it is well known that Vercingetorix was a meaningful name for ancient Gallic society.

In the real life of most societies on earth, still today and above all in the past, opaque, unmotivated, unchangeable (throughout one's lifetime) personal names do not exhaust onomastics at all. Rather, their adoption satisfies mainly the administrative and revenue purposes of our recent Western society. Even in Europe's recent past, in rural areas, in the Alps, or for particular human groups (such as Jews or Gypsies), the proper name could change during the life of the individual, who could also be identified with different names by his family and by his contemporaries. In Rumania, a few decades ago, the "true name" of a man was kept well hidden (only God knew it!) from fear of harm by his enemies.

Another linguistic aspect of naming, ignored by logicians and philosophers, for example, is that in certain areas, 50 percent of place names derive from personal names (see the maps of Italy made by the Army-IGM, Istituto Geografico Militare, and Zonabend's studies in Minot). Logicians and philosophers usually just do not know these things; linguists are just speaking of different things.

As regards onomatopoeia and its being an archaic trait of human language, the Finnish folklorist Martti Kuusi, in his work on popular sayings and about sun-showers (published in *Quaderni di Semantica*, the review directed by Caprini and Mario Alinei) maintains something that could be used as an epigraph for the present Convention: he affirms that nursery and forest lie not just behind us, and around us, but they also live in us. What Kuusi says could be of help in studying the origin of human language.

Pellegrini's metaphorical explanation for the presence of "cats" and "dogs" in a large number of plant names is unsatisfactory. The linguistic map of the Romance names of the caterpillar attests that most of these names refer precisely to "cats" and "dogs" (for example Fr. *chenille* < Lat. *canicula* "petite chienne" or It. *gatta* "female cat"); but, granted that a caterpillar may sometimes look like a furry animal, it is impossible to say the same thing of horses, horsemen, and land-surveyors that equally lend their names to caterpillars in the Romance domain. Perhaps some other solution, different from metaphor, is to be sought.

*

Glauco SANGA agrees with Caprini. If the same plant has animal names that seem to have contrasting characteristics, or if (as Gnerre pointed out) several rivers are called Anaconda, then we should not speak of realistic motivations.

Returning to the question of animal names, Sanga wonders what the original name of animals is: if we observe the fact that the animal (a) is used to denominate plants, (b) can have a taboo name, (c) can have a kinship name of a relative (as Alinei showed), what then is the real name of the animal? Taboo names, hunting names, being substitutive cannot be the original ones; then, is the original name the kinship one? Or maybe an onomatopoeic denomination (as we see in Indo-European languages, where both Lat. *lupus* and Eng. *wolf* derive from a **ul-kw*, acoustically similar to a howl)? Animal names are a central problem, for they seem to precede even plant names: we have a limited number of animal names that identify ten times as many plants, and this is probably related to the ideological centrality of hunting, with respect to the gathering, typical of hunting-gathering societies.

*

Against Berlin's hypothesis about the phonosymbolism of *a* (large) and *i* (small), Barbara TURCHETTA (Professor of Linguistics at Tuscia University, Viterbo) suggests an explanation based on graphic symbolism, at least as regards the response of the students from Georgia. People used to Latin script perceive *a* as larger than *i*, since *i* is simply smaller. It would be interesting to make an experiment with students whose alphabet has different graphic signs for our *a* and *i* (e.g., Arabic alphabet); the results would help us understand whether the symbology of our alphabet bears on our investigations when doing research into classifications.

*

Brent BERLIN asks Turchetta what her expectations would be, supposing the experiment were to be performed verbally with the students of Georgia, or were to be performed with illiterate people, and also asks her to give a theoretical explanation for her intuition.

*

Barbara TURCHETTA thinks that the results of the experiment could be very different for students using other alphabets, as well as for illiterate people. For example the Italian ideophone for a loud laugh would be *ha ha ha* and its equivalent for the Ewe from Western Africa is *iki*.

*

According to John TRUMPER, the key to "sound imagery" and "soundscaping" is not any particular vowel one may choose to use, but is a question of mapping maximally contrasting semantic sets on to a maximally mapped-out perceptual-acoustic plane, an intersection involving both the semantic and acoustical which will allow a maximum of contrast on all the pertinent levels of linguistic organization. One might actually reverse one's choice of particular vowel, i.e., *i* or *u* = "largeness," *a* = "smallness" instead of *i* = relative "largeness," as long as one exploits maximal perceptual contrast, in strictly phonetic terms, in maximally distinct divisions of semantic space.

Revel's "soundscaping" image seems very appropriate as far as hunting societies are concerned, but the farther you get away from them and into pastoral and agricultural societies, soundscaping gradually disappears from the organization of the environment. As very few hunting societies survive, it becomes more and more difficult to prove and actually define what soundscaping is and what its precise mechanisms are. Moreover, the fact that few hunting communities survive renders the category less and less appropriate, in an analytical sense, as we move towards a basically"deafer"type of society, deafened by a particular type and use of"noise" rather than of"sound"and soundscaping possibilities.

*

As regards hydronyms in Amazon, Maurizio GNERRE points out that those rivers that were renamed four or five centuries ago probably will not be changed again, because the new generations are focusing their attention on other areas of territory that are becoming more relevant to their culture (such as villages and places along the road).

According to Gnerre, philosophers are interested in onomastics, but they are not concerned with toponomastics.

It is true, as Sanga affirms, that"River Anaconda"is a toponym that is repeated more than once, probably because the importance of the anaconda is symbolic and not real; yet we cannot maintain the same explanation as regards Pajának and Antraenza. These toponyms, in fact, refer to something real and observable: the former is"river that cannot go into the big river,"which is true because its own tide goes upstream; the latter"useless river,"which is a negative, realistic notation.

Hunting ideology is very important, and a great many hydronyms are names of animals that people hunt. Maybe it would be possible to investigate the economic conditions of Indo-European peoples at the moment they arrived on the continent by considering those river names of ancient Europe that remained unchanged for centuries.

Turning to Turchetta, Gnerre observes that, apart from *ha ha ha*, he often hears the onomatopoeia *hi hi*, for a small laugh: in this case, therefore, the symbolism seems to work.

*

Mario ALINEI underlines that the importance of hunting ideology revealed by the prominence of the names of the animals entails a further complication. Kinship names also include a great many insects that show a wide onomastic, kinship, magico-religious variety. A Ph.D. thesis by a student from Zaire on the names of insects and small animals (see Tshimanga Kutangidiku,"A propos des motivations socio-sémantiques à l'origine de la création lexicale des noms des petits animaux chez les Bantu," *Quaderni di semantica* 34, a. XVII, n. 1, 1996), confirms the hypothesis that insects are part of the archaic diet; as a matter of fact, those insects with kinship and magico-religious names are included in the diet of the traditional populations of Zaire. As women gather insects, it is necessary to distinguish the male ideology of hunting from the female ideology of insect gathering.

As regards the real names of animals, if the supposedly real name (which could be the one that Indo-European scholars reconstruct on the basis of comparison) is contrasted with the taboo names, the real name should date from the prereligious period. Religion is generally held to have begun with the practice of the burial of the dead. This makes it difficult to imagine that the real animal-name could be a kinship name, since this presupposes a religion or a mythical conception of the animal. The kinship name is a variant of the many taboo names, the motivation of which is mainly descriptive: one moves away from the real name to indicate the animal by a common characteristic: color, habitat, behavior or even its being an uncle, a cousin or a grandfather and so on. This means the real name would be the only one not yet made taboo.

*

John TRUMPER observes that, at least in historical Europe where hydronyms are justly considered the most archaic and conservative elements, the instability of river names and the other toponymic elements depending on them always seems to have implied catastrophe from the demographic point of view. An Italian example is that of the river Amato and the Lamezia plain that takes its name from the river. When the pre-Italic Νάπη, as the Greeks wrote it, probably from an Indo-European root *[n-] ap- "water," decidedly distinct from its Greek homograph νάπη "grove" or "vale" [< Indo-European *nem- "bend,""bending," whence Gr. νέμο/νάπη, Lat. nēmus, in Celtic: Gallic inscriptions νεμητον, Erse neamhed, Welsh nant etc.], and which gave the Greek adjective Ναπητῖνος, Latin Napētīnus etc. (cf. Strabo, *Geography*, Italy bk. VI, 4, 12–15: "Εστι δ' αὐτὸς ὁ ἰσθμος ἑκατὸν καὶ ἑξήκοντα στάδιοι μεταξὺ δυεῖν κόλπων, τοῦ τε Ἱππωνιάτου, ὃν Αντίοχος Ναπητῖνον εἴρηκε, καὶ τοῦ Σκυλλητικοῦ" [This very isthmus is 160 stadii between two gulfs, the gulf of Vibo, which Antiochus called Napetinus, and the gulf of Squillace]), became the Lāmātus river (< lāma, therefore "muddy creek," < "mud"), modern river name Amato, and which the Greeks took from the Brutii and hellenized as Λαμῆτος > Lamezia, there was an implication that indigenous populations had been replaced by others, invaders. In other words, Italic tribes such as the Oscans [Brutii] had arrived sweeping away or submerging the pre-Italic Indo-European or even pre-Indo-European populations, while subsequently Greeks arrived and forced symbiosis between Greeks and Oscans, occupying a part of their territory. It would be interesting to have Gnerre's comments on the present instability of hydronyms and its implications in the case of the Jivaros.

*

Maurizio GNERRE explains that the Shuar arrived in the Upper Amazon five or six centuries ago, maybe sweeping away the former population or occupying a territory that, for some reason, was uninhabited. They had to change the river names since they could not retrace the old names: in any case, the linguistic rupture was dramatic.

*

Jane HILL points out that, in considering toponyms and especially hydronyms, it is necessary to control variables such as length of occupancy and ideology. In the case of the Shuar, very transparent toponyms seem to represent a fairly recent occupation of the soil, but there might be a different linguistic and ideological distinction between different parts of the world. For example, in the United States, European invaders and Native Americans had different sociolinguistic attitudes, since the former borrowed a great many river names from Native American languages, whereas the latter never used each other's names, or very rarely.

*

Glauco SANGA, coming back to the problem of the arbitrariness of the sign, claims that words are at first meaningful, even more strongly than in the models proposed (onomatopoeia, phonosymbolism and so on). The fact is that words become opaque and arbitrary, but only later; and they become opaque precisely in order to work: to be really used, words have to be opaque, and therefore transparent ones are a small minority. Sanga asserts that languages must be opaque, therefore we can use words automatically, as objects, without considering them in their "real" meaning every time we pronounce them. If we throw a stone, we normally do not calculate the trajectory mathematically: we just throw it, instinctively. Now culture is precisely what transforms meaningful events into instinctive ones; and language transforms meaningful words into instinctive ones, arbitrary words. But this is a long way from claiming that words are truly arbitrary.

Thought

Daniel Fabre

The Symbolic Uses of Nature

It would be interesting to orient contributions in this section towards three related themes, wide-ranging enough to allow for an open and topical discussion. Rather than engaging upon a critical discussion of the general problems proposed, I am asking contributors to reformulate the following according to their own point of view and their own experiences, in the hope that a fruitful dialog will arise from the confrontation between the various positions.

1. What does anthropology mean when it talks about thought? It would appear that this term is used not merely to indicate the combined activities of the spirit, both conscious and unconscious, but also the place from which these activities emanate. In short, we can say that the anthropologist does not locate thought inside the spirit,—for example, in the flow of psychic events—but outside, in the public use of signs, in symbolic exchange. It would appear, therefore, that for the anthrolopologist there is no spirit that is not objective. Thoughts are filled with content in a historic context, in the greater or lesser *hic* and *nunc* of a tradition, customs, institutions. This general position poses the problem of the relationship with what at present is by common consent called cognivitism, and it is here that a real debate may arise.

2. That which we define as thought can perfectly well be replaced with what we recognize as knowledge, for example local nature knowledge. Emphasis has often been placed on the problem posed by the empirical approach to objects of science on the part of ethnoscientists; in particular since ethnobotanical, ethnozoological, ethnometeorological categories etc. are the result of the dismemberment of ancient natural history and precede more abstract redefinitions based on physical/chemical categories or on methods of observation and analysis. In fact, in anthropology, the object is not the botanical, zoological or meteorological knowledge of a group, but the contents of thought, of which these areas of practical knowledge are merely occasions and supports. Furthermore, in ethnoscientific questionnaires there is often a point that is not expressed, or badly defined with respect to man himself, who uses and produces knowledge and thought, and who, as a bearer of customs, institutions, traditions, should thus be the only totalizing object of the enquiry and the analysis. This suggests that we should examine the theoretical status of the microspecialties produced by the plural notion of"knowledges."

3. Anthropology has not generally given up the possibility of passing from the local to the general, from particular societies to man as such. What has happened, today, to the program of universals? It remains a horizon that is generally shared and it can be challenged or widened, investigating the frontier of animality and the founding features of mankind. Nonetheless, the solutions proposed are connected in a somewhat contradictory manner with the debate that we have summarized above. In fact, to define universals as logical constrictions inherent to the human

spirit or as obligatory perceptions that give shape to the archaic themes of savage thought, means relocating thought inside the spirit, naturalizing or desocializing it. At this point one could justifiably ask whether, to observe thought in action, there is not a more direct route than that of ethnographic research into all sorts of different societies. Which means, however, depriving anthropology of that on which its very specificity is based.

CHAPTER 12

Thought of Nature and Cosmology

Jean-Pierre Albert

This session on "Thought. Symbolism: nature as model and metaphor" invites us to explore the meanings of the notion of "thought" that go beyond those dealt with so far, especially as regards problems of classification. An initial direction to be explored (not systematically considered by any of the participants) could be the study of the most explicit forms of symbolism, those using natural realities as "models" or "metaphors" for social realities. Indeed nature supplies political thought with models in the twofold sense of organizational forms and rule-prescribing references, if we bear in mind the importance of comparing society with living organisms made by thinkers from Plato to Auguste Comte. As regards "naturalistic" metaphors, we need look no further than the extremely rich literary genre of the fable. Without going into the problems raised by the possibility of self-conscious symbolism, I should like to focus on two ideas. Firstly, these kind of references to nature map out completely heterogeneous cultural areas. To my mind it is pointless to attempt to unite in one "mythical thought" all the uses of the metaphor (in the widest sense) so different in intention and modality. Secondly, as in naturalistic models of political philosophy, in the fable we also find the play of a vision of nature as a whole. In other words, general principles govern the representation of relations between phenomena or different parts of reality: the principle of finality in organicist models of society, and the principle of equilibrium or compensation in the "wisdom" of the fable. These general models—I will continue with an analysis of them later— also express a "thought of nature" that is different from particular forms of "knowledge"; this thought is to them what philosophy is to the sciences.

Two questions (at least) arise concerning these models: the question of their relation with religious representations (bearers of cosmological frameworks) and the question of their possible universality. I will deal with the first question at greater length since the answer to it has a direct bearing on how we handle the second.

Thought of Nature and Religious Thought

One of the aims of this conference is to stimulate interest in interdisciplinary exchanges. I hope I will not betray this aim in describing the rather specific reasons inducing me to reflect on forms of knowledge about nature without being a specialist on this subject. My main research concern is in fact the anthropology of religions. And that is why I am involved in the ongoing debate concerning knowledge about nature. The very idea of religion (the possibility of giving a single definition of cultural phenomena empirically labeled as "religious") has been the subject of considerable criticism over the past few decades. On one hand, the supposed universal nature of the determinants most often invoked in constructing such a definition (the ideas of the sacred, supernatural, the very general presence of ritual forms such as sacrifice, etc.)[1] has explicitly been attacked. On the other hand—and this mainly concerns the work of the French anthropologists—the vision of Claude Lévi-Strauss and the work it has inspired tend to class the religious with the "symbolic" or "cultural" in general.[2] In both cases the specific dimension of the religious is challenged.

These trends actually link up with the earliest assumptions of social anthropology, especially the relativist tendencies already present, for example, in Durkheim. Thus on the subject of the "primitive man" in *Les formes élémentaires de la vie religieuse*, he claims: "the rites he [primitive man] enacts to guarantee the fertility of the soil or the fecundity of animal species he eats, are no more irrational to him than the technical procedures adopted by agronomists for the same purposes are to us. The powers he brings into play through different means have nothing particularly mysterious about them... For those who believe in them [these forces] are no more unintelligible than gravity or electricity is for today's physicist" (1985: 35). If these statements are true, then clearly there is no boundary between rite and technique, and irrational belief and empirical knowledge. Each culture produces in its own way ("for those who believe") a vision of the world, which rationally justifies the means employed for acting on the real. Given that most of these means have been ineffective and the ideas mistaken, we may well wonder how humanity has survived so much blundering and wasted energy. Hence the argument of the advocates of the universality and innate nature of certain cognitive abilities, those selected for their fitness in contributing to the survival of the species.[3]

Thus the debates between ethnosciences and symbolic anthropology have an exact parallel in the debate concerning the specificity of the religious: in both cases a general theory of "symbolism" and the unity of culture are used to challenge the existence (or importance) of specific cognitive mechanisms to which experience of the world corresponds and perhaps also the specific cultural fields themselves. The idea of culture divided into relatively independent areas which leads to an immediate mapping out of the boundaries between factual truth and fiction, knowledge and belief, etc., is accepted by all, provided we are dealing with the West. We have even less reason to deny "distant others" similar notions to such distinctions resting—at least in part—on the activation of universal mind-sets.

In short, the idea of specifically religious facts recognized as such (even in variable terms) in all cultures implies inversely the construction of a stable ordinary experience of the world. Choosing the first option, as I have done, inevitably implies the second. Indeed the results of cognitive psychology have produced a decisive argument to be added to the controversial debate on the specificity of the religious. I feel justified in seeking, as Pascal Boyer does in his book *La religion comme phénomène naturel*, cognitive mechanisms, again, universal, underlying the position of a "world of religion."

Cosmology and Religion

So far nothing of what has been said enables us to set a clear-cut boundary between knowledge of nature and religious ideas. In particular the notion of "mythical thought" often invoked to describe all cultural ideas outside the field of rational or empirical knowledge confuses rather than illuminates the debate. Although, for example, folk stories are permeated with "mythical textures," both narrators and listeners clearly recognize that they are pure fiction and not part of a religious representation of the world.[4] Similarly, if we agree with Pierre Bourdieu (1982) that Montesquieu's "theory of climates" is a genuine mythology, nothing induces us to consider such texts within a kind of religious conception of reality. This problem arises, in fact, with everything that may be described as "erudite mythologies" or "indigenous theories": any propositions within an apparently rational cognitive project—ideas about the world marked by the ways of thinking and perhaps also the objectives of myth.

The register of natural realities lends itself particularly well to such ideas. Even excluding the most erudite forms, we easily find general principles at work in certain fields of knowledge. These principles correspond to ideas about nature in general, its most fundamental laws, and which therefore we may describe as cosmological. For example, popular wisdom about weather, found in the countless sayings gathered by folklorists, have a very clear principle of equilibrium: the French proverb "Christmas on the balcony, Easter at the hearth" sums up the idea of a necessary equilibrium or a obligatory ultimate rebalancing for momentarily occurring anomalies.

Predictions of this kind are of course very shaky. Although the most they suggest are effective meteorological possibilities, they tend to give the impression of an imaginary control over very random processes by setting them in a deterministic framework. At the same time claiming that sooner or later it will be cold in winter is simply comfortably embracing the certainty that there is an order conforming to how the world should be. On the other hand, the facetious story of the "rainmaker" who, claiming to outdo his creator, upsets everyone, shows that genuine wisdom is concealed in the apparent whims of the meteorological phenomena. This lesson links up with the more usual moral of fables.[5]

Similar constructions are found in more elaborate forms of knowledge about nature. Thus in Pliny the Elder's *Natural History*, we find the idea that each animal (or at least each powerful or dangerous animal) has a rival capable of overwhelming it: thus the elephant is opposed by the dragon, the asp by the mongoose, etc. In his work we also find a hierarchy of natural realities. The principle of equilibrium is a consequence of the principle of hierarchy: the balance of forces between animal species helps keep each one in its place. We could give other examples of these kinds of schemes. More or less explicit and more or less elaborate, they are applied both to nature and social phenomena. There is thus—to my mind—a kind of powerful correlation between the principles of a "limited good," especially perceptible in the interpretation of the inequalities of chance and wealth, and the more widespread scheme of "finite worlds" or "closed worlds" underlying theories such as astrology, the thought of signatures, and more generally the image of the universe as a finalist whole. Lastly, we note the totalizing potential of the "physics" theory of the Four Elements (with their determination according to the pairing of hot/cold and dry/wet) and its biological counterpart, the theory of humors as illustrated by Francis Zimmerman in chapter 16.

I am aware of the unusual character of the list I have just illustrated. It is in fact an attempt to redistribute the terms according to two criteria: according to their greater or lesser rationality (or empiricality) and their more or less "naturalistic" character. We would thus create alongside an almost homogeneous list of prescientific theories, a borderline area in which cosmological models would become indistinguishable from strictly religious hypotheses: the idea of an ordered world merges with that of Providence, the horizon of the closed world with cosmological ideas underlying sacrificial practices, the supposed influence of the stars with figures of magical or religious causality, etc. This would lead us to deny, at least in the second group, the specificity of religious representations in relation to other kinds of constructions of reality, and thus re-introduce in a different form the theory I claimed to be insufficient earlier.

To clarify this difficult issue I can see only two solutions. The first is to adopt a Kantian attitude (later also adopted in more or less the same form by Auguste Comte) consisting in returning cosmology to the side of metaphysics on the grounds of a restrictive view of the limits of experience and the verification methods of scientific knowledge. Moreover, if we accept the Nietzschean point of view that metaphysics is simply an offspring of religious thought, we must conclude that cosmological speculations are themselves an offspring of religion, a pseudo-rational form of knowledge, and even more so when they claim to solve questions of meaning, value and origin.[6] The second solution is to take into account the indigenous points of view that distinguish between that which concerns religion and that which concerns knowledge about nature. Under the label of "natural history," Pliny, for example, provided an often fanciful vision of nature but continually distinguished his subject from the reverie of "magicians" and the "lies of the Greeks" (i.e., their mythical beliefs). Similarly, in the Middle Ages, a writer like Raymond Lulle was very careful to situate his *Treatise of Astrology* in the register of natural sciences, to avoid the heretical implications of a fatalist determinism, which might suggest a magical-religious conception of astrology. The same rationalist need is found in the field of "natural questions," when it comes to proposing a naturalistic explanation for strange or frightening phenomena—lighting, volcanoes, earthquakes—which popular opinion attributes to the action of the gods.[7] This means that indigenous thought or more generally the resources of myth may be enlisted in intellectual projects claiming only to be situated in the sphere of the religious. These may be self-evident ideas but inevitably further augment the above-mentioned confusion between the symbolic and the religious.

In fact, the two points of view just described are not actually contradictory. The first is an external judgment, while the second takes into account indigenous representations. However it is conceivable—without contradiction—that a logic objectively influenced by a religious vision of the world can subjectively be enlisted in an argument against certain explicit aspects of a religious view of the world. This would be enough to make manifest the existence in a given cultural area of a division between what is religious and what is not. Would we be conceding too much to relativism if we admitted that this division, even if always applied, is still vague as regard its content? As we have seen, the question does not arise for the thought of natural beings concerns only elementary classification operations, ontological categorization, etc. as identified by cognitive psychology and ethnosciences: I have admitted that they create a "horizon of the natural" (and similarly of pragmatic effectiveness) which never wholly fits into the symbolic (or the cultural), the universality of their effects being the criterion of their logical priority in relation to everything set in a cultural form. Does the probable existence of cosmological universals enable us to pursue a similar line of reasoning? The answer lies in investigating the origins of these ideas and the visions of "nature" that they imply.

Whence Cosmological Universals?

Let us return to the evolutionist argument in favor of the thesis of universals in knowledge of natural realities. Its strength lies in the idea that beyond certain cultural developments, there is a core of representations more or less faithfully reflecting the real, thus inspiring effective behavior from the adaptive point of view. Hence the theory that cognitive abilities to build such representations is the outcome of the evolution of the species and is related to its genetic pool. Thus initially the idea that certain cognitive schemes are innate is derived from a vital need for truth. To support this logic using an independent argument, we have the experimental results of cognitive psychology showing the existence, for example, of an "innate idea" of the animal which does not contradict that suggested by contemporary biology. The existence of universals is thus related to the claim of their innateness and the approximate truth of the representations they imply, this approximate truth being the condition for effective action and consequently their adaptive value.

Supposing that this theory is correct, we may still make the following three remarks: (1) Alongside the innate aptitudes to build a more of less exact representation of nature, there could be equally innate aptitudes to present a mistaken version—for example, a religious version. (2) False ideas about reality may inspire effective practice and, as is much more often the case, neutral uninfluential action from an adaptive point of view. (3) When applied to humanity, the notion of adaptation cannot only be seen from a biological point of view but must also take into account the needs of life in society i.e., the classic hypothesis of sociological functionalism, which is far from necessarily being compatible with the naturalist functionalism mentioned so far.

These three observations have a direct bearing on our subject. Firstly, we may suppose that the more general constraints on social life (at least, the fact that a social order is always conditioned by obeying rules and not a pure product of objective determinisms) introduce an aspect of universality. Similarly, societies of the same size, level of technical and economic development, etc. have analogous solutions in their way of setting up and legitimating their rule-making procedures. Secondly, we note that the religious dimension is usually implicit in rule-making procedures. It is significant that the authority for social rules—by essence rationally unjustifiable[9]—rests on the mismatch between the representation of the world and that of our immediate experience: religion may well be a sociocognitive arrangement intended to accredit the unbelievable, i.e., statements in contrast with the world of religion to which we spontaneously surrender our individual cognitive abilities.[10]

At the same time the most general condition of knowledge and action is to suppose that the correlatives of our awareness are objective in form. Inevitably, then, the realities postulated by religion, although distinct from natural realities, also tends towards the form of objectivity. This means that religion gives an objective form (or origin) to principles of action, obligations, rules for circulation and trade, etc. These are laws, then, and not actions. The religious illusion thus lies in describing a world in which the factual realities (gods, souls of the dead, *mana*, etc.) would also be laws or sources of obligation. The sphere of laws, however, is all the more clearly distinguished from the sphere of actions (including those in the religious area) the closer it is to being a social model of authority: a personal god, for example, appears more as a maker of laws than a law in of itself. This is not necessarily the case in governing behavior when supernatural entities take the form of impersonal principles, such as *mana* and *hau* studied by Marcel Mauss (1950). These principles, if we follow indigenous logic, tend to act as laws of nature. Similarly, nothing seems more common in the most varied mythological and rituals systems than the idea of continuity, or a community of essence between the order of nature (cosmic order) and social order. Indeed, the two areas of rules can be joined simply by exchanging some of their features: the social order gains in terms of presuming necessity and universality; nature, on the other hand, becomes a moral entity, a world of rules and not laws (in the sense of deterministic laws). In plainer words, nature tends to become a projection of society, while its supposed laws at cosmic level are a reflection of the more general constraints of social life.

We thus understand how cosmological universals may exist, not as expressions of innate mindsets, but as correlatives of the more general constraints of social life. The fact that several of the cosmological principles mentioned above—the principle of equilibrium, and the principal of finality—directly evoke social ideas corroborate this hypothesis. More specialized theories, such as that of the Four Elements and the Humors also fit into this pattern, since they are inseparable from the thought of equilibrium. Nature conceived in this way lays down duties, but it, too, has duties to be respected. Its existence as an organized (or finalized) totality means that the sphere of facts is inseparable from the sphere of values. Having become a quasi subject, nature acts, calculates, plays with its own rules and at times even goes so far as to violate them.

You may all well have guessed what I am driving at. When defined in this way, is nature not a religious-type operator? This was Auguste Comte's feeling concerning the uses of the notion of the "metaphysical state" by philosophers.[11] Similarly, we may easily demonstrate that the cosmological principles mentioned so far are a superficially secularized version of religious ideas. In short, we can claim that the idea of "nature in general" (in those cultures where it exists) is *never* a naturalist representation. Equally we cannot ignore that the *naturalization* of principles whose relevance is mainly of a social order (for example, a principle of balanced exchange or of compensation) refers to a representation of nature as the place of the immutable, of order and of necessity: characteristics that can be easily deduced from the most rudimentary classification operations, ontological imputations, etc., as the most immediate constraints on effective action. In this sense nature is a twofold model: it enables us to think of general features of reality and action, and it is valid as a law insofar as phenomena conforming to the criteria of the natural (order and necessity) are more valid from a practical point of view than random and disordered phenomena. As P. Boyer has stressed, it is not surprising that the supernatural world includes many features from the natural world. Only on this condition can it interact with humanity and become an expedient, because what it shares with nature renders it to a certain extent predictable and manipulable.

What conclusions may be drawn? I have tried to show that admitting the consistency of a "horizon of the natural" allows us through inverse logic to suppose the existence of a specific area of the religious. This general proposition based on the idea of an innate aptitude to distinguish between ontological types then seemed to be undermined by the existence of universal cosmological principles. But the identity of such principles—religious? naturalist?—is still vague. These principles could be brought closer together more on grounds of their common social origin and the way they work, than because of explicitly religious representations. Equally however certain features of the natural—given at the same time as our most elementary experience of knowledge and of action (for example, a certain idea of determination)—are part of the construction of a normative vision of reality (both natural and supernatural). We thus grasp much better the expedient of legitimacy underlying all naturalization of the social sphere. As for the controversial claim of an "sphere of nature" set against the area of the supernatural, it represents an immanent possibility in that construction whose most radical expressions are also the most unlikely: far from the naturalist thought of nature, the straightforward naturalization of the supernatural seems in most cases to be enough to curb all the disturbing power of the gods.

Notes

1. On the sacred, see Isambert 1982; for the supernatural Pouillon 1993; and for sacrifice Detienne 1979.
2. Summing up Lévi Strauss' contribution, M. Izard and P. Smith comment in the preface to the collected essays *La fonction symbolique:* "The religious is thus of the contingent and local, whereas symbolism is sealed by the necessary and the universal" (1979: 13).
3. See, for example, D. Sperber 1996.
4. On this subject see the issue of *Ethnologie française,* (I, 1993) with the proceedings of the workshop on mythical textures.
5. For other examples of cosmological reflection in ordinary thought, see M. Albert-Llorca, (1991: ch. VIII).
6. This is more or less the argument in my article "La ruche d'Aristote: science, philosophie, mythologie" (1989).
7. This polemical aim is very obvious in the sixth book of *De rerum natura,* in which Lucretius tackles a list of "natural questions," very much like that dealt with by Seneca in *Naturales quaestiones.*
8. These remarks are not really objections. They are in any case taken into account by the advocates of evolutionist theories.
9. This point requires lengthy discussion. I would simply like to point out that one of the most consistent attempts to rationalize political authority, made by Rousseau in *Du contrat social,* ends with a long chapter on "civil religion" and the sacrality of the law.
10. For a more complete account of this theory, see J.-P. Albert 1997.
11. See the first lesson of the *Cours de philosophie positive,* (1830: 28). The idea of nature, as used by the nineteenth-century philosophers, is explicitly described in this work as one of an "intermediary conception of a crossbred character," suitable for making the transition from theological philosophy to positive philosophy.

References

Albert, J.-P. 1989. La ruche d'Aristote: science, philosophie, mythologie. *L'Homme.* 110: 94–116.

———— 1997. *Le sang et le Ciel. Les saintes mystiques dans le monde chrétien.* Paris: Aubier.

Albert-Llorca, M. 1991. *L'ordre des choses. Les récits d'origine des animaux et des plantes en Europe.* Paris: Editions du C.T.H.S.

Bourdieu, P. 1982. *Ce que parler veut dire. L'économie des échanges linguistiques.* Paris: Fayard.

Boyer, P. 1997. *La religion comme phénomène naturel.* Paris: Bayard Editions.

Comte, A. 1830. *Cours de philosophie positive* (première leçon). Paris: Antrhropos (réimpression: 1969)

Detienne, M. and J.-P. Vernant, eds. 1979. *La cuisine du sacrifice en pays grec.* Paris: Gallimard.

Durkheim, E. 1985. *Les formes élémentaires de la vie religieuse.* Paris: (1ére éd.: 1912) PUF.

Isambert, F. A. 1982. *Le sens du sacré. Fête et religion populaire.* Paris: Editions de Minuit.

Izard, M. and P. SMITH. 1979. *La fonction symbolique. Essais d'anthropologie.* Paris: Gallimard.

Mauss, M. [1950]. *Sociologie et anthropologie.* Reprint Paris: 1985. PUF.

Pouillon, J. 1993. *Le cru et le su.* Paris: Editions du Suil

Sperber, D. 1996. *La contagion des idées.* Paris: Odile Jacob.

CHAPTER 13

Symbolic Anthropology and Ethnoscience: Two Paradigms

Marlène Albert-Llorca

Daniel Fabre's second question asks contributors to consider the status of the ethnosciences (here the term stands for disciplines focusing on the study of indigenous knowledge: ethnobotany, ethnogeology, etc). The initial idea is that there is no "exact correspondence between what is described as 'thought' and the notion of 'knowledge,' such as local nature knowledge." I would like to focus on this issue.

First, I must point out that I take the notion of "thought"— as suggested by the subtitle of the session "nature as symbol, model and metaphor"—to mean "symbolic thought." I also take for granted the concept of symbolism proposed by C. Lévi-Strauss that there are no "natural symbols." An object assumes a signifying value only because it is correlated and opposed to another, and its meaning cannot be rendered without highlighting the system of relations in which it is included. Similarly I share—*cum grano salis*—the idea that "indigenous thought" rests on a "science of the concrete," i.e., thought responsive to the sensible features of human and natural reality. Does that mean that we should distinguish between ethnosciences, or to use the French equivalent of that notion, *savoir naturalistes* (nature knowledge), and symbolic elaborations? I should like to try and show that this distinction cannot be taken for granted. Indigenous or popular knowledge of nature is not always the simple empirical record of observable data. It may also be the outcome of the play of symbolic thought itself. This then makes employing the notion of ethnosciences problematic.

Ethnoscience

The notion of ethnoscience, like that of nature knowledge, is not neutral. It presupposes that there is no divide between "indigenous" knowledge and the sciences of nature or rather Western natural history. Talk of ethnosciences suggests, on the one hand, that people have more than a purely utilitarian relationship with nature, that the interest in itself also has a speculative dimension. On the other hand, it implies that forms of knowledge produced by "common sense" and naturalists rests on a mental outlook common to the whole of humanity (Atran 1986). Hence the importance of work that clarifies the "indigenous" classification of fauna and flora, such as the theory advanced by B. Berlin: 1). Such classifications rest on "observable affinities of the species themselves and not on the current or potential cultural importance of these species" (see B. Berlin, ch.1 in this volume); distinctions and groupings created in all cultures rest on morphological and/or ecoethological criteria. 2). These classifications always take the form of taxonomy, i.e.,"a hierarchical classification so that at a given level of hierarchy all the categories are mutually exclusive…: no animal for example can belong to two species, no species can belong to two genera, etc."(Sperber 1975: 12).

The importance of this theory—a radical break with the relativist concepts defended by B.L. Whorf (1956)—illustrates that research in the ethnosciences has tended to focus on highlighting local ethnotaxonomies. But this tendency leads to an interest in only one part of the"indigenous"knowledge of nature.

We may allow (although it is far from self-evident)[1] that all societies classify natural reality according to morphological and ecoethological criteria. But we must equally admit that knowledge of flora and fauna is not limited to knowing about this type of feature. Knowing an animal or plant species also implies knowing how to use it: how and when to hunt an animal, if it is a wild animal; knowing how to feed and look after a domestic animal; knowing if and how it is edible and how to cook it, etc. Narratives concerning the origin of animals and plants that I have studied in Europe (Albert-Llorca 1991) are very revealing about the importance and status accorded to this kind of knowledge. Many insisted on explaining the morphological, ethological or ecological features determining the animal or plant species (why rosemary has blue flowers, why whales live in water, etc.) but others justify in a similar way cultural uses. Thus, for example, a Catalan story explains why hemlock is poisonous—and that is true—but another claims that you will fall ill if you eat jay meat in the period from Holy Week to St. John's Day (Amades 1994: 140; 1988: 189). The first proposition has an empirical basis; the second is obviously based on a cultural influence, but the harmfulness of jay meat is still included in the narrative in terms of a"natural"feature of the species in question. The statement about jay flesh is not"nature"knowledge, but it is difficult to see how it can be left out of an inquiry into the "ethnosciences."This kind of knowledge is not unrelated to symbolic thought. I should like to take another example from European Christian culture to illustrate this point.

Knowledge and Symbolic Thought

Most narratives concerning the origin of the bee collected in Europe define it as an animal producing honey "for use by men,"and wax "to honor God."In fact in Christendom wax has very important ritual uses. Candles are burned to saints when an act of grace is being asked for; they are lit around the dead; canon law demands that two candles at least partly made of beeswax should burn on the altar during the Eucharist; lastly, from Easter to Ascension a large candle, called the Paschal candle, studded with five grains of incense representing the wounds of Christ, is burned in churches. These uses are justified by a metaphorical relation between wax and the body of Christ described explicitly in the medieval ritual texts: according to Guillaume Durand,"wax represents Christ because of three properties. The snuff stands for the soul, the wax the body and the light the divinity of Christ (1854: 142)."We see, therefore, why a candle is burned during the period when the resurrected Christ is bodily present on earth and why two candles are burned on the altar during the ritual re-enacting his sacrifice. Christ is the only victim that may be offered as a sacrifice to God because he is absolutely pure. His "natural" equivalent, the wax, must therefore be equally pure. This led medieval theologians to turn to ancient naturalists for the idea, on the one hand, that wax is "begotten" by bees, and on the other the idea that bees reproduce without copulating. Bees were said to be born either from a seed gathered on flowers or from the corpse of a dead ox. The equivalence between wax and Christ can thus be taken to its logical conclusion: Christ was made flesh in the body of an absolutely pure woman (the Virgin Mary); wax is engendered by a virgin, the bee. The purity of wax and consequently its liturgical uses are thus entirely justified.

The beekeepers' thought is less explicit than that of the medieval theologians. It is expressed by actions rather than words, but its coherence, as I have tried to demonstrate at greater length elsewhere (Albert-Llorca 1995: 143–63) is also based on the religious importance of wax in Christian culture. Here is the crux of the issue: the bee is considered to be a sacred animal (the "fly of the Good Lord," as the saying goes in the Morvan); it is forbidden, for example, to kill it or treat the bee as a "beast" because it produces a substance intended for God. This also explains why beekeeping was often practiced by hermits, monks or priests. Of course there were also lay beekeepers, but they were not all so well qualified for the task. It was actually claimed that bees attacked drunkards, the debauched, and red-haired people "who have a special smell," due—as Y. Verdier has pointed out—to the impure blood of those who stained them: they were said to have been conceived during the menstrual period (1979: 47). This horror of impurity would be puzzling without reconstructing the metaphorical background to the association of wax with Christ and the bee with the Virgin Mary.

Is this knowledge a "belief"? The answer is yes, if we include it in the category of statements never subject to scientific experiment or which contradict the principles of scientific thought. Claiming that bees attack debauched people means investing them with moral sentiments that an insect cannot possibly have. Claiming bees dislike certain smells is a hypothesis never tested by rigorous experimentation. However one of these beliefs does function as a form of knowledge: namely, beekeepers' stories mentioning the bee's sensibility to smell, which they take into account in their practice.

At this point the notions of ethnoscience or nature knowledge seem to be problematic, since they suggest that this kind of knowledge we have been describing should be left out of research or at least appear to isolate it arbitrarily from other forms of knowledge. The example of the bee seems to demonstrate that there is not a theology of the bee, on one hand, and an ethnozoology of the bee, on the other, but rather a set of forms of knowledge—whether true or false—based on the nature of things or on cultural-type presuppositions. These forms of knowledge are only coherent when seen as part of symbolic thought about the bee. Conversely, this does not mean that symbolism is outside the world, as the distinction between "symbolic and mundane classifications" suggests. The association between the Virgin and the bee rests on social practice and at the same time induces social practice.

In other words, knowing what a bee is implies not only knowing that it is an insect that builds hives and produces honey and wax, but also knowing that it is"a creature of the Good Lord," i.e., knowing its place in the world conceived as an ordered hierarchical whole.[2] The idea of the world is not the same as the idea of nature, which in order to develop had to carefully isolate natural history. Claiming that a bee is an insect means situating it in the animal kingdom. Saying it is a creature of God means giving it a place—which we might describe schematically as a mediator—in a universe also containing men and in which the existence of God is posited. At this point we are no longer dealing with a naturalist thought of nature but a cosmology, which as the philosopher Kant stressed is something completely different: natural history belongs to knowledge—it can be confirmed or refuted by experiment. Cosmology depends on thought, that is on the attempt—bound from the outset to defeat but inevitable, because intrinsically linked to the nature of human Reason, according to Kant—to unify the wholeness of experience. No more than Western metaphysics, the societies we study have never managed to produce entirely coherent cosmologies. Unlike what Lévi Strauss claims, there is no"dynamic and global taxonomy" incorporating"botanical and zoological taxonomies" (1962: 15). But I feel that the anthropologists will not be doing their duty if they fail to take into account the efforts made by people to construct a world endowed with a certain degree of coherence.

Notes

1. On this point see P. Descola's critique (1986: 103ff).
2. In his contribution (see ch.2 in this volume), Roy F. Ellen stresses that"it is impossible, for example, to make sense of Austronesian terms and categories for'bird' and for'tree' without considering utilitarian and symbolic criteria."I completely agree with this idea.

References

Albert-Llorca, M. 1991. *L'ordre des choses. Les récits d'origine des animaux et des plantes en Europe*. Paris: Editions du C.T.H.S.

_____ 1988. Les servantes du Seigneur. L'abeille et ses oeuvres. Terrain n.10: 23–36. In *Nel paese del tempo. Antropologia dell'Europa cristiana*, edited by G. Charuty. Naples: Liguori editore.

Amades, J. 1988. *L'origine des bêtes. Petite cosmogonie catalane.* trans. by M. Albert-Llorca. Carcassonne: GARAE.

_____ 1994. *Des étoiles aux plantes. Petite cosmogonie catalane* trans. by M. Albert-Llorca. Carcassonne: GARAE / Press Universitaires du Mirail.

Atran, S. 1986. *Fondements de l'histoire naturelle. Pour une anthropologie de la Science.* Paris: Editions Complexe.

Descola, P. 1986. *La nature domestique. Symbolisme et praxis dans l'écologie des Achuar.* Paris: Fondation Singer-Polignac et editions de la M.S.H.

Durand, G. 1854. *Rational, ou Manuel des divins offices.* Trad. C.Barthelémy. Paris: LouisVivès 5 vol.

Lévi-Strauss, C. 1962. *La pensée sauvage.* Paris: Plon.

Sperber, D. 1975."Pourquoi les animaux parfaits, les hybrides et les monstres sont-ils bons à penser symboliquement ? *L'Homme* XV (2): 5–34.

Verdier, Y. 1979. *Façons de dire, façons de faire.* Paris: Gallimard.

Whorf, B. L. 1956. *Language, Though and Reality.* New-York: Wiley & Sons.

CHAPTER 14

Doing, Thinking, Saying

Giulio Angioni

It is difficult to deny that in everyman there are differences in ways of knowing and thinking (Horton and Finnegan 1973), in addition to cultural differences and differences depending on the different stages in human life (Piaget 1970). Here I wish to address the question of the way of knowing and thinking that develops through actions, what is expected from such actions, and how they are coordinated. We shall not neglect images and concrete symbolism, nor conventional systems of signs, notably verbal language, by far the most important for humankind. We must point out, however, that even among anthropologists who study naturalistic knowledge there appears to be too much emphasis on an approach that privileges linguistic means of knowing and ordering, even to the point of considering them the only ones. Thus, naturalistic knowledge seems to be overly reduced to the operations of naming, classifying, and categorizing.

We must of course recognize the fundamental role of language in focalizing, preserving, and rescuing knowledge. But we reject here (for example with Piaget) the idea that language is the foundation of all thought, and the consequent idea that thought is only what can be uttered, with thought expressed in language considered to be the best, most complete and effective form of thought. As concerns anthropological studies, it is sufficient to mention the strong theories of linguistic-cultural relativism (Whorf 1956) to understand the risks of the misleading exaggeration of a pansemiological or panlinguistic vision of culture and human life or simply of the thought and visions of the natural world (Cirese 1984; 1988).

The existence and importance of the traditional knowledge implicit in "doing" and the more or less great and lasting "pretechnological" operating traditions are well known and self-evident. Here, I mean by "technology" the complex whole of notions and comments on the practical application of specialized scientific know-how that characterizes modern life (Angioni 1986). We are dealing with experiences, abilities, and knowledge incorporated in the individual as a single person, as part of a social class or group or an entire society. This represents both a knowledge and a capacity to do that direct and guide the performance of certain jobs, or which, more generally, also constitute a common foundation of notions and abilities, a "humus of tradition," a preliminary and common background necessary for individual "specializations," even those having to do with sex, age, caste, social class and so on (Leroi-Gourhan 1945; 1964).

This "common background" shows, more clearly than any specializations, its implicit nature, its belonging to a stage preceding theorizing, and explains the difficulty in describing it in words experienced by those directly involved; it shows up just as clearly both in primitive societies, where the common fund of knowledge and abilities tends towards omniscience (in the sense that all members of the group know everything there is to know, with distinctions evident only between the sexes and the different age groups), and in industrialized societies.

Despite innumerable differentiations and specializations, all members know and can do certain things before receiving formal education and becoming specialized: if, in fact, in a primitive society all adult individuals of the same sex can, let us say, hunt or gather, in our societies everyone is capable of tying shoelaces, of stopping and going at traffic lights, and of turning on the television set, even when they still cannot read or write. These are all things that count; they allow us to live our daily lives and forge individuals into normal members of a group. And even today, as in pre-industrial societies, this patrimony of abilities and elementary notions is learned and becomes incorporated mostly by impregnation, inference, and experience. In our societies, almost nobody consciously teaches others or learns from others how to behave as a generic male or female individual, despite the scholastic teaching of what is called sex education.

To sum up, informal knowledge and know-how carry more weight in pretechnological societies, but they are still important even in postmodern societies in which formal education overshadows and tends to become even more so. Institutional education is still far from canceling out the knowledge acquired spontaneously and implicitly, by impregnation. In traditional European societies, both urban and rural, operational knowledge and abilities had and still have as their most important characteristic this non-formal aspect, this acquisition by impregnation and in the lack·of explicit communication.

Even the most advanced modern technology would not have practical meaning were not for the application of man to machine and the consequence of the human use of machinery: even this knowledge of the practical use of machines is, and for a long time to come destined to remain, informal to a great extent, because one cannot learn and teach fully except through practical experience, as everyone discovers when learning to drive a car or ride a horse and, even before that, in learning to stand erect, something our species and every individual learns at the beginning of life (De Martino 1977). All these things are for the most part unutterable, or at least are not to be found completely in semiosis. For example, everybody knows the uselessness of manuals and lectures about computers, as compared to learning their use through trial and error. Regarding technical, and thus also naturalistic knowledge the world we call traditional and pre-industrial is to a greater or lesser extent immersed in and confined to this single dimension of practical knowledge which, however, still remains the most important and, in certain aspects and cases the truest form of practical knowledge, even for us.

Basically, man in general does not perform only semiotic operations such as naming and classifying: together with these, and before these, he acts to satisfy needs; he thinks of nature so as to "do" and not only and not principally so as to "say." Of course, in complex societies in all times and places there are and there have been categories of thinkers who speak and categories of thinkers who do; this is but another way of expressing the division between manual and intellectual work. And this inveterate division also explains to a certain extent the propension

to consider thought more as a question of"saying"rather than as one of"doing,"
more as"intellectual"rather than"manual."

These truisms appear obvious to those who must seek an"informer"to speak
explicitly about his knowledge or know-how concerning nature, and is even more
evident when the knowledge or know-how is obsolete but one is attempting to
reconstruct it through oral memory.

Let us examine an exemplary case. Generally speaking, when informers are
asked for information on things they no longer do, or when they are asked to speak
explicitly about notions and operations that they still perform in their normal
trades, what becomes clear is the lack, or considerable scarcity, of representation
and comment, of verbal or written formulation or any other means for the con-
veyance of knowledge: this is knowledge implicit in practical doing, and operators
prefer to explain by showing, rather than by attempting a verbal description or an
explanation through means that are not operational and ostensive (Goody 1977).

We are thus dealing with a kind of knowledge that is first and foremost operational, technical, non-verbal and to a great extent impossible to express in words,
a knowledge that is implicit in"doing"and finds its explanation in doing. Most
often this knowledge was learned by doing or through the doing of others, with
little room for verbal explanations, formalized reflection, predisposition of the
modus operandi. In fact, when recollection of a certain knowledge is stimulated in
an attempt to recover and communicate a dimension belonging to a person's past,
the problems to be faced are not much different from those that one comes up
against in trying to reproduce gestures and technical procedures with the use of
instruments and raw materials that are still available. Whether we are dealing with
obsolete trades or ones that are still practiced, the solution to the problem is all but
an obligatory one: that is, to explain, through doing, those implicit, incorporated
notions and operational capabilities that, almost exclusively in doing and by doing,
have been learned, perfected and have become manifest and explicit in the action,
often becoming things or modifying things.

It is also significant that the closer the notions and techniques are connected
with a specialized, albeit traditional trade, as is the case of the artisan with respect
to the generic farmer or worker, the more the informer succeeds in being clearer
and more explicit, sometimes with drawings and similar aids.

In the most common cases, we should speak of embodied knowledge and body
memory rather than abstract notions and oral memory. Embodied knowledge and
body memory are dimensions that are rarely considered the subject of reflection
and explicit and conscious speech but, since they have become mechanical concatenations of trains of thought and gestures; they are a kind of second nature,
almost a part of instinct an outcrop in our conscious minds almost exclusively in
the case of accidents, difficult situations, when something disturbs normal utilization. We tie our neckties without having to pay conscious attention and the same
thing can be said for tying our shoelaces, but not when one of them breaks and not
in the case of an adolescent trying to tie his tie for the first time and not even when
we try to tie it after a season spent without wearing one.

This explains why we all do nothing but utter confused words about things we know how to do, and do perfectly well without thinking, like tying our shoelaces when we are still half-asleep in the morning (but perhaps we all remember how difficult this was to learn when we were children). The fact is that these notions and techniques, which usually go under the name of "poor," have developed autonomously and independently from the kind of knowledge and language that we call scientific or technological. They are not learned institutionally as part of our formal education, from teacher to learner, but by implicit inference, by impregnation.

Science and techniques, especially in the wake of the industrial revolution, have developed side-by-side, with the sciences making a strong contribution to the development of techniques, thus leading to what we now term technology. But this must not make us forget that for millions of years knowledge and know-how—practical human activities in general—have developed and progressed through spontaneous leaps and bounds; as André Leroi-Gourhan would say, they have gone ahead to become more and more effective along an itinerary of practical rather than theoretical rationality, in a process of inventions and acquisitions accumulated by doing and in doing, following implicit and almost obligatory tendencies.

Technique is a practical line of reasoning. And technical reasoning is truly explicated only through technical action and becomes concrete as the result of operating. In that it is a line of reasoning, it comes to an end with its product. And it is often difficult to explain, even when it is perfectly feasible on the practical plane.

The technical line of reasoning is rather a way of acting, a way and a means for acting on nature: it teaches us nothing about nature, but it is the aggregate of notions that have become means for acting and operational concepts. That is why I do not feel that Marshall Sahlins (Sahlins 1980) is altogether wrong when he states, albeit a bit too peremptorily, that the intelligent efforts of the operator have been far more important than the tool in the history of humanity, and that the product of labor has increased and improved thanks more to the technical knowledge of producers than to the perfection of their instruments.

If what we have said up to now has any sense, then we must underscore the importance of a documentary effort concerning implicit notions and capabilities acquired in doing and stored in the body memory, in that sort of operational memory that sees to it that the body can operate in performing work without close and continuous control over the mind and the will, like the woman who knits, nurses her baby and watches TV, without paying constant attention, with that sort of memory which, on second thought, is what allows us to operate as effectively as an animal acting by instinct, as in the case of a musician who can perform a piece effectively only when his body, especially his hands and feet, is capable of "doing it on its own," and performs operations that no musical annotation, however perfect, can ever describe adequately.

We are often dealing with a true *recherche du temps perdu*, but which can be recalled to mind, not so much to the memory of speech, to the oral memory, as to the operational memory, the body memory. In the case of crafts, both old and new, those that were practiced fifty, two hundred, perhaps even two thousand years ago, are often still practiced, at least on an sporadic basis, although with the concretions of knowledge and know-how that time has slowly accumulated on them, while others are, or appear to be, completely new.

That knowledge and habits pertain essentially to the body and are embodied is one more reason for interest in them. Indeed, practical doing, more than other human activities, shows up the combining of things and ideas, of doing and knowing how to do, of action and conscience, of body and soul, of the abstract and the concrete, of the symbolic and the practical, of the corporeal and the incorporeal, of the sign and the manual, of language and its extralinguistic referent.

Rather than being knowledge and an immediate and empirical act, in all cultures knowledge and technical action are perhaps, above all, that which imparts sense, besides usefulness, to things, and thus to the operator himself, who sees himself as an operator and feels that he is realized in his work and represented by what he produces, which is thought transformed into something useful.

For those who tend to uphold the primary position of the word, of the superiority of *Homo loquens* over *Homo faber* (Cirese 1988), and who believe that true and complete thought is only that which performs and explicates itself semiotically, an especially important and complex duty is that of elucidating the notions and representations of one's own work, and of oneself as a"knower"and"operator."

Perhaps the greatest conquest of researchers in the field of traditional knowledge, especially the part that is implicit in doing and which is explained through doing, that knowledge that makes up body memory, thus still thought elaborated for doing and not for saying, would thus be that of succeeding in getting the persons directly involved to disseminate knowledge and perhaps even to formulate their knowledge explicitly in some form of language or representation, and to represent their know-how in semiotic form. And then, those who go in search of cultural universals will be able to find them perhaps more easily in the very field of thought-for-doing, than in the field of thought-for-saying, or in any case of knowledge formulated and organized in the various natural languages, in short more readily than in semiotic thought: break, cut, pound, tear, throw and dig, besides techniques of the body (Mauss 1936): how to run or jump, get dressed or inhabit, are cultural universals (besides being precultural universals) more than to name, classify, categorize or elaborate other forms of semiotic knowledge concerning nature.

This line of thinking engenders many problems, to being with the recognition of a thought and knowledge that are not immediately dependent on language and not entirely utterable in a language of signs. This entails elucidating the resemblances and differences between the two kinds of cognitive activity: that for doing and that for saying, or the differences, resemblances and interconnections between body memory and linguistic memory. In traditional and ethnological societies those specialized in some form of doing often have a language and a specific metalanguage, a trade language and even trade secrets. How much is actually expressed and to what extent explicitness is actually achieved in such cases, if it is true that such secrets are expressed or alluded to using jargon and cryptic language, if it is true that when they become explicit at the same time they also wish to hide and camouflage themselves?

References

Angioni, G. 1986. *Il sapere della mano: saggi di antropologia del lavoro*. Palermo: Sellerio.

Cirese, A. M. 1984. *Segnicità fabrilità procreazione*. *Appunti etnoantropologici*. Roma: CISU.

———— 1988. Homo faber, homo loquens. In *André Leroi-Gourhan ou Les Voies de l'homme*. Paris: Albin Michel.

De Martino, E. 1977. *La fine del mondo: contributo all'analisi delle apocalissi culturali*. Torino: Einaudi.

Goody, J. 1977. *The Domestication of the Savage Mind*. Cambridge: Cambridge University Press.

Horton, D. and R. Finnegan. 1973. *Modes of Thought*, London: Faber & Faber.

Hallpike, Ch. R. 1979. *The Foundations of Primitive Thought*, Oxford: Oxford University Press.

Leroi-Gourhan, A. 1971. *L'homme et la matière*. Paris: Albin Michel [1943].

———— 1973. *Milieux et techniques*, Paris: Albin Michel [1945].

———— 1964. *Le geste et la parole*. 2 vols. Paris: Albin Michel.

Lévi-Strauss, C. 1962. *La pensée sauvage*. Paris: Plon.

Mauss, M. 1936. Les techniques du corps. In *Journal de psychologie* XXXII, nos. 3–4

Piaget, J. 1970. *L'épistémologie génétique*. Paris-Neuchatel: Delachaux & Niestlé. Eng. ed. 1972.

Sahlins, M. 1980. *Stone Age Economics*. Ital. trans. Milano: Bompiani.

Solinas, P. G. ed. 1991. *Gli oggetti esemplari*. Montepulciano: Edizioni del Grifo.

Whorf, B. L. 1956. *Language, Thought and Reality*. J.B. Carroll ed. New York: MIT Press.

CHAPTER 15

Thought, Knowledge, and Universals

Jack Goody

In this chapter I respond to Daniel Fabre's comments on thought in relation to nature, and I do so in three parts following that prolegomenon. I discuss the way that the concept of thought has been used by social scientists not only in relation to individual cognition but also as a system attached to a social group or to a type of social group (e.g., "primitive" or "savage"). I point to the difficulty of determining either level and to the desirably of discussing the issues in terms of more specific cognitive processes.

Regarding nature, this topic can best be considered in terms of systems of knowledge, which are clearly related to thought but in complex ways. I see those systems of knowledge as influenced by specific cultural constellations, for example, as well as by more general factors such as the means of communication (for example, literacy), the means of production (for example, pastoralism) and the use of the means of destruction (for example, how war alters perceptions about nature and the environment, as with the fields and valley of the Somme in 1914/18, the pesticides in Vietnam, the land mines and the barbed-wire worldwide). But there are also yet more general factors that influence perceptions of nature, not only the built-in ones discussed by the cognitive sciences, but problems intrinsic to the interface of language-using animals with their environment, an interface which I suggest produces, inter alia, doubts, ambivalences and contradictions that become more explicit in societies with writing, but which are already present in embryonic form in ones without.

Thought: Systems of Thought and Mentalities

The problem for most social scientists is that thought is necessarily internal, a happening inside the black box, which is silent and invisible to the outside observer. The anthropologist cannot examine thought inside the human mind, he does so in signs, in symbolic exchange. Signs have to be used as the only way we have of getting at what is inside. One can only deduce what may be happening in that box if one assumes that outward events represent inner ones, if one utilizes signs to indicate, in the words of the initial statement for this conference, the spirit, l'esprit humain. However when we as social scientists talk about the thought of a particular people, or even of a particular person (whether at the thought of Karl Marx or at the mentalities of the French or of the Asante), we are usually looking not at specific mental events, but at more general dispositions. The thought of particular cultures, at the most abstract level, resembles of course the thought of other cultures of the same general kind, so that we find a further level of discussion, in the shape of the modes of thought, of say, "primitive,""'savage" or "simple" societies usually in opposition to that of "modern,"""advanced" or industrial ones. This particular topic has of course had a long-standing and controversial history, especially in the writings of Lévy-Bruhl, Evans-Pritchard, Lévi-Strauss, not to speak of many psychologists. In more particular, less globalizing terms, the topic has emerged in the use of the concept of mentality in historical research. Both these notions imply some set of mental predispositions that continue over time, the latter being a way of summing up psychological tendencies to act in one way rather than another. The concept of mentality, the use of which has often been questioned, but which has played a considerable part in the work of some writers of the Annales group, appears to be used to include affect as part of thought (for example, in the work of Ariès and Stone on childhood and the family). But while the concept involves thought, it does not necessarily imply thinking, which is an active conscious process of review, whereas thought as mentality is an established form that approaches knowledge, confirmed belief. If confirmed belief is shattered, say, by a new educational regime, then "thought, mentality," changes. And both are contentious in that the specification of modes of thought and mentalities results in generalizations that are hard to substantiate and fail to account for the quite rapid changes over time that societies and individuals experience.

That is not to say there isn't a grain of truth behind these highly generalized discussions, but it seems better to focus attention on cognitive operations at a more specific level. For example, Lévy-Bruhl's observation about the non-recognition of the idea of contradiction by "primitives" can be rephrased in the light of Evans-Pritchard's work, by giving a consideration to the means of communication. Simple societies certainly do have an understanding of and a feeling for contradiction, but when one is limited to oral communication alone, its perception is more difficult. One cannot put different versions of a myth side-by-side but only sequentially, often with long intervals in between. In any case contradictions are in a sense implicit in religious beliefs that often run counter to the understanding of ordinary life (rising from the dead, visits to the other world, being here and elsewhere, eating souls). These apparent contradictions with daily thought, which in this case are perhaps better called incompatibilities and exist in all systems, may differ in different religions and be rejected in their entirety by secular approaches. In any case, the recognition of contradiction that Lévy-Bruhl saw as absent in the simpler societies was derived from written dialectical inquiries of early philosophers, whose formalizations were often far removed from the workings of ordinary minds.

The same can be said of Lévy-Brulh's other characterization of primitive thought as "prelogical." Again, as Evans-Pritchard showed, the belief systems of simple societies had their own form of logic. What they did not have was the formalized type of "logic" based on the constructed syllogism that was developed in writing by Aristotle and other philosophers, a formalization that was largely a consequence of the written mode.

We need therefore to look much more specifically at thought processes in terms of cognitive operations. Such operations we can associate with particular cultural factors that can by definition be altered (sometimes with difficulty). That approach saves us from the resort to some general feature that is seen as a fixed characteristic of a certain range of societies, embedded in their culture. There is a third factor in this discussion which concentration upon classificatory schema, whether built-in or culturally generated, fails to bring out. That factor is thinking about nature, as well as thought and knowledge. I tackle this question because it seems essential for us to reconcile structure and process, to introduce a dynamic or generative element, and to consider the individual as well as the social.

In the process of thinking about nature as one can at least describe the way that this seems to happen in ourselves, a dangerous approach that one is sometimes forced to adopt because of the nature of the black box that is our mind, our sensitive processor of mental operations. But to try that presents far greater difficulties. As Shakespeare remarked,

There is no art
To tell the mind's construction in the face.

So let me move to the other extreme, since extremities are easier to deal with, namely that of knowledge, in which we are dealing with thought that has achieved a degree of social acceptance. Here we know what we mean when we are dealing with knowledge recorded in the *Encyclopaedia Britannica*, which is not simply the result of a so-called modern Western scientific outlook since we find a parallel process earlier on in China in the encyclopedias of the Song dynasty. But acceptance is not so easy to determine in oral cultures or even in the oral traditions in literate cultures, for different groups may accept different systems of classification and measurement, as with carpenter's space and sailor's space, or hunter's time and herdsman's time, or male and female categories of plants and animals.

Leaving that aside I will return to thought as in so-called primitive thought (another dangerous category but not to be too readily disposed of by a politically-correct culturalism, as Lévi-Strauss has affirmed) or more acceptably in the historian's notion of mentalities, which is a way approaching cultural specificity in a wider framework, or the thought of say Francis Bacon, for which it is important to note (as with encyclopedias) that we have as our data base a written corpus rather then a spoken one. In each of these contexts a central aim of the social sciences and humanities is to produce a unitary, homogeneous account, often looking for homologies, to serve as some sort of explanation, or backdrop, for considering the more flexible, more pluralistic discourse that is closer to the process of thinking, or thought in the dynamic sense. In fact the homogeneous account often conceals the active process.

I want to stress this dynamic element, although it often goes against the general thrust of anthropological and linguistic analysis, and even those dealing with historical shifts over time, as with the mentality school, who often see their data as subject to revolutionary change rather than evolutionary process. In using this phrase I am employing a term often seen by contemporary social scientists as politically incorrect, but I do so in the social rather than the genetic sense, as did

Childe in speaking of social evolution, which was a differentiating as well as a common process of development.

When we describe different taxonomies of natural objects, we also have to be prepared to present mechanisms of how such differences emerged, since they were not all created by God at his original baptism of things in Genesis. That is to say, we have to envisage change, not only the change we see in every part of the world today with people (individuals and cultures) abandoning existing ways of behaving in favor of international crosscultural ones, but there is also the long-term evolution of different schema among neighboring peoples.

We can admit, and here the so-called cognitive approach has done a lot of good in combating culturalist fundamentalism, that taxonomies have universal elements that this school attributes to wired-in dispositions. I would not reject this possibility but would always demand some satisfying proof. However for me universals can have some other explanation and I would often attribute them to the situation of language-using animals facing their environment, in other words to transcultural experiences (which are not always universal but widespread). At the same time we need to be able to account for those features that distinguish one culture from another, which we can see emerging in historical records of literate cultures.

If we had the observational data over time we could also see this same generative process in oral cultures, the so-called traditional ones often seen as self-perpetuating in their conceptualizations. Of course they are not. Continuity has to be fought for, to be won, negotiated as some would say (but I prefer the conflict metaphor to the commercial one). We see this process of change in West Africa in the religious sphere where at present we constantly find a turnover of what have been called "medicine shrines." Some have associated this turnover with the impact of Europe and its colonial expansion. However I would regard turnover as being an intrinsic feature of such shrines since they offer solutions to human problems that often fail to work, and are perceived in this light, hence are often abandoned in favor of new gods.

Taxonomies are more enduring but even they change over time and in any case are not always as explicitly formalized as written accounts suggest. That is one of the problems for the social sciences, or indeed for any forms of written knowledge in the humanities, where we are trying to generalize and make sense of human behavior, especially linguistic discourse. To establish order out of the varied and abundant material with which we are faced, we select a limited number of aspects, of the phenomena to deal with, to re-represent. We create lists and tables, which are invaluable for certain purposes. But for many others they are highly misleading and fail to catch the contextual, metaphorical, constantly changing use of language, which as Barth has documented for New Guinea leads to changes in cultural norms. Culture in other words responds to human agency, to the constant problem-solving engendered in face-to-face confrontations of man and his environment, natural and social. That continual interaction between what we can crudely call the social (or the cultural) and the individual, between the realms of sociology and psychology, implies a generative approach to culture and a cultural approach to psychology, to the mental, to thought (in the first sense).

There is the same problem with attempts to fit "thought," concepts, into taxonomies. Primarily as a result of the work of Lévi-Strauss, efforts were made to categorize concepts into binary pairs (polarities) and to link these in columns (as analogies). Such schemas were conceived as underlying thinking in all domains of a particular culture (that is, as homologies). In the first place, such schemas were based upon a fixed usage of a single-stranded kind. Black was opposed to white and homologous with "night, evil, infertility, the left hand, women." The schemas excluded ambivalence, metaphor, even many "normal" usages; they raised what I have called the black-is-beautiful problem. Where did that obvious sentiment, widespread in African cultures, fit in? If one sees such usages as contextual there is no problem. If one presents a taxonomy, which is almost inevitably single-stranded, the question is insoluble.

Continuity/Discontinuity

Even where taxonomies remain the same at one level, the reference of the individual items may change radically with changing cultural circumstances. There is no meaningful sense in which a cow is the same for me as a cow for our Anglo-Saxon forefathers. In a limited sense the word is the same but a critical feature of the cow (as Saussure insisted) was its opposition to the Norman "boef" and to the English "beef" for the dead but edible animal. As in English folk perceptions which achieved a written form in Walter Scott's *Ivanhoe*, that division related to the class/ethnic distinction between Saxon and Norman, whereby the former looked after the living animal (the cow) while the latter were the ones who ate the meat on the table (beef). However for me, the cow has moved out of the edible category. Cows are now scarcely domesticated because they are "mad," the carriers of disease. The fact that we use the same classifying term, with the same morpheme in the same taxonomic system, seems not very meaningful compared to the changing position of the cow in my cognitive space. I'm no longer involved in the class/ethnic relations of my English ancestors; the terms developed in that original context have remained but with quite a different practical and symbolic meaning.

It is necessary to elucidate and examine these various concepts because in so-called traditional societies (as in ours), thinking and thoughts about nature are not simply about utilizing concepts, category systems, ideas that have been handed down, either because they are wired-in or because they are culturally transmitted. Human agency manipulates, plays with those schemes for its own purposes and in so doing, may well modify them by an endogenous process giving rise to a plurality of classification schemes.

I want to make one other point about category and symbolic systems. There is no doubt, as many have shown, that you can *elicit* hierarchies of living kinds and such hierarchies if not universal are certainly transcultural. Whether they are wired-in seems a moot point; they could also arise in an interactive situation or possibly in the process of language formation when inclusive categories (families) were presumably the first to emerge, later to be broken down into species and varieties. Such specification is certainly subject to exogenous social pressures—in early medieval China at the capital of Loyang the Empress Wu produced many varieties of the peony, all of which (unlike wild species) needed to be given names, perhaps of the kind of Princess Di or Eleanor Roosevelt. That proliferation of nature resulted not just from neolithic domestication but from Bronze Age surplus, that is from a culture of luxury. In addition the explicit formulation of category systems was greatly promoted by the advent of writing (and of literate scholars) and such explicit categorizations took the form of bounded lists, rather than say Venn diagrams, which posed problems on the boundaries. Is the tomato to be included under fruit or under vegetable?

Let me return to the question of thinking, thought and knowledge.

Knowledge

Knowledge is an outcome of thought, of individual mental processes, and the two clearly interact. But knowledge is an explicit, social product, more or less codified, that in literate societies can be produced in books and added to, in the way Song Chinese did to their encyclopedias; whereas thought is personal, largely unspoken, and has to be deduced from outward signs; thought is talking to oneself but often in a way that is based on "custom," or learned behavior. The relation between thought and knowledge is complex. Take nature. In oral societies, knowledge about nature can be seen as thoughts about that topic which have gained some kind of social recognition or acceptance; there has been some process of filtration to form, say, an ethnobotanical system. Those thoughts have been externalized in some way, in a category system, in customary practices, in established beliefs. In some sense all this knowledge is open and belongs to a collectivity (though that collectivity may not be societywide); it is a social fact.

What is there to say specifically about thoughts of nature, about "nature knowledge"? The subtitle for this section of the conference runs "Nature as a model and metaphor" and in a previous version of the communication the word "symbolism" was inserted before nature. The notion of symbolism does not seem to have much force in this context since all verbal usage is symbolic. The subsequent references to model and metaphor suggest we should address those contexts in which knowledge about nature is used to illumine some other aspect of life, for example, "my love is like a red, red rose." I am not sure there is anything general that one can say about such usages except that they are common in all human cultures, though they may vary in particular ways that are not easy to establish.

I find some difficulty with the notion of symbolic classifications of nature as distant from pragmatic (utilitarian) or necessary ones. The difficulty arises from the fact that all language use is symbolic, involving re-presentation. By symbolic some people mean arbitrary elements, in which all systems of naming or classification partake in varying degrees. Others refer to the kind of usage that crosses domains, as in "my love is like a red, red rose," in other words the use of metaphor or simile that links different spheres.

My plea here is that we should not over-formalize such linkages. We can build analysis on the cultural links between sex and flowers only to find Marvell writing that:

> Our vegetable love will grow
> Vaster than empires and move slow.

So we now have to revise the domain linkages to include love and legumes. But should we be doing one or the other on a cultural basis, outside specific contexts? The use of metaphor of this kind is inherent in the plurality of domains, serves to breakdown the boundaries and is generative both on an individual level (not only of course with poets) but on a cultural, or better, social one, since as Barth has shown for New Guinea, individual usages that are in a sense potentially present as metaphors or similar usages may lead to cultural changes if they are "selected," not in any mechanical or utilitarian sense, but because of generational drift, or not just drift but a desire to distinguish oneself as a new generation.

Herein lies a major problem. If our understanding is of nature dominated by the taxonomies of ethnic science or by homologies between systems, whether these are seen as built-in cognitive givens or as cultural products, they are in a sense knowledge as opposed to thought (in the sense of active thinking about). Yet the two co-exist. In the final analysis knowledge derives from thought. We know that one of the major characteristics of humanity is the way it can adjust to situations, by utilizing knowledge (culture) and at the same time adapting it (and perhaps more permanently changing it). One of the ways this is done, as Marlène Albert-Lorea suggests (in ch.13 of this volume), is to manipulate/operate a plurality of taxonomies, so that the tomato is a vegetable in one context and a fruit in another. And the choice between these has to be up to the actor and the context. That does not make it an anomaly, the focus of a taboo; it represents a constant feature of such systems facing the encounters of actors with their environment, which makes them respond to actual situations with a range of tools, or on another level, quite different ways of classifying the world. For example, the LoDagaa of northern Ghana are not conscious of the family, species, variety system; they classify animals implicitly into edible and inedible, and explicitly into wild and domesticated (*wedun* and *yirdun*, the first constituting the herds of the beings of the wild) and into black and white, referring to the ritual danger they may carry.

One of the problems about analyses of mentalities as systems of thought is the Eurocentric way this matter has been approached. While there may be some basis to the notion that systems of thought changed significantly in Renaissance Europe, the general ways in which this shift has been phrased are open to objections. Ideas about children may have changed as Ariès claimed, but to talk of the invention of childhood is quite mistaken as one sees from Chinese paintings or even from African toys and games. It is the same with the idea of nature. Not only were there earlier developments in Europe of the classical period (where doubts were felt about the slaughter of animals by Porphyry and others) about man's domination of nature, but the Chinese too showed their appreciation of nature in their painting of winding rivers and craggy mountains as early as the Han dynasty, indeed at a time when Europe virtually confined its representation to holy images, setting aside not only landscape painting but even restricting the development of botanical illustrations for scientific purposes. The Byzantine illustrations of Dioscorides continued to be used until replaced in the Renaissance. In other words, the growth of knowledge about plants was held back by the presence of an iconoclastic trend that was virtually only over-ridden for religious purposes.

These examples bring out an essential aspect of both thought and knowledge which totalizing, holistic approaches often neglect. On the European scene, as we have seen, accounts of the Renaissance and the Enlightenment in Europe often stress as significant achievements the invention of childhood or the emergence of a new respect for nature. As far as the recent history of that continent is concerned, social scientists may not be altogether wrong in taking that view. But that perspective on "modernization" neglects the fact that on the one hand other cultures had different views and on the other the Middle Ages had gone through a phase during which much earlier secular thought and knowledge, as well as artistic activity in painting, sculpture, theatre and in literature, had been set aside in favor of the pre-eminence of a religious (specifically Christian) worldview. In this sense, mentalities had undergone a radical reversal that affected processes of thought and knowledge in ways that were only overcome as the result of the secularization and growth of learning (knowledge) in the Renaissance and the Enlightenment. China did not have to go through such a transformation while classical Europe had seen perhaps the birth itself (rather than the rebirth). Neither had to experience an enlightenment since they did not experience the dark ages. China held the totalizing Buddhist Church at bay by encouraging paganism, which by definition avoided fundamentalism in instituting a measure of pluralism, in thought as in institutions. One of the results of the fall of the Roman Empire and the drastically reduced role of the (secular) state under "feudalism" was the dominance of the church in worldviews, in art, as in many economic contexts. Political decentralization allowed for the hegemony of religion with dramatic results for the history of culture and of ideas.

One particular facet of nature that I myself have recently been interested in as a "symbolic" and as a "natural" system, has been the use of flowers, principally domesticated, as gifts and for decoration. One relevant aspect for the present discussion was the development of an elaborate Language of Flowers in France in the early 1800s. Many societies allocate meaning to floral species—the red rose for ardent love, for example. But the Language of Flowers produced an exhaustive set of equations for all floral species and created a complex language so that when flowers were combined together in a bouquet they sent a translatable message. It explained, formalized, and greatly simplified the range of meanings; in this context, the red rose had to have one meaning and one alone; it could not represent the blood of martyrs, political or religious, as well as passionate love, since the listing did not permit such contextual flexibility. This highly artificial and restrictive system flourished in the urban environment of nineteenth-century France and spread to all European nations shortly thereafter. It represented a limited form of "knowledge" (extra-scientific) created for a specific purpose of heterosexual communication of a courting kind. It was invented at a particular moment in time and for a particular type of bourgeois urban space. Popular for a period, its popularity then declined, though even today it still maintains a shadowy existence. The recent unparalleled employment of flowers in Britain on the occasion of the funeral of Diana, Princess of Wales, where the country ran out of supplies, involved no such specific meanings, only a general idea that any floral tributes would be an appropriate way of showing grief.

Even that use of flowers cannot best be described in terms of modes of thought or mentalities unless one makes room in some cases for an element of affect, and unless one is prepared to see these beliefs and practices as being more subject to change than analytic frameworks usually allow. Moreover, the use of flowers is subject to change not only as the result of "cultural drift" along the lines of linguistic drift, nor yet because of the adoption of the ways of one's neighbors. Rather the more radical changes of attitude occurred because their use involved implicit cognitive contradictions about natural objects, the result of thoughts which had to do with doubts about the use of flowers in general or for religious purposes in particular. For neither Judaism nor Islam condones the use of flowers in places of burial, since such offerings would seem to be asking the dead for favors rather than appealing to the omnipotent All-mighty. Many early Christians took a similar view and even the later Catholic Church, committed to their use as decorations, did not in principle countenance them as offerings. The Protestants were far more severe and the objection of Puritans was not simply to flowers as offerings to God but to the waste of expenditure on any ephemeral objects, expenditure that could better be devoted to other worthy causes. It was an objection to flowers as luxuries, as inessential to human survival, the same objection that emerged in Maoist China where they failed the criterion of filling the peasant's rice bowl.

The underlying contradictions worked on two levels. First, gifts of flowers to the dead or to God himself seemed to challenge his omnipotence; and in secular terms the use of flowers could be seen as a unnecessary luxury which neglected the difficulties of the poor in getting the essentials for their livelihood. Such challenging thoughts were not limited to specific cultures, specific traditions: they were potentially present in all theistic religious systems as well as in all luxury (heavily differentiated) cultures. Hence their potency when they are brought into the open under conditions of religious or social change.

Thought, Thinking and Universals

Such a state of affairs regarding flowers involves taking into account a dynamic element, a swing from one set of thoughts to another, about representation or about nature, for example. These particular shifts are not random in that they arise from the existence of contradictions, or ambivalence, within the human situation, for example, about the extent to which representations can present the original, or are they always a misrepresentation, a lie, as Plato argued on somewhat different grounds? Such contradictions are rooted in the life of a language-using (i.e., representation-using, symbol-using) animal. For example the essential exploitation of nature raises potential doubts about its exhaustion, its non-renewability, about the gradual disappearance of animals through slaughter, about the very process of cutting the stalks of grain and killing the spirit or god who has to be reborn the following year. Such doubts, ambivalences and contradictions mean that actual beliefs and practices may switch over time (or between groups) from one set to another, providing the dynamics of social change from, for example, figurative to abstract art (which is seen as pure and cannot lie).

This is one layer of thought process in human cultures. There are others. Firstly there are those that are "culturally" specific, whatever the cultural unit may be. Secondly, there are those that are specific to certain more widespread sociocultural situations, for example, those turning on the presence of the kind of cognitive mechanism that some have associated with the coming of writing (the making of lists and tables, such as those used in listings of fruits and vegetables which enforce a binary choice between inclusion and exclusion). Like other changes in the mode of communication, writing is of fundamental importance not only in enabling humanity to accumulate (by providing new recording techniques) and to organize thoughts and knowledge about the natural world, but in developing ways of thinking about it, ways that bring out the implicit thoughts and sensitivities of oral cultures, for example, against the slaughter of animals among Buddhists, Jains, Benedictine monks and secular vegetarians. And there are those cognitive procedures that people have rightly or wrongly associated with particular ways of gaining a livelihood, sometimes with modes of production as distinct from modes of communication, as in the perceived characteristics of thinking in a pastoral way of life or under capitalist production with its continual calculation of the profit and the loss and its "Protestant ethic."

Thirdly, there are the universal elements referred to in the introductory remarks and which have recently been viewed as built-in features of the human mind, following Chomskian notions of the generation of language. That trend has been manifest in the work of Atran on category systems in botany, on the supposedly universal division between artifacts and natural kinds that has been the subject of psychological investigations in children's perceptions by Keil (1979, 1989) and others. The notion that these features are in-built, that universals provide logical restrictions inherent in the human mind, in human thinking, has been seen as an alternative to the more cultural approaches we have outlined above. It is not. Whatever may be attributed to in-built notions, the cultural differences (not by any means all randomly distributed) that we have discussed above have to be differently accounted for. Even universals, as I have suggested earlier, may result not from features built into the human mind genetically, but from the universal situation of human beings reflecting in linguistic terms about their position in the universe, with regard to themselves, to other humans, and to the external environment, itself mainly "natural," with which they are faced.

I want to finally refer to some extensive material that I have gathered, namely the many versions I have transcribed (with K. Gandah) of the Bagre myth and which offer in my eyes a more profound entry into the thoughts or rather the speech of the LoDagaa about nature, since we are concerned not so much with the study of individual words in relatively decontextualized taxa or lists as with the wider linguistic structure of the sentences, sectors (e.g., paragraphs) or complete units (standard oral forms), as well as in a sociological context that treats words in the context of their total usage (of presence as of absence).

I'm not going into the Bagre in any detail; it is about the creation of the world of humans, not of the natural world that seems less problematic. I want to stress that it offers a very complex view of nature in which different versions emphasize different aspects; some include references, for example, to the creation of the world where we find both the big-bang notion of creation on the one hand and on the other the gradual process of evolution and continuity between man and animals. These two contradictory notions represent the widespread dilemma that faces people of recognizing the continuity of man with the other animals (we are all living kinds) but the discontinuity that man's development of speech creates within the animal kingdom, a discontinuity that calls for some supernatural explanation. Secondly, this dilemma represents an instance of a wider set of doubts, skepticism about the world, the existence of which certainly runs against many earlier views of the thought systems of oral cultures, savage mind, or primitive mentality.

I have referred to the Bagre because I wanted to draw attention to certain aspects of the problem of trying to give an account of thoughts, mentality, about nature. First we need a large body of "natural" material and it is not clear from where this should be drawn. I have spoken of the Bagre as a unique collection of oral material made possible by the portable tape recorder and its analysis by the computer. The plurality of versions reveals ambivalent attitudes that arise from the cognitive contradictions inherent, in my view, in the very process of a language-using animal facing his or her environment. The existence of such contradictions, which are not built-in but created in interaction, provide a generative component in culture for thought, for metaphor as well as for classification.

I regard these doubts as being an embryonic form of the critical views of the world that were later developed in writing, beginning with the Greeks, as some see it, with the Enlightenment as others do, and which contribute to the rise of modern science with its own category systems. There is both cognitive continuity and discontinuity, so-called folk systems are precursors of learned ones. There is often however an automatic assumption that this scientific development was a Western one. To a comparative cultural historian, that notion appears fundamentally wrong and highly ethnocentric; it has been countered by the encyclopedic work of Joseph Needham in *Science and Civilization in China*, where he shows that the numbers of botanical species listed in Chinese treatises about 400 BC was the same in number as in Theophrastus of the same time in Greece. And in the ensuing period the numbers in the records of botanical species in works such as the extensive encyclopedias of the Sung (centuries before the *Encyclopaedia Britan-*

nica!) were greater than in Europe until the work of the German botanists in the fifteenth century. During this period, for reasons connected with the dominance of the church, despite all its many achievements, there was a great loss of knowledge about nature.

To return to the Bagre, the third element it brings out is the ambivalent attitude towards the exploitation of nature, though this attitude is also present in some rites as in the case of homicide. On the one hand, it is necessary to slaughter animals, to shed their blood, in order to provide ourselves with food, and more importantly to give to the supernatural agencies, since virtually no meat is consumed except after an offering to the gods (as in Islam and Judaism). On the other hand, the sacralization of the flesh, and I would argue the existence of totemism, which is of course much more than classification (a notion that in a sense trivializes the institution) is necessary because in order to be able to kill animals (and this comes close to the shedding of human blood), you have also to conserve them. Besides, they are all God's creation (as well as the products of a kind of evolution). In other words, conservation is in no sense an invention of the West (any more than is science, except for "Western science" by definition). There is already an ambivalence about killing animals, and indeed often about harvesting the grain, which may be seen (as in the works of Sir James Fraser) as the killing of the spirit of the corn.

This chapter has attempted to address the relationship between thinking, thought (in the general sense, of anthropologists and others when they talk for example, of "la pensée sauvage," or of historians when they talk of mentalities, though this latter concept includes affect as well as cognition) and finally knowledge, which I take to mean widely accepted, formalized, validated thought, perhaps in the hands of sub-groups.

In their attempt to characterize these processes anthropological, linguistic, and social scientific approaches often adopt techniques (including the very fact of recording them in writing) that tend to freeze the situation, including terms and their meanings (when origins and usage are sometimes conflated), including metaphoric and idiomatic usage. Whether in the form of culturalist or so-called cognitive approaches, these tend to neglect generative factors. I refer to so-called cognitive approaches in the plural because it amazes me, in the intellectual world generally, how the concepts of cognition and evolution have been hijacked by the biological sciences. That must be very limiting to a rounded analytical approach.

I have not had time fully to elaborate the point or to outline the generative mechanisms, but I wanted to end by returning to one of these, that deriving from the cognitive contradictions that are involved in the process of re-presentation which is intrinsic to the use of language. And language use involves, as Plato implied, a measure of deception; a horse is never a "horse." In simple societies individuals are possibly more aware of the arbitrary element in language than we are (especially Americans and English). Ethnic groups are relatively small; there are always boundary problems that involve communication with other groups for whom a horse is not a "horse" but a "cheval" or a "pferd." The arbitrariness of terminology is more apparent than in monolinguistic communities and raises doubts about the relationship between the signifier and the signified. It is to those doubts, those ambivalences, about nature to which I have tried to draw attention in an effort to introduce a generative element into ethnoscientific analysis.

CHAPTER 16

Bodily Humors in the Scholarly Tradition of Hindu and Galenic Medicine as an Example of Naive Theory and Implicate Universals

Francis Zimmermann

This chapter, in which I address myself to theoretical issues currently debated between cognitivists and the proponents of symbolic anthropology, is based on long-standing ethnographic and historical researches into the scholarly tradition of Ayurvedic medicine, in South India, and European parallels in Galenism. I shall be focusing on the three topics delineated by Daniel Fabre in his part introduction, namely, (1) the rather strained relations between cultural anthropology and cognitivism, (2) the most recent developments of ethnoscience and area studies of local knowledge regarding Nature, and (3) the controversies over universals.

Current debates have been shaped into a quarrel between the Ancients and the Moderns, and, at least in France, they resulted in a division into two camps, a divide between two intellectual worlds ignoring one another. Cognitivists, on the one hand, and cultural anthropologists, on the other hand, pretend not to read one another. Cultural anthropologists, personifying the Ancients (with whom I belong), are inveterate enthusiasts of local color. They tend to take refuge in glossy paper and color-plate journals like *Terrain*, or *Ethnologie française*, or in anthropological journals intended for historians like *Etudes rurales*, and they spread themselves thin in thousands of small rustic or exotic localities. Cognitivists playing the role of the Moderns, to the contrary, have a strategy of distancing themselves from ethnographic fieldwork by setting up a fence of concepts and epistemology between field studies and scientific argument, and they proceed by importing the metalanguage and problematics of British and American philosophy of mind and of child psychology into the field of anthropology. An example of such imports will be given in the following pages, when we trace the concept of *ontology* introduced in cognitive anthropology by Pascal Boyer and Dan Sperber back to psychologist Frank Keil and philosopher Gilbert Ryle. In this paper, I would like to suggest that a rapprochement between cognitive and symbolic anthropology would be most fruitful to expose the cultural construction of nature knowledge, as I have attempted to do in the anthropological study of scholarly traditions. A scholarly tradition like humoral medicine may be defined as a combination of local knowledge and of scholarly procedures of knowledge transmission. The Hindu tradition of Ayurvedic medicine, for example, is both a body of local knowledge that can be documented ethnographically, as I have done through long-standing fieldwork in

Kerala (South India), and a body of learned practice based upon the transmission and scholastic commentary of Sanskrit texts, parallels of which are to be found in Greek and Latin texts (Galenic medicine). We could not do full justice to the sophisticated knowledge of nature encapsulated in the discourse, clinical gaze, and medical prescriptions of a local Ayurvedic practitioner, if we were not going beyond ethnography and symbolic analysis to address the broader theoretical issues of language and cognition.

Daniel Fabre emphasizes one of the fundamental tenets of Durkheimian anthropology, when saying that "anthropologists have never located thought in the privacy of mind, or in the flow of experience for that matter, but always outside the mind in the public use of signs and in the public space of symbolic transactions." One would be mistaken to object that the concept of public space here is an anachronism recently introduced in anthropology. From an anthropological point of view, there does not exist any mind that would not be objectified. Thoughts can only be seized by ethnography, and modes of thought can only be described, insofar as they are historically contextualized, there and then, in a given set of manners, customs, and institutions. Such an approach to the mind through its objectified products in public space (*le milieu*) was advertised by Lucien Lévy-Bruhl for example in the opening pages of his book on the soul, *L'Ame primitive* (1927), and it was labeled as "the indirect approach" (*la voie indirecte*). The method of indirect approach to the mind through its objectified products was borrowed from Auguste Comte, along with the idea of the collective mind of an epoch and society objectified in its intellectual public space, or *milieu*. As far back as then, already a new brand of philosophy of mind, the philosophy of the positive mind (*l'esprit positif*), served for a base of anthropology. Was it actually very different from the philosophy of mind invoked today by anthropologists who participate in cognitive sciences and the mapping of the mind? This is not so sure. One clue to a rapprochement between cognitive sciences and cultural studies might very well be found in the Durkheimian classical arguments, and a reassessment of Lévy-Bruhl's texts, revolving around the articulation between cognition and symbolism.

How to break the ice and take up with the cognitivists again? As far as I for one share with them the most serious misgivings about cultural relativism and the basic tenets of universalism, conceived of as a guiding principle in our researches into nature knowledge, I am readily adhering to Dan Sperber's phrasing of the issue, when he writes: "The question now is whether there exists in anthropology a body of knowledge worth retaining, which can only be retained at the expense of our recognizing that irreducible cultural types do exist" (Sperber 1996: 26). In other words, after one hundred and thirty years of publications that have built the discipline we use to define as social and cultural anthropology, we may question the value of this heritage, because any piece of anthropological knowledge that seems to be worth retaining as scientifically valid is likely to be infected, as it were, by relativism (the idea of irreducible cultural types). But for all that I am not prepared to share Sperber's *Naturalism*, a biologistic ideology that pushes him to add immediately that anthropology's "so-called concepts" (*sic*), as long as they are not articulated with population genetics and evolutionary biology, do not have any precise ontological implications.

On the contrary, I would like to defend a seasoned practice of symbolic anthropology and comparative ethnology in which the *categories* selected for (comparative) survey and (symbolic) interpretation have been chosen precisely for their ontological implications. A distinctive feature of these categories of collective thought is that they are located halfway on the scale of abstraction. Neither too concrete nor too abstract. Neither specific (related to one particular language and culture) to the point of their being untranslatable, nor abstract to the point of their being unable to refer specifically to one particular semantic domain. Bodily humors like Bile and Phlegm in Galenic and Ayurvedic medicine seem to perfectly exemplify the kinds of concepts and names that are endowed with ontological implications.

A number of appealing cognitive models have been developed in the last fifteen years to account for the ways of knowing natural objects, and to describe mental processes brought into play in the perception and categorization of natural objects—plants, animals, diseases, emotions, bodily organs and tissues. Each of these models has evolved in connection with a different set of psychological assumptions taken from child psychology. I shall limit myself to a brief presentation of only one of them, the mind-mapping model based on the theory of specific domains and the concept of naive theories, which seems to me extremely appropriate to the anthropological study of Humoralism, and to the description of perception and categorization procedures in local knowledge.

Over the last twenty years, psychologists have developed the view that cognition is domain-specific. Cognitive abilities are specialized to handle specific types of information. Another assumption was more recently grafted upon the idea of domain specificity, according to which our early conceptions of natural objects, naive as they are, take the form of intuitive theories. Freed from its biologistic prejudices and turned upside down, as it were, this model can be fruitfully applied, in ethnoscience studies, to the description of the cultural construction of nature in any given body of local knowledge.

The Concept of Naive Theory

Hundreds of papers have been published in the last twenty years in the most established journals of medical anthropology on the hackneyed concept of *the Hot-Cold syndrome,* meaning, the complex of ideas revolving around the division of foods, diseases and complexions into Hot and Cold. This is the popular version of humoralism that ethnographers are most likely to meet with in the field. For, the medical aetiologic concept of the humors, in its current meaning, can be defined as the concept of a polarity between opposite qualities—hot and cold, dry and wet, bile and phlegm, and so on—that is constitutive of complexion and producive of disease. Whenever cultural anthropologists have attempted to explain the birth of this naive biological theory in a given cultural context, they have always been trapped into the most conventional assumptions of diffusionism. The Hot-Cold syndrome was seen as an outcome, transmitted orally, of the degradation and popularization of learned cosmologies and sophisticated theories laid down in the Sanskrit, Greek, and Latin classical texts. Telling examples of this kind of diffusionism recently applied to the history of humoral theories are to be found in South America (George Foster) and in South Asia (where Sanskrit pundits tend to trace vernacular medical practices back to brahminic knowledge). I shall skip the details of my argument, here, since one may refer to *Généalogie des médecines douces* (Zimmermann, 1995) where ethnoscientific studies of humoral systems of medicine have been put back in the context of current epistemological and political debates in anthropology.

Recent developments in child psychology, however, might give us a clue as to the way biological concepts and naive theories have emerged in local knowledge. Psychologists have argued that, even at the youngest age, emergent concepts in a child's mind are embedded in sets of naive assumptions already structured like theories. Children seem to embed their concepts and interpret their properties in arrays of systematically connected beliefs rather than in an atheoretical similarity space. According to conventional wisdom—one could speak of an empiricist obsession among British and American philosophers—young children are pure phenomenalists. Slavishly tied to representations, they merely compute similarities over a set of familiar instances to reach conceptual knowledge. Reacting against these unwarranted assumptions of the late 1970s, Frank C. Keil, an eminent child psychologist affiliated with Cornell University, devised tests to detect the most fundamental conceptual cuts a very young child can make in the world, such as those between animals and plants, artifacts and animals, and the like. His work suggests that even preschoolers are sensitive to certain ontological contrasts, and he argued "strongly against an account portraying the young child as an unrepenting phenomenalist slavishly bound to computing correlations among salient characteristic features" (Keil, 1989: 214). Emergent concepts, to the contrary, seemed to be embedded in certain intuitive theories. Note that Keil, who is a psychologist, phrased his conclusions in the conditional and shaped his argument against the empiricist's *tabula rasa* into a set of new philosophical assumptions. "If young children were complete, unrepenting phenomenalists," he argues, "how could they ever make the transition to more principled biological knowledge? It seems more reasonable to suggest that even at the youngest ages the children might have started to build intuitive theories concerning the true nature of biological kinds and that, if these theories could be accessed, the children might well be willing to overrule characteristic features" (1989: 195). I am not concerned with the psychological side of the question, here, but only with the logical and cultural aspects of emergent categorization, and I limit myself to the biological domain, that is, the properties ascribed to living things as soon as they are "accessed"—to use Keil's phrase—in the context of food consumption and medical practice.

Some of the distinctions through which living things are sorted out and assigned a distinct domain of cognition (the biological) are known only to sophisticated scientists, but others may be much more universal. Quite a few anthropologists would be prepared to access them in the catalog of cultural universals. Let us take an example. In a brilliant paper illustrating the birth and nurturance of basic categories and naive theories in the domain of biological thought, Keil (1994: 236–7) devised a list of seven distinctions, none of which were strictly true of all living things and untrue of all else, but which seemed to capture the biological world. In a somewhat artificial way, however, the psychologist formulated them as if they could be apprehended by the reader through the eyes of a very young child. The second distinction in the list is worth quoting, since it gives me an opportunity to show how close the *naive* theory is to the *learned* tradition:"Biological kinds have a complex, heterogeneous internal structure. Except for broad axes of symmetry, if you chop them up into pieces, the pieces tend to be different from each other as well as not being simply smaller versions of the original. Gold and water clearly don't behave that way ... ."(Keil, 1994: 236). I shall skip over the epistemological critique that could be addressed to the psychologist, insofar as his constructs are based upon some ancient philosophical assumptions. Let me just point to the Aristotelian origin of the distinction between gold or water, on the one hand, and biological kinds, on the other hand, which Aristotle called *anomoiomeric* (consisting of unlike parts).

In Keil's formulation of the distinction, the dividing line is drawn between minerals (gold, water) and living things, which makes it look very rough and naive. But the criterion in itself, by which the distinction is made, is a sophisticated conception. If we trace it back to Aristotle's *Parts of Animals*, where the scholarly formulation of the distinction is to be found, the dividing line is drawn within the realm of living things, to separate the anomoeomeric ones (the organs in a living body) from the homoeomeric ones (the tissues). The same distinction thus comes in two versions. Aristotle's version is the strong or sophisticated one. The weak or naive version is that construed by Keil and attributed by him to kindergartners and pre-school children. Now, I have shown elsewhere (1989, 1995) that Aristotle's distinction between tissues (homoeomeric) and organs (anomoeomeric) was unknown to physicians in the scholarly traditions of humoral medicine. Humoralism, in Ayurvedic and Galenic medicine, is actually based upon the confusion of tissues and organs and the valuation of bodily fluids that are conceived of as anomoeomeric. Humoralism truly deserves to be called a naive theory, since it is both an elaborate combination of theoretical assumptions underlying a scholarly system of pathology and therapeutics, and a very rough and naive imagery of juicy saps and vital fluids that contrast with cold metals, dry minerals and flat water.

Keil's table of the seven basic categories of naive biological thought would not have the slightest relevance if the humors were not taken into account. I met with them indeed, but in an unexpected guise. Let us forget the medical aetiologic concept of the humors in its most popular usages described earlier in this paper (the Hot-Cold syndrome), to focus on its scholarly usages, as when Thomas Sydenham in the 17th century spoke of *the peccant humors*,"morbific entities,"as historians of Ayurvedic medicine use to say in translating the Sanskrit word *dosha*, a"fault," a "flaw,"hence a vitiated and pathogenic bodily fluid, a"humor."Actually, the medical theory of humors postulates the existence of invisible physiological realities—

vital fluids—that are the causes of visible pathological manifestations offered for inspection to the clinical gaze of the medical practitioner. We do find this ontological category of the humors in Keil's list of seven basic categories as number five:"Typical phenomenal properties are usually diagnostic of underlying nonphenomenal ones. We assume not only a kind of essence, but also a rich set of causal links between that essence and the merely phenomenal ... ." (Keil, 1994: 237). Humors belong with this category of essences underlying phenomenal properties of excreta and tissue alteration. The good point about Keil's phrasing is that it emphasizes the central role inferences (diagnostic) will play, even in the most naive of all biological conceptions, from visible effects to invisible causes. Physiological and pathological phenomena are observed—excreta, bodily marks and complexion, behavior, etc.—from which the prevalence of a given humor is inferred. Clinical manifestations implicate a pathogenic essence hidden in the depths of the body.

Implicate Universals

To my knowledge, the word *ontology* was introduced in psychology by Keil in 1979. His work on domain specificity in cognition is thus at the origin of later developments in anthropology—with Dan Sperber, Pascal Boyer, and Philippe Descola, to limit my references to French authors—and of current anthropological researches into "natural ontologies." In *Semantic and Conceptual Development. An Ontological Perspective*, Keil presented the first psychological investigation of the developing child's ontological knowledge and theories. The function of a theory is to specify the nature of the knowledge that underlies predications. Keil assumed that ordinary concepts and predicates attributed to natural things in usual conditions of perception carry a number of implicit ontological categories such as *event, object, living thing, animate, human*, and the like, which are the keys to a categorization of everything. In all the most naive processes of categorization, some classes are found to be more natural than others, and only certain pairs of predicates can be combined sensibly or naturally. Naturalness thus is one of the major working hypotheses tested in cognitive psychology over the last two decades, the naturalness of classes and the naturalness of copredications. What was at first a sheer working hypothesis, however, has come to be a philosophical assumption, when naturalness was ascribed to the very possibility of predication. Predicability, according to psychologists, is the phenomenon that only certain predicates can be combined with certain terms in a natural language. In such a case, the naturalness of classes and copredications is warranted by the implicit recognition of ontological categories. Rooted in nature itself, as it were, this ontological knowledge is the basis of a person's understanding of what sorts of things there are in the world and how they relate to each other. This description of children's natural ontologies is most akin to anthropological presentations of taxonomies in exotic cultures. Exotic modes of categorization documented in ethnoscience studies also imply a framework of ontological categories.

I shall set aside an aspect of Keil's description that appears obsolete at first glance to anyone having worked on polythetic classification—a mode of thought which has been intensively studied by symbolic anthropologists for the last thirty years or so. Deriving their working hypotheses from classical Western philosophy, child psychologists are prisoners of Aristotelian modes of hierarchical classification per genus and species, as it can be seen, for instance, on the predicability trees and ontological trees that illustrate the principle of naturalness in Keil (1979: 15–16). They assume that categories commanding a given domain, categories like *event*, *object, living thing, animate, human*, for example, that command the specific domain of living things in child cognition, are organized in a hierarchy. This particular point merits further consideration, but a discussion here would be beside the question. I shall set aside the issue of hierarchical arrangements, to concentrate upon the issue of ontological implications.

Children submitted to experimental research proved able to make spontaneous ontological assumptions about words of the natural language when naming things they were still incapable to conceptualize. Keil showed that even young children make surprisingly fine-grained ontological distinctions, between, for instance, living things and artifacts. In order to achieve this level of discrimination, they must elaborate an intuitive theory covering a specific domain of experience long before the corresponding repertoire of categories becomes available to them. Keil ventured to tread on the dangerous ground of culture and symbolism only once, in the mid-1980s (re. Keil, 1989: 197), in a paper demonstrating that even when authors such as Ovid and the Brothers Grimm invented metamorphoses in their stories, the transformations tended to occur more often within ontological categories—animals into animals, plants into plants—than across them—as when fish are transformed into stones, porcupine into cactus, or lizard into stick.

A *domain* is a body of knowledge that identifies and interprets a class of phenomena assumed to share certain properties. It should not be confounded with sheer semantic fields (a linguistic concept) or given sets of schemata in a typology (a psychological concept), although all of these mental structures are ways of achieving conceptual interconnectedness and mental economy. The language of philosophy of science is more appropriate than linguistics or psychology to describe the concept of domain specificity. Scientific paradigms, research programs, and scholarly traditions define domains of inquiry, and they do so by providing an ontology of the fundamental phenomena to be explained. Similarly, cognitivists argue that human reasoning is guided by a collection of innate domain-specific systems of knowledge. We may entertain misgivings about the presupposition of *innate ideas* and the obvious ethnocentrism of the above-mentioned examples and references from Aristotle to the Grimms, but these misgivings do not affect the core of the argument. In contrast with sheer schemata or linguistic categories, the set of principles that define the entities covered by a specific domain and support reasoning about those entities is entrenched in the ontology they determine. It is this *entrenchment* that characterizes the mental structures described in the present paper under the names of "specific domains," "intuitive theories," and "ontological categories." I have attempted to strip the cognitivistic concept of *ontology* of its biologistic connotations by clearly dissociating this observable process of entrenchment from the presupposition of innate ideas.

Let us try to ascertain the meaning of the word *categories* in the classic texts of our discipline, which spoke of "cognitive and linguistic categories" *(les catégories de pensée et de langue)* and assigned the anthropologist the task of making an exhaustive inventory. Says Mauss ([1924] 1950: 309): *Il faut avant tout dresser le catalogue le plus grand possible de catégories; il faut partir de toutes celles dont on peut savoir que les hommes se sont servis.* Categories are not built around defining features but around central members of a class or prototypes that exemplify them. Prototypes are described as consisting of collections of correlated attributes. They do not always refer to a corresponding essence and they are liable to apply to different ontological domains. In other words, not all categories are ontological categories, and there is something more in (ontological) domains than in (ordinary) prototypes. For want of time I cannot do justice to the distinction here, and I can only make an allusion to the form it takes in the scholarly tradition of humoral medicine (re. Zimmermann, 1987, 1989, 1995). Our earlier definition of specific domains stressed the expectation that there would be some specific property of the mental representations pertaining to a given domain accounting for their distinctive cognitive role. We said that they were ways of achieving conceptual interconnectedness, but this does not exclude arrays of correlated attributes that constitute *categories* without constituting *domains*. Pairs of concepts like Hot and Cold, for example, are too narrow to be domains. The distinction between categories and domains is most illuminating when applied to the humoral theory. In the scholarly tradition of humoral medicine, it explains the interplay of two different sets of biological entities constituting physiology and pathology: 1) the sensory qualities—hot and cold, dry and wet, sweet and sour, etc.—that play a fundamental role in Galenic pharmacy as well as in Ayurvedic medicine although they constitute a tabulation of correlated attributes devoid of any ontological implication, and 2) the humors proper—bile, phlegm, etc.—that, on the contrary, constitute an ontological domain. In the weak or naive version of humoralism (the Hot-Cold syndrome), as if it were the result of a process of degradation and popularization of the scholarly tradition, the humors have been stripped of their ontological complexity and conflated with the sensory qualities. *The Hot-Cold syndrome* is more or less synonymous with *the dialectics of Bile and Phlegm*.

I suggested elsewhere (1995) that the full-fledged conception of the humors constituting an ontological domain represented an interesting instance of culture universals. My approach to controversies over universals and my argument in favor of culture universals were influenced by the reading of linguist Joseph Greenberg and hellenist Geoffrey Lloyd. The notion of *implicate,* or *implicational universals* was invented by Greenberg. In addressing myself to the question of universals in the context of comparative ethnology with special reference to nature knowledge, I am trying to find a way out of the dilemmas of cultural relativism. In the field of medical anthropology, for example, cultural relativism, which has been the dominant paradigm for the last two decades, resulted in our juxtaposing all the medical systems of the world with one another, on an equal footing, as so many pieces of local knowledge good for transmission, and so many ethnic commodities good for consumption. It is high time we broke with the prejudices of relativism. But far from setting out to look for absolutely universal or primary concepts, as one use to say in the realm of cognitive sciences, I am interested in some arrangements, images, figures of speech, and specific theories, several occurrences of which are found in different cultures, so that they may be said to be endowed with a conditional universality (if I am permitted this oxymoron). The theory of humors and the humors as ontological categories, which are found in various medical systems all over the world, represent excellent examples of conditional or implicate universals."It may be and has been objected that it is strange to call something a universal when it does not occur in all languages," Greenberg ([1980] 1990: 316) writes as a justification for the phrase he coined, adding that"what is universal [in the concept of implicate universals] is the logical scope of our statement."In other words, these categories have *universal implications*, just as categories commanding a specific domain of experience and cognition; in the cognitivist's parlance, they have *ontological implications*. We need not trace the presence of these universals in all cultures and localities to ascertain their existence, they impose themselves upon the anthropologist as guiding principles in the comparison of cultures.

Since a distinctive characteristic of implicate universals is their patterning power in all its ontological implications, we should not be surprised to rediscover them at the core of the new"Naturalism,"in cognitive anthropology. I can only make allusive references to Pascal Boyer's analyses of"empty notions,"in his earlier books, and the primacy of *episodic* memory over *semantic* memory in traditional categorization (1990: 43), of what he calls *assumptions* in his more recent book (1994a), and eventually (1994b), Boyer's adhesion to the new model of cognition based on the concepts of domain specificity and ontological categories. A new response has developed over the past decades to immemorial controversies over universals. Anthropologists of this new persuasion will argue that human reasoning is guided by a collection of entrenched domain-specific systems of knowledge. Each system is characterized by a set of core principles governing all processes of perception, inference, and categorization in this particular domain of experience. Cognition, on this view, is a process of entrenchment, the entrenchment of categories in ontological domains of which they are the specific keys. Such categories no longer resemble the language universals that cognitivists of the older generation were looking for. Universals entrenched in specific domains with ontological implications, if they exist, will be embedded in emotion, sensibility, the body, the unconscious. Times are ripe for a rapprochement between cognitive and symbolic anthropology.

Afterword

In the ensuing discussion, Professor Giorgio Solinas rightly suggested that the method of approach to intelligence built in objective systems of thought that was delineated in this presentation took its inspiration from Louis Dumont, who was my teacher, and from the Durkheimians, and he wondered whether its implementation in the context of a highly literate society did not entail some limitations. I did agree that I had worked in South India within an exclusive and highly learned cultural milieu completely organized by writing. The literate tradition represented only one of the many facets of indigenous taxonomies and of the cultural construction of categorization. On the other hand, I could avail myself of a direct access to the Sanskrit and Malayalam texts and catch the local modes of thought in a pincer movement by addressing them from the two sides of philology and ethnography.

References

Boyer, Pascal. 1990. *Tradition as truth and communication. A cognitive description of traditional discourse.* Cambridge: Cambridge University Press.

———. 1994a. *The naturalness of religious ideas: Outline of a cognitive theory of religion.* Berkeley: University of California Press.

———. 1994b. Cognitive constraints on cultural representations: Natural ontologies and religious ideas. In *Mapping the mind. Domain specificity in cognition and culture* edited by Lawrence A. Hirschfeld & Susan A. Gelman. Cambridge: Cambridge University Press.

Greenberg, Joseph H. [1980]. 1990. Universals of kinship terminology. Their nature and the problem of their explanation. In *On language. Selected writings of Joseph H. Greenberg* edited by Keith Denning & Suzanne Kemmer. Stanford: Stanford University Press.

Keil, Frank C. 1979. *Semantic and conceptual development. An ontological perspective.* Cambridge MA: Harvard University Press.

———. 1989. *Concepts, kinds, and cognitive development.* Cambridge MA: MIT Press.

———. 1994. The birth and nurturance of concepts by domains: The origins of concepts of living things. In *Mapping the mind. Domain specificity in cognition and culture* edited by Lawrence A. Hirschfeld & Susan A. Gelman. Cambridge: Cambridge University Press.

Lloyd, Geoffrey E. R. 1990. *Demystifying mentalities.* Cambridge: Cambridge University Press.

Mauss, Marcel. [1924]. 1950. Rapports réels et pratiques de la psychologie et de la sociologie. In *Sociologie et anthropologie* edited by Marcel Mauss. Paris: Presses Universitaires de France.

Sperber, Dan. 1996. *La Contagion des idées. Théorie naturaliste de la culture.* Paris: Odile Jacob.

Zimmermann, Francis. 1987. *The Jungle and the Aroma of Meats. An Ecological Theme in Hindu Medicine.* Berkeley: University of California Press.

———. 1989. *Le Discours des remèdes au pays des épices. Enquête sur la médecine hindoue.* Paris: Editions Payot.

———. 1995. *Généalogie des médecines douces. De l'Inde à l'Occident.* (This book includes a detailed bibliography on Humoralism, from a comparative and historical point of view) Paris: Presses Universitaires de France.

Session III: Thought

Edited by Gabriele Iannàccaro

Maurizio GNERRE underlines the connection between Marlène Albert-Llorca's principle of equilibrium, also studied by cognitive anthropology (see Michael Leighton, *On the Principle of Symmetry*), and the idea of polarity mentioned by Zimmermann. Their fundamental idea of equilibrium and symmetry can also be combined with Giulio Angioni's reference to all the knowledge languages normally do not express or we do not think we have to express through languages.

All this suggests that language is a mechanism that is always in tension with extralinguistic knowledge: what is made explicit through language and what is not is regulated through time by a sort of balance. According to Goody, this balance always implies a dynamic relationship. Therefore the emergence and the development of human language can be considered as a progressive *claim* towards explicitation, towards making more and more explicit what is implicit in much traditional knowledge. Different historical and sociocultural contexts entail different standards and degrees of polarity in the complex balance between the knowledge of action and the knowledge made explicit through language.

*

Commenting on Albert-Llorca's paper, John TRUMPER points out a problem of directionality in the relation between Latin Apis "bee" and the plant-name Apium = Apium graveolens (L.), Petroselinum hortense (Hoffm.), Smyrnium olusatrum (L.), Fumaria officinalis (L.), as well as some Ranunculi, though in most cases Pliny, Latin Dioscorides, Pseudo-Dioscorides and others add modifiers. Peudo-Apuleius gives us Cxix. herba apivm (Interpolationes ex Diosc. II 76, 11 sqq. "Græci selinon, melissofillon, profetæ hema Horu, Latini apium. Omnibus notum est acerrimæ uirtutis esse atque mictualis."). For "Horus" (blood) "see J. André *Notes de lexicographie botanique grecque* [Paris, Champion 1958: 17 aima (Wrou with comments). J. André, *Les noms de plantes dans la Rome antique* [Paris, "Les Belles Lettres" 1985: 20] sees no problem here and would derive Apium directly from Apis "herbe aux abeilles." But the problem cannot be resolved in such an apparently simple manner because we are dealing with a complex semantic and cognitive configuration that involves Insects (bees), Products (honey, wax) and Plants (Apium sp. etc.), with no particular solution that gives precedence to either the plant or the insect in the construction of a Cognitive Model, since Apium, as a plant, was also culturally central, from the culinary and medicinal point of view: it was far more central and meaningful in ancient culture than its Romance descendants or congeners (French, S. Italian, etc.) are or have been. This chicken-or-egg problem may well have to be rethought.

A first proposal might be the following. Although it is obvious that APIUM has nothing to do with the colour of the plant, unlike the alternative Greek name a[rgion (J. André [1958: 19], evidently from *arg- Pokorny, *Indogermanisches Etymologisches Wörterbuch* Vol. 2: 64, the "white" or "shining" base, from the color of its small flowers), nobody has yet succeeded in demonstrating that, apart from BEE associations, it does not have to do with some property of the plant Apium graveolens (L.), e.g., Pokorny id. 780.1 *op- "strong," "be strong," "revivify," given that, as André (1958:17) testifies, this plant was associated with Horus and the cult of the dead. It was used as a garland entwined around the necks of mummies, perhaps to accompany them in their "new" life by virtue of its potency, perhaps because of associations with the Horus cult. In this last sense, the plant may also have long-term negative associations: it is used rurally, in infusion, in European folk culture to provoke abortions. An *o/ *a alternation in Proto-Latin or Italic—it is possible to postulate something like Italic *op- / *ap- = Oscan *ap-, more rarely *up- = Latin *ap- in relics, more generally *op-—is not too problematic for us to be able to argue the case, as Emilio Peruzzi brilliantly taught. Trumper asks to know the speaker's views on this problem.

Goody's criticisms on current "static methodology" are not generally justified, since sociolinguists and/or ethnolinguists have been trying to come to terms over the last forty years with questions like intrinsic variability and "areas of existence" with basically mathematical properties, which go beyond determining the acoustic and perceptual "areas" of vowel existence, though rather similar in practice. Although the question has not yet been completely resolved in problems of meaning, and a great deal has still to be achieved at the interdisciplinary level, linguists are working seriously and dynamically.

As regards the Middle Ages, Albert is right in maintaining that they have been grossly and seriously underestimated for too long. A great deal of responsibility for this lies with Augustine and his relative intellectual and cultural weight throughout the Middle Ages; however, the Middle Ages are not just Augustine, and the Latin West has to be measured against the Greek East, which included a large chunk of Italy; both should be rethought and re-appreciated. Boethius and Cassiodorus may have some responsibility in this. Even though the Latin Pseudo-Apuleius contains only 130 species of plants, Pseudo-Dioscorides (ca. 400 AD) contains about as many as Theophrastus, i.e., ca. 400. It is true that later authors never discuss as many plants as Dioscorides writing in 50–60 AD (ca. 600 plants), though to say that the Greek East loses ancient botanical knowledge seems exaggerated. Even Albertus Magnus' *De Vegetabilibus* in the thirteenth century, demonstrates that a good deal is still known about plants even in the West [he knows at least 375 species, while modern scholarship has not successfully identified thirty-five of his plant names, so we are dealing with potentially 410 species in all], while, on the other hand, western Europe did benefit from the odd eastern Emperor's occasional generous gift of copies of Dioscorides. In fact, the Greek-speaking world did keep alive, and benefit from, the tradition in Damascenus' Peri; Futw'n (originally first century BC), which throughout the Byzantine Middle Ages was known as Pseudo-Aristotle (the Loeb edition still published it in 1980 in Aristotle XIV. Minor Works, following Becker's early nineteenth century title), as well as the later Dioscoridean tradition. Exchanges of knowledge continued too, at first only between Greeks and Arabs, much later, with Gundisalvus and Geraldus [of Cremona], between the Latin and Arab world. The Middle Ages only lost knowledge about particular things, for example Fish and the Sea. One of the reasons may well be Augustine's perplexity on the ambiguity of the relevant passages of Genesis 2 (19–21) where Adam gives names to other "beasts" but not to fish, cf. Augustine's De Genesi ad litt. libri XII (Migne's *Patrologia Latina* vol. 34). There are others: Zug Tucci has a long essay (H. Zug Tucci, "Il mondo medievale dei pesci tra realtà e immaginazione," in *L'uomo di fronte al mondo animale nell'Alto Medioevo* [Atti delle Settimane di Studio del Centro Italiano di Studi sull'Alto Medioevo XXXI, Spoleto 1985: 291–360]), which detractors do not seem to have read, that demonstrates that in certain northern European and certain northern Latin areas there is relatively little fishing and fish culture for a great variety of reasons, whereas the Mediterranean in general still represents a strong fishing culture: Cassiodorus at the end of the sixth century talks about this in his Variê and, as a Mediterranean, born at Squillace in Calabria, introduces the "piscarium" into monastic culture and practice, perhaps a little-known fact.

It has been objected that Calabria was not central to the transmission of Greek knowledge, but it might just be remembered that Cassiodorus, Calabrian, was an Emperor's secretary, a Roman Consul and Senator whose main preoccupation was mediating between his own Latin culture and his Emperor Theodoric's obtuse Germanic (Gothic) mentality: he had also spent some time in Constantinople and demonstrates in the Variæ a more than passing knowledge of Greek, even though occasionally not well understood (Aristotle's e[naima vs. a[naima functional and classificatory opposition is used to etymologize Latin animal!), like Boethius but unlike Augustine who seems to have known hardly any. Under the Byzantines (700–1100 AD) Calabria produced copies of Dioscorides and compendia of strictly lay knowledge as well as the *Codex Purpureus* and Eastern Liturgies; Nilus of Rossano, Calabrian, re-introduced an awareness of Greek culture and its importance at Monte Cassino, in Grottaferrata and in Rome in the tenth century, while his disciple the Calabrian Fantinus the Younger acted as a trait d'union with the Greek culture of Greece where he is still venerated as a saint, though not in Italy; Ioannis Italos, cofounder with Michael Psellos of the Byzantine "Renaissance" at Constantinople, was Calabrian, born in Reggio Calabria, Barlaam (his lay surname was Massari, amply demonstrating his "Latin" origins), a Calabrian who risked becoming the Patriarch of Constantinople but fortunately got the worst of an argument with Gregory Palamas, was Petrarch's Greek mentor as well as a renowned writer of Greek theological disputations, while his pupil the Greek Simon Atamanus became his successor as Greco-Latin bishop of Gerace in Calabria, and another, Leonzio Pilato, became Boccaccio's mentor and translated Homer in Florence, anticipating the Renaissance [A. Pertusi, *Scritti sulla Calabria medievale* [Rubbettino, Soveria Mannelli 1994: 233] "…con il Kristeller, si può affermare senza tema di smentite che già gli italo-greci insegnarono la filologia ai nostri primi umanisti"), etc. Just to stay within the eleventh century, one might remember that Leo, Metropolitan Archbishop of Reggio Calabria at the beginning of that century, was a renowned grammarian and historian who wrote treatises on grammar and a continuation of Theophanes' Chronographia (see Migne's Patrologia Greca vol. 120 col. 177 et seq.); his successor Archbishop Nicholas was author of a number of important Scholii (*Codices Vat. Gr.* 1650–1658), some in an extremely anti-Latin vein; the Calabrian physician Philip Xeros was author of a number of important medical tracts (Codex Vat. Gr. 300 folios 267–273, 299–300 etc.); Constantine the Protosecretarios, Calabrian, probably from Reggio, was a well-known translator of Arab authors into Greek, e.g., Abu-Gafar Al-Gazzar's works (*Codex Vat. Gr.* 300, folios 11–267), that the so-called Suida or Suda Lexicon of that century was composed by Calabrian monks (S. G. Mercati, in *Byzantion* XXVI–XXVII, 1956–57 fasc. 1, A. Pertusi op. cit. 228), that the᾿ Ἐρωτήματα of Byzantine grammatical tradition, which later become the Renaissance model of grammars, was first projected as a teaching method by Italo-Greek monks in Calabria and Apulia, at least according to Anna Comnenus' private correspondence, etc. It would be tedious to go into further details or take the question into earlier centuries. One might therefore say that Southern Italy and Calabria were far more important culturally before the thirteenth to fourteenth centuries than ever after it, and this in a strongly "Greek" cultural sense.

*

Glauco SANGA points out that Zimmermann's theory of humors shows many points of contact with the ideology of illness (or sickness), understood and named through animals. According to this ideology we have quiet animals inside our own body, which provoke illnesses when they start moving (e.g., worms). Coming now to the *kinship* names of animals, in Sicily we have an illness with a *kinship* name, the *matrazza* (= *mother*) "the womb," which is thought of as an animal, a sort of polyp, that can move and cause psychological pathologies.

*

Pier Giorgio SOLINAS asks Zimmermann whether his interesting suggestion of reading Ayurvedic medical learning as a practice of knowing refers either symmetrically or inversely to Bourdieu's *Théorie de la pratique*. Solinas shares the intelligent character of the practice, or the active logos, the knowledge of the hand. He asks Zimmermann whether it is possible to rediscover the mechanism of thinking in terms of thinking that studies itself: metathinking, which identifies itself for a while and works within the object-thought. This he considers very classical, as is Angioni's reference to this embodied intelligence. It would be interesting to know from him how it would be possible to regenerate in thought what is not corporeally reproducible; ethnographic observation of an action in fact cannot be the repetition of that same action. By what means can we think of the action, see it, have it reproduced?

*

Addressing his remarks to Marlène Albert-Llorca, Roy ELLEN states his skepticism about those theories that claim that we should conflate all the meanings of nature all of the time. Rather, when we engage in generating classifications, in thinking about nature and in articulating knowledge, we systematically repress, or forget, or ignore certain characteristics and associations of particular natural things. Sometimes this results in more naturalistic classifications, sometimes in more symbolic ones, or a combination of the two. The ability to cope with those cognitive contradictions that Goody may have been alluding to is perhaps itself some cognitive universal of humankind.

*

On cognition, Tim INGOLD underlines the inconsistency of separating the cognitive activity based on neurological hard-wired mechanisms from the manifold varieties of bodily activities in the world. Dan Sperber, Scott Atran and others have pointed out that, even if a significant component of knowledge is not innately "wired-in," certain innate acquisition devices are needed to enable the intergenerational transmission of representations and beliefs. However, Giulio Angioni has referred to the knowledge revealed in skills like riding a bicycle or knotting a tie, which is built into long-term memory and is part and parcel of the process of our bodily development, but which seems to be virtually impossible to represent in the form of rules or representations. This kind of knowledge is not given from the start, nor is it added on to a preformed bodily constitution. Is it thus innate or acquired?

The results of some experiments in knot-tying conducted in the Department of Social Anthropology at the University of Manchester showed that it was impossible to distinguish the cognitive task of problem solving from the practical activity of the body. Moreover, since the detailed written instructions (supported by diagrams) that should help students to perform the operation really made sense only when the knots were completed, the skill of knot-tying turned out to be impossible to represent in terms of any precise algorithm.

In her book *Androgynous Objects* (Chur: Harwood Academic, 1991), Maureen MacKenzie supplies an interesting example of literally embodied knowledge, a knowledge incorporated into the modus operandi of the human organism. She describes the way Telefolmin girls from Central New Guinea, without explicit instructions, learn from their mothers how to make string bags. If we cannot understand string-bag making as the output of some kind of acquired cognitive program, then we cannot understand a weaverbird's nest construction as the output of a genetic program. The conventional distinction between innate and acquired skills, or between universal capacities and culturally particular competencies is therefore quite unstable.

Maurizio Gnerre's distinction between non-linguistic knowledge, which is implicit, and linguistically articulated knowledge, which is explicit, is not convincing, since it seems to rest on the idea that language is a vehicle for making inner ideas explicit. Rather, speech is an activity, like any other kinds of activity in the world, and therefore there is not such a clear distinction between these two kinds of knowledge. The American anthropologist A. Irving Hallowell, in his article "Ojibwa ontology, behavior and world view" (in *Culture in History: Essays in honour of Paul Radin*, Editor S. Diamond [NewYork: Columbia University Press, 1960]) supplies two interesting anecdotes about the Ojibwa, who, like most hunter-gatherer people, have no concept of "nature" that corresponds to ours. They consider a feature of nature as inanimate, animate or a person according to the context. For example, in the Ojibwa language, a grammatical distinction is made between animate and inanimate objects, and stones are sometimes counted as linguistically animate. They are only alive in a context of power, because life is not an intrinsic property of objects but rather a property of relationships.

Another characteristic of the Ojibwa has to do with the ability to speak. Speech, for them, is not the expression of inner ideas but a way of being. Thunder, like a person, can speak, because it exists in its kind of sound, just as the waterfall exists in its kind, and human persons exist in the sounds they make in their song and speech.

*

Renzo GIACOMINI, physician, observes that body memory is connected with the front brain-lobe and when the right brain-hemisphere is damaged, patients cannot move the left side of the body, but they do not realize it. Describing the indirect relationship between our thinking and our topical representation of the body in terms of psychodynamics, Roberto Saggioli affirms that images, mental figures and ideas tend to reproduce the corresponding physical conditions and external actions. This means that we tend to transform our mental figures either into movements, such as walking and making gestures, or into spoken actions.

*

Gherardo ORTALLI observes that in the Middle Ages there was a notable downswing of botanical knowledge compared to the classical age (Pliny). In fact men knew different things for different purposes: they read natural phenomena in a symbolic Christian way. Indeed, a paradox of Christianity is the desacralization of nature brought about by its strong anthropocentrism; today our secular world considers nature in a much more sacred way, due to the influence of enviromentalist theories.

*

Alessandra PERSICHETTI (anthropologist) asks Jean-Pierre Albert whether the systematization of universals and naturalistic and scientific knowledge are privileged and have a greater epistemological supervision, and if it would be possible to consider religious thinking too in a Foucaultian perspective: that is, whether it might have replaced a former kind of thinking, like the two other levels of knowledge.

She asks Zimmermann whether his statements on complex systems of thinking about nature start from the basic supposition that in the various social systems contrasting theories about nature are present, while other systems work on a single elementary level. If, on the other hand, Zimmerman simply agrees with Goody in thinking that the complex system of thinking about nature is nothing but the faculty of those societies endowed with writing and thus have a greater possibility of formalization, his assumption would not explain why also in societies without writing there are contrasting theories about bodily humors and substances.

*

As regards animals and naming of animals according to a kind of theory of signatures, Francis ZIMMERMANN responds to Sanga that the inferences about humors could be easily compared to inferences of that kind.

Answering Solinas, he confirms that he is a disciple of Bourdieu and that he follows the guidelines of Dumont, and structuralism. His work on classification in India concerns only a learned cultural tradition completely organized by writing, therefore it is the exploration of a single aspect of the problem.

*

Marlène ALBERT-LLORCA agrees with Roy Ellen in thinking that different kinds of classifications work in parallel in different societies; for example, in Europe plants are classed sometimes according to morphological criteria (size, flower colors, etc.), sometimes according to pragmatic criteria (edible or not), or according to whether they are wild or cultivated. This difference, moreover, cannot be reduced and combined by symbolic or morphological classifications. Nevertheless, symbolic thinking is worth the attention of ethnotaxonomy since it is an essential aspect of the relationship between human beings and their environment.

The ethnography of apiculture confirms Trumper's remarks; as a matter of fact Virgil's description of bees as organisms reproducing without copulation suggests that bees were considered to have a constitution that falls somewhere between the animal and the vegetable. The supposed affinity of bees with the vegetable kingdom is also attested in W. Frantze's *Historia animalium*, published in 1665, where they are said to "sprout" to form a new swarm; in a similar way, Virgil uses the word bunch" to refer to a swarm hanging on a branch.

*

As regards John Trumper's question about the Middle Ages, Jack GOODY underlines the enormous loss of secular knowledge involved. For example, the Roman town where he grew up in Britain had a theatre in 300 AD and never had another for 1500 years. As far as botanical knowledge is concerned, the whole tradition of Roman naturalistic painting disappeared virtually until the *primi lumi* of the Italian Renaissance, until we come across the background in early Renaissance painting. It would be interesting to know why Dioscorides' illustration continued to be used for more than 1000–1500 years.

The rejection of the natural image-iconoclasm is reflected in botanical work during the earlier Middle Ages. Iconoclasm characterized early Christianity and was an attitude present in the Old Testament from which the Muslim's objection to graven images derived. The concern with the truthfulness of representation underlying iconoclasm, something stressed by Plato, ran very deep in Judaism and still runs deep in Islam and at times in Christianity. Protestants, for example knocked off all the heads of the statues in the cathedral of Ely.

Iconoclasm is not simply a Western European phenomenon, it also relates to Buddhism and Confucianism. It arises from the contradictions implicit in the process of representation. According to Goody, they are part of the problem of language-using animals facing the world which they have to classify when they are interacting with it on a linguistic basis. This creates problems in understanding the world, involved in all classification, and leads to Goody's own point about cognition and action. However, he considers it extreme to claim that there is no aspect of understanding the world that does not involve action, or involves it only minimally or in the circular sense that mental processes themselves involve action.

*

Giulio ANGIONI maintains that there exists a coherent thinking that renders itself explicit in action, and that in action grows upon itself and realizes itself. Although trying to make it explicit through internal means is important, it is not enough, and anthropological or scientific explicitation through means of representation and discourse is equally worth carrying out. Studying the differences between what he roughly schematized as "thought-of-doing-for-doing's-sake" and "thought-of-saying-for-saying's-sake" would be interesting.

PART IV

Use

Antonino Colajanni

How have we come to use Nature, from a practical point-of-view?

The nature knowledge of native peoples—about the different eco-systems in which they have long lived, and their capacity to manage, transform and use their resources—has received increasing attention over recent decades: both as subjects of study and research, and also as relevant items of knowledge in programming and carrying out projects revolving around environmental-agricultural change. Students in this field and operators in the development field need to investigate more closely the connection between classificatory and conceptual studies of nature on the one hand, and studies of techniques of use and their feasibility in the context of development projects on the other. In this way one might hope that the influence of *knowledge* on *practical action* will grow both quantitatively and qualitatively. The list of studies on indigenous knowledge and the processes of economic development is now such a long one—beginning with the pioneering anthology edited by D. Brokensha, D. M. Warren, O. Werner (1980)—that it is worth stepping back to make a critical and historical assessment of the various aspects, problems, and solutions proposed. This could also help to promote new kinds of research and strategies of negotiation with the decision-making contexts of the development programs. The most recent collections of essays and reflections on practical experiences (Dupré 1991; Warren, Slikkerveer, Brokensha 1995) reflect the belief that it is only through ethnographic research and an intelligent fusion between the classificatory-conceptual studies and those on techniques of use and on technological efficiency and efficacy, that success can be guaranteed in this new field of study. These recent works also constitute practical instruments of information on the immense body of work that is being carried out in numerous research and experimentation centers, many of which are based in countries outside Europe.

The Dynamic Nature of Indigenous Knowledge Systems

The first problem to tackle in an international meeting of specialists on the subject would seem to be that of methodology, strategy, communicative logic and the search for "efficacy" in the process of the *practical application* of nature knowledge and techniques of use in well-defined operations (development projects and programs). This is a complex process of communication and encapsulation of knowledge in practical action, in which indigenous knowledge may constitute the central nucleus, the starting point and at the same time the most efficacious instrument in methodological and operational terms to achieve the aims of the projects. Comparisons between projects that have taken account of local knowledge and others that have ignored it have given sufficient evidence of the expediency and efficacy of this tendency, but it is worth investigating and discussing yet further the methodologies of communication with the technicians and politicians of development.

A second problem, which is often ignored in discussions of the issue, is that of the *dynamic and adaptive nature of indigenous knowledge systems and the practical use of resources,* which prove susceptible to modification, incrementation, and energization in response to external stimuli. Local systems often have a long history behind them, of which it is important to know about. The processes of change they have undergone in the past are thus pertinent to the solutions that may be chosen for their future. The capacity of knowledge systems to adapt themselves, to lend themselves to use over the centuries, to reconstitute and resystematize themselves (which often engenders reclassification), in the context of long decades of encounters and clashes between native societies and the various fronts of Western penetration, reveal that native societies possess their own forces of response and reaction: they are, in short, active and not passive subjects, also in the field of technique and natural knowledge. They are thus able to respond efficaciously to the proposals—often "radical" and "unconditional"—of development programs. It will therefore be indispensable to devote some time to considering and discussing the process of *technical domestication* that the West has exercised over native systems; it will be important to deal with the active and reinterpretative reactions to innovations, the forms of resistance, "technical syncretism," and the functional readaptation of elements from outside. The loss, disappearance, and dispersion of knowledge and techniques of use should therefore be studied alongside the relationship between innovation and response, between selection and adaptation, between acceptance and rejection, between integration and exclusion.

Local Technical Knowledge as a Resource for Development

Recent studies and practical experience have now shown—and it is accepted by the most important international agencies responsible for the promotion of economic development—the strategic importance of "Local Technical Knowledge" as a fundamental *resource* for the correction of the often perverse effects of many development initiatives; this not only serves the interests of local populations but also renders projects more efficient and efficacious by increasing local participation. Indigenous knowledge of the natural world and the indigenous systems of use can thus be considered a valuable *installed social capital;* they act as the very hub of the local context, constituting a driving force, an essential source of creative and constructive energy, essential to ensure that the projects of planned change are mastered by the local society. There is a general shift in the attitude of development technicians (agriculturalists, foresters, agrarian economists, engineers, managers etc.) towards the resources of knowledge and the strategies of use of local populations. This progressive recognition of the potentiality of knowledge systems and of indigenous action is producing—and will increasingly continue to produce—a beneficial effect on the processes of creation and recreation of local *social and cultural identities,* which, as is well-known, constitute an important aspect of social change. If debate among specialists in the subject were to result in the proposal of a common basic methodology for the introduction of homogenous and recurrent elements into all development projects—whether agricultural, forestal or environmental—that treat "knowledge and local-use capital" as an indispensable element, it might be possible to make a major change in the planning of development initiatives, which could become decisive in terms of theory and methodology.

The Great International Institutions for Development and the Indigenous Knowledge

In discussing the topics listed above it would be impossible to omit a careful assessment of the interest that the great international institutions for development have shown over the last ten years in these arguments. The World Bank and the F. A. O., for example, have recently organized seminars and discussion groups on the theme of local knowledge of the environment and its resources, involving many specialists on the subject, and they have begun to introduce certain methodological innovations in the direction suggested here (see, for example, Warren 1991; Davis Ebbe 1995). It is well worth discussing and constructively criticizing—with the aim of improving—this new line adopted by the great development institutions. And this can be done on the basis of concrete experience in work and research, and with the contribution from indigenous organizations and the research centers created over the last few years in many developing countries. It is worth considering whether this new interest on the part of international institutions constitutes a real innovation in methodology and theory, one that will modify the traditional attitude towards local knowledge and local power-structures in the development-process, or whether their interest merely represents an obligatory and unconvinced concession to pressure from the periphery of the world and from certain persistent scholars, and a consequent modification of language, having little or no operational effect on the various development programs underway in the world at the moment. It will be interesting, in this sense, to make a collective assessment of any concrete results (both in terms of theory and practice) that have been achieved by the effort to coordinate and exchange experiences that began in 1993 with the publication of the influential magazine *Indigenous Knowledge and Development Monitor,* published by the Center for International Research and Advisory Network (C.I.R.A.N.) at the Hague, in collaboration with the network "Indigenous Knowledge Resource Centres," which now includes twenty-six Centers, most of which have branches in developing countries.

Indigenous Knowledge and the Different Ecosystems

Finally it will be useful to record the opinions and specific conclusions drawn from the experience of various researchers and experts on the relations between local knowledge, forms of use and development contexts, in the various ecosystems inhabited by human societies, each of which presents its own characteristics. The general problems presented above may assume special features according to their specific contexts, whether they be *tropical rain forests* or the *arid and semi-arid regions of the Sahel*, to cite just two classic examples. By relativizing the discussions of the general themes to specific regions and ecosystems or to the different activities of the local populations (agriculture in arid zones, nomadic sheep-rearing, coastal and river fishing, etc.), it may well be possible to render the discussions more specific and, at the same time, more concrete.

References

Brokensha, D., D. M. Warren and O. Werner, eds. 1980. *Indigenous knowledge systems and development.* Lanham-New York-London: University Press of America.

Davis, S. H. and K. Ebbe, eds. 1995. *Traditional knowledge and sustainable development.* Environmentally Sustainable Development Proceedings Series, no. 4. Washington: The World Bank.

Dupré, G., ed. 1991. *Savoirs paysans et développement.* Paris: Ed. Karthala/ORSTOM.

Warren, D. M. 1991. *Using indigenous knowledge in agricultural development.* Washington: World Bank Discussion Papers, no. 127.

⎯⎯⎯ , L. J. Slikkerveer and D. Brokensha, eds. 1995. *The cultural dimension of development. Indigenous knowledge systems.* London: Intermediate Technology Publications.

CHAPTER 17

Indigenous Knowledge: Subordination and Localism

Giulio Angioni

Also in the case of traditional knowledge and skills (whether of nature or not) "using" involves practical activities and planning. Consequently, it must deal with politics and, especially, political economy and geopolitics. More generally, it has to confront power, authority, prestige and the use of force, even in a world in which the distinctions between local and global are diminishing and cultural phenomena are increasingly widespread and potent (Milton 1996; Robertson 1996).

There must be a greater focus, therefore, on two often overlooked aspects of knowledge commonly described as indigenous or traditional: subordination and localism. Roughly speaking subordination means that traditional indigenous forms of knowledge are more or less subjugated and lack authority and prestige compared to other existing hegemonic systems of knowledge. Their localism reflects the fact they have less of a global vision in time and space than modern official knowledge. They have no long-term ecological strategy, their practical use is limited to a special ecological niche, and their planning scales are of no more than a few agricultural years. This localism closed in limited spatial-temporal strategies is a frequent condition of traditional nature knowledge, despite the popularity in the West of the stereotype of the "savage ecologist" (Ellen 1986). And yet this more or less closed localism has not hindered the development of cosmogonies and comprehensive mythological escatologies and, more importantly, it has not prevented the almost universal development of the careful recycling of all residues from processing, consumption and biological assimilation. This has taken place since the so-called Neolithic revolution. Indeed, recycling and the shrewd management of scarcity are founding aspects (although rather neglected in our studies) of agrarian and pre-agrarian cultures from all times and places.

There has, however, been some critical focus on localism, despite the fact that it has been categorically attributed only to traditional indigenous knowledge, whereas official hegemonic modern knowledge is not often enough seen as being relative to a given place and period, i.e., to what we call the modern Western world. The Western form of localism lies in the powerful idea that its own knowledge is not local or even historically determined but has qualities of absolute certainty and excellence, reflecting the well-known modes of ethnocentrism or rather of Eurocentrism. But it is above all the subordinate features of traditional knowledge compared to hegemonic official knowledge that has led to them being played down, written off, or left in the background as irrelevant—or perhaps even as too politically embarrassing.

Compared to official knowledge and skills, traditional or indigenous or practical knowledge and skills are more or less tacit, implicit, intuitive, informal, non-coded and so on. But compared to "official" hegemonic knowledge, they are also more or less subordinate or subjugated—when not simply misunderstood. This is true both within the developed societies we call Western and also in the societies of the so-called Third and Fourth worlds. Traditional knowledge has the power of being *able to do* but often they have little or no decision-making power over what to do, never mind the problems of more general economic means handled by different players than the "bearers" of "indigenous knowledge." From this point of view, the word "indigenous" is justified by the fact that they are typical and belong to people from a specific place. There are also forms of knowledge and power which are difficult to localize, although they can be grasped fairly well when we learn that the trends of the Singapore stock exchange influence parsley farming in Tipperary.

All the features identified as traditional or indigenous or informal knowledge are present and continue to be regenerated as such also in Western societies: practical daily informal knowledge only slightly influenced by explicit discourse and learning; the kind of knowledge that underlies the Italian proverb *vale più la practica che la grammatica* ("an ounce of practice is worth a pound of precept"). We may thus conclude that what really distinguishes indigenous or traditional knowledge is a lack of authority and power—their subordination to what we describe as scientific knowledge and its holders. Anthropologists know better than most that traditional indigenous knowledge is no more or less categorically subordinated together with its bearers, but that these forms of informal knowledge are still widely and deliberately ignored by the holders of hegemonic knowledge. They tend to consider peasants, shepherds, fishermen or Third or Fourth World craftsmen as "untaught" or *tabula rasa* in terms of any kind of knowledge, and not only compared to hegemonic formal or scientific knowledge in a specific place at a given time.

As a provocative addition to the meticulous terminological research to describe adequately the subject of our discourse concerning traditional or indigenous knowledge, it may be interesting to note how Antonio Gramsci seriously used, also for these things, the discredited old term "folklore"—in the strictly etymological meaning of the "people's knowledge."

In the first of his *Prison Notebooks* (henceforth PN), at the date of 8 February 1929, Gramsci included the "concept of folklore" at Point 7 of the "principal topics" (PN, 5). Gramsci believed that folklore, and especially the concept, was a "principal topic" and in the *Prison Notebooks* he further explored folklore, which "must not be seen as bizarre, strange, ridiculous or picturesque, but must be conceived as a very serious matter, to be taken seriously" (PN, 90).

Gramsci was fully aware that he was looking at things in an unusual way also in the case of folklore. He began by categorically refuting its irrelevance and declaring that folklore should be studied as "a vision of the world and life ...of certain strata of society determined in time and space," and that is "people should be seen as the set of subordinate and exploited classes of all societies to date" (PN, 89,

2311–17). Thus there is a need "to change the spirit of folklore research, as well as deepen and widen its scope" (PN, 2314). Obviously this also applies to folklore visions of the natural world.

This shift in perspective and values is the result of Gramsci's overall vision of reality contemplated in his political and theoretical commitment. It also raises the problem of the political use of folklore. Gramsci introduces the problem in his lesson on folklore of why, despite being been studied so much, it had not produced— even less in the common meaning of the left—that radical reassessment of the term and its subject as serious study matter, which has continued to be "at most something picturesque."

While in prison Gramsci dedicated much of his reflections on working-class culture, i.e., folklore seen as the study "a vision of the world and life of the subordinate and exploited classes," and six years later in a second draft (1935) of the *Notebooks* in Notebook 27, entitled *Observations on folklore* (PN, 2311–13), he reiterated that it was "a very serious matter, to be taken seriously." In the *Notebooks* he dealt at length with a host of topics: spontaneous philosophy, common sense, and the co-existence of these forms of social awareness (or, as we would say today, of different cultures), language and dialect, literature, especially popular literature, spontaneity and deliberate action along with important aspects of his vision of the intellectual class, hegemony and historical blocs, and the *questione meridionale*— the southern Italian question. All of these subjects are best understood—and, in some cases, can only be understood—with reference to his views on the features of folklore.

We must thus return to Gramsci's notes on the "vision of the world and life of the subordinate and exploited classes," that is to folklore and popular culture in which he believed it possible often to see "an unassimilated agglomeration of fragments of all the conceptions of the world and life that have occurred in history." But "even the people is not a culturally uniform community. It has many cultural layers, combined in various ways, whose pure nature cannot always be identified in given historical popular collectivities" (*PN*, 2312). Consequently, "it is the greater or lesser degree of historical isolation of these collectivities that permit a certain identification." But in folklore "we must distinguish between several strata: thus there is a fossilised strata, reflecting conditions of past life and as such conservative and reactionary; strata which are a series of innovations, often creative and progressive, determined spontaneously by life forms and conditions in ongoing development and are in opposition to, or only different from, the morality of the ruling strata" (*PN*, 2313). There are also some memorable sketched notes on the strength and solidity of popular beliefs, the tenacity of the moral tradition and, above all, on the civil and political duty to progressively acquire awareness of one's own historical character by the masses who, after such long periods of subordination and exclusion, have never even guessed that it is possible and worth the effort.

Gramsci himself came from those subordinate strata of an isolated marginal island (Sardinia), made up of people "who do not even guess that their history can be of any importance or that there is any point in leaving documentary traces."

Understanding this for Gramsci means making a commitment to raise—in the majority of the mass—"the awareness of your own historical character" and not rush into some "group adventure in the name of the mass."

Thus the discredited term "folklore" is used not only as an empathic feeling of nostalgia and regret, or of the typical. It involves understanding that the "re-appropriation of one's own historical individuality" is not achieved by idolizing true or presumed traditions, but only by grasping the time to give to the masses (as Gramsci had always claimed) the task of that "theoretical awareness" creating historical and institutional values, or founding states" (*PN*, 1041).

This is anything but sentimental indulgence and arcadian folksiness. If the task still seems utopian and distant, we can only say that Gramsci's lesson has still to be understood, digested and assimilated so that it may become common sense, or that kind of "popular belief" that "often has the same energy as a material force."

This brings us—*quod demonstratum erat*—to a typically Gramscian issue: the question of the relations between the intellectuals and the masses (today the same thing would be conveyed with formulas such as "cultural institutions and administrative area") and between spontaneity and conscious action. This is expressed or was expressed recently in a more confusing manner in terms of needs, participation, representation, and new players… The issue is still relevant and not only because it appears to be remote from thinking on a grand scale and collectively, from civil commitment. It marks a return to the importance of the private, that is, the spontaneous and elementary, previously artificially opposed to the elaborate and the "politically aware and organised"—to use Gramsci's words, again from the *Notebooks*.

Today we must reckon with new aspects of common sense, a new folklore, new spontaneous philosophy to be taken into account just as much as the old peasant-origin folklore. Although we must never relinquish making a "more cautious and accurate calculation of the forces acting in societies," like the old folklore, this new folklore may have the tenacity of material forces, despite the fact it does seem more ephemeral and subject to passing fads.

Reflecting on Gramsci as a folklore scholar does not mean the usual old rushing off into the past to seek out the fine things from the good old days. It is an aspect of long-term work aimed at preventing the structural and subjective conditions of marginalization and subordination from growing further. It is also a way of ending certain deep-rooted reactions, such as the Manichean attitude tending to see what is or appears to be indigenous as wholly good and everything outside as bad, or vice versa. Against the temptation to simply look back nostalgically, Gramsci himself reminds us that "the historian with all the due perspective manages to establish and understand that the always harsh and stony beginnings of a new world are better than the decline of a dying world and the swan song it produces" (*PN*, 1377).

For the first time in history, the feeling of belonging to the human race and to the planet Earth—our (endangered) common land—is now a strong and widespread persuasion. The decline in Western forms of nationalism and patriotism are probably also due to the obvious fact that the world of today is increasingly one, and that the same is true of its inhabitants, even though the ecumene is roughly divided into rich, hegemonic countries and areas (the West, the First World) and poor, subordinate countries and areas (the Third and Fourth Worlds), thus creating two new macro-identities—rich and poor—on a planetary and ecological scale.

It is a trite generalization to claim that various aspects of ethnic belonging are no longer positive when they come into conflict with wider communities and the need for solidarity, or the feeling of belonging to humanity as a whole in a increasingly smaller interdependent world threatened by individual and collective selfishness. Cultural assimilation and standardization are neither good nor bad per se. Similarly, it is not always and not in every case good to preserve and strengthen ethnic, national, tribal or other characteristics and sentiments: cultural variety and cultural standardization have both caused strife and brought benefits. If, as Lévi-Strauss held, cultural differentiation is, and has been, a condition for progress, it has also been a source of strife, although not as such and not directly. There would not have been the horrors of five centuries of European colonialism had there not been such enormous cultural differences that were, and are, also differences in economic, political, military, ideological, religious, and linguistic power.

Many other characteristic aspects of Western identity may be revealed in the Western way of experiencing the world. But one essential aspect would be overlooked if we did not bear in mind the underlying, fundamental ontological conviction of the superiority and excellence of what we call the West, which is continually measured against the diversity of the "rest of the world," for the purposes of consolidating itself. Thus the Western ways of being and experiencing (Euro -Western, White, civilized heir of the best that has been done in the world) are accompanied by the notion that the rest of the world and humanity are vague, fluctuating, muddy, often confused and even contradictory things. But the vagueness of ideas and feelings are not per se the cause or sign of less strength or poor functionality.

The Western vision, the inveterate modern and Western way of seeing the world and life, of perceiving one's own place in the world, is based on a solid and proteiform racist-type outlook. In the ethnocentric view, non-Westerners are inferior in a world becoming increasingly Western. This is tantamount to saying, for example, that no Westerner can conceive of the world without positing the West as its foundation, as the developer of industrial civilisation, as being in a position of excellence, while all the others occupy more or less inferior positions, according to their differences with the Western way of life.

This way of making sense of the world and the Western way of living is more solid in so far as the Westerners' feeling of superiority rests on the fact that the West does effectively dominate the world, especially recently in a period in which part of the West embodies this sense of security for having won the "Cold War." This sentiment is now so subdued and mild that it hardly ever needs to be expressed as intolerance or violence, now much rarer than in the past. In fact, the counterweight to the unassailable sense of our superiority is expressed by charity towards the Third World in the form of development aid, especially in the field of knowledge and expertise about the natural world.

Of course the vision and sentiments of Western security are more or less implicit or lurking beneath the surface, ready at any time to become explosive and even aggressive. There are no longer those who believe that democratic industrial Western white superiority has biological roots and causes, that white Western superiority is inevitable because of a genetic and psychophysical superiority. This idea has lost currency and is even discredited, although for around two centuries it was part of the West's indigenous nature knowledge, together with the idea of geographical and, especially, climatic determinism.

Few would dare feel or claim to be superior from a biological, psychophysical, genetic or climatic point of view. This explains the sincere recurrence of the common claim: "I'm not racist, but..." Culturally, however even the great minds feel and at times proclaim Western superiority, just as in the past. Today's reason is historical and cultural and no longer biological. Giving up the idea of biologically based superiority does not imply that the historico-cultural racist conceptions are any less solid, although less arrogant and rarely violent. Indeed, historico-cultural racism is subtler and more persuasive, more up-to-date and less crude. It does not pass judgement or explain; it simply states its own superiority, considered too self-evident to be called into doubt either within or outside the West.

Moreover, today's mild racism wishes to account for diversity, and especially diversity demonstrating the incontrovertible Western superiority, primarily in cultural terms, as technology—the human tool making use of the world for its own purposes. The West sends probes into the cosmos, while the Third World is unable to feed itself, when "left to its own devices." From the point of view of basic needs, consumerism is objectively preferable to starvation, even though Christian culture values the paradox of the moral and spiritual superiority of the poor, and especially of anyone who becomes poor among the poor. Thus on the subject of material culture, and especially elementary nature knowledge and capacities required to guarantee survival through food, there is often strong cultural unease in many Westerners when confronted with the cultural history of everyday things such as American Indian potatoes, tomatoes, peppers, beans, cocoa, maize, pumpkins, aubergines, vanilla, tobacco, turkey, or Asian rice, tea, silk, or coffee and Middle Eastern citrus fruits and so on—not to mention the pizza originally only consumed by the subordinate Neapolitan plebe.

The sense of the other usually generates reactions that swing from defence to aggression, often with the development of feelings of superiority. Whether called racism, intolerance, or ethnocentrism, it is as old as the sense of belonging and identity. And perhaps the anthropologist is most aware that the balance between self-esteem and the way of relating to the other is varied and very difficult to assess historically. There is a recurrent tendency to reduce differences to inferiority, so that the other becomes something worse and dangerous, often a scapegoat. There may also be a tendency to assimilate the other by denying differences, so that equality involves minimizing differences to make people identical. Both attitudes—one aggressive and the other incorrigibly charitable—have been present in

our civilization from at least the origins of what we call the Modern Age, symbolically opened by Columbus and the advent of the great waves of extra-European
colonialism (Gliozzi 1997).

Anthropologists are unwittingly responsible for spreading the idea that the
racist attitude considering the other as inferior, or ethnocentrism in general, is
something universal, something biological, genetically inherited, which, moreover, we supposedly have in common with other animals. It is thus everyone's
duty, educators and educated, to play down and try to eliminate this drive in the
human species to consider the other as inferior. Racism should thus be overcome
just as children learn to control their sphincters. Even the fact of having understood this, and having developed attitudes of tolerance, is attributed by some to
the superiority of Western culture or civilization, hastily presented as the only culture capable of re-educating itself against the dangers of ethnocentrism and
racism.

In some writers this argument becomes very sophisticated (De Martino 1977;
Cherchi 1997), but at least one fallacy is often present: ethnocentrism is neither
inevitable nor universal. And even less so its racist extremes. History and ethnography reveal almost just as many cases of not considering the other as inferior as
vice-versa. Indeed the attitude of considering the other, the stranger or foreigner,
as superior is not at all rare. This phenomenon of "superiorization" of the other
underlies the sudden defeat of the Aztec and Inca empires at the first contact with
Cortez and the Pizzarro. But without resorting to exotic examples, this is also a
common feeling on my native island of Sardinia, where history ensured that—as
often happened to peoples subordinated by external powers—those from outside
are instinctively perceived as better, superior, and more capable in this or that
sphere, while the locals are described as "small, dark and stupid." Too often
strangers came to Sardinia in arms, as the dominator, boss, lord, cock of the
walk—and never the hen—only later to be pushed down the pecking order by a
new arrival. Now, however, foreigners also come to the island as the homeless,
gypsies or Africans. And they no longer come with an exotic aura—as a chalk
statuette to place on the mantelpiece. So the Western feeling of superiority is also
felt in Sardinia, although being fully Western and European at times becomes
problematic, especially when there is an identity crisis because of stronger pressure in marginal Western places, like Sardinia, to stay in the West, since they are
faced with the danger of sliding towards nearby Africa and the Third World.

Those who believe that racism today is mainly found in the deeds of young
neo-Nazi skinheads and the like, fail to understand that as Westerners they participate in a new form of racism making them feel in a position of obvious unchallenged superiority. When that superiority is challenged, it is usually only as a way
of criticizing contemporary mores—along the lines of the bourgeois Enlightenment critique in Montesquieu's *Persian Letters*.

Today's sense of Western superiority can thus be mild and condescending, with
no violence or intolerance. It is often collaborative, charitable, ecumenically Christian, Third Worldist, relativist, merciful, philanthropic, exoticlike, and ethnic as well
as obviously being anticolonial. In short, it holds its hands out to the difficulties of

the poor South. But basically this is not dissimilar to when evangelical conversion was the white man's burden, and Barbarians and savages had to be civilized through various forms of colonialism. The West incorrigibly believes it basically has something to give and teach. This is felt as its mission in the world, even though the majority of educated Westerners would now smile at the three "Cs" in seventeenth- and eighteenth-century Europe's mission to the world: Christianity, Civilization, and Commerce. Or in a word, progress. And progress was better, because based on the idea that a non-European, savage, primitive, colonized people was incapable of joining Europeans at the top of the evolutional ladder. This idea was then overtly proclaimed and violently pursued by twentieth-century fascism at all latitudes.

It is as if the new kinds of historico-cultural racism—with its ancient origins—were the constituent genes of Western culture and the Western way of being and experiencing the world. And every Westerner absorbs it from the earliest age. But it is an unchallenged supposition which when called into question creates unease.

Today, as a result of standardization, a large part of humankind lives on resources produced in few regions by a small number of farmers, while greater numbers of people re-ly on recent varieties of a few crops. In contrast, most of the world's crop genetic resources are concentrated in the poo-rest regions of the earth (Brush and Stabinski 1996). Modernization has led to ecological specialization, with only a few crops grown or breeds of livestock raised, while traditionalism generated a wealth of genetic resources together with poverty.

Will this always be so?

The challenge today is to conserve land and maintain live-stock diversity, in situ or ex situ, in order to preserve the biological and social processes of crop and livestock evolution, ecosystems, natural habitats and native know-how and skills (Warren, Sikkerver and Brokensha 1995).

The role of the anthropologist is to disseminate the idea that indigenous peoples are not a *tabula rasa*, but know how to do some things better than agronomists, engineers and other specialists in the fields of conservation and also how to address the challenge of what to do with the knowledge and skills of natives most directly affected by these problems.

However, natives have not yet assimilated the concept of the use of natural resources within a modern eco-logical framework: in most traditional native ecological con-ceptions, time and space are measured in terms of one or a few years and of a local ecological niche. But present-day eco-logical conditions and conceptions and the world market require open-mindedness and plans for production and trade on a planetary scale. This necessarily generates contradictions, contrasts and incomprehension, especially when planners (e.g., of nature parks) neglect local indigenous skills and knowledge. On the other hand, indigenous skills and knowledge can also be useful in conservation in situ, as well as for conservation in general.

Clearly, for example, a conservation policy of recovering and extending woodlands in Mediterranean regions must take into account the population's habits, which are several thousands of years old, and be based on effective consensus. In many Mediterranean regions country-people have ways of doing and thinking that almost suggest that Mediterranean peasants and shepherds often by tradition do not like woods, hold trees in little consideration, unless immediately useful for wild fruit, bark, wood or timber. And there is also a long tradition of short term competition over land use between peasants, shepherds and woodcutters. Sometimes, as in Sardinia (Angioni 1989), this attitude is explicitly declared as hostility and uneasiness towards huge wooded areas. I believe it would be difficult to find such an indifferent and ignorant culture as regards the ecological functions of trees and plants—even for the uses that Sardinian farming itself makes of land as fields or pasture. Here understanding the importance of trees, if not love for the wood or Mediterranean brush, could be the outcome of good schooling and of a publicly financed awareness campaign. But, more importantly it would be the inevitable consequence of immediate material interest in the economic advantages of carefully managed forestry. The woods must also be used in the short term, and not only for urban people on Sunday trips. This is the key practical point. Forestation must be economically viable first and foremost for country people. Vaguely ecological forestation dictated and conducted only in the name of combating desertification would certainly be a failure. The spirit and way of conducting it would be alien to the thousands of year of customs and the equally ancient and rooted ways of feeling of the Mediterranean people, who value a tree as being economically useful and can measure it on the yardstick of daily and seasonal concerns of shepherds and peasants. This attitude may well be objectionable and outmoded, but we cannot deny that it has a certain strength and justification.

The worldwide battles of the ecologist are not won by ignoring or mocking ways of looking at the animal and plant world endorsed by thousands of years of experience. Those thousands of years may not even have produced a fraction of modern knowledge about ecological balances. But they did produce knowledge, attitudes, and outlooks that will not be changed overnight and especially not through witchhunts or disdain. The rural masses are used to "cutting a long story short," of "making virtue of a necessity," and therefore—to use other Italian proverbs—of "making a stack from every straw" or a fire from every piece of wood, if the animals die and the cold never lets up. It is very hard to stop look at the trees in order to see the wood. And what seems clear to planetary ecologists is obscure for many of those directly affected. And quite rightly this must be studied and taken into account for possible use.

The role of the anthropologist studying indigenous knowledge and relevance of the idea of planetary utility will only become easier when planetary utility is no longer tantamount to planetary inequality.

References

Angioni, G. 1989. *I pascoli erranti: antropologia del pastore in Sardegna.* Naples.

Brokensha, D. W., D. M. Warren and O. Werner eds.1980. *Indigenous knowledge systems and development.* Lanham MD.

Brush, S. B. and D. Stabinsky. eds.1996. *Valuing local knowledge:Indigenous people and intellectual property rights.* Washington D.C.

Cerchi, P. 1996. *Il peso dell'ombra: l'etnocentrismo critico di Ernesto De Martino e il problema dell'autocoscienza culturale.* Naples.

Cirese, A. M. 1976. *Intellettuali, folklore, istinto di classe.* Turin.

De Martino, E. 1977. *La fine del mondo: contributo all'analisi delle apocalissi culturali.* Turin.

Ellen, R. F. 1986."What Black Elk Left Unsaid: on the Illusory Images of Green Primitivism"*Anthropology Today,* 2: 6.

_____ and K. Fukui, 1996. *Redefining Nature: Culture, Ecology and Domestication.* Oxford.

Gramsci, A. 1975. *Quaderni del carcere.* 4 vols.Turin.

Gliozzi, G. 1977. *Adamo e il Nuovo Mondo: la nascita dell'antropologia come ideologia coloniale.* Florence.

Milton, K. 1996. *Environmentalism and anthropology: Exploiting the role of anthropology in environmental discourse.* London.

Robertson, R. 1996."Globalization: Time-Space and Homogeneity-Heterogeneity"In *Global modernities* Featherstone M., Lash S. and Robertson R. Sage, London.

Warren D. M., J. L. Sikkerveer and D. Brokensha. 1995. *The cultural dimension of development: Indigenous knowledge systems.* London.

CHAPTER 18

Indigenous Environmental Knowledge, the History of Science and the Discourse of Development

Roy Ellen and Holly Harris

There are a number of recent books extolling the virtues of "indigenous knowledge," by which we mean here indigenous *environmental* knowledge. The concept—or its terminological cognates—is now widely employed in development studies, environmental conservation programs, and in the political rhetoric of international funding agencies and national governments. It is also being increasingly adopted by "indigenous" minorities and regional movements throughout the developing world as part of a defense against the expropriation of intellectual or cultural property. The paper, of which this is a summary,[1] does not seek to demonstrate the superiority, or even the complementarity, of local knowledge over dominant scientific knowledge in particular instances, nor does it seek to enter into the polemical discourse suggesting the converse. We are already persuaded that indigenous environmental knowledge (hereafter IK) can be advantageous in development contexts, and has applications in industry and commerce as well. Most professional anthropologists also accept that the claims made for the environmental wisdom of native peoples have sometimes been ignorant and naive, replacing denial with effusive blanket endorsement presenting an "ecological Eden" to counter some European "world we have lost." Rather, we try here to present a dispassionate examination of the status of, and claims made for, IK in the discourse and practice of different academic disciplines, and as found in different political contexts (ranging through environmental movements, states, NGOs and indigenist activism), and in the transfer of ideas between these.

What is meant by "indigenous knowledge" is by no means clear, and part of the purpose of the longer versions of this paper has been to examine the variable uses of the terminology and concept through its East-West and local-global cultural refractions. IK is increasingly criticized for its lack of organizing themes, as well as for the apparent ease with which its proponents are prepared to extract the part from the whole: the tendency to remove knowledge from its particular context and turn it into generalizable and empirically applicable lessons. Moreover, the distinction indigenous non-indigenous has many highly specific regional and historical connotations that are not always appropriate to particular Western or Asian contexts. Some even argue that the category of IK is wholly compromised by the "hegemonic opposition" of the privileged *us* to the subordinated *them*, and therefore is morally objectionable as well as practically useless.

The West often assumes that it has no IK that is relevant, in the sense of "folk" knowledge; that it once existed but has now disappeared, and that somehow science and technology have become its indigenous knowledge. Certainly, there is plenty of evidence showing that the existence of, for example, codified pharmacopoeias, displaced local knowledge and oral tradition extensively in Europe and the Mediterranean from the early modern period onwards. But Western folk knowledge is arguably just as important as it ever has been, just different: informed by science where appropriate, and located in different contexts (e.g., in situations as varied as pigeon-fancying, computer-person interfaces and high-tech deep sea fishing). The folk are no less creative. Moreover, in parts of Europe, urbane folk actively seek out the authoritative knowledge still regarded as being present in their own peasant traditions, as in truffle-hunting, geese-rearing or the preservation of rare breeds of sheep. Peasant or rural knowledge becomes, in this latter context, Europe's own inner indigenous other.

In Asia, by comparison, indigenous knowledge is variously that of some great tradition (e.g., Ayurvedic medicine), or more often that of myriad little local traditions. Where the two merge is unclear, and as in the European case there is historical evidence to suggest, for example, that the great Asian herbalist traditions have been systematically absorbing and then replacing local folk knowledge. IK is constantly changing, being produced as well as reproduced, discovered as well as lost, though we may describe it in different ways. What passes for scientific knowledge is often simply the knowledge of culturally dominant agencies of government. We ask if it is possible to effectively define the shifting boundaries between science and folk knowledge, and whether the distinction is in any way helpful. We examine what the differences between *indigenous, local* and *folk* knowledge might amount to (if anything), and whether there is a difference between folk knowledge and folk science. Is there just good and bad science, or is science qualitatively different in its underlying cognitive organization? Is it all applied common sense, the only difference being that one is practiced by the folk and the other by professionals? In other words, is it an outcome of some division of intellectual labor? Alternatively, is folk knowledge hopelessly embedded in particular symbolic patterns of thought, while *real* science is a distinctive kind of uncommon sense, driven by a logic that often results in demonstrating its counter-intuitive character? At this point, of course, the trail leads us into the familiar anthropological thicket of the rationality and relativism debate. In our paper we have deliberately avoided getting too entangled in these issues, though we believe it is necessary to acknowledge the intellectual link.

In the developing world, contemporary reliance on indigenous knowledge has been a combination of economic necessity and tradition. In many countries, the state sector and NGOs have, however, moved from colonial hegemonic denial towards the positive acceptance of the utility of local knowledge in medicine and sustainable development, partly for political and economic reasons. Individual native peoples, though less so in Asia than in, say, the Americas, have seen indigenous knowledge as part of their own identity. Both LDC states and Western NGOs have sought to protect indigenous rights to such knowledge, and this has given rise to a whole set of new issues in merging the philosophies, legal traditions, and discourses of East and West, North and South.

Another major theme that emerges is the often contradictory and changing scientific and moral attitudes towards indigenous knowledge. Much Western science and technology emanates from European folk knowledge (e.g., herbal cures) and knowledge acquired in a colonial context. The nomenclature and classificatory schema employed by Linnaeus, for example, depended extensively on Asian folk knowledge as this was absorbed into the writings of colonial naturalists working in the seventeenth and eighteenth centuries. During the nineteenth and twentieth centuries such knowledge was increasingly tapped and codified, acknowledged and yet somehow denied. Explicit and full recognition, together with the rights that are deemed to accompany this, has only come in the West with the quest for appropriate and cheap technologies for development, and the rise of ethnobotany in the pharmaceutical industry—at a time when the environmental movement has become morally committed to the notion of indigenous environmental wisdom. No wonder then that at this precise historical moment, when IK (via intellectual property rights), and the rights of "indigenous" peoples in more general terms, are higher on the political agenda than they have ever been before, "indigenous" as a label is being reclaimed by the protagonists themselves in pursuance of their own interests.

A final problem we face is what some recent writers have called "time-space compression." What, for example, does the implicit distinction between West and "other"—used throughout our paper—encode? More particularly, global-local distinctions are now blurring and we are told that we inhabit a world of "transcultural discourse." However sloppy some may find the conceptual apparatus offered to cope with these issues, it does directly address the question as to whether it is still possible to regard local knowledges as something discrete, even less pristine, or whether we are trapped by the representations of such in global (Western) media and their reformulation by indigenous people who learn it from, and who raise their consciousness of it through, Western sources. Should we continue to try to separate local knowledge from global knowledge on the assumption that one or other is superior in a particular context, or should we give preference to the mixture of local and global which most indigenous peoples now rely on? It is this blurring that results in neologisms such as "glocal," and a new analytical emphasis on how people shift the geographical context of their knowledge. For Piers Vitebsky the problem is encapsulated in the historical simultaneity of shamanism expiring on the tundra just at that moment that it is taken up by new agents. Can indigenous knowledge survive such appropriations? We take the view that IK, in the sense of tacit, intuitive, experiential, informal, uncodified knowledge, will always be necessary and will always be generated, since, however much we come to rely on literate knowledge that has authority, the validation of technical experts, and what is systematically available, there will always be an interface between this kind of expert knowledge and real-world situations. Such knowledge will always have to be translated and adapted to local situations and will still depend on what individuals know and reconfigure culturally, independent of formal and book knowledge.

Notes

1. The original published paper of which this is a summary appeared as *Concepts of indigenous environmental knowledge in scientific and development studies literature: a critical assessment*, (APFT Working Paper No. 2, October 1997) University of Kent at Canterbury: Avenir des Peuples de Forêts Tropicales, (See also http://lucy.ukc.ac.uk/Rainforest/SML—-files/Occpap/indigknow.occpap—TOC.html). A modified version by the same authors was also published in 2000 as "Introduction," in *Indigenous environmental knowledge and its transformations: critical anthropological perspectives*, edited by Roy Ellen, Peter Parkes and Alan Bicker Amsterdam: Harwood Academic Publishers, Studies in Environmental Anthropology, vol. 5.

CHAPTER 19

Two Reflections on Ecological Knowledge

Tim Ingold

I have written this paper in two relatively independent parts. Each, so far, is little more than a sketch of ideas that still need to be properly thought out. The first is about organisms, persons, and ecological relations. Here I try to replace the conventional idea of organisms and persons as distinct, substantive entities with a view of the organism-cum-person as a position or nexus—situated within an unbounded field of relations—where growth is going on. This leads me to suggest that the kinds of relations we are used to calling "ecological" are not really *between* organisms and their environments, as though each were initially "given" independently of one another, but rather constitute the very existential foundation from which organisms grow. The second part is about knowledge. I argue against the idea that the knowledgeability of practitioners committed to making a living within a certain environment lies in the accumulation of received mental content—that is, of rules and representations that are available for transmission in advance of their application in practice. Instead, I maintain that the essence of practitioners' environmental knowledge lies in skills, that is in developmentally embodied capacities of awareness and response built up through a history of involvement with the land and its inhabitants. Thus the growth of knowledgeability is an aspect of the growth of the organism-person in his or her environment.

There is a connection between the two parts, though it is not yet worked out as fully as I would like. It is most easily seen in terms of the link between the respective views *against* which I argue in each part. If the organism or person is regarded as a discrete entity, and the living world as an aggregate of such entities, then every one of them must be specifiable in its essential nature—as a "thing-in-itself"—independently of, and prior to, their mutual involvement. Whence, then, come the components of these specifications? The general answer is that they are "passed on" by way of some mechanism of inheritance. This passing on, in turn, defines recipients as descendants of those from whom the components were received, in genealogical succession. Thus organisms are said to receive a set of biological specifications through genetic inheritance, while persons receive a parallel set of cultural specifications through a "second track" of cultural inheritance or traditional transmission. The first guarantee that the genetic offspring, for example, of human parents will themselves be human beings, regardless of the circumstances of their life in the world. The second furnish these offspring with additional information which "closes the gap," as Clifford Geertz once put it, "between what our body tells us and what we have to know in order to function."[1] Knowledge, then, consists of a package of information, coded in words or other symbolic media, that is passed from generation to generation in an ancestor-descendant sequence.

My argument, to the contrary, is that knowing is not a matter of being in possession of information handed down from the past, but is rather indistinguishable from the life-activity of the organism-person in an environment that has itself been, and continues to be, fashioned through the activities of predecessors and contemporaries. It follows that knowledge is perpetually generated, rather than applied, in practice. This generative process is tantamount to the growth of the organism as it reaches out, along the lines of its relationships, into its surroundings. But if we take this view, it is no more possible to regard the organism than it is to regard the person as a substantive entity whose nature is fixed in advance of its encounters with other entities of the same or different kinds. Nor is it possible to regard the person as a carrier of "cultural" information additional to the "genetic" information bequeathed to the organism. Organism and person are, in effect, one and the same: in the words of the biologist, Brian Goodwin, they are not so much things as sites of "pure, self-sustaining activity," not creatures of nature but places where creation is going on (Goodwin 1988: 108). Perhaps no one has expressed the point better than a Cree man who explained to the ethnographer, Colin Scott, that to be a person is to *live*, and that life is a process of "continuous birth" (Scott 1989: 195).

Part I: *Fungal Persons*

I was talking recently with the mycologist Alan Rayner, who complained bitterly about the extent to which the entire field of biological science has been dominated by a view of the organism whose source lies in zoology. Most animals, as Alan pointed out, can be readily identified as discrete, externally bounded entities, capable of moving around in their environments under their own steam. It is not like that with fungi. What we see above ground may indeed look like an object with its own distinctive characteristics, albeit fixed to the ground and in that sense more resembling a plant than an animal. Look under the ground surface, however, and what we find is a complex and indefinitely ramifying network of strands or fibers, technically known as a *mycelium*. What we thought was the "fungus" was actually nothing more than a fruiting body, situated at one particular node of the network. Above ground, we perceive individual fungi as things, rather like street lamps in a city, and we might even pretend to count them. Below ground, however, they are not things at all but places, positions, or points of emergence of the total mycelial network. What, Alan wondered, would biological science look like if the fungus, not the animal, had been taken as the paradigmatic instance of a life-form?[2]

It is possible, I replied, that recent debates in social science might—by way of analogy —suggest an answer. I explained that social scientists, and social anthropologists in particular, have been much exercised by the problem of how to establish a general notion of the person that could be used for comparative purposes, and that did not end up being conflated with the Western notion of the individual. There is a certain similarity between this latter notion and the classic zoological concept of the individual animal. The human individual, like the animal, is supposed to have a certain constitution, integrity, and boundedness that is given quite independently, and in advance, of its involvement with others of its kind. But some anthropologists, myself among them (Ingold 1997), have been arguing that we need to understand the person not as a substantive entity in that sense, but as a point of growth or emergence within a wider field or network of social relationships. Listening to Alan speak, I could not fail to be struck by the parallel between what he was saying about fungi and what I wanted to say about persons. I am not, of course, alone in my view. Perhaps, then, the kind of rethinking he was calling for in biology is already well under way in social anthropology.

I do believe this is the case. Yet precisely because biology has not yet caught up with anthropology in this respect, most anthropologists continue to assume that even though as *persons* humans are a bit like fungi, as *organisms* they fit the standard zoological model of the animal. Hence they conclude that the person and the organism are two quite different aspects of human being, respectively social and biological, existing on different levels of reality and calling for quite different methods of study and analysis. However Alan's point was not that fungi are quite unlike animals (though on one level, that is obviously the case). It was rather that when we come to address the most fundamental questions about living systems, we need to think of animals, too, not as things in themselves, but as points of emergent growth within a relational field. What makes the fungus such a useful model is just that the pathways of relationship are manifestly traced out in its fibers (or *hyphae*). With animals they are more difficult to see—but they are there nonetheless. And they are there for humans too. Thus with the "fungal model," if I may call it that, being a person is not necessarily any different from being an organism. In what follows I want to consider the implications of this view for the way in which we might understand the relations—conventionally called "ecological"—between organisms-persons and their environments.

We run into trouble right away, even when we begin to think about the notion of "environment." An environment surrounds, and therefore presupposes something—an organism—to be surrounded. One's immediate inclination is to represent this diagrammatically by drawing a figure like a circle, or a sphere if you could do it in three dimensions. The line or surface then represents an interface, between what is "inside" (the organism) and what is "outside" (the environment). But suppose instead that you were to draw a network of lines, branching out and coming together at various points, and ramifying indefinitely. What sense could it make to speak of the "surrounding" of such a figure? The best one could do would be to draw an arbitrary boundary around some region within it, simply in order to delineate a focus of interest. If you were a neurologist you might want to draw this boundary so as to coincide with the inner surface of the cranium, and concentrate on the stupendously complex network of neural pathways making up the brain. But the human brain is not "joined on" to the rest of the body but fully part of it. The neural net extends throughout the body space, as far as the surface of the skin:

thus if we consider the brain to be such a net, we would have to conclude—not unreasonably—that the body is one big brain. But then, there is nothing absolute about the skin, it is just one of a number of points along a sensory pathway where impulses traveling in one form are transformed into equivalent impulses in another form.

Gregory Bateson made precisely this point by way of the example of a man cutting a tree with an axe (Bateson 1973: 433). Here, the interface between the hand and the handle of the axe is as nothing compared with that between the blade and the trunk of the tree. For it is the impact of the former on the latter that the woodsman feels when he wields his axe. Moreover this tactile pathway is intimately linked to another, visual one, by virtue of the fact that the man watches as he works, continually adjusting his bodily movements in response to a monitoring of the evolving form of the cut. To understand what is going on here we have to consider the dynamics of the entire man-axe-tree system. But again, any boundaries we may draw around this system are entirely of our own making. The tree, after all, stands in a forest, where it competes with other trees for sunlight and provides shade for the plant and animal life beneath its boughs. The man has a life history of involvement with other people, and the axe, too, has its own history of manufacture and use. One way of thinking about this, which is the way Bateson actually adopted, is to imagine a nested hierarchy of systems of progressively more inclusive scale. Thus the visual and sensorimotor systems of the man are included within the system comprising his whole body, which is in turn included in the man-axe-tree system, which is included in the wider ecological system consisting of a human social group in its environment.

For my part, I do not accept this hierarchical model. The problem with it is that it presupposes a certain view of part-whole relations, according to which units of smaller scale are constituted independently of the larger systems in which they participate. In other words, the hierarchy is defined by the following property: that higher-level units "include" lower-level ones, but lower-level units do *not* include higher-level ones. This way of thinking is very deep-rooted in the Western tradition, and it lies behind the incessant controversies surrounding "reductionism." Whether this is something that one is for or against, the very notion of reduction implies that larger wholes are put together from smaller parts. And this, in turn, fosters the impression that it is possible to specify in some sense what an organism *is*, independently of the totality of its environmental relations, and likewise that it is possible to say what human persons are, without regard to their involvement in the wider sphere of social relations. The very same logic underwrites the way we often think about places: as though somehow every place existed in itself with a clear horizon around it, setting it off from other places in a territorial mosaic. To view the world from a place is certainly to be "somewhere looking about," but it is not to be "down there looking up." And our knowledge of the wider world is forged in the passage from place to place, expressed in narratives of movement and changing horizons, not in the ascent from a narrowly constricted "local" view to a panoptic, "global" one (Ingold 1993: 40–1).

Evidently, while it succeeds in relativizing boundaries, the hierarchical model is unable to eliminate them altogether. Rather, they have a "now-you-see-them-now-you-don't" character. Moving from the inside out, in a nested series, the boundaries seem to dissolve, one after another. But moving in the reverse direction, outside-in, they promptly reappear. It is not, then, easy to see where an organism ends and an environment begins, yet if we adopt the perspective of the "system as a whole"—that is, of the ecosystem—the individual organisms that are bound within it stand out sharp and clear. Likewise, in a social relation it is impossible to tell precisely where the "self" ends and the "other" begins; vis-à-vis the social system, however, human individuals are revealed as the units of which it is composed. And the same goes, too, for places in an environment. As you travel from one place to another, there is no boundary that you have to cross in order to pass from place A to place B. But to conceive of the environment "as a whole" is immediately to conjure up the image of a totality composed of discrete, spatially demarcated places. Indeed it seems that the problem of boundaries is sure to persist so long as we continue to think of organisms, individuals or localities as together constituting "wholes" that are greater than the sum of their parts. For as totalized entities, such wholes will necessary appear bounded from the perspective of the next level up the hierarchy, just as their components appear so from the perspective of their own level.

What is missing from all this is any recognition of time and process. I have spoken of organisms and persons as points of growth within a relational field—that is, as places where certain pathways merge in the creation of new ones. But it is important to emphasize that these pathways, or lines of relationship, are not laid down in advance for beings to follow. Rather, their laying down is itself tantamount to the process of growth. Here again, the image of fungal *hyphae* is appropriate. Wherever growth is going on, organisms are coming into being, enfolding into their own constitution, and through their histories of becoming, the constellations of pathways within which they emerge. Thus the relationship between the organism and the wider field is better viewed as one of enfolding and unfolding than one of part and whole. Or in Bohm's (1980: 177) terms, the order of life is *implicate* rather than *explicate*. Every being, in its characteristic patterns of movement, awareness and response, enfolds the totality of life from a certain position within it, and projects it forward through its own actions. With this view, the hierarchical logic of part-whole relations clearly collapses.

How, adopting this very general point of view on life, should we understand an "ecological" relationship? It cannot be an interaction between one thing and another, for that would be to suppose that they existed, as discrete entities, in advance of their mutual engagement. If organisms, in general, "issue forth" along the lines of their relationships, then each organism must be coextensive with the relationships issuing from a particular source. It is not possible, therefore, for any relationship to cross a boundary separating the organism from the environment. If the concept of environment is to mean anything at all, it must refer to the *interpenetration* of organisms. This is perhaps easier to see in the case of persons, where we are used to using the word "social" to denote the condition of interpenetrability. But just as we need to be careful not to reify the social as an exclusive, higher-order domain going by the name of "society," we also have to avoid reifying the interpenetrability of organisms as a domain that exists apart from them, and with which they can interact—namely "the environment." In short, organisms no more interact with the environment than do individuals with society. Rather, ecological relations—like social relations—are the lines along which organisms-persons, through their processes of growth, are mutually implicated in each others' coming into being.

Part II: *Traditional Knowledge*

I am writing this in Tromsø, northern Norway, where I have just been talking to one of the researchers in the Institute of Social Anthropology at Tromsø University, Bjørn Bjerkli.[3] Bjørn has been studying the traditional pattern of using common land among a group of farmers, mostly Saami-speaking, inhabiting a small valley not far from here. He is concerned because this pattern has recently brought them into conflict with the authority of the state, which asserts that they have no right to use the land in the way they do. The case will shortly come to court. The difficulty is that in order to justify in law their claim to use the land as a commons, local people will have to demonstrate that this use has been subject, from time immemorial, to clearly articulated (if heretofore unwritten) rules and procedures. So far they have been unable to do this, and have only come up with very vague statements about "the way we do things here." They can talk about what went on in the past, about people and events, good years and bad years, and so on. But they cannot formulate explicit principles of "traditional" land use. Bjørn's worry is that if they are forced to enunciate such principles in order to win their claim for land, then in the future they will be required to regulate their land use according to these principles. And the effect of this regulation could be to destroy the very tradition they seek to sustain.

The crux of the problem lies in the concept of "tradition," to which lawyers, bureaucrats, and politicians appeal just as often and freely as do local people. But they clearly mean different things by it. Below, I want to suggest what some of these differences are. It also seems to me, however, that in much of the anthropological discussion about so-called "traditional ecological knowledge," the sense in which the concept of tradition is used comes closer to that adopted by the state administration than to that understood by local people. One solution might be to drop the concept of tradition altogether: to regard it as so tainted by its conventional opposition to modernity that it can give only a distorted view of people's real lives, one that is flattened in time and devoid of any sense of history. But this solution is hardly satisfactory in a situation where local people value, and want to continue with, what they themselves see as a traditional form of life, and where they have to contend with administrative authorities for which "tradition" is taken to be an essential ingredient in the management of local affairs. A better solution, I think, is to find a way of talking about tradition that chimes more accurately with local sensibilities, and to reconstruct our theory of "traditional ecological knowledge" around this. The following remarks are motivated by this aim.

I am going to contrast two understandings of traditional knowledge, one embedded in the modernist discourse of the state apparatus, the other in the everyday life of local people. For simplicity, I shall call these MTK (traditional knowledge in modernist conception) and LTK (traditional knowledge in local conception) respectively. Now one of the characteristic things about MTK is that it is very closely bound up with what I call the "genealogical model." This is based on the idea that the elements that go together to constitute a person are *passed down*, along one or several lines of descent, from that person's ancestors, independently and in advance of his or her life on the land, in an environment. Just this kind of model is implicit in the standard anthropological convention for drawing kinship diagrams, where the lines represent channels for the transmission of substance. The substance may be in part material, providing the recipient with a component of "biology," and in part mental, providing a complementary component of "culture." MTK, then, is unequivocally located within "culture." It amounts to a corpus of ideal rules, recipes and prescriptions, transmitted from generation to generation as a kind of heritage, in parallel to the transgenerational passage of bodily substance ("blood").

In this view, however, the environment (which I take to include the land along with its animal and plant life) plays no part whatever in the constitution of persons. It is simply the backdrop of nature against which a certain way of life is played out. That is why, in the past, administrations have often seen no principled objection to moving "indigenous" people off the land. It did not occur to them that such displacement might rupture the continuity of tradition or cut people off from their pasts. So long as the stuff of tradition could be passed along, like a relay baton, from generation to generation, it made no difference *where* the people were. Yet clearly, when local people say "that's the way we do things here," they are referring to knowledge that only makes sense in the context of their involvement in a familiar environment. Perhaps they might say that a really traditional person is one who knows the country "like the back of his hand." This doesn't mean that he carries it in the form of a cognitive map inside his head, but it does mean that through having grown up there, he has learnt to "know" it rather as an experienced craftsman might be said to know his raw material. That is, he is acutely sensitive to its forms and textures, can respond creatively to its variations, and is ever alert to the possibilities these afford—and the hazards they present—for pursuing different kinds of tasks.

Now knowledge of this kind, namely LTK, is not really "passed down" at all. Rather, it is continually generated and regenerated within the contexts of people's skilled, practical involvement with significant components of the environment. This means that LTK is not cognitive: it does not lie "inside people's heads," as opposed to "out there" in the environment. It lies, rather, in the mutually constitutive engagement between persons and environment in the ordinary business of life. Or to put the contrast in another way, whereas MTK consists of items of knowledge that are stored in memory, from which they may be accessed and expressed in practice, LTK subsists in practical activities themselves, activities that may also be understood as ways of remembering. Consider, for example, the knowledge of procedures for herding goats on a mountainside. If these exist in the form of MTK, then the herder faced with a certain situation "on the ground" has first to perform a purely mental operation (to retrieve the relevant items of information from memory), followed by a physical or behavioral one (to implement the retrieved instructions in practice). However the goatherd equipped with LTK remembers how to do it as he goes along: confidently negotiating his herd through the terrain he is not so much applying his knowledge in practice as knowing by way of his practice.

Still more fundamentally, the distinction between MTK and LTK hinges on the difference between thinking of tradition as a kind of *substance*, and thinking of it as a type of *process*. Basic to the modernist conception of the person are the metaphors of container and content. Equipped by nature with universal capacities, human beings are viewed as containers for the culturally variable, substantive content that specifies traditional knowledge in its diverse spheres of application. (There is much debate among cognitive scientists about whether there are few or many containers, or "mental modules," per human being, and about whether—or to what extent—these containers constrain what is acceptable by way of content, but I shall not go into these issues here.) Regarded as substantive mental content, it is in the nature of MTK that it should stay the same, from generation to generation. In reality, of course, it does not, but this is attributed to "mistakes" in the transmission process—somewhat analogous to genetic mutations—whereby knowledge is passed from one container to another. Some stuff may be lost; other stuff gained: overall, then, traditions change.

With LTK, by contrast, tradition is understood as a process—as continually going on. This process is none other than that of people's practical engagement with the environment. The important thing, for them, is that the process should keep on going, not that it should lead to the replication of identical forms. Indeed "keeping the process going" may involve a good measure of creative improvisation, rather like keeping the music going in a jazz band. The crucial point here is that, regarded as a process, tradition can be continuous without taking any fixed form. There is no opposition in LTK between continuity and change. Change is simply what we observe if we sample a continuous process at a number of fixed points, separated in time. The growth of an organism, for example, is continuous, but if we compare its appearance at different times it will appear to have changed. So, too, the growth of LTK is an aspect of the growth of persons, in the contexts of their relations with one another and with the environment. Just because people are doing things differently now, compared with the way they did them at some time in the past, does not mean that there has been a rupture of tradition. What would really break the continuity, however, would be if people were forcibly constrained to replicate a fixed "traditional" (*sensu* MTK) pattern. The effect would be similar to a gramophone needle getting stuck in the groove of a record. One could not keep the music going. Likewise, any attempt to "traditionalize the traditional," as Bjørn Bjerkli so beautifully put it (1996: 18), would rupture the continuity of life.

In essence, LTK is a kind of knowledge that might be better denoted by the concept of *skill*. To explain what I mean by this, I need to make three general points about skill.[4] First, skills are not properties of the individual body considered, objectively and in isolation, as the primary instrument of a received cognitive tradition. They are rather properties of the whole system of relations constituted by the presence of the agent in a richly structured environment. Thus the study of skill demands an ecological approach, which situates the practitioner, right from the start, in the context of an active engagement with his or her surroundings. Secondly, skilled practice is not just the application of external force but involves qualities of care, judgment and dexterity. This implies that whatever the practitioner does *to* things is grounded in an active, perceptual involvement *with* them, or in other words, that he watches and feels as he works. Thirdly, skills are refractory to codification in the programmatic form of rules and representations. So it is not through the transmission of any such programs that skills are learned, but rather through a mixture of imitation and improvisation in the settings of practice. What happens, in effect, is that people develop their own ways of doing things, but in environmental contexts structured by the presence and activities of predecessors.

It would be wrong, then, to say of LTK that it is "cultural" rather than "biological," or in the head rather than in the body. It is rather a property of the whole human organism-person, having emerged through the history of his or her involvement in an environment. Having recognized this, it is possible to understand local people's inability to formulate their traditional knowledge in anything other than the most vague and general terms. This fact has often been judged by outsiders, including government officials, as a measure of its inadequacy or inauthenticity. But in reality, the vagueness or elusiveness of their formulations is the very source of their strength. They are, in effect, *rules of thumb*, very general notes of guidance that may be drawn on as resources for action, but that in no sense govern its course (Suchman 1987: 52). They cannot do so, since the normal environment for any kind of practice is never quite the same from one moment to the next, and the essence of dexterity lies in being so attuned to these variations as to be able, continually and fluently, to respond to them.[5] To the extent that procedures are precisely specified, or movement "pre-programmed," dexterity is compromised. In short, the vagueness of LTK's formulaic prescriptions is a condition for the sheer *precision* of the actual movements of knowledgeable practitioners.

I would like to conclude with just two more points. The first is about the use of knowledge. Clearly MTK is conceived as knowledge that can be applied. It is supposed to exist as an objective body of rules and representations, in principle separable from the agency of those persons who can use it. But in what sense, if any, could LTK be used? It seems that skill, rather than being something that agents can harness to their practice, is constitutive of their very agency. Thus we have to think of use not as something that happens when we put two previously separate things together—namely an agent, on the one hand, and instrumental knowledge on the other—but as a habitual pattern of involved activity (what one is "used" to doing) in which the agent becomes effectively one with the task in hand. The separation of the agent and his or her knowledge is only an artifact of subsequent analysis, or perhaps a consequence of the breakdown of activity, when existing skills prove insufficient to the task and things go wrong.

The second point concerns the constitution of persons. I have shown how MTK is linked to a genealogical model according to which every person is constituted from two substantive components, respectively biological and cultural (mental and material), passed down from predecessors. In the framework of LTK, by contrast, the person is not conceived as a substantive entity, but rather as a locus of growth and development within a field of relationships. And by the same token, the contribution that other people make to one's own knowledge—often represented in the idioms of kinship—is not one of substance but rather one of setting up the conditions in which growth can occur. Persons, in short, come into being not by having stuff passed on to them along lines of descent, and against the backdrop of nature, but through growing up in an environment. That is to say, personal powers of perception and action are developed through the immediate experience of sensory participation with human and non-human components of the dwelt-in world. And the knowledge born of this experience, though commonly dismissed as "intuitive," must necessarily form the bedrock for any system of science or ethics that would treat the environment as the *object* of its concern.

Notes

1. See Geertz (1973: 50). In the same article, Geertz clearly distinguishes between cultural knowledge—"plans, recipes, rules, instructions"—and its overt manifestation in "concrete behavior patterns" (44). Thus to know is one thing, to function on the basis of that knowledge is another.
2. Rayner has elaborated on this idea in a recent book (1997). The extraordinary difficulties he experienced in finding a publisher for this volume says much about currently entrenched attitudes in biological science.
3. See Bjerkli (1996). I am most grateful to Bjørn for a stimulating discussion. He is not responsible, however, for the shortcomings of what I have written here.
4. See Ingold (1996) for a fuller discussion.
5. This point was made by the Russian neuroscientist Nicholai A. Bernstein in an essay entitled "On dexterity and its development" written some fifty years ago but only recently published. The essence of dexterity, Bernstein argued, lay not in bodily movements themselves but in the "tuning of the movements to an emergent task" (Bernstein 1996: 23).

References

Bateson, G. 1973. *Steps to an ecology of mind*. London: Granada.
Bernstein, N. A. 1996. On dexterity and its development. In *Dexterity and its development*, edited by M. L. Latash and M. T. Turvey. Mahwah, NJ: Lawrence Erlbaum Associates.
Bjerkli, B. 1996. Land use, traditionalism and rights. *Acta Borealia* 1, 1996: 3–21.
Bohm, D. 1980. *Wholeness and the implicate order*. London: Routledge and Kegan Paul.
Geertz, C. 1973. *The interpretation of cultures*. New York: Basic Books.
Goodwin, B. 1988. Organisms and minds: the dialectics of the animal-human interface in biology. In *What is an animal?*, edited by T. Ingold. London: Unwin Hyman.
Ingold, T. 1993. Globes and spheres: the topology of environmentalism. In *Environmentalism: the view from anthropology*, edited by K. Milton (ASA Monographs **32**). London: Routledge.

_____ 1996. Situating action V: The history and evolution of bodily skills. *Ecological Psychology* 8: 171–82.

_____ 1997. Life beyond the edge of nature? or, the mirage of society. In *The mark of the social*, edited by J. B. Greenwood. Lanham, MD: Rowman and Littlefield.

Rayner, A. 1997. *Degrees of freedom: living in dynamic boundaries*. London: Imperial College Press.

Scott, C. 1989. Knowledge construction among Cree hunters: metaphors and literal understanding. *Journal de la Société des Américanistes* 75: 193–208.

Suchman, L. A. 1987. *Plans and situated actions: the problem of human-machine communication*. Cambridge: Cambridge University Press.

CHAPTER 20

Indigenous Knowledge
and Cognitive Power

Pier Giorgio Solinas

The issue at hand, our "use" of knowledge, involves a question that I find most pressing, even though it is one that has become so obscured by a veil of archaism as to be vain perhaps to hope to lift it. The question is that of the relationship between knowledge and its application (or rather that between knowledge, application, and development). In his part introduction Colajanni assumes that this series is hierarchized and that the path from knowledge to development is tied to a consecutive order dependent upon the very nature of the thing in question. Nobody could say he is wrong. In fact, if one accepts that "knowledge" exists and that it must be liberated and allowed to work, then it follows that each phase is the condition of the other: first one knows and then one acts; application presupposes knowledge.

The fact is, however, that the history of knowledge and its application provides us with little proof of the existence of this ordered relationship, especially in ethno-economic history. If we look, for example, at what has happened to farming techniques, we discover that development has been dependent not upon accumulation of indigenous cognitive capital, but upon forced and perhaps inevitable conversions. How many people have "eaten" the forests for generations, burning and fertilizing the land by means of systematic destruction, without the least doubt as to the goodness of their system?

As far as I know, the slash and burn method of cultivation declined significantly when it met with a definite will for change, which succeeded in adopting alternative, if not opposing, techniques. The more prudent fireless methods of crop rotation predominated only after suffering fierce competition from the traditional concurrent techniques. Of course, this does not mean that the new knowledge and uses (that C. refers to as the Western domestication of techniques) were without their damaging effects. In many cases, domestication did none other than substitute more itinerant methods of farming with the plantation system.

I find it interesting that this example of development illustrates, if only metaphorically, the paradigm we can label as critical transition. How many critical transitions, how many "Neolithic revolutions" have been induced, kindly or harshly in third-world societies in the past two or three centuries? The West's irreversible technical evolution has lasted a couple of millennia. The technology that European civilization has exported and replicated abroad, by means of education programs on economics or forced labor, has been assimilated in just one or two generations.

In the whirlwind of all this change, indigenous knowledges are swept away. In order to achieve predominance, the new technical systems cannot tolerate the presence of two different kinds of knowledge. Of course after, but only after, complete deculturation has taken place, those born into the new society may acknowledge and research their lost patrimony. However, the results are often prolix or artificial: a laborious revival ritual performed by a mimical mind.

My aim, then, is to insist that the recovery and re-use of indigenous knowledge is not as positive as it seems according to some edifying versions. It is not a matter of "recovery"or restoration. It is not a matter of discovering a treasure of buried knowledge and putting it back into use.

*

Knowledge of nature does not involve only plants, animals, water, and the sky. Consciousness and ideas about nature are expressed in many other fields. Among these, the nature of humanity is of primary importance. What is a human being? What takes place inside his body? How does what he does, eats, and communicates to others influence his nature? What does she become after death? How are new human beings born?

Let's take reproduction, for example. Theories that attempt to explain the transmission of life from one generation to the next deal with a complicated problem. They must take into consideration those who do not yet exist (the individuals who will be born) in the context of that which now exists, but will cease to exist in the future. In other words, they must attempt to reconcile life and death, future and past. As we know, a decisive part of indigenous theories assume the existence of reincarnation. The dead are reborn; their person is converted and returns to the earth in the form of an embryo. (Although differing in form, the idea of an infinite, repetitive cycle of existence is found in totemism, in Melanesian theories, among the Eskimos and in many other cases too well known to require illustration).

That which we must ask ourselves is: what kind of knowledge are we dealing with here? And then: what are the consequences that ideas on reproduction and reincarnation have in real life?

*

I think that first of all, one must briefly examine the place of this type of knowledge with respect to the knowledge that we normally consider to be parallel to the ethnosciences (ethnobotany, ethnoastronomy, etc.) Indigenous theories that deal with the transmission of life are neither entirely speculative nor entirely technical. They are at the same time *descriptive* and *prescriptive*; that is, while they state what they describe as natural facts, at the same time they prescribe them as having a kind of ontological value. The word is the founder of the truth, one could say. However, the word is not a coded message surrounded by a sonorous magic. It is a matter of something quite different: the subject who learns is defined by the categories he is learning; thus knowledge is ascriptive and neither the being nor the value is negotiable.

As to the effects that a system of belief has on life and action, we must clarify whether the ideas encompassed by the system must be understood as conditions of being and acting, or whether the two things—the systems of action and systems of thought—are parts of an integrated whole in which neither is dependent on the other. If customs depend on their founding ideology, then a hierarchy exists in which ideas have priority over actions. We might say that such and such a ritual is a response to a need and is justified by a belief; for example, since it is believed that the sun could burn out, sacrifices must be offered to restore its heat; since the body of a dead man could contaminate, relatives should segregate themselves from other people until the danger has passed; since the newborn is the embodiment of his deceased grandfather, it is necessary to pay attention to signs of distress in the infant, and so on.

I think that the hierarchy between systems of ideas and procedures of action is rather ingenuous. But I also believe that it is not wrong to try, as much as possible, to discover consistencies between the two levels.

*

In India, the country in which I am studying "tribal" culture ("civil tribe," according to the current terminology, not just the administrative language), I often come across questions that have to do with the transmission of genetic identity and the integrity of the person. The group to which I am referring, the Santals, have quite definite ideas on the transmission of identity. Their system of kinship relations is rigid, almost rigorous. Descent rules are undeniably patrilinear; any connection to the maternal line is canceled before it can even take form. Women lose the name of the paternal clan as soon as they are included in the husband's family (to the point that many are even unable to remember the paternal name); they do not belong to the *gustì* (to tell the truth, neither to the paternal *gustì* they have left, nor that of the husband into which they are never fully integrated) and do not have the right to inherit the father's property. (All this, needless to say, is Santal tradition; the changes introduced by modern state laws are quite different.)

In spite of this, the biological perception of the Santals does not correspond to their giuridical theories on patrilinear descent, as far as the blood, the bones, and the flesh are concerned. One would think that the "natives" would stick to some theory in which man is all and woman is nothing (as ideas of this type are not uncommon). Unexpectedly, however, they recognize that the blood of the mother and of the father contribute equally to the formation of a newborn, and that resemblance, dreams, and other transcendent ties between parents and children and between grandparents and grandchildren come as much from one side as from the other.

At the same time, they affirm that maternal blood (the many that are mixed together in the family tree) is dispersed or diluted while that transmitted along the agnatic line prevails. This, however, seems to be the result of some vague and mysterious observation of frequency rather than an actual doctrine about life. This asymmetrical vision would be explained in this way: yes, women bestow their blood, but this blood is no other than the blood of the paternal *gustì*, which, once mixed with that of the husband, loses its ability to endure. (There is no distinction between masculine and feminine blood. There is only one blood, which is replicated through male descent.)

It seems that the principle of continuity is inseparable from the principle of integrity. Common identity is a physical fact as well as symbolic and a marker of ethnicity. (The political and intercultural implication is marked. The tribe members, or *adivasi*, feel surrounded by the Hindus, whom they consider foreigners to be resisted.) Defense of the community and the common identity means maintaining the purity of blood. Mixed marriage is considered sacrilege and those guilty of it—violation is extremely rare—are brutally deprived of all rights to membership in the community. Curiously, it is usually the woman who marries a non-Santal who receives banishment. (It's not so surprising: a woman giving birth to the child of a foreigner would introduce biological corruption to the integrity of the community; while the marriage of a man with a non-Santal woman would have less dangerous effects. The question of blood is further complicated by the bone/flesh dichotomy (thus, the hard and enduring versus the soft and perishable) in the body. Not only are the bones and flesh opposing as masculine and feminine, but they alternate and are transformed in the cycle of life and death (Marine Carrin-Bouez has done very good work on these questions); bones resist decomposition while the flesh disintegrates.

*

At this point we could say that the "use" of this fluid body of ideas lies in this or that implication (in the first place, in the social structure: ethnic unity is seen as natural communion, community members instinctually feel that its conservation is necessary). I think, however, that to distinguish between "knowledge" and its "application" in this case is more a tactic device we resort to for the sake of categorization and order rather than an actual difference between knowing and applying in every day life. I doubt that it is possible to separate from practice and local norms that which we perceive as ideology (the theoretic part of practice).

The theory of identity of blood between family members that belong to the same line is so saturated by the very fact of being such, that it is problematic to establish rules that dictate the way they come into being. In any case the sequence seems backwards with respect to that which we assume when we put the theory before the application. Here the ethnography must struggle to extract the theory from practice and not the opposite. Ideology is not a cognitive premise to which we add a pragmatic extension. That which is between one and the other is perhaps a firm link of analogical coincidence, but not necessarily a stringent consistency. When a Santal has to explain why blood transmitted on a paternal line prevails over that of the maternal line, he will attempt to defend his own supremacy of belief rather than demonstrate the validity of his thesis.

The fact is that the thesis in itself is proposed neither as an axiom nor as a provable statement, and in all probability is not limited to the circle of things that can be affirmed or negated; it is rather an accessory, an ethical and mental institution at the same time. Is it "knowledge" in the sense that our discussion intends? Even if we convinced ourselves of the fact that it is, notwithstanding everything, knowledge, it would be necessary to immediately intervene on the word that we adopt and clarify that it is a matter of fragile knowledge, highly subject to getting lost or corrupted when barely transferred into other contexts and codes.

Perhaps I am risking a rather integralist position, but I have the impression that reality does not allow much space for maneuver on this point. Maybe the only possible destiny for cognitive data extracted from "native" worlds, once deprived of their natural context, is ethnography. That is, a sort of conservation under glass, where the glass represents the impossible passage of life between the outside and the inside.

To escape from this condemnation, I believe that one must work *within* the ideologies and knowledge, taking them for what they say and also for what they are unable to say, operating inside the thought-object. "Use" is not an instrument or a resource to adapt to programmed needs, but a cultivable space, like a space in which we ourselves can make experiences. A paradigm still hospitable for this particular approach is that of Dumont, which, in fact, copes with ideologies by adopting their categories within a formal syntaxis, and which as Dumont sometimes says, has them produce more then what they themselves have insofar produced.

CHAPTER 21

The Role of Indigenous Knowledge Systems in Facilitating Sustainable Approaches to Development: an Annotated Bibliography

D. Michael Warren

A few days after the end of the conference we heard of the sudden death, in Nigeria, of the anthropologist Michael Warren. This is a very sad loss, and we would like here to commemorate both the scientist and the man, and to pay tribute to the vigorous contribution he made to the conference.

The first use of the term "indigenous knowledge" dates to a 1979 publication by Robert Chambers—who used the term "indigenous technical knowledge"—and a 1980 book co-edited by Brokensha, Warren, and Werner. Both publications presented indigenous knowledge (henceforth IK) as an overlooked—and sometimes maligned and denigrated—but potentially powerful ally in a new era of development moving away from a transfer-of-technology (TOT) top-down approach to a more participatory approach that could facilitate more sustainability. The term was used to replace what had earlier been called "traditional knowledge" in order to avoid the underlying stereotypes of simple, static, savage/primitive that the term "traditional" carried with it in many quarters. What initially appeared to be a highly rational and cost-effective approach to development was quickly perceived by many development professionals as a threat to the TOT approach. Those of us involved in presenting IK to the development community soon realized that we were involved in a major paradigm shift that would take nearly two decades for development agencies to accept. This acceptance is only now taking place, albeit far more slowly than many of us would like to see happen.

Except for a few detractors who may not have carefully read the massive amount of literature that has accumulated during the past two decades, it is now generally accepted in both the academic and the professional development community that IK represents localized—sometimes ethnically-based or community-based—knowledge that has evolved within a micro-environmental context. It is also accepted that localized knowledge domains have counterparts in the global knowledge systems generated by the network of universities and research laboratories. Indigenous knowledge—like its global counterpart—has comparative strengths and weaknesses that can be identified from both emic (insider's) and etic (outsider's) perspectives. IK can be compared and contrasted with its global counterpart. IK is always dynamic, reflecting indigenous approaches to innovation and experimentation that result from the changing sets of problems faced by any community at a given point in time. IK is likewise variable according to gender, age, and occupation. We now differentiate core knowledge for any given knowledge domain that allows all members of a community to communicate at a basic level on virtually any topic important for the community. It addition to the core knowledge, there is also more specialized shared knowledge that cuts across related occupational areas, as well as highly specialized knowledge limited to persons involved in the same occupational niche.

Twenty years ago we viewed IK as the end point in our attempt to understand the ways that a given community, through its language, organizes physical, social, and ideational phenomena through a process of identifying, categorizing, and classifying. Those of us working in international development soon came to realize that the IK serves as a convenient starting point in the IK Cycle (Warren and Pinkston 1997: 158), IK providing the foundation for local-level decision making at the individual and community levels. We also realized that equally invisible to the development professional was the wide array of indigenous organizations within even small towns and villages, both formal and informal types of associations, many of which have local-level development functions and serve as forums for the community to identify and prioritize their problems and seek approaches to their solution. Often these forums stimulate the indigenous creative process leading to experiments of various types as well as innovations. The results of the experimentation are usually evaluated, often through indigenous approaches to communications; the results deemed useful and cost-effective at a given point in time being added into the IK. These four points on the IK cycle are well-represented by a host of case studies in *The cultural dimension of development: Indigenous knowledge systems* (Brokensha et al. 1995), the first volume published by Intermediate Technology Publications in its IT Studies in Indigenous Knowledge and Development book series that already numbers eight.

In the 1979 and 1980 publications, the majority of the contributors came from the disciplines of anthropology and geography. In the 1995 volume there are contributors representing more than twenty disciplines ranging from agronomy to veterinary medicine, from botany to range management, from entomology to fisheries science, forestry, and plant pathology. The volume of publications now emerging in virtually every discipline imaginable is enormous. In order to share information on the growing number of unpublished and published case studies that are sent to CIKARD, we now have a very fast key word search engine that allows anyone with access to the CIKARD Home Page (http://www.iitap.iastateedu/cikard/cikard.html) to search the bibliographical citations and abstracts for the entire collection, currently numbering about 5,000. The Home Page is managed and supported by the new International Institute for Theoretical and Applied Physics (IITAP) established at Iowa State University by UNESCO as a counterpart to its center in Trieste.

The development community has emerged in the past few years in support of our argument that participatory approaches to development that include both the clientele group and the development professionals will be greatly facilitated by recognizing the comparative strengths and weaknesses of existing contemporary indigenous knowledge systems and indigenous organizations. It has been argued that both short and long-term sustainability of development efforts will be greatly enhanced by working with, building upon, and strengthening the capacity of indigenous knowledge and organizations. On 3 December, 1990, Michael Cernea invited me to present a seminar at The World Bank on "Indigenous Knowledge and Development."That presentation was published in 1991 as *Using indigenous knowledge in agricultural development* (World Bank Discussion Papers, No. 127). A survey of project outcomes for projects that ranged from those that ignored (usually inadvertently) indigenous knowledge to those that systematically built the project on existing structures has demonstrated the cost-effectiveness and sustainability of the latter approach. The paper has now been translated into Vietnamese, Spanish, and French (the latter two available in their full texts on the CIKARD Home Page).

By definition participatory approaches to development build upon indigenous knowledge. A vast set of experiences in participatory approaches is now available, primarily through the efforts of Robert Chambers (1979, 1983, 1994, 1997) and his associates. Development agencies such as those of UNESCO (1994a and b), FAO (Saouma 1993; den Biggelaar and Hart 1996; Herbert 1993), UNEP (Dowdeswell 1993), the Consultative Group on International Agricultural Research (CGIAR 1993), Directorate General for International Cooperation of the Netherlands Ministry of Foreign Affairs (support for publication of the *Indigenous knowledge and development monitor*), The World Bank (Davis 1993; Pichon et al. 1998; Davis and Ebbe 1995; Lutz et al. 1994), UNDP—which recently established the Indigenous Knowledge Programme (Rural Advancement Fund International 1994), United Nations Research Institute for Social Development (Hviding and Baines 1992), International Board for Plant Genetic Resources (1993), International Plant Genetic Resources Institute (IPGRI) (Eyzaguirre and Iwanaga 1996; Guarino 1995), International Development Research Centre (IDRC 1993; Inglis 1993), ORSTOM (Dupre 1991), and UK Overseas Development Administration (ODA), now the Department of International Development (Sillitoe 1998; Willcocks and Gichuki 1996) are all currently involved in activities related to the use of indigenous knowledge in development.

Other important signs of the recognition of the role of IK in facilitating sustainable approaches to development are recent publications by the United States National Research Council on sustainable development (1991) and the role of IK in the global concern with biodiversity (1992a) as well as key examples of very effective cases of indigenous knowledge related to the Neem tree and Vetiver grass that have disseminated across much of the globe from South Asia (1992b and 1993). International and national conferences on the topic have been conducted in the Philippines (CIRAN 1993a and b), Sri Lanka (Ulluwishewa and Ranasinghe 1996; Warren 1996a), and China (Center for Integrated Agricultural Development 1994). Periodicals devoted to indigenous knowledge and development are now widely disseminated in both hard copies and through the internet. The *Indigenous knowledge and development monitor* published in The Hague by CIRAN goes in hard copy to recipients in about 130 countries, while all back issues are available at http://www.nufficcs.nl/ciran/ikdm. The *Honey bee*, devoted to indigenous experimentation and innovations, is published in a dozen languages (Gupta 1990).

In order to make the growing number of methodologies developed primarily by academicians for recording indigenous knowledge systems more user-friendly, one now finds an excellent manual published by the Regional Program for the Promotion of Indigenous Knowledge in Asia (REPPIKA) at the International Institute of Rural Reconstruction (IIRR 1996). A second guide that is focused on indigenous natural-resource management is currently being field tested by the Canadian International Development Agency (CIDA) (Centre for Traditional Knowledge 1997). Gary Martin (1995) has a new manual for ethnobotany. Other useful guidelines are provided by Mathias (1995, 1996) and Kothari (1995).

A major value shift has also occurred recently with indigenous knowledge now being viewed as cultural capital (Berkes and Folke 1992) and a major national resource (Aboyade 1991). Like the global concern with the decline in biodiversity, there is a growing concern about the continuing extinction of languages, the primary vehicle for the oral transmission of indigenous knowledge from one generation to the next (Hunter 1994; Warren 1998). Another major issue is the exploitation of recorded indigenous knowledge as reflected in the debate about intellectual property rights (Brush and Stabinsky 1996; Posey and Dutfield 1996).

CIKARD provides a variety of mechanisms to provide access to the growing number of case studies on IK. The Studies in Technology and Social Change series of monographs now includes seventeen volumes on IK (e.g., Titilola 1990; Leakey and Slikkerveer 1991; Atte 1992; McCorkle 1994; Rajasekaran 1994; and Warren et al. 1997). The Bibliographies in Technology and Social Change series includes five volumes devoted to IK related to ethnoveterinary medicine (Mathias-Mundy and McCorkle 1989), tree management (Mathias Mundy et al. 1992), livestock production (Slaybaugh-Mitchell 1995), farming systems (McCall 1995), and naturalized knowledge systems in Canada (Winslow 1996). In addition to key word searches possible through the CIKARD Home Page, research associates at CIKARD have also packaged bibliographies available on the Home Page for agroforestry (324 citations), aquatic resources (109), biodiversity (61), ecology (356), ethnobotany (199), natural resource management (160), rainforest resources (69), soil classification and management (297), sustainable agriculture (354), terrestrial vertebrate wildlife (82), traditional medicine (158), and water management (186).

Another major effort has begun recently that focuses on Indigenous Knowledge and Education with a growing global network of individuals and institutions developing teaching modules for use in primary, secondary, and tertiary educational institutes based on exciting case studies that reflect the contribution to global knowledge by communities around the globe, some representing indigenous peoples, ethnic groups, and minority groups that have frequently been devalued by majority groups. Draft teaching modules recently added to the CIKARD Home Page include natural resource management by the Cree of Canada, management of natural resources in the intensive agricultural system of Bali, ethnobotany of the Tzeltal of Mexico, the holistic therapeutic system of the Bono of Central Ghana, and the classification and management of soil by the Yoruba of

Nigeria. A USAID-funded workshop was conducted at the University of Ibadan in Nigeria in December 1996 to discuss how IK case studies can be introduced into the existing educational curricula in Nigeria, curricula that do not yet adequately reflect the knowledge generated by its own citizens (Warren, Egunjobi, and Wahab 1996). It is anticipated that numerous students will be stimulated to use the methodologies for recording IK to conduct their own research on the knowledge generated within their own communities. These newly recorded systems will be added to the IK special collections already initiated by the indigenous knowledge resource centers and indigenous knowledge study groups that have been established in many countries. The centers reflect an important opportunity for the development agencies. With a modicum of annual support, these centers can provide an enormous resource for development. In Nigeria, for example, the indigenous soil classification and management systems for Nigeria's 250 ethnic groups could be recorded by a growing number of interested students in Nigeria's 30 universities and archived for use by development professionals at the IK centers located at the Nigerian Institute for Social and Economic Research (African Resource Centre for Indigenous Knowledge) in Ibadan, three resource centers at the universities in Zaria, Nsukka, and Makurdi, and study groups at four institutes in Ibadan and Ile-Ife.

The global network of indigenous knowledge resource centers continues to expand rapidly. Currently there are thirty-three centers located in North America (Canada, 2 in the USA), Europe (2 in the Netherlands, Russia, Georgia, Greece-Crete), Latin America (Mexico, Venezuela, Brazil, Uruguay), Africa (Burkina Faso, Sierra Leone, Madagascar, 2 in Ghana, 4 in Nigeria, Cameroon, South Africa, Tanzania, Kenya, Ethiopia), and Asia (2 in India, Bangladesh, Sri Lanka, Indonesia, 2 in the Philippines), with another 20 in various stages of becoming formally established.

Within the past several years the exploration of indigenous knowledge has extended into the realm of biotechnology with both a major text (Bunders et al. 1997) and a USAID-funded conference conducted at Obafemi Awolowo University in Ile-Ife, Nigeria (Warren 1996b). The published case studies of IK in a growing number of disciplines are expanding at an exponential rate. Many of these involve the use of the IK in a development context. Since this literature is growing so rapidly and often is found in disciplinary journals it is important to provide an overview of some of the prominent recent publications according to discipline, understanding that this presentation represents only the tip of the proverbial iceberg. The largest number of publications is found in the area of *production agriculture/farming systems/agroforestry* (Adams and Slikkerveer 1997, Cashman 1989, Chambers et al. 1989, den Biggelaar 1991, den Biggelaar and Hart 1996, Haverkort et al. 1991, Systemwide…1997, Van Veldhuizen et al. 1997, Warren 1989a, 1991b, 1994, Roling and Wagemakers 1998, Sumberg and Okali 1997, Warren and Rajasekaran 1993, Hiemstra et al. 1992, Niemeijer 1996, Innis 1997, Moock and Rhoades 1992, Leakey and Slikkerveer 1991, McCall 1995, McCorkle 1994, McCorkle and McClure 1995, Prain et al. 1998, Pretty 1991, 1995, Reijntjes et al. 1992, Rhoades and Bebbington 1995, Richards 1986). Other areas include *crop breeding* (Ashby et al. 1989, 1996, Berg 1996, de Boef et al. 1993, Eyzaguirre and Iwanaga 1996, Bunders et al. 1997); *horticulture* (Aumeeruddy 1995); *economics* (Barreiro 1992, Titilola 1990; Mazzucato 1997); *entomology/pest management/plant pathology* (Bentley and Andrews 1991, Thurston 1992); *ethnobiology* (Berlin 1992, Martin 1995); *indigenous organizations* (Blunt and Warren 1996); *environmental science* (Davis 1993); *soil science* (Lutz, Pagiola and Reiche 1994, Hecht and Posey 1989, Pawluk et al. 1992, Dialla 1994, Rajasekaran and Warren 1995, Roach 1997, Warren 1992); *soil and water management* (Carney 1991, National Research Council 1993, Ulluwishewa 1994, Reij 1993, Willcocks and Gichuki 1996, Critchley 1992, Critchley et al. 1994); *natural resource management* (DeWalt 1994, Pichon, Uquillas, and Frechione 1998, National Research Council 1992b, Quintana 1992, Rajasekaran, Warren, and Babu 1991); *environmental sanitation* (Foster et al. 1996); *forestry* (Castro 1995, Everett 1995, Mathias-Mundy et al. 1992, Messerschmidt 1995, Rusten and Gold 1995, Posey 1985); *off-farm technologies* (Adegboye 1990, Gamser et al. 1990); *indigenous knowledge conservation/cultural capital* (Rural Advancement Fund International 1994, Berkes and Folke 1992); *biodiversity conservation* (Gadgil et al. 1993, National Research Council 1992a, Prain and Bagalanon 1994, Saouma 1993, Warren 1995a, 1998, Warren and Pinkston 1997); *human health* (Green 1994, 1996, 1997a and b, 1998, Green et al. 1989, Slikkerveer and Slikkerveer 1995, Warren 1989b, Warren, Egunjobi and Wahab 1996); *ethnoveterinary medicine* (Mathias-Mundy and McCorkle 1989, McCorkle et al. 1996, McGregor 1994); *water resource management/irrigation* (Groenfeldt 1991, Lansing and Kremer 1995); *arid lands management* (Warren and Rajasekaran 1994); *wildlife management* (Gunn et al. 1988); *fisheries/aquatic and marine resource management* (Hviding and Baines 1992, Morrison et al. 1994, Price 1995, Ruddle 1994, Pauly et al. 1993); *ecology* (Inglis 1993, Johannes 1989); *animal science* (Slaybaugh-Mitchell 1995); *communications* (Mundy and Compton 1995); *extension* (Rajasekaran 1994, Roling and Engel 1989, 1991, Scoones and Thompson 1994, Warren 1991a); *grain storage* (Phillips 1989); *food policy/nutrition* (Rajasekaran and Whiteford 1993); *town planning/housing/architecture* (Wahab 1997, Wiltgen and Wahab 1997); *famine relief/refugees* (Walker 1995); *development management/development planning* (Selener et al. 1996, Warren, Adedokun and Omolaoye 1996); *education* (Warren, Egunjobi and Wahab 1996, Semali 1997, Kreisler and Semali 1997, Kroma 1995); *local government administration* (Warren and Issachar 1983); *co-management of natural resources* (Pinkerton 1989); *craft production/aesthetics* (Wolff and Wahab 1995, Warren 1990); and *gender and development* (Davis 1995, Fernandez 1994, Appleton and Hill 1994, Ulluwishewa 1994, Simpson 1994).

Warren, Adedokun and Omolaoye (1996) have demonstrated that the Yoruba of Nigeria have an extensive lexicon for problem identification, directed change, development planning, and development management. Once regarded as the domain of the Western world, more recent surveys are finding that similar vocabularies exist in many other ethnic groups. Although this has not been systematically explored, if communities have these concepts for development and operationalize them through indigenous organizations and local-level experimentation, this further strengthens our argument that indigenous knowledge and organizational structures provide the most cost-effective and humanistic approach to participatory decision making in sustainable approaches to development.

References

Aboyade, Ojetunji. 1991. *Some missing policy links in Nigerian agricultural development*. Ibadan: International Institute of Tropical Agriculture.

Adams, W. M. and L. J. Slikkerveer, eds. 1997. *Indigenous knowledge and change in African agriculture*. Studies in Technology and Social Change, No. 26. Ames: CIKARD, Iowa State University.

Adegboye, R. O. and J. A. Akinwumi. 1990. Cassava processing innovations in Nigeria. Pp. 64–79. In *Tinker, tiller, technical change*, edited by Matthew S. Gamser, Helen Appleton, and Nicola Carter. London: Intermediate Technology Publications.

Appleton, H. E. and C. L. M. Hill. 1994. Gender and indigenous knowledge in various organizations. *Indigenous Knowledge and Development Monitor* 2 (3): 8–11.

Ashby, J. A., T. Gracia, M. del Pilar Guerrero, Carlos Arturo Quiros, Jose Ignacio Roa, and Jorge Alonso Beltran. 1996. Innovation in the organization of participatory plant breeding. In *Participatory plant breeding*, edited by Pablo Eyzaguirre and Masa Iwanaga. Rome: IPGRI.

———, C. A. Quiros and Y. M. Rivers. 1989. Farmer Participation in Technology Development: Work with Crop Varieties. In *Farmer first: farmer innovation and agricultural research*, edited by Robert Chambers, Arnold Pacey, and Lori Ann Thrupp. London: Intermediate Technology Publications.

Atte, O. D. 1992. *Indigenous local knowledge as a key to local level development: possibilities, constraints, and planning issues*. Studies in Technology and Social Change, No. 20. Ames: CIKARD, Iowa State University.

Aumeeruddy, Y. 1995. Phytopractices: Indigenous horticultural approaches to plant cultivation and improvement in tropical regions. In *The cultural dimension of development: Indigenous knowledge systems*, edited by D. M. Warren, L. Jan Slikkerveer and D. Brokensha. London: Intermediate Technology Publications.

Barreiro, J. ed. 1992. Indigenous economics: Toward a natural world order. Special Issue of *Akwe:kon Journal* 9: 2.

Bentley, J. W. and K. L. Andrews. 1991. Pests, peasants, and publications: Anthropological and entomological views of an integrated pest management program for small-scale Honduran farmers. *Human Organization* 50 (2): 113–24.

Berg, T. 1996. The compatibility of grassroots breeding and modern farming. In *Participatory plant breeding*, edited by Pablo Eyzaguirre and Masa Iwanaga. Rome: IPGRI.

Berkes, F. and C. Folke. 1992. A Systems perspective on the interrelations between natural, human-made and cultural capital." *Ecological Economics* 5: 1–8.

Berlin, B. 1992. *Ethnobiological classification: Principles of categorization of plants and animals in traditional societies*. Princeton: Princeton University Press.

Blunt, P. and D. M. Warren, eds. 1996. *Indigenous organizations and development*. London: Intermediate Technology Publications.

Brokensha, D. W., D. M. Warren, and O. Werner, eds. 1980. *Indigenous knowledge systems and development*. Lanham, MD: University Press of America.

Brush, S. B. and D. Stabinsky, eds. 1996. *Valuing local knowledge: Indigenous people and intellectual property rights*. Washington, D.C.: Island Press.

Bunders, J., B. Haverkort, and W. Hiemstra, eds. 1997. *Biotechnology: building on farmers' knowledge*. Basingstoke, UK: Macmillan Education.

Carney, J. 1991. Indigenous soil and water management in Senegambian rice farming systems. *Agriculture and Human Values* 8 (1+2): 37–48.

Cashman, K. 1989. Agricultural research centers and indigenous knowledge systems in a worldwide perspective: Where do we go from here?." In *Indigenous knowledge systems: Implications for agriculture and international development*, edited by D. M. Warren, L. Jan Slikkerveer, and S. O. Titilola. Studies in Technology and Social Change, No. 11. Ames: CIKARD, Iowa State University.

Castro, P. 1995. *Facing Kirinyaga: A social history of forest commons in southern mount Kenya*. London: Intermediate Technology Publications.

Center for Integrated Agricultural Development (CIAD). 1994. *Indigenous knowledge systems and rural development in China: Proceedings of the workshop*. Beijing: Beijing Agricultural University.

Center for Traditional Knowledge. 1997. *Guidelines for environmental assessments and traditional knowledge*. Ottawa: Centre for Traditional Knowledge. (2nd Draft Prototype for Field Testing).

CGIAR. 1993. Indigenous knowledge. In *People and plants: The development agenda*. Rome: Consultative Group on International Agricultural Research.

Chambers, R. 1979. Rural development: Whose knowledge counts? Special issue of *IDS Bulletin* (Institute of Development Studies, University of Sussex) 10, 2.

———— 1983 *Rural development: Putting the last first*. London: Longman.

———— 1994 The origins and practice of participatory rural Appraisal. *World Development* 22 (7): 953–69.

———— 1997 *Whose reality counts? Putting the first last*. London: Intermediate Technology Publications.

————, A. Pacey, and L. A. Thrupp, eds. 1989. *Farmer first: Farmer innovation and agricultural research*. London: Intermediate Technology Publications.

CIRAN. 1993a. Background to the international symposium on indigenous knowledge and sustainable development. *Indigenous knowledge and development monitor* 1 (2): 2–5.

———— 1993b. Recommendations and action plan. *Indigenous knowledge and development monitor* 1 (2): 24–29.

Critchley, W. R. S. 1992. *Indigenous soil and water conservation—Prospects for building on traditions*. Overseas Division Report OD/92/12. Silsoe, UK: Silsoe Research Institute.

————, C. Reij and T. J. Willcocks. 1994. Indigenous soil and water conservation: A review of the state of knowledge and prospects for building on traditions. *Land Degradation and Rehabilitation* 5: 293–314.

Davis, D. K. 1995. Gender-based differences in the ethnoveterinary knowledge of Afghan nomadic pastoralists. *Indigenous Knowledge and Development Monitor* 3 (1): 3–4.

Davis, S. H., ed. 1993. *Indigenous views of land and the environment*. World Bank Discussion Papers, No. 188. Washington, D.C.: The World Bank.

———— and K. Ebbe, eds. 1995. *Traditional knowledge and sustainable development*. Environmentally Sustainable Development Proceedings series, No. 4. Washington, D.C.: The World Bank.

de Boef, W., K. Amanor, and K. Wellard, with A. Bebbington. 1993. *Cultivating knowledge: Genetic diversity, farmer experimentation and crop research*. London: Intermediate Technology Publications.

DeWalt, B. 1994. Using indigenous knowledge to improve agriculture and natural resource management. *Human Organization* 53 (2): 123–31.

den Biggelaar, C. 1991. Farming systems development: Synthesizing indigenous and scientific knowledge systems. *Agriculture and Human Values* 8 (1+2): 25–36.

———— and N. Hart. 1996. *Farmer experimentation and innovation: A case study of knowledge generation processes in agroforestry systems in Rwanda*. Rome: FAO.

Dialla, B. E. 1994. The adoption of soil conservation practices in Burkina Faso. *Indigenous Knowledge and Development Monitor* 2 (1): 10–12.

Dowdeswell, E. 1993. Walking in two worlds. Address presented at the InterAmerican Indigenous People's Conference, 18 September, Vancouver.

Dupre, G., ed. 1991. *Savoirs paysans et developpement*. Paris: Karthala/ORSTOM.

Everett, Y. 1995. Forest gardens of Highland Sri Lanka—An indigenous system for reclaiming deforested land. In *The cultural dimension of development: Indigenous knowledge systems*, edited by D. M. Warren, L. Jan Slikkerveer, and David Brokensha. London: Intermediate Technology Publications.

Eyzaguiire, P. and M. Iwanaga, eds. 1996. *Participatory plant breeding*. Proceedings of a workshop on participatory plant breeding, 26–29 July 1995, Wageningen, the Netherlands. Rome: International Plant Genetic Resources Institute (IPGRI).

Fernandez, M. E. 1994. Gender and indigenous knowledge. *Indigenous Knowledge and Development Monitor* 2 (3): 6–7.

Foster, L. M., S. A. Osunwole, and B. W. Wahab. 1996. Imototo: Indigenous Yoruba sanitation knowledge systems and their implications for Nigerian health policy. In *Alaafia: Studies of Yoruba Concepts of*

Health and Well-Being in Nigeria, edited by F. Fairfax III, B. Wahab, L. Egunjobi, and D. M. Warren. Studies in Technology and Social Change, No. 25. Ames: CIKARD, Iowa State University.

Gadgil, M., F. Berkes and C. Folke. 1993. Indigenous knowledge for biodiversity conservation. *Ambio* 22 (2+3): 151–56.

Gamser, M. S., H. Appleton, and N. Carter, eds. 1990. *Tinker, tiller, technical change.* London: Intermediate Technology Publications.

Green, E. C. 1994. *AIDS and STDs in Africa: Bridging the gap between traditional healing and modern medicine.* Boulder: Westview Press.

———. 1996. *Indigenous healers and the African state.* New York: Pact Publications.

———. 1997a. Is there a basis for modern-traditional cooperation in African health promotion? *Journal of Alternative and Complementary Medicine* vol. 3 (forthcoming).

———. 1997b. The participation of African traditional healers in AIDS/STD prevention programmes. *Tropical Doctor* 27 (l): 56–59.

———. 1998. *African theories of contagious disease.* Newbury Park, CA: Altamira/Sage Press.

———, B. Pedersen and D. M. Warren. 1989. *Strategies for the establishment of cooperative programs involving African traditional healers in the control of diarrheal diseases of children.* Report for WHO. Washington, D.C.: WHO Ad Hoc Working Group on Traditional Medicine and Diarrheal Disease.

Groenfeldt, D. 1991. Building on tradition: Indigenous irrigation knowledge and sustainable development in Asia. *Agriculture and Human Values* 8 (1+2): 114–20.

Guarino, L. 1995. Secondary sources on cultures and indigenous knowledge systems. In *Collecting plant genetic diversity: Technical guidelines,* edited by L. Guarino, V. Ramanatha Rao, and R. Reid. Wallingford, Oxon UK: CAB International on behalf of the International Plant Genetic Resources Institute in association with the FAO, IUCN, and UNEP.

Gunn, A., G. Arlooktoo and D. Kaomayok. 1988. The contribution of the ecological knowledge of Inuit to wildlife management in the Northwest Territories. In *Traditional knowledge and renewable resource management,* edited by Milton M. R. Freeman and Ludwig N. Carbyn. Edmonton: Boreal Institute for Northern Studies.

Gupta, A. 1990. *Honey Bee.* [A Quarterly Journal Devoted to Indigenous Innovations.] Ahmedabad. India: Indian Institute of Management.

Haverkort, B., J. van der Kamp and A. Waters-Bayer, eds. 1991. *Joining farmers' experiments.* London: Intermediate Technology Publications.

Hecht, Susanna B. and Darrell A. Posey. 1989. Preliminary results on soil management techniques of the Kayapo Indians. *Advances in Economic Botany* 7: 174–88.

Herbert, John. 1993. A mail-order catalog of Indigenous Knowledge. *Ceres: The FAO Review* 25 (5): 33–37.

Himestra, Wim with Coen Reijntjes and Erik van der Werf. 1992. *Let farmers judge—experiences in assessing agriculture innovations.* London: Intermediate Technology Publications.

Hunter, Phoebe R. 1994. *Language extinction and the status of North American Indian languages.* Studies in Technology and Social Change, No. 23. Ames: CIKARD, Iowa State University.

Hviding, Edvard and Graham B. K. Baines. 1992. *Fisheries management in the Pacific: Tradition and the challenges of development in Marovo, Solomon Islands.* Discussion Paper No. 32. Geneva: United Nations Research Institute for Social Development.

IIRR. 1996. *Recording and using indigenous knowledge: A manual.* Silang, Cavite, Philippines: REPPIKA, International Institute of Rural Reconstruction.

Inglis, Julian T., ed. 1993. *Traditional ecological knowledge: Concepts and cases.* Ottawa: International Program on Traditional Ecological Knowledge and International Development Research Centre.

Innis, Donald Q. 1997. *Intercropping and the scientific basis of traditional agriculture.* London: Intermediate Technology Publications.

International Board for Plant Genetic Resources. 1993. Rural development and local knowledge: The case of rice in Sierra Leone. *Geneflow* 1993: 12–13.

International Development Research Centre. 1993. Indigenous and traditional knowledge. Special Issue *IDRC Reports* 21 (1).

Johannes, Robert E., ed. 1989. *Traditional ecological knowledge: A collection of essays.* Gland, Switzerland: International Union for the Conservation of Nature (IUCN).

Kothari, Brij. 1995. From oral to written: The documentation of knowledge in Ecuador. *Indigenous Knowledge and Development Monitor* 3 (2): 9–12.

Kreisler, Ann and Ladi Semali. 1997. Towards indigenous literacy: Science teachers learn to use IK resources. *Indigenous Knowledge and Development Monitor* 5 (1): 13–15.

Kroma, Siaka. 1995. Popularizing science education in developing countries through indigenous knowledge. *Indigenous Knowledge and Development Monitor* 3 (3): 13–15.

Lansing, J. Stephen and James N. Kremer. 1995. A socioecological analysis of Balinese water temples. In *The cultural dimension of development: Indigenous knowledge systems*, edited by D. M. Warren, L. Jan Slikkerveer and David Brokensha, London: Intermediate Technology Publications.

Leakey, Richard E. and L. Jan Slikkerveer, eds. 1991. *Origins and development of agriculture in East Africa: The ethnosystems approach to the study of early food production in Kenya.* Studies in Technology and Social Change, No. 19. Ames: CIKARD, Iowa State University.

Lutz, E., S. Pagiola, and C. Reiche. 1994. The costs and benefits of soil conservation: The farmers' viewpoint. *The World Bank Research Observer* 9 (2): 273–95.

Martin, Gary J. 1995. *Ethnobotany: A methods manual.* London: Chapman and Hall.

Mathias, Evelyn. 1995. Framework for enhancing the use of indigenous knowledge. *Indigenous Knowledge and Development Monitor* 3 (2): 17–18.

_____ 1996 How can ethnoveterinary medicine be used in field projects? *Indigenous Knowledge and Development Monitor* 4 (2): 6–7.

Mathias-Mundy, Evelyn and Constance M. McCorkle. 1989. *Ethnoveterinary medicine: An annotated bibliography.* Bibliographies in Technology and Social Change, No. 6. Ames: CIKARD, Iowa State University.

_____ , Evelyn, Olivia Muchena, Gerard McKiernan and Paul Mundy. 1992. *Indigenous technical knowledge of private tree management: A bibliographic report.* Bibliographies in Technology and Social Change, No. 7. Ames: CIKARD, Iowa State University.

Mazzucato, Valentina. 1997. Indigenous economies: Bridging the gap between economics and anthropology. *Indigenous Knowledge and Development Monitor* 5 (1): 3–6.

McCall, Michael K. 1995. *Indigenous technical knowledge in farming systems of Eastern Africa: A bibliography.* Bibliographies in Technology and Social Change, No. 9. Revised edition. Ames: CIKARD, Iowa State University.

McCorkle, Constance M. 1994. *Farmer innovation in Niger.* Studies in Technology and Social Change, No. 21. Ames, Iowa: CIKARD, Iowa State University.

_____ , Evelyn Mathias, and Tjaart W. Schillhorn van Veen, eds. 1996. *Ethnoveterinary research and development.* London: Intermediate Technology Publications.

_____ and Gail McClure. 1995. Farmer know-how and communication for technology transfer: CTTA in Niger. In *The cultural dimension of development: indigenous knowledge systems*, edited by D. Michael Warren, L. Jan Slikkerveer, and David Brokensha. London: Intermediate Technology Publications.

McGregor, Elizabeth, compiler. 1994. *Indigenous and local community knowledge in animal health and production systems—gender perspectives: A working guide to issues, networks and initiatives.* Ottawa: World Women's Veterinary Association.

Messerschmidt, Donald A. 1995. Local traditions and community forestry management: A view from Nepal. In *The cultural dimension of development: Indigenous knowledge systems*, edited by D. M. Warren, L. Jan Slikkerveer, and David Brokensha. London: Intermediate Technology Publications.

Moock, Joyce Lewinger and Robert E. Rhoades, eds. 1992. *Diversity, farmer knowledge, and sustainability.* Ithaca: Cornell University Press.

Morrison, John, Paul Geraghty, and Linda Crowl, eds. 1994. *Science of Pacific island peoples.* 4 vols. Suva, Fiji: Institute of Pacific Studies, The University of the South Pacific.

Mundy, Paul A. and J. Lin Compton. 1995. Indigenous communication and indigenous knowledge. In *The cultural dimension of development: Indigenous knowledge systems*, edited by D. M. Warren, L. Jan Slikkerveer, and David Brokensha. London: Intermediate Technology Publications.

National Research Council. 1991. *Toward Sustainability: A plan for collaborative research on agriculture and natural resource management.* Washington, D.C.: National Academy Press.

_____ 1992a *Conserving biodiversity: A research agenda for development agencies.* Washington, D.C.: National Academy Press.

_____ 1992b *Neem: A tree for solving global problems*. Washington, D.C.: National Academy Press.

_____ 1993 *Vetiver Grass: A thin green line against erosion*. Washington, D.C.: National Academy Press.

Niemeijer, David. 1996. The dynamics of African agricultural history: Is it time for a new development paradigm? *Development and Change* 27 (1): 87–110.

Pauly, D., M. L. D. Palomares, and R. Froese. 1993. Some prose on a database of indigenous knowledge on fish. *Indigenous Knowledge and Development Monitor* 1 (1): 26–27.

Pawluk, Roman R., Jonathan A. Sandor and Joseph A. Tabor. 1992. The role of indigenous soil knowledge in agricultural development. *Journal of Soil and Water Conservation* 47 (4): 298–302.

Phillips, Adedotun O. 1989. Indigenous agricultural knowledge systems for Nigeria's development: The case of grain storage. In *Indigenous knowledge systems for agriculture and rural development: The CIKARD inaugural lectures*, edited by Paul Richards, L. Jan Slikkerveer, and Adedotun O. Phillips. Studies in Technology and Social Change, No. 13. Ames: CIKARD, Iowa State University.

Pichon, Francisco, Jorge Uquillas and John Frechione, eds. 1998. *Traditional and modern approaches to natural resource management in Latin America*. In review.

Pinkerton, Evelyn, ed. 1989. *Co-operative management of local fisheries: New directions for improved management and community development*. Vancouver: University of British Columbia Press.

Posey, Darrell Addison. 1985. Management of tropical forest ecosystems: The case of the Kayapo indians of the Brazilian Amazon. *Agroforestry Systems* 3 (2): 139–58.

_____ and Graham Dutfield. 1996. *Beyond intellectual property: Toward traditional resource rights for indigenous peoples and local communities*. Ottawa: IDRC Books.

Prain, Gordon and C. P. Bagalanon, eds. 1994. *Local knowledge, global science and plant genetic resources: Towards a partnership*. Los Banos: UPWARD.

_____ , Sam Fujisaka and D. M. Warren, eds. 1998. *Biological and cultural diversity: The role of indigenous agricultural experimentation in development*. London: Intermediate Technology Publications. (Forthcoming).

Pretty, Jules N. 1991. Farmers' extension practice and technology adaptation: Agricultural revolution in 17th–19th Century Britain. *Agriculture and Human Values* 8 (1+2): 132–48.

_____ 1995. *Regenerating agriculture: Policies and practice for sustainability and self-reliance*. London: Earthscan.

Price, Thomas L. 1995. Use of local knowledge in managing the Niger river fisheries project. In *The cultural dimension of development: Indigenous knowledge systems*, edited by D. M. Warren, L. Jan Slikkerveer and David Brokensha. London: Intermediate Technology Publications.

Quintana, Jorge. 1992. American Indian systems for natural resource management. *Akwe:kon Journal* 9 (2): 92–97.

Rajasekaran, Bhakthavatsalam. 1994. *A framework for incorporating indigenous knowledge systems into agricultural research, extension and NGOs for sustainable agricultural development*. Studies in Technology and Social Change, No. 22. Ames: CIKARD, Iowa State University.

_____ and D. Michael Warren. 1995. Role of indigenous soil health care practices in improving soil fertility: Evidence from South India. *Journal of Soil and Water Conservation* 50 (2): 146–49.

_____ , D. M. Warren and S. C. Babu. 1991. Indigenous natural-resource management systems for sustainable agricultural development—A global perspective. *Journal of International Development* 3 (4): 387–401.

_____ and Michael B. Whiteford. 1993. Rice-crab production: The role of indigenous knowledge in designing food security politics. *Food Policy* 18 (3): 237–47.

Reij, Chris. 1993. Improving indigenous soil and water conservation techniques: Does it work? *Indigenous Knowledge and Development Monitor* 1 (1): 11–13.

Reijntjes, Coen, Bertus Haverkort and Ann Waters-Bayer. 1992. *Farming for the future: An introduction to low-external-input and sustainable agriculture*. Leusden: ILEIA.

Rhoades, Robert E. and Anthony Bebbington. 1995. Farmers who experiment: An untapped resource for agricultural research and development. In *The cultural dimension of development: Indigenous knowledge systems*, edited by D. Michael Warren, L. Jan Slikkerveer and David Brokensha. London: Intermediate Technology Publications.

Richards, Paul. 1986. *Coping with hunger: Hazard and experiment in an African rice-farming system*. London: Allen and Unwin.

Roach, Steven A. 1997. *Land degradation and indigenous knowledge in a Swazi community.* MA thesis. Ames: Department of Anthropology, Iowa State University.

Roling, Niels and Paul Engel. 1989. IKS and knowledge management: Utilizing indigenous knowledge in institutional knowledge systems. In *Indigenous knowledge systems: implications for agriculture and international development*, edited by D. M. Warren, L. Jan Slikkerveer, and S. O. Titilola. Studies in Technology and Social Change, No. 11. Ames: CIKARD, Iowa State University.

———— 1991. The development and concept of agricultural knowledge information systems (AKIS): Implications for extension. In *Agricultural extension: Worldwide institutional evolution and forces for change*, edited by W. M. Rivera and D. J. Gustafson. New York: Elsevier Science Publishing Company.

———— and A. Wagemakers, eds. 1998. *Facilitating sustainable agriculture: Participatory learning and adaptive management in times of environmental uncertainty.* Cambridge: Cambridge University Press.

Ruddle, Kenneth. 1994. *A guide to the literature on traditional community-based fishery management in the Asia-Pacific tropics.* FAO Fisheries Circular, No. 869. Rome: FAO.

Rural Advancement Fund International. 1994. *Conserving indigenous knowledge: Integrating two systems of innovation.* New York: United Nations Development Programme.

Rusten, Eric P. and Michael A. Gold. 1995. Indigenous knowledge systems and agroforestry projects in the central hills of Nepal. In *The cultural dimension of development: Indigenous knowledge systems*, edited by D. M. Warren, L. Jan Slikkerveer, and David Brokensha. London: Intermediate Technology Publications.

Saouma, Edouard. 1993. Indigenous knowledge and biodiversity. In *harvesting nature's diversity.* Rome: FAO.

Scoones, Ian and John Thompson, eds. 1994. *Beyond farmer first: Rural people's knowledge, agricultural research and extension practice.* London: Intermediate Technology Publications.

Selener, Daniel with C. Purdy and G. Zapata. 1996. *Documenting, evaluating and learning from our development projects: A systematization workbook.* New York: International Institute of Rural Reconstruction.

Semali, Ladislaus. 1997. Cultural identity in African context: Indigenous education and curriculum in East Africa. *Folklore Forum* 28: 3–27.

Sillitoe, Paul. 1998. The development of indigenous knowledge: A new applied anthropology. *Current Anthropology* 39 (2). (Forthcoming).

Simpson, Brent M. 1994. Gender and the social differentiation of local knowledge. *Indigenous knowledge and development monitor* 2 (3): 21–23.

Slaybaugh-Mitchell, Tracy L. 1995. *Indigenous livestock production and husbandry: An annotated bibliography.* Bibliographies in Technology and Social Change, No. 8. Ames: CIKARD, Iowa State University.

Slikkerveer L. Jan and Mady K. L. Slikkerveer. 1995. Taman Obat Keluarga (TOGA): Indigenous Indonesian medicine for self-reliance. In *The cultural dimension of development: Indigenous knowledge systems*, edited by D. M. Warren, L. Jan Slikkerveer and David Brokensha. London: Intermediate Technology Publications.

Sumberg, James and Christine Okali. 1997. *Farmers' experiments: Creating local knowledge.* Boulder: Lynne Rienner Publishers.

Systemwide Programme on Participatory Research and Gender Analysis. 1997. *A global programme on participatory research and gender analysis for technology development and organisational innovation.* AgREN Network Paper No. 72. London: Agricultural Research and Extension Network, UK Overseas Development Administration (ODA).

Thurston, H. David. 1992. *Sustainable practices for plant disease management in traditional farming systems.* Boulder: Westview Press.

Titilola, S. O. 1990. *The economics of incorporating indigenous knowledge systems into agricultural development: A model and analytical framework.* Studies in Technology and Social Change, No. 17. Ames: CIKARD, Iowa State University.

Ulluwishewa, Rohana. 1994. Women's indigenous knowledge of water management in Sri Lanka. *Indigenous Knowledge and Development Monitor* 2 (3): 17–19.

———— and Hemanthi Ranasinghe. 1996. *Indigenous knowledge and sustainable development: Proceedings of the first national symposium on indigenous knowledge and sustainable development, Colombo, March 19–20, 1994.* Nugegoda, Sri Lanka Resource Centre for Indigenous Knowledge, University of Sri Jayewardenapura.

Stet UNESCO. 1994a—Traditional knowledge in tropical environments. Special Issue *Nature and Resources* vol. 30, no. 1.

_____ l994b – Traditional knowledge into the twenty-first century. Special Issue *Nature and Resources*, 30, 2.

Van Veldhuizen, Laurens, Ann Waters-Bayer, Ricardo Ramirez, Deb Johnson and John Thompson, eds. 1997. *Farmers' experimentation in practice: Lessons from the field.* London: Intermediate Technology Publications.

Wahab, Waheed Bolanle. 1997. *The traditional compound and sustainable housing in Yorubaland, Nigeria: A case study of Iseyin.* Ph.D. dissertation. Edinburgh: Department of Architecture, Heriot-Watt University.

Walker, Peter J. C. 1995. Indigenous knowledge and famine relief in the horn of Africa. In *The cultural dimension of development: Indigenous knowledge systems,* edited by D. M. Warren, L. Jan Slikkerveer, and David Brokensha. London: Intermediate Technology Publications.

Warren, D. Michael. 1989a. Linking scientific and indigenous agricultural systems. In *The transformation of international agricultural research and development,* edited by J. Lin Compton, ed. Boulder: Lynne Rienner Publishers.

_____ 1989b. Utilizing indigenous healers in national health delivery systems: The Ghanaian experiment. In *making our research useful,* edited by John van Willigen, Barbara Rylko-Bauer, and Ann McElroy. Boulder: Westview Press.

_____ 1990. *Akan arts and aesthetics: Elements of change in a Ghanaian indigenous knowledge system.* Studies in Technology and Social Change, No. 16. Ames: CIKARD, Iowa State University.

_____ 1991a. The role of indigenous knowledge in facilitating a participatory approach to agricultural extension. In *Proceedings of the international workshop on agricultural knowledge systems and the role of extension,* edited by Hermann J. Tillmann, Hartmut Albrecht, Maria A. Salas, Mohan Dhamotharah, and Elke Gottschalk. Stuttgart: University of Hohenheim.

_____ 1991b. *Using indigenous knowledge in agricultural development.* World Bank Discussion Papers, No. 127. Washington, D.C.: The World Bank.

_____ 1992. *A preliminary analysis of indigenous soil classification and management systems in four ecozones of Nigeria.* Ibadan: African Resource Centre for Indigenous Knowledge and the International Institute of Tropical Agriculture.

_____ 1994. Indigenous agricultural knowledge, technology, and social change. In *Sustainable agriculture in the American midwest,* edited by Gregory McIsaac and William R. Edwards. Urbana: University of Illinois Press.

_____ 1995a. Indigenous knowledge, biodiversity conservation and development. In *Conservation of biodiversity in Africa: Local initiatives and institutional roles,* edited by L. A. Bennun, R. A. Aman, and S. A. Crafter. Nairobi: Centre for Biodiversity, National Museums of Kenya.

_____ 1995b. Indigenous knowledge for agricultural development: A Keynote speech. Workshop on Traditional and Modern Approaches to Natural Resource Management in Latin America, The World Bank, April 25–26, 1995.

_____ 1996a. Keynote address. *Indigenous knowledge and sustainable developments: Proceedings of the Sri Lanka National Symposium on Indigenous Knowledge, 1994,* Rohana Ulluwishewa and Hemanthi Ranasinghe, eds. Nugegoda: SLARCIK.

_____ 1996b. The role of indigenous knowledge and biotechnology in sustainable agricultural development. In *Indigenous knowledge and biotechnology.* Ile-Ife, Nigeria: Indigenous Knowledge Study Group, Obafemi Awolowo University.

_____ 1998. The role of the global network of indigenous knowledge resource centers in the conservation of cultural and biological diversity. In *Language, knowledge and the environment: The interdependence of cultural and biological diversity,* edited by Luisa Maffi. In preparation.

_____ , Remi Adedokun and Akintola Omolaoye. 1996. Indigenous organizations and development: The case of Ara, Nigeria. In *Indigenous organizations and development,* edited by Peter Blunt and D. Michael Warren, ed. London: Intermediate Technology Publications.

_____ , Layi Egunjobi and Bolanle Wahab, eds. 1996. *Indigenous knowledge in education: Proceedings of a regional workshop on integration of indigenous knowledge into Nigerian education curriculum.* Ibadan: Indigenous Knowledge Study Group, University of Ibadan.

———— 1997. *Studies of the Yoruba therapeutic system in Nigeria*. Studies in Technology and Social Change, No. 28. Ames: CIKARD, Iowa State University.

————, Layi Egunjobi and Bolanle Wahab. 1996. The Yoruba concepts of health and well-being: Implications for Nigerian national health policy. In *Alaafia: Studies of Yoruba concepts of health and well-being in Nigeria*, edited by Frank Fairfax III, Bolanle Wahab, Layi Egunjobi, and D. M. Warren. Studies in Technology and Social Change, No. 25. Ames: CIKARD, Iowa State University.

———— and Joe D. Issachar. 1983. Strategies for understanding and changing local revenue policies and practices in Ghana's decentralization programme. *World Development* 11 (9): 835–44.

———— and Jennifer Pinkston. 1997. Indigenous African resource management of a tropical rainforest ecosystem: A case study of the Yoruba of Ara, Nigeria. In *Linking social and ecological systems*, edited by Fikret Berkes and Carl Folke. Cambridge: Cambridge University Press.

———— and B. Rajasekaran. 1993. Indigenous knowledge: Putting local knowledge to good use. *International Agricultural Development* 13 (4): 8–10.

———— 1994. Using indigenous knowledge for sustainable dryland management: A global perspective. In *Social aspects of sustainable dryland management*, edited by Daniel Stiles. New York: John Wiley.

————, L. Jan Slikkerveer and David W. Brokensha, eds. 1995. *The cultural dimension of development: Indigenous knowledge systems*. London: Intermediate Technology Publications.

Willcocks, Theo. J. and Francis N. Gichuki, eds. 1996. *'Conserve water to save soil and the environment': Proceedings of an East African workshop on the evaluation of indigenous water and soil conservation technologies and the participatory development and implementation of an innovative research and development methodology for the provision of adoptable and sustainable improvements*. SRI Report No. IDG/96/15. Bedford, UK: Silsoe Research Institute.

Wiltgen, Beverly and Bolanle Wahab. 1997. The influence of Alaafia on the design and development of Yoruba housing: A case study of Ibadan and Iseyin. In *Studies of the Yoruba therapeutic system in Nigeria*, edited by D. M. Warren, Layi Egunjobi, and Bolanle Wahab. Studies in Technology and Social Change, No. 28. Ames: CIKARD, Iowa State University.

Winslow, Donna. 1996. *An annotated bibliography of naturalized knowledge systems in Canada*. Bibliographies in Technology and Social Change, No. 10. Ames: CIKARD, Iowa State University.

Wolff, Norma H. and Bolanle Wahab. 1995. Learning from craft taxonomies: Development and a Yoruba textile tradition. *Indigenous Knowledge and Development Monitor* 3 (3): 10–12.

Session IV: Use

Edited by Gabriele Iannàccaro

Brent BERLIN asks Ingold how the results of their respective endeavors in producing a local flora would differ if he and Ingold were to set out working with the same people, each using their own different approaches. He also asks Ingold to comment upon what would need to be done in a situation where traditional knowledge risked being lost for posterity.

*

Tim INGOLD answers the first question by saying that while they might end up with results that would look very similar, the difference would lie in their understanding of how the flora relates to the knowledge of the people themselves, as well as to their own understanding of what knowledge is. Berlin would assume that the artifact produced—an inventory of plants and their distinguishing characteristics—corresponds to something that already exists in some mental representational form within the native mind, while Ingold would not.

As regards the second question, the project of catching a window of time and documenting it is important. However, once again, there would be a difference between Berlin's and Ingold's approaches. It might lie in how they conceive the relationship between the documents they would produce and the past and future of the people whose knowledge is supposed to be recorded therein. In other words, it would hinge on how, according to them, people relate these documents to processes of memory, on how the act of writing knowledge down affects the future practices of the people concerned, and how this, in turn, influences their understanding of the relationship between knowledge and practice.

*

John TRUMPER points out that both Indonesia and Ghana are culturally influenced by Islam and underlines the triadic complication of Islamic culture. The Classic and Middle Greek diglossia concept resurrected in the 1950's by Ferguson and others in relation to Islamic culture is not exhaustive, since Muslims are people of the *book*; they have a highly literary knowledge, but they have, as well, a highly oral-literary knowledge, because Koranic schools *are* oral. Although schooling deals ostensibly with primarily written materials, mnemonic formulae are oral and literary at the same time. Moreover, Muslims have a strictly *oral orality* that is distinct from the other two types. It would be interesting to investigate what effect this particular type of complication has on the transmission of indigenous knowledge.

*

Roy ELLEN confirms that there is—in a sense—a free market in ideas, and certainly it is right to point out that medical therapies, for example in parts of Indonesia, are heavily influenced by ideas imported from Islam from the near East, and also from Hinduism, from India. Sometimes these techniques are maintained relatively separately by people who are specialists in certain kinds of system; in other cases, there is clearly hybridization. One of the consequences of this juncture is in the legal area, in particular in discussions about intellectual property rights. These rights generally do not work as well in a country like Indonesia, where there has been much more hybridization; if a certain people in particular locality claim rights over certain knowledge, then that is going to have implications for people next door, who have exactly the same knowledge.

*

Michael WARREN confirms Islamic influence in Ghana. Travelling by road from Ghana to Hybrid in Nigeria thirty-six years ago, he was able to observe that Islam penetrated all the way down to the coast; moreover, Islam was already there when the first Portuguese arrived in the 1490s. The Yoruba indigenous therapeutic system is based on *Alaafia*, a sort of greeting. It comes from Hausa, and Hausa derives from Arabic; it means peace, well being. It ends up being a focus point for a very complex holistic approach to medicine.

*

Francesco AVOLIO, dialectologist, praises Ingold's paper for stressing very clearly that continuity and change are not opposed, but, on the contrary, can be identified, since we cannot imagine the former without the latter and vice versa. In Italy, these two elements were often conceived as a pair of contraries, as in some sociological, anthropological and "meridional" essays, published during the 1970s or even before, in which the continuity of "peasant civilization" (as it was often called) was a sort of constant factor existing outside of time and history, and based, probably, on hasty interpretations of novels such as *Cristo si è fermato a Eboli*, by Carlo Levi, or *Fontamara*, by Ignazio Silone. Ingold made it clear that the accusation of "archaeology"—frequently addressed to those forms of anthropological and dialectological research that are interested in re-establishing traditional cognitive networks (i.e., an aspect of continuity)—is excessive, and, therefore, without a real justification.

Regarding Warren's paper, the important data that he discusses merits more attention in concrete international politics, and it would be interesting to know if there are, nowadays, international occasions of comparison among anthropologists to try to obtain a more prominent and sharper role in decisions concerning the development of the countries they study.

*

Michael WARREN points out that the major transformation that has occurred in the last fifteen years happened when Michael Cernea was brought into the World Bank to increase the number of non-economist social scientists working there—which was close to zero when he came. There have also been significant attempts even on the part of U.S. Aid: they had anthropologists in Washington and regional offices in Abidjan and Nairobi for example.

*

Antonino COLAJANNI adds that in the United States, the Bingenton Institute for Development Anthropology has also done research-work, consulting, and studies, mostly in collaboration with U.S. Aid. In the Bingenton Institute Review, *Development Anthropology Bulletin Network* (one of the numbers issued on its tenth anniversary), David Brokensha, some years ago, wrote that, in the last twenty years, studies by anthropologists have been considered much more seriously. The Institute for Development Anthropology has published volumes attesting the influence of anthropologists on the programs of USA Aid. Still, political pressure on the World Bank and other big institutions has necessarily lowered the intensity of knowledge accretion, sometimes causing difficulties in their relationship with the academic world. Working more closely with the world of research would perhaps convince other anthropologists of the value of development anthropology and would be of help in preventing the perversity of development. Warren could (was able to?) explain what obstacles hinder the alliance today.

*

As a fellow member himself, Michael WARREN maintains that the Institute of Development and Anthropology, founded by Mike Horowitz with Theodore Scadder and David Brokensha, has been extremely successful in using anthropology as a basis for getting World Bank, UN and U.S. contracts. He points out that in the 1999 April edition of *Current Anthropology* there is a major article by the British scholar Sillitoe, a major critique in fact of why more anthropologists are not involved in international development. Sillitoe points to the anthropological academic enterprise, particularly in the U.K. and the United States, where those scholars belonging to the theoretical area have often looked down on those who do practical things. He also focuses on the fact that many anthropologists do not have cross-disciplinary sensitivity—an interest in understanding the history and development of a given community; furthermore, they lack the basic development-planning and management skills and tools needed to carry out a significant short-term development assignment.

Many of our academic enterprises designate people with a fairly narrow approach to the matter, while what we really need is to open up and look at the wealth of insight and perspectives offered by the academic world and to study how they should be used.

*

Diego MORENO appreciates Ingold's mention of the ecological paradigm. His own research on flora in the European area attests that the whole floristic and vegetable heritage of a place is an artifact, and this is the theoretical premise for the work of historians, archaeologists, and anthropologists interested in the history of resources, in the history of the knowledge that changed it. As Ingold points out, nature knowledge is a relational kind of knowledge and today ecological analysis is relational. The demands for new classifications advanced by the zoologist Minelli and the anatomo-pathologist Fassina in the first session of the Conference have not been taken up.

*

Tim INGOLD equally considers anthropology and archaeology to be essentially indistinguishable. From a British anthropological perspective, it is paradoxical that while it has been normal and indeed fashionable to claim that there is very little difference between anthropology and history, and while history has an obvious connection with archaeology, anthropologists have had very little to do with archaeology. This paradox owes much to implicit assumptions about the fundamental ahistoricity of nature, and about the incommensurability of timescales (particularly the distinction between historical time, which anthropologists feel they can handle, and evolutionary time, which they cannot). Archaeology, like anthropology, stands rather precariously on the division between the natural sciences and the humanities, a division which it should be anthropology's mission to bridge. Michael Warren is right in stressing the importance of interdisciplinarity, although it is a very difficult thing to achieve. It can only be done properly if it is backed by significant theoretical work. It is essential to sort out the conceptual structures that different disciplines bring to their tasks, to see whether or not they can be reconciled, and to look critically at their underlying assumptions, before their insights can be integrated into some grand synthesis. Otherwise interdisciplinarity is bound to remain something of a hodgepodge rather than leading to real intellectual advance.

*

Brent BERLIN says Moreno is absolutely incorrect.

*

In anthropologist Alessandra PERSICHETTI's opinion, there is still some discrepancy between the clear-sightedness of the prospects proposed by Warren and by the World Bank and the frail local knowledge that Solinas discusses. In the Arabic-Islamic world, it emerges in conflictual forms; in Italy as well, the knowledge of Arabic immigrants is assimilated and dissolved by our knowledge, which is richer and equipped with better instruments.

*

Pier Giorgio SOLINAS points out that, while it is possible to transpose parts of knowledge through hybrizidation so that they can flourish in new grounds, the innermost aspects of social structure and symbolism risk being destroyed in order to *save whole populations economically*. As a matter of fact, the knowledge that is necessarily abandoned is a frail kind of knowledge, in which the relationship between being and meaning is so close that changing the former necessarily entails changing the latter.

*

Michael WARREN offers two examples of exogenous linguistic elements somehow incorporated locally without people knowing it. The first refers to the Alaafia, spreading from Hausa and ultimately from Arabic into the Yoruba therapeutic system. Only few Yoruba speakers fluent in Hausa or familiar with Arabic know that the word *Alaafia* derives from Hausa, the others are unaware of this fact. The second example concerns over a hundred words in American English: they come from a stretch of landscape from the Wolof speakers in Senegal all the way down to Kimbundu and Angola, but American people are not conscious of the origin of these borrowed words.

*

Antonino COLAJANNI points out that there are processes of dynamic reconstitution everyday and we can observe them. Trying to use less frequently the categories of disgregation, destruction, and exhaustion and more frequently those of reconstitution would be sometimes advisable, because experience proves that mixing and reconstituting on a particular basis is possible. A society, if not too violently undermined, can develop and build up its own future counting on 50–60 percent of its own knowledge.

For example, a native Tucan from Colombia has been working for months to invent a hydraulic grater for bitter manioc to save his wife 50 percent of work. His invention is based wholly on a Tucan linguistic and symbolic-ritual apparatus, but at the same time is an innovation.

In Moko-moko, Bolivia, international aid is supporting the reconstruction of the ancient pre-Inca system of *andenes*, terraced land with water inside, which with solar warming emanate four to five degrees of warmth, preventing potatoes from freezing. Traditional technologies are obviously altered, but the changing process is organized through the perceptive, knowing, linguistic, and cultural categories belonging to the society of that area.

*

Paolo SEGALLA, anthropologist involved in development and recently working with an approach based on indigenous knowledge, expounds some objections people make, asking Warren to comment on them.

The first one concerns the inadequacy of interventions based on indigenous knowledge, which are small with respect to the major challenges of the present day; for example, the desertification of the Sahel. The second objection comes from the field and it is often expressed by the small farmers themselves: they are convinced that what they need is not indigenous culture, but modern materials such as fertilizers, or machinery, or whatever modern technology has to offer.

*

Michael WARREN admits that indigenous structure can play a role in improving things only in small-scale enterprises where participatory decision is involved. In large-scale enterprises, like the building of gigantic tunnels with American equipment and Italian technical assistance in Libya, or the construction of gigantic dams in India and China, there are no possibilities of decision and participation for local communities.

*

As regards variability and discreteness, John TRUMPER explains that linguists in the 1960s (Labov et al.) rediscovered a great debate that started off in the 1870s on variation. Bailey and others in the 1970s and 1980s reformulated this debate on what is discrete and what is continuous in languages, and by reflex in cultures. It might even be the sort of problem newborn babies are faced with. They have to find out where they end and where Mummy begins, as it were. We are born, it appears, into a continuum and we have to discover our discreteness; later on, growing up intellectually, we have to discover that we are both discrete and continuous, in different ways in differing contexts. We are, so to speak, context-sensitive as well. Nature is therefore both properties, as are we, something beyond Aristotelian accidence (characteristic) and substance (definitory). The same question crops up in the reification or not of environments. An environment is a succession of discrete rēs, though at the same time every environment is also a continuum. So mushrooms, plants, animals etc., just like language, which is also a natural historical product, just like human beings themselves and their cognitive processes, are all simultaneously discrete and continuous phenomena. Therefore whatever analytical model we exploit has to accommodate both these phenomenological aspects; otherwise it is methodologically a *non-model*.

*

Tim INGOLD points out that the condition of continuity is, in some sense, ontologically prior to, and a precondition for, the objectification of nature as a domain of things, as Merleau-Ponty states in assuming that we have what he calls a "pre-objective" awareness of our environment in which perception is embedded, and only because of that can we go on to say, for example, that we have a body, that there are things out there, and that we can study them scientifically. This is not in any sense to reject science, or to set up an opposition between science and humanistic inquiry. It is merely to plead that we give due recognition to science's pre-objective foundations.

*

Marta MADDALON does not consider scientific knowledge to be a guide for the study of nature knowledge.

PART V

Conservation

Cristina Papa

What does it Mean to Conserve Nature?

The Conservation of Environmental Resources

This part introduction offers a brief survey of the debate and the main anthropological topics surrounding the conservation of environmental resources. We begin with the very concept of conservation itself. Here conservation is understood in terms of a type of human action that, albeit implying manipulation, transformation, and the domestication of environmental resources, allows these resources to be conserved and renewed rather than destroyed. The importance of focusing on the relationship between human action and nature conservation is a reaction to the negative effects that have arisen from considering them as distinct from one another and distinguishing between conservation of environmental resources and human knowledge (Ingold 1992). Neither environmental resources nor human knowledge can be considered as abstract separate entities, and it is in this that an initial difference lies with respect to those who consider human action and conservation to be irreconcilable.

My approach, then, is in line with the principles outlined in the 1991 International Union for Conservation of Nature (IUCN) document. It defines conservation as "the management of the human use of organisms and ecosystems capable of ensuring that such use is sustainable." Within this overall approach different analytical categories are adopted according to different criteria, which obviously influence the analysis of and judgments about the behavior of entire populations as regards their ability to conserve. For example, a number of well represented positions in the US debate of the 1970s tended to associate some cultural and behavioral models with the objective of guaranteeing sufficient food in the future for its group and thus of conserving resources. This was the purpose of food taboos, male dominance or war. In this context, a price was paid by certain individuals in order to avoid the extinction of the prey and thus guarantee hunting for future generations (Harris 1974; McDonald 1977; Ross 1978). In Western culture these analyses have strengthened the stereotype of the "savage ecologist" (Ellen 1986) and the "ecologically noble savage" (Redford 1993). In their view, the indigenous populations cut off from the Western world pursue an automatic conservation strategy by living in harmony with nature without threatening either animal or plant species, thus avoiding any changes to their own ecological environment and culture. The critique of the "savage ecologist," seen as the self-critical invention of the West, arose at the same time as the adoption of narrower definitions of conservation. This is the case with the definition advanced by Alvard, whereby decisions for conservation are "subsistence decisions with a short-term cost for those implementing them but with the objective of increasing the sustainability of harvests in the long term" (Alvard 1995: 790). This definition attempts to distinguish between a self-conscious strategy implying short-term costs to guarantee the long-term stability of the harvest, which Alvard calls "conservation," from apparent conservation described as "epiphenomenal" (Hunn 1982; Hames 1991). In the second case conservation results are not the outcome of conservation strategies but rather the consequence of a positive relation, for example, between low population density (partly due to territorial mobility, illness, famine or war) and the wealth derived from exploited resources.

This distinction is based on an analysis of the aims of and the degree of self-awareness about the effects of using resources rather than their results. On the basis of this definition, for example, Alvard shows that the Piro Indians on the eastern coast of Peru are not conservationists. In fact this population has unlimited access to hunting wild game and has never introduced any limit in hunting the most vulnerable and easily-captured prey. Alvard believes the best way to understand the Piro hunting strategy is through the "optimal foraging theory," whereby hunters tend to optimize the results of their activity. This would explain why they do not protect prey that should be safeguarded from a conservation point of view (young animals or pregnant females), but hunt where the prey is most abundant, thus saving time and energy and avoiding places where prey is scarcer. Critics of the optimal foraging theory stress that utilitarianism is not always the basis of human action. Often action is motivated by reasons and emotions that contradict utilitarian criteria and, moreover, is dependent on conditions linked to the social and environmental context that may lead to a change in behavior. The risk involved in Alvard's theory is that of pigeon-holing peoples by replacing one stereotype—the "noble conservationists"—with another, the "savage destroyers" (Puri 1995).

Like other critics, Puri attempts to bring out the full complexity of our behavior towards natural resources, ranging from the utilitarian dimension to that of representation and social relations. These meanings can hardly be reduced to individually identifiable wills and reasons, or given causes. Smith, for example, although stressing the importance of adopting intentionality or design as a distinctive criterion, points out that it cannot be a clearly separate factor because overtly conservationist ideologies may conceal selfish actions (Smith 1995). Moreover, he does not believe it necessary to assume cost as a condition when speaking of conservation. The criterion of cost is of no use in eliminating "epiphenomenal" conservation. In fact a costly behavior resulting in conservation as a food taboo may have other functions and meanings, such as that of acting as an ethnic marker.

The concept of conservation obviously raises the problem of what and for whom. From this point of view, behavior appearing to be destructive from one perspective may be conservationist from another. As Hames points out "clearing a forest to make arable fields is not what we would commonly called conservation. This is because an ecosystem has been destroyed and replaced by an anthropogenic form. But it may be useful to think of the fields made arable as a system of conservation for those inheriting them" (Hames 1995: 805). Hames also suggests that a sufficient condition for speaking of conservation is Roger's notion (1991) of "security of long-term return."This condition presupposes the existence of social mechanisms guaranteeing the possibility through a system of rules or an authority governing the access to and use of natural resources. That is why agricultural systems controlled by community authorities are more frequently associated with the practice of conservation than economic systems based on hunting or gathering with more fragmentary and mobile social organizations. The question of managing resource use from this point of view seems to be crucial in studying conservation practice.

This relation between decision-making about resources and their use was tackled by Hardin in his famous book *The tragedy of the commons* (1968). This work highlights the conflict between collective ownership of pastures and private ownership of flocks. Hardin came to the conclusion that the free access to resources of which no-one assumes the responsibility combined with private use leads to degradation. But, as Beckerman and Valentine claim, when everything is collectivized, both the resources and their uses, we find conservationist behavior associated with resources (Beckerman and Valentine 1996), just as happens when both access and use of resources take place at the individual level. Collectivization of resources and consumption may be found in several Amazon peoples such as the Tucuna, Arawak, and Curripac. They inhabit areas with poor land along the banks of rivers significantly called the "rivers of hunger." For although these people consume a small part of their food in private, in general they communally eat most of their catch from hunting and fishing, despite the conflicts over the share, each family group should contribute to collective use. But in general the individual good and collective good tend to coincide and give rise to moderation in hunting and fishing. As Bekerman and Valentine point out, this means establishing a relation between conservationist behavior and the more general conditions that frame them and give them meaning.

The use and management of resources is in any case a sphere of potentially conflictual relations. As many works have shown, conflicts concerning the conservation of natural resources tend to create opposition between social and environmental players with local populations and institutional authorities at various levels: national, supranational, and regional. They are the bearers of different interests and points of view concerning which resources and in what quantities these resources should be used. In opposition to our view is the particularly significant position that conceives of conservation as excluding human use. Indeed, linking conservation with use runs counter to the notion of conservation as the preservation of a wild humanless environment, considered as the only way to fully protect individual species and environments in a world where human activity and nature are basically considered to be antagonistic.

Many nature parks in the United States have been created on the basis of this principle, a principle that defines a park as a place "where a man is a visitor and must not remain" (Gomez-Pompa and Kaus 1992: 271). This position has at times led government bodies and environmental associations to propose and encourage moving traditional local inhabitants away from zones designated as nature parks, often to inadequate small areas, with serious consequences for the populations in terms of living standards and health (Scudder and Colson 1982; Turton 1987), often resulting in their extinction and environmental decay. In his book *The mountain people,* Turnbull (1972) describes the consequences of the expulsion of the Ik, a population of hunters and gatherers from their homelands in Uganda destined to be the Kidapo National Park. Forced to take up farming in poor areas near the park, the Ik had to abandon their way of living without ever adapting to a new one. Decimated by hunger and poverty, the survivors turned to poaching, begging, and prostitution. But biodiversity conservation can benefit from human presence. For example, the ecosystem of the Serengeti fields survives thanks to the presence of the Masai and their cattle. The expulsion of the Masai from this territory would lead to the fields being overrun by bush, with a consequent reduction in food for the antelopes (Adams and McShane 1992).

In any case these problems only have a limited effect on protected areas that today are mostly inhabited. According to recent estimates, 70 percent of the world's protected areas are inhabited, while a 1985 IUCN report estimates that 86 percent of the protected areas in Latin America are inhabited (Amend and Amend 1992; Kempf 1993). These figures also suggest that the use of natural resources in protected areas and their conservation in practical terms cannot be separated (Pitt and McNeely 1985; McNeely and Miller 1984).

Conservation of Environmental Resources and Local Cultures

The conservation of environmental resources has long been considered a question to be tackled only in political terms with the backing of the natural sciences. Only recently has there been a focus on the indispensable role that local populations and their culture have played and can continue to play, and the role of anthropology. Thus while the study of the indigenous and technical knowledge of a population in its own environment is one of the classical themes of anthropological research on individual local communities, only in the past few decades has there been an attempt to find a practical application for the results of such research: a delay that results from the fact that the usefulness of indigenous and technical knowledge of nature has been underestimated. These populations were seen more as pupils to be taught than as teachers from which something could be learned. More recently there has been a greater need to involve them in development schemes that include many environmental resource conservation programs. But although the importance of local cultures is now recognized, they are often interpreted in an ideological and abstract way.

Such attitudes are found in the statements of the special bodies for biodiversity and environmental associations or even the social sciences. Despite stressing the close relationship between the conservation of local varieties and the environments and cultures in which they were produced or conserved, they seem to view local cultures as closed static entities immune to history and the effects of socioe-conomic trends and cultural exchanges.

There has been a tendency to ignore the fact that in the same society there exist several levels of knowledge which also translate into conflicts between the social groups that hold them. These groups are the bearers of specific interests in the use of environmental resources. One example is the conflicts between arable farmers and herdsmen in rural European areas (Angioni 1989; Amiel 1985), or between farmers and hunters. A complex system of relations has also taken root in some European countries between local farmers and new farmers, especially young peo-ple of urban extraction who have moved to the country to work the land, as sev-eral studies reveal (Buttel and Newby 1980; Kayser 1990; *Etudes rurales* 1994) or immigrants from non-EU countries employed more or less permanently in farm-ing or sheep-raising. An intense debate has also sprung up around the relation-ship between ideology and practical action in the natural world. This has mainly focused on challenging the static vision of culture, whereby certain philosophical or religious traditions or cosmologies based on respect for the natural environment are automatically claimed to be ecological, ignoring the fact that the spheres of ideology and practical action are not comparable (Callicot 1982; Brightman 1987; Hames 1987; Ellen 1995).

The real problem however is one of meaning and awareness with respect to order and disorder in the world, and the relation between humanity and its sur-roundings, and the other living species whether plants or animals, taken individ-ually or within the ecosystem. This problem is tackled in a different way in different cultures, and it may or may not include the concept of conservation. The study of the technical relations between humanity and its environment cannot be undertaken without taking into account the overall ecosystem. At the same time the framework used to represent the set of relations which will in turn influence technical action cannot be ignored. This does not necessarily mean that the repre-sentation mirrors the actual relationship. On the contrary, in Western history we find a series of gaps and contradictions. In his contribution to this volume, Ortalli illustrates for example how in the Middle Ages, from the twelfth century on, there were detailed widespread rules about environmental protection, as demonstrated by town statutes, but from the ideological point of view this form of conservation was not represented or elaborated conceptually (Ortalli 1997). There was thus a gap between practical modes of protection, worked out at an everyday level, and ideological interest. Today the opposite is true: there is a good deal of refined ide-ological elaboration about conservation and great difficulty in effectively imple-menting environmental protection.

Thus a careful analysis of the history of the West, but also of other world regions, reveals the presence of gaps and contradictions rather than logical sys-tems. In this context the opposition between one civilization, that of the West based on the separation of body and mind, subject and object, nature and culture, and the other civilizations that do not separate the human being from the natural environment—but make him as an integral part of the surrounding nature with a conservationist behavior—often says more about Western self-criticism than the reality of the "other" cultures. This is nothing new. In Western culture Montaigne and Rousseau, to mention only two illustrious names, adopted similar points of view which were more the outcome of a critical approach to Western institutions rather than the result of an in-depth analysis.

The ideological import of this approach is also clear in the very definition of "other" cultures and the term "indigenous." As Arne Kalland (1994) has pointed out, the concept of indigenous is ambiguous. In the wider meaning, indigenous—often interchangeable with the terms "native" and "aboriginal"—is used to indicate the first known inhabitants of an area. In a more limited sense, however, the term has come to denote often small groups of populations descending from non-Europeans without a state apparatus, and organized on the basis of non-market economy with unsophisticated and simple technology, and often refers to people who have been oppressed by invaders for decades. Indigenous knowledge has become synonymous with non-developing static traditional knowledge, which, as we have seen, fails to account for trends of cultural circulation between social groups and different cultures as well as forms of cultural transmission.

The category of "indigenous knowledge" is used rhetorically by the indigenous populations themselves in their struggle against the governments of their territories with the aim of obtaining recognition as populations with their own culture, the rights to self-determination and to exploitation of the resources in their area. In this struggle, the indigenous populations thus often use Western visions of their environmentalism to reach the objectives of self-determination. Anthropologists have long studied the indigenous claims on territories, the conflict between states, local populations, movements and political trends between the various players (Conklin and Graham 1995), and they have often been on the side of local populations and ethnic minorities in relation to the policies for the protected areas (Turton 1987; Clay 1988).

Opposed to those claiming that indigenous populations are naturally conservationist, some people believe that their habitual activities must be greatly limited if not stopped altogether in order to guarantee conservation. It is on these grounds that in some cases populations are expelled from protected areas or certain activities are stopped or developed in the direction wanted by the authorities. This is the case with the Inuit whalers who must prove in annual meetings with the International Whaling Commission that they have not sold too much whale meat (Kalland 1993). Since 1931 they have been forced to continue using traditional boats without engines or firearms (Freeman 1993). Whaling has been a classical case of conflict between environmentalist associations and local populations whether "indigenous," like the Inuit, or social groups from Western countries, like the Icelandic whalers. In this case the sustainable use of resources was obviously not an objective fact but rather a tool to put forward a point of view—that of Greenpeace—which had chosen the whale as a key symbol in its own political initiative (Einarsson 1995) and that of the fishing populations who saw the curbs on whaling as the cause of emigration and lost jobs.

A pragmatic-type position (Brush and Orlove 1996), which takes into account the complexity of interests of the interacting players and also the problems of managing protected areas stripped of local populations (Gibson 1995), seems more suitable for highlighting the need to reconcile the requirements of resident populations with those of conservation on the basis of what is described as a "scale of threats."This perspective was used in a study on Alaska which underlined the fact that the threat to environmental conservation from fishermen and hunters was infinitely smaller than that occasioned by outside fishermen and hunters. Their destructive role, in turn, is much less than that of the oil companies (Childers 1994).

Conservation of Genetic Resources *in situ* and *ex situ*

One of the most fiercely debated questions concerns the conservation of plant genetic resources, which account for a very large part of overall biological diversity. Until recently plant genetic resources were conserved *ex-situ* in germplasm banks. The collections made by public institutions in industrialized countries have amassed a large quantity of genetic material in the germplasm banks, which contain most of the germplasm gathered worldwide. As Brush (1996) points out, citing data given in Pluknett et al. (1987): "The relative abundance of germplasm in public institutions lessens the possibility that breeders will purchase crop germplasm from primary sources. National gene banks in industrialized nations and those at international agricultural research centers of the CGIAR system (e.g., IRRI, CIMMYT, CIP) control a large percentage of the world's collected germplasm reserves.... Germplasm from these collections is more attractive than germplasm from uncollected landraces for several reasons—including seed health, biological identification, and characterization of agronomic traits. Since much of the use of genetic resources is within public programs, many users are likely to expect a public source for breeding material. The abundance of collected germplasm thus undermines a market for crop genetic resources." (Brush 1996: 425).

Despite the fact that this abundance reduces the probability"that users will pay for unknown germplasm when they can obtain it without cost from international and open collections,"in recent years there has been a growing awareness that *ex-situ* conservation of cultivated plants is not enough and must be accompanied by forms of *in situ* or on-farm conservation. The latter are the only kind of conservation that permit the preservation not only of genetic resources but also of agroecosystems and the individuals belonging to them with their attendant heritage of knowledge, techniques, and social and cultural forms. Moreover, only *in situ* or on-farm conservation allows the germplasm to be preserved within evolutional processes so as to obtain future variability."Perhaps the most profound change in the context of genetic resources is political. The existing institutions and policies for conserving and utilizing genetic resources were established under the mantle of common heritage, in which resources were collected, preserved and exchanged as public goods." (Brush 1996: 419).

Brush considers two important international events, the Convention on Biological Diversity and the 1992 Agenda 21, as key stages in identifying the conservation of agroecosystems as an indispensable form of biodiversity conservation and as hailing the end of the view that considers these resources as a public asset inherited by humanity. There is now a vast literature on the question of producers' rights, highlighting the various positions (Greaves 1994; Brush 1992; 1994; 1995; 1996). At the same time the literature stresses the need to recognize these rights, since the importance of exploiting genetic resources and their economic spin-off is so great that the producers must be given some form of economic support. The problem is more about how to give this support: one possible form is direct compensation and moves in the direction of acknowledging intellectual property rights or contracts; another is indirect compensation of investments in farm developments meant to improve *in-situ* and on-farm conservation as well as the living standards of farmers.

This shift in perspective has been accompanied by a different way of focusing on local cultures and the prospects of the loss of biodiversity. The conservation of genetic resources in germplasm banks is based on the premise of the inevitable loss of the world's genetic diversity and assumes that the only way of conserving knowledge, seen basically as information related to the uses and techniques of manipulation and cultivation, is in repertories and other forms of documentation. In this picture, the problems of cultural change and its dynamic relation with the social structure become fairly insignificant, since the documentation and conservation of such knowledge is linked to the prospect of their inevitable disappearance. As Warren, Slikkerveer and Brokensha (1995) stress, work on biodiversity conservation must also be supported by research on indigenous knowledge, by trying to combine the efforts of preserving the germplasm of plants with that of documenting "the human knowledge accumulated about this plant material, particularly that which is at high risk." (Warren, Slikkerveer and Brokensha 1995: XVIII). The objective is thus to ensure that the information is "systematically deposited and stored for use by development practitioners" (XVII). This means storing knowledge in isolation from the situations that produced, reproduced or transformed it, as well as from the bearers, for the sake of genetic improvements in research, medicine, pharmaceutics, the food and agriculture industry and their consequent use on the market.

Unlike with *ex-situ* intervention, in the case of *in-situ* conservation, which implies the involvement of local populations, the complexity and dynamism of culture become indispensable reference points. In fact *in-situ* projects are about sustainable development. In pursuing conservation their objective is controlled change aimed at increasing income and yield, without replacing the local genetic resources. Obviously the cultures (techniques, representations, and forms of learning), and the social and productive systems that have made the conservation of these varieties possible cannot be seen as static situations in the projects but rather as agents subject to change as part of more general transformations. Thus cultural change, instead of being exorcised, must be oriented towards a sustainability so that knowledge of nature is developed and its transmission encouraged. In reality, as has often been pointed out, the emphasis on developing local culture is not accompanied by a real commitment to use it within development programs, nor an adequate evaluation of its complexity. In particular, as Ellen and Harris point out in their contribution to this volume:"the result is an ambiguous representation of IK [indigenous knowledge] as 'indigenous science,' rational knowledge or empirical knowledge. Current literature on IK presents it as largely separate from the cultures in which it originates. At best, reference is made to certain ritual and symbolic factors which could be considered, but any consideration of whether and how indigenous knowledge and culture might differ is ignored." (Ellen and Harris 1997).

Another important limit in the documentary collections of so-called popular knowledge of nature lies in their incapacity to take into account the complex original contexts due to the abstract way in which the knowledge of nature is described. Especially in Western Europe, where there is no lack of historical sources, the analysis and planning of biodiversity conservation cannot ignore the fact that the natural environment is the product of historical transformation and cultural exchange between geographical areas and social classes. From this point of view the conservation of genetic resources and the cultural contexts in which it takes place have as their main subject the rural populations materially exploiting the resources of an area by animal rearing, hunting, or arable farming. But this perspective also implies past intervention by other agents: politicians, the holders of political power at various levels for the land in question; economists, landowners and experts; intellectuals, botanists and agronomists who may act in agreement with the other agents or may be in conflict with them, as is illustrated in Ambrosoli's work (1992) on the rise of the"new agriculture"in Europe from the fifteenth to the nineteenth century or in Amiel's work on the relation between scientific and popular knowledge and the local vine-growers in the Beaujolais area (Amiel 1985). A strong emphasis on cultural diversification and circulation in relation to popular knowledge of nature also emerged in the 1983 workshop entitled *Les savoirs naturalistes populaires* at Sommières and in the issue of *Terrain* (1986) also dedicated to popular knowledge of nature.

The development of environmental archaeology also suggests that the environmental features of a given place are historically determined by the uses and knowledge of the populations inhabiting them over time. An historical approach failing to give due weight to the local scale would thus underestimate the technical action on the environment and its repercussions on the process transforming the real, seen only schematically within long-term processes from the Neolithic to the Industrial Revolution (Moreno 1989 and 1990).

In the West, more than anywhere else, the full historical dimension can be restored to research into the relationship between environmental resources and local cultures by seeing the signs of social and economic trends in social practices and techniques, such as arable farming or animal rearing.

Biodiversity and Traditional Products

In Western countries specific questions are raised by the relation between local cultures and the conservation of neglected local varieties: kinds of oil, fruit trees, minor cereals, legumes (cow-pea, lentils) or local animal races or local products made by processing local genetic resources (cheeses, hams and salami, etc.), which are rediscovered and put on the market.

In very different forms than those of the poor areas of the world, in the West the conservation of on-farm biodiversity by local producers is not carried out so much for direct local consumption as for the needs of the Western urban market. Indeed, some of these food products are destined for the quality end of the market. It is for the needs of this market that local varieties become "traditional products," reflecting the transition from direct consumption and local markets to "global" markets. Laurence Bérard and Philippe Marchenay (1995; 1998) have cast light on how the very procedures stipulated by the European Union and individual states to obtain quality denominations defining products as traditional necessarily imply a reinterpretation of the cultural aspects of the production and use of those products.

The registration of a trademark gives producers the exclusive rights to use the registered designation, which then becomes a full right of industrial property reserved for a collective holder. The entitlement is reserved to the group of producers who applied for it and is extended to all those in the related zone manufacturing the product according to the conditions established in the "regulation." This kind of protection prevents the commercial use of the designation by businesses failing to follow the conditions established by the "regulation" and by those producing similar products that the consumer might confuse with "designated" products. The fact that these products are individual local animal or plant varieties with more or less sophisticated processing has little influence when it comes to the procedures for obtaining "traditional-product" designation and the mechanisms introduced. Examples of such products are obsolete cereals, now only cultivated in circumscribed areas, such as Saracen wheat or unhusked cereals such as spelt (*T. dicoccum, T. monococcum, T. spelt*) or animal varieties such as the *poulet de Bresse bourguignonne* in France or other products, in which the processing is of greater importance, such as with oil, wine, salami and hams and cheeses.

On the one hand as these local products move from a local market, where their specific characteristics were known, to a wider market they require characterization. This happens with products unknown to the wider public, requiring information on product identity and possible use when they have to compete with similar products from other areas. The mechanisms behind the protection of these local products, however, lead to deep changes in the product itself and the local culture, influencing their variability on the one hand, and their changeable nature as living material on the other. "Local food products are situated in a complex world of relations involving the biological and the social. An animal race, a cultivated plant, and a product such as a sausage or cheese are the outcome of an accumulation of knowledge, practice, observation and adjustments that must be seen in relation to the way they are represented. In short, they are objects heavily invested with many processes and meanings. The living form, the state characterizing it, involves and brings into play many factors. It authorizes all kinds of manipulation at all levels and on all scales. Thus the living form conceals a considerable potential for evolution and variability which people may draw on. But it is also associated with a very short life span. The various forms of expertise shaping the living form are the basis of life itself and *a fortiori* of the conservation of the perishable ephemeral matter which must constantly be maintained and renewed to ensure the repeatability of production systems" (Bérard and Marchenay 1998: 163–64).

However the mechanisms of conservation and protection of the local product tend to modify considerably the product itself and the local culture, when the producers decide to work out specifications for the detailed characteristics of the product and its different phases of processing. This leads to a gap between the changeable features, the diversity of the product, and the characteristics needing to be defined in specifications based on opposed features, such as repeatability, conformity to a model and hygiene rules. As Bérard and Marchenay point out, because of this situation, diversity is automatically reduced:"we might imagine, for many of the local and traditional products, a solution consisting of'leaving things as they are,' but the commercial logic underlying our society also plays an influence through hygiene rules that tend to remove any possibility for them of surviving economically, since when they are sold they must respect the health regulations in force"(Bérard and Marchenay 1997).

This takes place through a mechanism that may be defined as the constitution of the heritage of these products. According to Bérard and Marchenay (1998), this has already happened in France. In Italy this is not yet the case. While local genetic resources are considered a heritage to be protected, the products from these local varieties have not yet taken on the status of a heritage. They are rather identity markers of the local community from outside, and as such have a function that is more than merely economic. Protection is invoked not to guarantee against loss, but against imitations of the product and those usurping their reputation to sell substandard products.

In this picture the view of local culture underlying production becomes ambivalent. On the one hand it is exploited for the sake of claims of typicality, or distinguished from similar products, on the other, it is manipulated in order to eliminate aspects held to be inconsistent with commercial success whether through the mechanisms for obtaining the trademark or market forces. The processes must thus be seen within a more general system of relations between individual local communities and the outside world, built up through tourism and trade. Within these trends, producers, but also individual communities and their institutions, represent themselves by offering a"tourist supply"including the possibility to buy and consume typical products, whether they have the quality trademark or not.

In this way individual products are described in the packaging texts or advertising leaflets in terms of the specific tradition or place of origin, and its physical features: climate, landscape, altitude, rivers, lakes and culture. This thus involves history, local cultures, cultural assets, urban features—in short, the human and natural environment in which they are produced. The players involved in building up this "typical character" as the conservation of a product and/or of a natural resource are not only the producers, as happens with industrial products not rooted in the local production area, but also many players active in the community:

producers, retailers, restaurant managements, hotels, members of tourist promotion associations, organizers of local festivities, regional and municipal authorities.
These players are aware of selling to tourists and outside consumers not only food
products but also an image and especially its typical character and links to the
area, of which it becomes an identifier. They build up the image by highlighting the
permanent nature of certain overall features, but at the same time concealing other
variable features, which they fear may undermine the product or are outside the
rules established by the specifications. In other words, a reduced simplified tradition is constructed in order to present a purported unchanging "overall" continuity. This product is then attributed potential legitimacy and social recognition.

The idea of a social construction does not mean that there are no differences
between these products and similar products of industrial-origin, on the market.
On the contrary, it is precisely this social construction that makes typical products
different from industrially produced consumer goods. Industrial products are relatively undifferentiated and impersonally uniform, while designated products are
loaded with an identity, a genealogy, the identities of place with the natural and
cultural features that produce it, the identity of the natural environment and a specific landrace.

In typical products the cultural dimension is exhibited, monetized and built up.
This emphasis tends to highlight the gap between knowledge incorporated in
expertise and use only as expressed in practice and catalogued constructed knowledge, which seems independent from its players and related contexts. We are thus
faced with a kind of protection that both conserves and reduces cultural—and
arguably—genetic variability. There has not been sufficient analysis of how the
emphasis on intangible aspects tends to form the basis of a relationship between
very different players founded on a cultural dimension and its expression. This
trend reflects a continuous cultural elaboration and innovation on the basis of the
comparison between different points of view.

As commodities, typical products therefore show a complex identity that
enables us to distinguish between two levels of traded goods: on the one hand, the
material nature of the product destined to satisfy the same needs as industrial
products of the same kind, and on the other, the local identity it bears. The product thus takes on the role of priceless symbolic capital, which becomes, thereafter,
an integral part of the world of traded commodities.

References

Adams, Jonathan S. and Thomas O. McShane. 1992. *The myth of wild Africa: conservation without illusion.*
London: W.W. Norton.
Alvard, Michael. 1995. Intraspecific prey choice by Amazonian hunters. *Current Anthropology* 36:
789–818.
Ambrosoli, Mauro. 1992. *Scienziati, contadini e proprietari.* Turin: Einaudi.
Amend, Stephan and Thora Amend, eds. 1992. *Espacios sin Habitantes? Parques nacionales de America del
Sud.* Gland (Switzerland): IUCN.
Amiel, Christiane. 1985. *Les fruits de la vigne. Representation de l'environnement naturel en Languedoc.* Paris:
Editions MSH.

Angioni, Giulio. 1989. *I pascoli erranti*. Naples: Liguori.

Beckerman, Stephen and Paul Valentine. 1996. On Native American conservation and the tragedy of the commons. *Current Anthropology* 37 (4): 659–61.

Bérard, Laurence and Philippe Marchenay. 1995. Lieux,temps et preuves. La construction sociale des produits de terroir. *Terrain* 24: 153–64.

―――― 1997. Diversity, protection, and conservation: local agricultural products and foodstuffs. Paper presented at the *International Congress "Saperi naturalistici/Nature knowledge."*, Venice 4–6 December 1997 [see in this volume Section V: Conservation].

――――. 1998. Les procédures de patrimonialisation du vivant et leurs conséquences. In *Patrimoine et modernité*, edited by Dominique Poulot, 159–70. Paris: L'Harmattan.

Brightman, Robert A. 1987. Conservation and resource depletion. The case of the boreal forest Algonquians. In *The question of the Commons: the culture and ecology of communal resources*, edited by B. M. McCay and J. M. Acheson, 121–41. Tucson: University of Arizona Press.

Brush, Stephen B. 1992. Farmers' rights and genetic conservation in traditional farming system. *World Development* 20: 1617–30.

―――― 1994. A non market approach to protecting biological resources. In *Intellectual property rights for indigenous peoples. A Sourcebook*, edited by Tom Greaves, 133–43. Oklahoma City: SfAA.

―――― 1995. *Providing farmers' rights through in situ conservation of crop genetic resources*. Rome: Food and Agricultural Organization of the United Nations.

―――― 1996. Valuing crop genetic resources. *Journal of Environment and Development* 5 (4): 418–35.

―――― and Benjamin S. Orlove. 1996. Anthropology and the conservation of biodiversity. *Annual Review of Anthropology* 25: 329–52.

Buttel, F. and H. Newby. eds. 1980. *The rural sociology of advanced societies*. London: Croom Helm.

Callicott, J. B. 1982. Traditional American Indian and western European attitudes towards nature. *Environmental Ethics* 4: 293–318.

Childers, R. A. 1994. Cultural protection: a link to tradition. *Forum Appl. Res. Public Policy* 9 (4): 79–83.

Clay, Jason W. 1988. *Indigenous peoples and tropical forests*. Cambridge: Cultural Survival.

Colchester, Marcus. 1994. *Salvaging nature. Indigenous peoples, protected areas and biodiversity conservation*. Gland (Switzerland): UNRISD—World Rainforest Movement—WWF.

Conklin, Beth A. and Laura R.Graham. 1995. The shifting middle ground: Amazonian Indians and eco-politics. *American Anthropologist* 97 (4): 695–710.

Dixon, John A. and Paul B. Sherman.1991. *Economics of protected areas: A new look at benefits and costs*. London: Earthscan Publications.

Einarsson, Neils. 1995. All animals are equal but some are cetaceans: conservation and culture conflict. In *Environmentalism. The view from anthropology*, edited by Kay Milton, 73–84. London: Routledge.

Ellen, Roy F. 1986. What Black Elk left unsaid. On the illusory images of green primitivism. *Anthropology Today* 2 (6): 8–12.

―――― 1995. Rhetoric, practice and incentive in the face of the changing times: a case study in Nuaulu attitudes to conservation and deforestation. In *Environmentalism. The view from anthropology*, edited by Kay Milton, 126–43. London: Routledge.

―――― and Holly Harris. 1997. Indigenous environmental knowledge, the history of science, and the discourse of development. Paper presented at the *International Congress "Saperi naturalistici/Nature knowledge."*, Venice 4–6 December 1997 [see in this volume Section IV: Use].

Etre étranger à la campagne. 1994. Special Issue of *Etudes Rurales* 135–36, July–December 1994.

Freeman, Milton M.R. 1993. The International Whaling Commission, small-type whaling, and coming to terms with subsistence. *Human Organization* 52 (3): 243–51.

Gibson, C. C. and A. A. Marks. 1995. Transforming rural hunters into conservationists: an assessment of community-based wildlife management programs in Africa. *World Development* 23 (6): 941–57.

Gomez-Pompa, Arturo and Andrea Kaus. 1992. Taming the wilderness myth. *Bioscience* 42 (4): 271–79.

Greaves, Tom, ed. 1994. *Intellectual property rights for indigenous peoples. A Sourcebook*. Oklahoma City: SfAA.

Hames, Raymond. 1987. Game conservation or efficient hunting? In *The question of the Commons: the culture and ecology of communal resources*, edited by B. M. McCay and J. M. Acheson, 92–107. Tucson: University of Arizona Press.

_____ 1991. Wildlife conservation in tribal societies. In *Biodiversity, culture, conservation and ecodevelopment*, edited by Margery L. Oldfield and Janis B. Alcorn. Oxford: Westview Press.

_____ 1995. Comment on: Alvard Michael, Intraspecific Prey choice by Amazonia hunters. *Current Anthropology* 36 (5): 804–805.

Hardin, G. 1968 The tragedy of the commons. *Science* 162: 1243–48.

Harris, Marvin. 1974. *Cows, pigs, wars, and witches: the riddles of culture*. New York: Random House.

Hughes, D. J. 1983. *American Indian ecology*. Texas: University of Texas.

Hunn, E. 1982. Mobility as a factor limiting resource use in the Columbia plateau of North America. In *Resource managers: North American and Australian hunter-gatherers*, edited by N. Williams and E. Hunn,. 17–43. Boulder: Westview Press.

Ingold, Tim. 1992. Culture and the perception of the environment. In *Bush base, forest farm: culture, environment and development*, edited by E. Croll and D. Parkin, 39–56. London: Routledge.

IUCN/UNEP/WWF. 1991. *Caring for the earth. A strategy for sustainable living*. Gland (Switzerland): IUCN.

Kalland, Arne. 1993. Whale politics and green legitimacy. A critique of the anti-whaling campaign. *Anthropology Today* 9 (6): 3–7.

_____ 1994. Indigenous knowledge. Local knowledge: prospects and limitations. Paper presented at the *AEPS Seminar on Integration of Indigenous Peoples Knowledge*, Reykjavik 20–23 September 1994.

Kayser, B. 1990. *La Renaissance rurale*. Paris: Colin.

Kemf, Elizabeth. 1993. *Indigenous peoples and protected areas: the law of mother earth*. London: Earthscan Publications Ltd.

Léger, D. and B. Hervieu. 1979. *Le retour à la nature*. Paris: Seuil.

Les hommes et le milieu naturel. 1986. Special Issue of *Terrain* 6, March 1986.

Les savoirs naturalistes populaires. 1985. Actes du seminaire de Sommières, 12–13 December 1983. Paris: Editions MSH.

McCay, B. M. and J. M. Acheson, eds. 1987. *The question of the Commons: the culture and ecology of communal resources*. Tucson: University of Arizona Press.

McDonald, D. R. 1977. Food taboos: a primitive environmental protection agency (South America). *Anthropos* 72: 734–38.

McNeely, J. A., J. Harrison and P. R. Dingwall, eds.1994. *Protecting nature: regional reviews of protected areas*. Gland (Switzereland): IUCN.

_____ and K. R. Miller, eds. 1984. *National parks, conservation and development: the role of protected areas in sustaining society*. Bali: Washington Smithsonian Institute.

Moreno, Diego. 1989. A proposito di storia delle risorse ambientali. *Quaderni Storici* XXIV (3), December 1989: 883–96.

_____ 1990. *Dal documento al terreno. Storia e archeologia dei sistemi agro-silvo-pastorali*. Bologna: Il Mulino.

Ortalli, Gherardo. 1997. Forms of knowledge in the conservation of natural resources: from the Middle Ages to the Venetian "tribe". Paper presented at the *International Congress "Saperi naturalistici/Nature knowledge."*, Venice 4–6 December 1997 [see in this volume Section V: Conservation].

Pitt, D. C. and J. A. McNeely, eds. 1985. *Culture and conservation: the human dimension in environmental planning*. London: Croom Helm.

Plucknett, D. L., N.J.H Smith, J. T. Williams and N. M. Anishett. 1987. *Gene banks and "the world's food."* Princeton: Princeton University Press.

Puri Rajindra, K. 1995. Comment on: Alvard Michael, Intraspecific Prey choice by Amazonia hunters. *Current Anthropology* 36 (5), December 1995: 809–10.

Redford, Kent H. 1990. The ecologically noble savage. *Orion Nature Quarterly* 9 (3): 25–29.

Rogers, A. 1991. Conserving resources for children. *Human Nature* 2: 73–82.

Ross, E. B. 1978. Food taboos, diet and hunting strategy: the adaptation to animals in Amazon cultural ecology. *Current Anthropology* 19: 1–16.

Scudder, Thayer and Elizabeth Colson. 1982. From welfare to development: a conceptual framework for the analysis of dislocated people. In *Involuntary migration and resettlement: the problems and responses of dislocated people*. Edited by Art-Oliver Hansen and Anthony Smith. Boulder: Westview Press.

Smith, Eric A. 1995. Comment on: Alvard Michael, Intraspecific Prey choice by Amazonia hunters. *Current Anthropology* 36 (5): 804–805.

Turnbull, Colin. 1972. *The mountain people*. London: Simon e Schuster.

Turton, D. 1987. The Mursi and the National Park development in the Lower Omo Valley. In *Conservation in Africa*, Edited by D. Anderson and R. Grove, 169–86. Cambridge University Press: Cambridge.

Warren, D. Michael, L. Jan Slikkerveer and David Brokensha. 1995. *The cultural dimension of development*. London: Intermediate Technology Publications.

CHAPTER 22

Random Conservation and Deliberate Diffusion of Botanical Species: Some Evidence out of the Modern European Agricultural Past

Mauro Ambrosoli

The Historical Context of the Conservation of Available Resources

Individual European farmers put in practice some kind of selection and conservation of natural resources within the institutional context of early modern and modern society that caused other constraints upon the environment. From this point of view, I would like to discuss two cases that concern the random or deliberate diffusion of natural resources in the light of the problems, which rose between major political institutions and local society. In this context, upper and lower classes were much closer to each other than indigenous societies and the ever-present Western society are today. It is almost impossible to approach the social dimension of the knowledge and the employ of natural resources without referring to social groups, dominant and subject, élites and minorities, which have continuously rearranged their relations within the ever-changing European environment (e.g. Rackham 1986). The contrast between lower and upper social groups (or between indigenous and Western civilization) discussed in many of the papers at this volume, should be defined as a contrast between socially different groups with established interests in controlling and exploiting vast regional areas. These groups are in turn divided by regions, at times by social classes, without the two sets matching precisely. The use of goods and available resources always raises the question of their conservation as well as their ownership, exploitation, and political control. The uses of collective assets, pastures, woods, fenlands, and marshes provide the most common example of environmental exploitation by local social or dominant groups. Periods of accumulation or saving are necessary phases in the exploitation of a resource. The stop-and-go models, phase A—phase B, saving-consumption all refer to the neoclassical model of an attempt to strike a balance between consumption and accumulation (see, for example, the natural accumulation of fertility). "Change is the rule of nature" (Sauer [1952] 1969): historians cannot but share doubts about the over-simplification of the figure of the *ecologically noble savage* (Ellen 1986, Redford 1990). Even the natives, the indigenous peoples, exploit their territory according to the growth of the population

(Sahlins 1972) or under pressure from commercial agriculture. Ellen and Harris (in this volume) discuss the weakness of indigenous knowledge and its chance of oblivion in the context of European colonial exploitation that result in the mis-management of natural resources. The technological development of European or Western agriculture is characterized by the same ambivalence. It may be pictured as the shift from the (good or bad) *old* agronomy integrated by social groups in the historically determined environment that lasted till the early nineteenth century, to a (good or bad) *new* agronomy of selection and choices that built the *new* European countryside. We are still in a very complex system of data and situations: *good* and *bad* are not absolute terms. Colajanni (in this volume) speaks of technical domestication (on the one hand the natural sector, an unspoiled databank, and, on the other, a technological culture that acquires data and inputs). In the history of European commons there are only rare examples of social groups (large or small) that managed a natural resource without abusing other local interests and con-sciously carried out some environmental conservation, not for the sake to guard their own rights over their territory. Hunting (a feudal privilege) had the effect of defending wild game against the interests of dominant social groups; woods and pastures controlled by the "original" inhabitants excluded those who immigrated more recently in the region. Woodcutting was not a right for all. The protection of plant and animal varieties was random and often contradictory. For an example, the European aristocracy enforced legislation against poaching to protect certain wild animals particularly representative of the social hierarchy, whilst at the same time local pressure groups cut down much of European woods and forests either for fuel or for timber (Thompson 1977; Thirgood 1981). On the other hand, the process of the adoption of elements from other sectors of the same society can be pointed to, which thus became innovations. The acquisition and spread of novel elements was facilitated by practical experiences stored in the individual or social memory and in some instances even supported by government. The result is that peculiar forms of historical development of technology continued to occur because their seedlings were embedded within the historical-cultural dominion exercised by Western society. The social capital of information and knowledge which histor-ically developed in the low sector was often under attack by the institutions of the upper classes which acted with no long-term plan and on the basis of contingent motivations either of individual or limited ruling groups. The gap between local knowledge and high culture was ambiguous: it was enforced by local groups such as craft guilds that also proved to be dominant political channels within the local society. They concealed their technological experience in a culture of "secrets." High technical culture in Europe was often divided according to crafts, sectors, regions, and regional specialization (cod fisheries, soap making, silk throwing, wool weaving) (Rapp 1976). The market economy leveled out productive systems, regional differences, basic goods, and typical products, and was organized around the advanced economic system (De Vries and van der Woude, 1997). The simplifi-cation of social relations of production to yield more money from money limited the combination of productive factors for the natural environment. From being an asset, land became a means of production. European agrarian institutions in the sixteenth century raised a set of problems, connected to the rising population, which were addressed immediately but solved only partly in the eighteenth and nineteenth centuries, thanks to the general change of the agricultural systems. In the widespread process of privatization of common lands, which occurred in Western Europe throughout the eighteenth and the nineteenth centuries, the conservation of natural resources was more random than deliberate. Nevertheless all major instances of conservationism find their roots in the changing attitudes towards the natural world, which developed in certain milieus of the European society alongside with the more general process of the privatization of common lands and natural resources (Thomas 1983: 173–80, 254–86). Far from being a pure individual choice, conservation or diffusion of botanical species operated within the institutional options available in early modern and modern European regions. It is helpful to look at today's problems concerning the management and the property rights of natural resources with a particular attention to some instances, which occurred in our European past and leave remains even today. Ingold in this volume quotes the lawsuit of a Sami community against the state to defend their rights to common lands in Norway today. Moreover, European farmers have always practiced some kind of selection and improvement of their sown species: that was sometimes a personal choice, whilst on other occasions they drew fruitful opportunities from major institutions. However, individual farmers faced selection problems, which are similar to those of today: species and cultivars were part of the European regional floras, whose description had been attempted, in the main botanical textbooks of the Renaissance (e.g. Bauhin 1623, 1651) before Linné.

The Dilemma of Specialization: Selection, Weed Control and Social Consumption of Food

The dilemma of specialization is that selection weakens the system. The botanical history of European agriculture (but not only European agriculture) is played out through the contrast between species and variety. In a relatively short period (compared to the history of farming from the Neolithic revolution to the present day), the highest yielding varieties per unit were selected. The deliberate selection of better seeds out of the farmer's crop led to the elimination of the worst cultivars in favor of the better developed plants without obvious diseases (this is true both of European wheat and American beans). Unfortunately this process of selection caused a loss of good germplasm that happened to be discarded alongside bad seeds. To give one example, we must remember Rev. W. Wilks, who described very accurately in 1880 the whole process of selection of what was later called *Papaver rhoeas*. He was proud to declare that the best garden poppies of the world descended from one batch of seeds grown in one flower box of his vicarage. It was a lucky event that the combined efforts of spontaneous growth and of numerous growers, who considered their poppies as weeds, continued to produce millions of the specimens that Rev. Wilks had required to develop his *P. rhoeas*.

This loss was balanced out by a continuous exchange of seeds from distant regions, from poor lands to fertile lands (never vice versa), which brought mountain seeds to the plains, and seeds from open fields to the enclosed ones. The conscious exchange of seeds was practiced on a European and then world scale from the sixteenth through the nineteenth centuries thanks to human curiosity. That was supported by the work of individual botanists, the early Renaissance botanical gardens, individual travelers and trade networks increasingly attuned to the needs of farmers (Ambrosoli [1992]1997; Woodward, ed.1984; Goody 1993: ch. 7). I have illustrated elsewhere the fundamental role of European clover seeds in the British agricultural revolution (1650–1850) (Ambrosoli [1992] 1997: ch. 7). Seed selection undoubtedly encouraged greater yields in England (Allen 1992) and Europe in early modern centuries. This remained a positive feature as long as there was a highly diversified peasant agriculture producing on a continental scale the seeds required for the development of the capitalist sector in agriculture. The market price of seeds was a scanty reward for the toil and labor required in producing them. And more importantly they did not alone guarantee the conservation of peasant agriculture on the European continent. Once again yesterday's European peasants gave and today's peasant throughout the world give similar answers to the introduction of crop innovations. In this volume Stephen Brush describes the question of genetic erosion of crop resources bringing in evidence from the center of genetic diversity of potatoes in an Andean valley. Today's peasant behaviours in storing seed varieties of wheat, potatoes, and maize explain certain secular delays in the adoption of new crops and so called innovations. Seed varieties maximized environmental advantages and lowered risks of crop failures caused by bad weather and plant disease.

Thus the theme of a reward for innovations and/or saving the species remains an unsolved question. Italian long lasting rural institutions proved to be not very flexible in front of innovations and Italian agrarian history is full of highly indicative cases of institutional restraints. The well-known example of Camillo Tarello illustrates the question in an appropriate manner. Between1565 and 1567 Tarello, a medium size landowner of Lonato, a village in view of Lake Garda, and a subject of the Republic of Venice, applied for and obtained a "Privilege (i.e. a patent) of agriculture" from the Venetian Senate to defend the agricultural method he had designed (Berengo 1975). A farmer of learning in the Latin agricultural writers and great practical experience, he suggested that, better tillage, thinner seeding, and the continuous crop rotation between clover and wheat produced a higher grain harvest. Not only did he never receive a pound for his patented method, but also his method went into soon oblivion and it was only rediscovered two hundred years later. Among other reasons, this happened, firstly because in his days the availability of clover seeds was poor and secondly because his system called for crop selection according priorities which were not shared by the mass of peasant producers (Ambrosoli [1992] 1997: 135–41). Therefore, the specialized clover-wheat rotation reduced the number of natural resources available to the average farm. In mid eighteenth century, when civil servants, pamphleteers, and landowners looked for new methods to come out of an everlasting downswing in agricultural productivity, they rediscovered Tarello's method. Despite of all good intentions, it remained without application for another century or so before Napoleons conquest and nineteenth century reforms cleared public and private lands of feudal constraints.

The second trend in the history of European agriculture was the elimination of symbiotic crops (two or more species sown at the same time, different from crops seeded together in rows). For example, seeds of rye and wheat were sown together: a crop known as *barbariato* in the Italian region of Piedmont, *maslin* in English, *paumelle* and *bisaille* in French. In France *sain foins* were a mixture of various forage plants with mainly *Onobrychis*, Lam., and/or luzerne-alfalfa: a crop known as *sanum fenum* in sixteenth century French, or *fenasse* today in Dauphiné valleys. Thus the will to crop selection annulled a farming practice that lowered the risk of a failed monoculture in particularly difficult years (cold or wet) and in poor soil. Wheat and rye, barley and oats for forage or grain, and sainfoins to keep the fields under control when the selection of seeds was difficult or impossible. This practice had one far from secondary advantage, such as control over weeds: since they could not be eliminated, they were kept down by sowing a range of mutually tolerant, similar useful seeds, leaving no room for true weeds. However, in order to maximize price gains, commercially oriented farmers preferred to select one major crop, dropping the least favored harvests that were, anyway, valuable from the genetic point of view.

Higher quality food consumption led to a revision of the practice of symbiotic crops (as did the later spread of weed-killers). Selected crops have advantages but are ambiguous: in early modern Europe good quality wheat fetched a better price in the market but was only suitable to the best soils. From the fourteenth throughout the nineteenth centuries *good* farming was synonymous with the production of high-quality wheat. In different ways, regional agrarian system in Western Europe tried to find the solution of feeding a rising population thanks to the reclamation of wet lands (the *polders* in Holland, the *marais* in France, the wet lands in the Venetian mainland, e.g.) and making permanent fields wherever it was possible. *Good* farming equaled land improvement that meant tillage and wheat in the first instance and secondly tillage and pasture, with one major product in mind, wheat. Yet, economic cycles between 1450 and 1750, with peaks in crisis years such as the mid seventeenth century crisis, turned former peasant owners into landless tenants, who became food consumers more than producers in many European regions. The result was an agricultural system biased to meet the needs of the landowners, aristocracy, church, and bourgeoisie, which had food priorities that were very costly in terms of soil requirements. This agricultural system left the majority of the consumers' population at risk of dearth crisis (such as 1764/66 in Italy or 1787/88 in France). Fortunately peasants in other parts of the world had developed two crops, that fitted in the European wheat cycle with success: maize and potato, which were grown for family consumption by those producers who required wheat as a cash crop. At a very great extent European agricultural system throughout early modern to modern centuries had to find a balance between upper classes food requirements and laboring classes' consumption.

So is *good* agronomy a science or a practice admitting the distinctions between classes of producers? Dominant cereal cultivation developed into a true wheat civilization, which classified grain consumption according to social class: was there any agreement between peasant practices and high agronomy? I mean that set of practices, which developed in Holland, England and France in the eighteenth century. They were later diffused in upper class circles throughout Western Europe mainly in the following century and were referred to by contemporaries as "nouvelle agriculture" or "high farming" and, according to the most common view, had been labeled "the agricultural revolution" by twentieth century historians (Overton 1996).

More than the crops, the state of the soil is the true yardstick of good farming. Desertification or complete infestation by weeds is the negative limit to which crops tend. High European agronomy from Agostino Gallo (1565) to Jethro Tull (1731) took into careful consideration the issue of soil, which ought to be kept in a regenerative condition from the point of view of fertility and the struggle against weeds. Three-quarters of the world's cereal growers still do weed control by hand today. Weak weeds like the easily uprooted poppy, which proliferate in cereal fields, has been replaced by much more resistant plants making the most of the free spaces between crops, the fertilizers spread on the land and crop care that benefits all biological forms (Mooney 1980, 1983). Crop selection brought up the number of plants, which begun to be regarded as weeds. Genetic erosion began to be a byproduct of the market approach to agriculture, although it went unnoticed as long as market oriented farmers were fewer in comparison to peasant owner-occupiers or tenants, who were providing Western European markets with a resourceful quantity of agricultural seeds. This trade had begun early in Western European history but developed regularly since the latter years of the seventeenth century bringing in an immense supply of new plants, which were sold by specialized dealers (Brockway 1979, Goodman and Redclift 1991, Ambrosoli [1992] 1997: ch. VII, Bush in this volume, Drayton 2000).

Deliberate Collection and Diffusion of Plant Inputs: the Napoleonic Museum of the European Flora

From 1550 to 1880 various forms of regional development of European agriculture took place mainly due to the multiplication of plant inputs. Renaissance scientific culture embarked on the enormous task of recognizing and using a huge number of plants (around 5,000 exemplars) from both the old and new worlds (Crosby 1972, 1986; Ambrosoli 1997 Drayton 2000). Simplifying, we can mention two models of plant diffusion in European history. On the one hand, in England (1550–1750), the land owners solved the problem of seed supply by themselves, required by the improvement of the productive resources of their estates and therefore of English agriculture in general. On the other hand there was the French model. First the monarchy and then the republic (l'*État*) provided the necessary structures for multiplying and distributing the plants required for developing and supporting agriculture throughout the country. While the English model is fairly well known (Allen 1992; Ambrosoli [1992] 1997), I should like to mention the main elements of the Napoleonic project and its implementation in the Department of the Po around 1810. These events reflect a precocious and significant awareness of the problem of conservation of plants, species, varieties, and cultivars for the development of resources and its insertion into a far-sighted policy program.

First I wish to highlight that the Napoleon's project to give order to the environment was enforced thanks to the centralizing tendency of the French State through administrative rationalization and served in the modernization of the French economy. Particularly significant was the Napoleon government's action to create an official network of nurseries and seed beds in every department and supply selected plants and seeds to French agriculture and further afield. The project aimed to construct, order, and regulate the urban landscape and the countryside in general. The Napoleonic Empire designed and implemented this project at great speed as part of a declared program of urban modernization and general action in favor of the environment. Using the new administrative structure of the Departments, from 1811 to 1813, a detailed network was created from the Calais to the Pyrenees (with the exception of the Catalonia) and from the Rhine to the Tiber. In every department an official nursery was created (the existing ones were reorganized) and appointed to grow the varieties considered necessary for improving the farm species or for their aesthetic value or their function in enriching the soil. These nurseries were linked to the great institutions of French natural science in Paris: the Jardin des Plantes, the botanical garden in the Palais du Luxembourg, the garden of Versailles and the experimental farm at Roulle. These institutions met local demand for exemplars required to improve the varieties and functioned as thoroughgoing banks of seed and botanical variety (Ambrosoli 1998). The political economy of the empire was organized in the name of regional integration according to new French interests. Because of the war with Britain, the French sought to be self-sufficient in economic terms (Crosby 1965).

Through institutions such as the Bureau de Statistique (but also the individual ministries), this policy centrally organized the gathering of specific data and their diffusion. In this context scientific information was used politically to solve the urgent problems of the French economy. Encouraging agriculture became one of the emperor's personal priorities and his policy followed the ideas and methods established by the eighteenth-century physiocrats.

I will center my attention to French administered Piedmont (1797–1814) to bring in detailed evidence necessary to make my point clear about the institutional support offered by the French state to a deliberate policy of diffusion of natural inputs. The case of the Department of the Po is significant not only from this point of view but also from that of exchanges of information about seeds. A centuries-old custom practiced by botanists, seed exchanges were the basis for the growth of research in the botanical gardens in the sixteenth throughout the eighteenth centuries (Ambrosoli 1997: ch. VII). Indeed Napoleon believed introducing agricultural innovations from above and his efficient beureaucracy was ready to comply with his orders. Thus, as early as the 19th of December 1806, the prefect Loyseh sent many Piedmontese varieties to meet a request by the Pepinière of Luxembourg for its botanical collection. In the Jardins du Luxembourg in1809, the Minister of the Interior organized what was rather pompously described as "the most complete existing collection of fruit trees in France." But there was no "cours pratique de culture" to multiply the thousands of botanical varieties to be used by France in enhancing their own agriculture and that of the imperial departments. Through the Administration Général des Colonies, the Ministry of the Navy wrote to Paris (on 29 June 1808) to organize experiments with "quarantine maize" as victuals for the navy and the colonies. The seed was not among those either in the Jardin des Plantes in Paris or the southern departments of France, where maize only just had begun to be grown (Braudel 1967). French bureaucracy supported capably and willingly the Emperor's project. It was the Prefect of the Po who made the connection with Signor Viala, of the Turin academy of agriculture, who sent 240 liters of maize seed to Paris (4 July, 19 September 1808). The interest in seeds from Piedmont did not diminish, and the Bureau d'Information Administrative et Statistique asked for "further information about growing maize" (Paris, 28 January 1811). On 20 June 1811 the Ministry wanted even more information about maize: how much seed per hectare, the average yield, and the uses of maize for humans and animals. A new wave of correspondence followed between Turin and Paris after 1813.

But to foreign eyes Piedmont meant above all raw and spun silk for the French manufacturers, and so silkworm breeding was also tried out in the Crocetta experimental garden in Turin. In the early nineteenth century the silk sector was still struggling to rise from a slump begun in 1767 with the arrival of Bengali and Chinese silk on the London market. The Piedmont silk mills and Lyons weavers, who were highly dependent on the latter, were unable to compete with white silks from Nanking. The traditional French departments dedicated to seed production had vainly attempted to rear Chinese worms. In March and April 1811, 1813 and 1814, the Nimes Division des Travaux Publiques, and the Gard prefect sent three hectograms each of silkworm eggs of white Nanking silk to Turin, which they had in turn received from the Ministry of the Interior (21 March 1811). Nuvolone and Lascaris attempted to grow them in the Crocetta garden in February 1812. After obtaining a reasonable quantity to guarantee the reproduction in March 1812, they sent a small amount of cocoons to the sub-prefectures of Pinerolo and Susa. The Ministry of the Interior assumed the responsibility of spreading the Piedmont experience in the silk sector and acquired the work on silkworm by Modesto Paroletti as soon as it came out in print in 1812. The French Ministry thus emulated the centralized control practiced by the House of Savoy on the quality of Piedmontese silk. In Paris, on 18 November 1813, the Ministry of the Interior, Second Bureau of Agriculture sent further information on the silk campaign of 1812. In Turin, too, the prefect stressed the importance that the Napoleons government attached to the cultivation of white Chinese-origin silkworms (7 April 1814).

Imperial Plant Diffusion and the Italian Connection:
G. B. Balbis director of the Botanical Garden of Turin

In Napoleonic times the driving force behind updating European flora and spreading it to Paris, from Turin and the rest of Italy was Giovan Battista Balbis (1765–1831), a greatly neglected figure. Like other civil servants at the time, as a young medical student, Balbis spent time in all the institutions of the day. With Jacobin leanings, he was first in exile in France, then a physician in the first Napoleon's army in Italy before finally becoming the remarkably active director of the Turin Botanical Garden. He had first been a student of the botanist Allione and then a member of the Imperial Academy, formerly renown as the Academy of Sciences of Turin. After Waterloo he lost his honors and position, which were then restored after a two-year suspension. Of interest is Balbis' role at the Botanical Garden in the context of Franco-Italian (but also European) scientific exchanges, and his own personal capacity to maintain a vast correspondence with the Napoleonic authorities and botanists dedicated to pushing back the frontiers of the discipline and supporting the government policy of investing in natural resources. The lists of plants grown in the Garden were a vital tool for exchanging seeds and plants and Balbis proved to the French government the importance of the Turin Botanical Garden, hastily renamed after the Empress Josephine in 1805. Three hundred copies were made of the catalogues published in 1813. They dealt with the exchange of seeds with other authors and gardens (Balbis 1813). Thus in his speech at the School of Agriculture and the *Calendario Georgico* in 1814, on the subject of horticultural and flower gardens, Balbis mentions the private gardens of San Sebastiano, Buttigliera, Rivoli, those of the vice-president of the Accademia Agraria of Turin, Signor Colla, and of Antonio Evasio Borsarelli, director of the experimental garden of the same agrarian society (Forneris and Pistarino 1990).

It is above all in Balbis' correspondence that we see the depth of his social and scientific relations and his wholehearted support for Napoleon's policy to develop tree and plant cultivation. From 1807 to 1809 alone, Balbis received over 200 letters from illustrious figures such as the Frenchmen Gouan from Montpellier, Jauvry from Grasse, Suffren, Curten from Lyons, Thoiin, the director of the Musée d'Histoire Naturelle in Paris, and Destonchamp from Paris. Then there were the Italians Armano from the Milan Botanical Garden, Bertolini from Sarzana, Savi from Pisa, Viviani and many others. Balbis was at the center of a network of exchanges on botanical information and books, but most importantly plant inputs, lists of seeds and plants, sent back and forward from Paris and France and from the rest of Italy. Here are some examples: on 20 April 1808, Bertolini sent a list of desired plants from Sarzana; on 29 April 1807, Giovanni Brignoli from Udine asked for 156 seeds or roots and sixty-four books dating from Mattioli's time to the end of the eighteenth century; on 23 April 1807, the Marchese Spino asked for fourteen plants for his own garden. The Marchese again wrote to Balbis because he wanted to create a nursery with "trees of all species" and asked his correspondents for seeds on 18 September 1807. Michele Tenore wrote from Naples on 16 March 1807 that the Minister of the Interior, Miat had ordered a small private garden and also wished to increase the botanical species in the public gardens, the so-called Villa Comunale. Within a year he was being congratulated for his enthusiasm and the success of the structure (Naples, 1 April 1808) and added a number of plants from the Kingdom of Naples, which arrived in Turin through the good offices of a friend, Signor Dephayes, a confidant (sic) of the Minister. On 12 January 1810 Savi sent a list of plants he wanted for the Pisa Botanical Garden, which comprise at least 364 species: vetchling, alfalfa, sweet clover, millet, plantain, sage, rye, clover, fenugreek, tare, acacia, agrostis, allium, artemisia, astragal, centaury, hedisarea, lettuce and other farm plants.

The main Parisian interest was in local varieties for the grand projects, to enhance the Luxembourg gardens, the Jardin des Plantes and Versailles. There is an unending list of ornamental plants and farm varieties: from Montpellier on 17 September 1807, Gouan offered dry plants for the herbarium and bulbs for his garden. From Grasse on 24 July 1807, Jauvry offered Balbis thirty-six plants that the Turin botanist still did not have. The illustrious Suffren wrote from Salen on 26 October 1807 asking for about 240 plants and mentioning his intention of collecting herbs at Cogne, Val d'Aosta, and in the neighboring villas. On 3 March 1807 Curten wrote from Lyons asking for fourteen plants (*alnus glutinosa, bignonia, celsia/celosia orientalis, hypericum balearium, pinus orientalis, quercus ceris et rubra,* and *robinia frutescens*). In March 1807 Thoiin sent 304 seeds and another 300 in February 1808, that were not to be found in the Turin Botanical Garden. On 9 March 1807 the Mayor of Dijon requested seeds for plants not present in the local botanical garden. Balbis made a note, "répondre le 29 avec envoi les graines par la diligence." On 21 February 1808 Merat requested 336 seeds from Paris. Although ill, at that time, Balbis made a memo: "répondre le 4 mars avec envoi des plantes par la diligence."

Destonchamps wrote several times from Paris in 1811 and 1812 requesting eighty-five varieties of "fraxinus, juglans, acer, celtis, ilex, laurus, platanus, pirus, robinia, ulmus" as well as peaches, willows, pines and firs. He duly received them all. The average time taken to meet the requests was two to three weeks at the most, plus the eight to ten days for the Turin to Paris postal service. Thus a month was required to supply the great imperial collections with inputs and/or broker with the small institutional or private Italian acclimatization gardens.

On the look out for local varieties, Balbis traveled from Turin to Milan, Novara and Saluzzo with the aim of developing new gardens or find novelties. Thus, for example, his trusty gardener Piota wrote to him in 1803: "we are now transplanting a species of onion from Bra in the garden, and I am told it is excellent in spring salads."

Conclusions

I have focused my former research on the deliberate and/or informal spread of plants thanks to the action of individual farmers, in connection with the theme of natural selection. In this paper I wish to stress that well before the present times major European governments provided the institutional background to both selection and diffusion. In the two instances presented here, the Senate of Venice supported the requests of single operators, such as Tarello in 1566–67, with trade by-laws and guilds and French government enforced the deliberate environmental policy of the years 1804–1815 thanks to the efficiency of Imperial bureaucracy. Men of science, such as Balbis or farmers such as Tarello benefited from institutional support: however the general conditions of their times, both agricultural and political, drew back the positive effects of their deliberate action of diffusion of plants. How did it happen?

In traditional practice selection or diffusion was pursued for each of those varieties with the best chances of economic success or by following a definite program, a genuine plant policy. From 1550 to 1880 the development of agriculture took place also through the selection and multiplication of plant inputs that had been the care of individuals, farmers and landowners who had a great respect of the natural world. After a centuries-long stage in which random selection provided a few farmers with the best cultivars, between 1550 and 1770 academic botanical gardens and privately owned gardens supported the personal interests of the most skilled and improving landlords. Later the conscious diffusion of trees and flowers (ca. 1650–1750), the physiocratic project of great cultivation, taken up by Napoleons imperial gardens, created institutions that provided systematical reserves and collections, spreading botanical inputs. The lower world, consisting of small farmers or natives, continued to supply local species and varieties, often in large enough quantities for the development of Western high farming and functioned as a germplasm reserve for all future developments (Mooney 1983; 1985).

It is well known that even behind the collection of Kew Gardens, London, there were considerable economic interests linked to the diffusion of colonial products, such as rubber and coffee, outside their original acclimatization areas. Scientific research often perpetrated thefts in appropriating species to be grown on a commercial scale (Brockway 1979; Grove 1995). The same happens today as regards the privatization of genes, the base of everyone's natural heritage, which should be managed by indigenous social groups in the interest of local communities and future generations. This might be the only way of assessing in economic terms the function and work of conserving the natural environment.

The problem remains of how to pay for this function, whose social value is much greater than the simple economic remuneration of the seeds commercially produced. Today the conservation and multiplication of endangered germplasm still raises the question of how to sustain economically viable social groups (not only ethnic ones) carrying out socially useful functions (protection of the land, plant species and animal species; Colajanni 1997 [in this volume]).

References

Allen, Robert C.1992. *Enclosure and the Yeomen, 1450–1850,* Oxford: Oxford University Press.

Ambrosoli Mauro. [1992] 1997, *The Wild and the Sown,* Cambridge: CambridgeUniversity Press.

_____ 1998, Alberate imperiali per le vie d'Italia: la politica dei vegetali di Napoleone Bonaparte, *Quaderni Storici,* 99,3: 707–38.

Balbis, Giovan Battista, 1806, *Flora Taurinensis sive enunciatio plantarum circa Taurinis nascentium*, Turin: Giossi.

———— 1813, *Catalogus stirpium Hortensium Academiae Tauriniensis ad annum MDCCCXIII*, Turin: Academia.

Bauhin, Caspard. 1623, *Pinax teathri botanici*, Basle: Rex.

Bahuin, Jean. 1650–51, *Historia plantarum universalis*, Yverdon: Tipographia Caldoriana.

Berengo, M., ed.1975 [1567]. *Camillo Tarello. Ricordo di agricoltura*. Turin: Einaudi.

Braudel, Fernand. 1967, *Civilisation matérielle et capitalisme. Economie et capitalisme, XVIe–XVIIIe siècles*, Paris: Colin.

Brockway L. H., 1979, *Science and Colonial Expansion: the role of the British Royal Botanic Gardens*, New York [u.a.]: Acad. Press.

Colajanni, Antonino. 2004. How have we come to use Nature, from a Practical Point of View? [in this book: 283–86].

Crosby, Alfred W. 1965, *America, Russia, Hemp and Napoleon*, Columbus: Ohio State University Press.

———— 1972, *The Columbian Exchange: biological consequences of 1492*, Westport (Conn.): Greenwood.

———— 1986. *Ecological Imperialism. The biological expansion of Europe, 900–1900*. Cambridge: Cambridge University Press.

Drayton, Richard. 2000. *Nature's Government: science, Imperial Britain and the improvement of the world*. New Haven-London: Yale University Press.

Ellen, Roy F. 1986. What Balck Elk left unsaid: on the illusory images of Green primitivism. *Anthropology Today* 2 (6): 8–12.

Forneris, Giuliana and Pistarino Antonio. 1990. Note biografiche e attività scientifiche di Giovan Battista Balbis (1765–1831): opere, erbario e documentazione bibliografica. *Museologia scientifca* VII (3–4), pp. 201–57.

Goodman, David and Michael Redclift. 1991. *Refashioning Nature. Food, ecology and culture*. London: Routledge.

Goody, Jack. 1993. *The Culture of Flowers*. Cambridge: Cambridge University Press.

Grove, Richard H. 1995, *Green Imperialism: colonial expansion, tropical island Edens and the origins of environmentalism*, Cambridge: Cambridge University Press.

Mooney Patrick R. 1980. *Seeds of the Earth: a private or public resource?* Repr. with rev. Ottawa: Inter Pares.

———— 1983. *The Law of the Seed. Another development and plant genetic resources*. Uppsala: Dag Hammarskjöld Centre.

Overton, Mark. 1996. *Agricultural Revolution in England. The transformation of the agrarian economy 1500–1850*. Cambridge: Cambridge University Press.

Paroletti, Modesto V. 1812. Recherches sur les maladies qui ont affectées quelques éducations des vers à soie. *Annali Accademia Agricoltura*, IX.

Rackham, Oliver. 1986, *The History of the Countryside. The classic history of Britain's landscape, flora and fauna*, London: J. M. Dent.

Rapp, Robert T., 1976, *Industry and Decline in 17th century Venice*, Cambridge (Mass.): Harvard University Press.

Redford, Kent H., 1990, The ecologically noble savage. *Orion nature quarterly* IX. 3: 25–29.

Sauer, Carl O. [1952] 1969, *Seeds, Spades, Hearths and Herds. The domestication of animals and foodstuffs*, Cambridge (Mass.): MIT Press.

Sahlins, Marshall. 1972, *Stone Age Economics*. Chicago: Atherton.

Thirgood, Jack V. 1981. *Man and the Mediterranean Forest: a history of resource depletion*, London: Academic Press.

Thomas, Keith. 1983. *Man and the Natural World. Changing Attitudes in England 1500–1800*, London: Allen Lane.

Thompson, Edward P. 1977. *Whigs and Hunters: the origin of the Black Act*. Harmondsworth: Penguin Books.

De Vries, Jan and van der Woude Ad. 1999. *The First Modern Economy. Success, Failure, and Perseverance of the Dutch Economy, 1500–1815*, Cambridge: Cambridge University Press.

Woodward, D., ed. 1984, *The Farming and memorandum Books of Herny Best of Elmswell, 1642*, London: The British Academy by OUP.

CHAPTER 23

Diversity, Protection, and Conservation: Local Agricultural Products and Foodstuffs

Laurence Bérard and Philippe Marchenay[1]

In recent years there has been a growing interest in local and traditional agricultural products and foodstuffs within European societies. This is a different trend from that which has been developing for almost two decades in the related sphere of fruit and vegetable enthusiasts. Indeed, unlike the "historical and traditional" local varieties in a context of conservation essentially based on unpaid activities, the so-called *terroir* products in France have from the outset been in a market perspective. The phenomenon received a great boost in 1992 following EC directives on protecting product names through geographical areas and traditions. This ruling produced various effects at different levels and at the least can be said to have led to a great stir in the world of producers, who have suddenly found themselves faced with having to take into account a number of previously unconsidered criteria. Behind the simple appearance of protecting a name, the product itself is also to be protected through its historical grounding, production modes, and—if necessary—the plant varieties or animal breeds recognized as being part of the characterization of the product. Other players—usually not very active in the economy of this family of products—also enter the game, such as agricultural advisory chambers and agencies, local and national communities, farming organizations, promotion committees and associations, including research institutions.

In this context of protection several cultural issues arise at different levels: how can we evaluate the historical nature of a product, its geographical links, but also the technical practices to which it is subject? What status must be given to indigenous forms of knowledge and what role can they be seen as playing? What becomes of the inherent diversity in this group of products and what are the relations between protection, local knowledge, and conservation? These issues will be addressed by looking at some practical examples that we have chosen from dairy and meat products and animal and plant production, mainly in France.

Living Organic Things: Diversity and Knowledge

Local food products are situated in a complex world of relations involving the biological and the social. An animal race, a cultivated plant, and a product such as a sausage or cheese are the outcome of an accumulation of knowledge, practice, observation and adjustments that must be seen in relation to the way they are represented. In short, they are objects heavily invested with many processes and meanings. The fact they are living means they are continuously evolving and brings into play many factors. It authorizes all kinds of manipulation at all levels and on all scales. Reproduction modes are not the same for a plant, an animal or cheese. Even within the plant world, there are varying degrees of complexity concerning the plant, depending on its life duration, biological reproduction factors and multiplication techniques. In other words, a fruit species and a vegetable species are managed in different ways. Vegetative reproduction and sexual reproduction have different consequences on the plant, its uses and its place in society. Grafting allows identical varieties to be reproduced, whereas sowing the seed of an apple or pear tree involves genetic modifications; moreover, the influence of the physical environment (for example, soil or exposure) on the expression of a plant (phenotype) must be considered.

When we come to consider scale, a new level of complexity is reached. Thus micro-organisms, which may be described as "invisible living beings," are vital in making cheeses, as in certain kinds of beverages or prepared meat products. These elements—ferments, yeasts and fungi—play a key role in the processes of fermentation, maturation and ripening, which may or may not be associated with technical practices—sometimes complex—or with various levels of intervention on living matter.

Thus the living involves a considerable potential for evolution and variability which people may draw on. But it is also associated with a very short life span. The various forms of knowledge shaping the living are the basis of the existence and *a fortiori* the conservation of the perishable ephemeral matter that must constantly be maintained and renewed to ensure the repeatability of production systems.

This cultural and biological diversity is found at various levels of local agricultural food products. We find an initial general diversity in the field concerning fruits, vegetables, cereals, animal breeds, and local products such as cheese, prepared meats, and fermented beverages. Several existing or ongoing inventories in France and various southern European countries highlight the great wealth still to be found and at times with unexpected features. Within this overall diversity, we find local variations intrinsic to the products which have a second level of diversity: a wide range of varieties (various cultivars) or even a population variety within the same production system. Such variants may even be numerous and closely circumscribed for cheese, meat products or equally be due to the scale of production, ranging from small craft operations to intermediate farm output and industrial level.

Some productions such as carp rearing in Dombes[2] or mountain cheese making in alpine pastures, called "Alpages" in French, are based on complex systems generating a specific landscape and resulting in distinct levels of biodiversity. In the first case the farming techniques involve a cyclical rejuvenation of the ecosystem, and thus the production of animal and plant biodiversity. In the second case, the techniques employed influence the flora through the management of the mountain fields, the choice of local bovine race, the micro-flora specifically associated with cheese-making techniques and with the places in which they ripen.

This abundant polymorphous biodiversity rests on knowledge and practice that directly concern the management of the living.

Protection and the Society of Merchants and Traders

The misappropriation and use of a place name, production method or historical feature in order to promote a product is considered unfair competition for producers making efforts to respect given rules. It is also misleading for the consumer. The advent of a European market with open borders providing the opportunity for producers from different EC countries to freely sell their commodities raises the problem of the future of "specific quality" products and more generally of products particular to each country.

This is the general background to two European Council regulations of 14 July 1992. The first concerns the protection of *geographical indications and designations of origin for agricultural products and foodstuffs* (no. 2081/92), while the second certifies the specific character for *agricultural products and foodstuffs (no. 2082/92)*. The objective is to provide a legal framework in which to establish and protect the relation between a product and a place or tradition.[3]

> The Protected Designation of Origin (PDO) and the Protected Geographical Indication (PGI) guarantee the protection of a connection with a geographical area. They both designate the name of a region, a specific place or, in exceptional cases a country, used to describe an agricultural product or foodstuff.
>
> In the case of the PDO, the given product or foodstuff originates in a specific region, place or country, and its quality or characteristics are essentially or exclusively related to a particular geographical environment with its inherent natural and human factors, and the production, processing and preparation take place in the defined geographical area.
>
> In the case of the PGI, the given product or foodstuff originates in a specific region, place or country and it possesses a specific quality, reputation or other characteristics attributable to that geographical origin and the production and/or processing and/or the preparation take place in the defined area.
>
> The Certificate of Specific Character (CSC) also called today Traditional Speciality Guaranteed (TSG) protects a tradition and does not refer to the origin. Specific Character is defined as a feature or set of features that distinguishes an agricultural product or a foodstuff clearly from other similar products or foodstuffs belonging to the same category. To register for a Certificate of Specific Character an agricultural product or foodstuff must either be produced using traditional raw materials, or be characterized by a traditional composition or a mode of production and/or processing reflecting a traditional type of production and/or processing. The name must be specific in itself or express the specific character of the product (see EC Regulation 1992).
>
> The European PDO was strongly influenced by the French *Appellation d'origine contrôlée* (AOC). Obtaining the PDO requires prior attribution of the AOC in France or the equivalent in other countries where this kind of protection exists.

This regulation of protection implies that the European Community countries are bound to a normative framework concerning the free circulation of goods. Control structures must check if products bearing a protected designation comply with the criteria in the specifications."As from 1 January 1998, in order to be approved by the Member States for the purposes of this Regulation, private bodies must fulfill the requirements laid down in standard EN 45011 of 26 June 1989"(EC Regulation 2081/92). This standard establishes the general criteria for organizations proceeding with the certification of products in all spheres of economic activity, from industry to health and security.

The specifications for a Protected Designation of Origin (PDO) or a Protected Geographical Indication (PGI) must include: the name of the product; a description including the raw materials, if appropriate, and principal physical, chemical, microbiological and/or organoleptic characteristics; the definition of the geographical area; evidence that the product originates in the geographical area; a description of the method of obtaining the product and if appropriate, the authentic and unvarying local methods; the details bearing out the link with the geographical environment or the geographical origin; details of the inspection structures.

The Certificate of Specific Character must include the name and elements on which to base the assessment of the traditional character of the product, the description of the production method and the minimal requirements in the control procedures of specificity.

Establishing the Specifications: Making Choices and Reducing Diversity

When producers choose to follow a protection process, they must work out together—it is a collective procedure—the specifications, describing in detail the various manufacturing phases of their product. Diversity is thus inevitably reduced, since one practice, one production method, and one ingredient must be selected rather than another. In France the principle of cheeses made from a mixture of milks (cow and goat milk, for example), is rejected but this practice was very common in the rural world and reflected the seasonal management of milk production of various dairy animals. Depending on the time of year, the percentage of milk from goats, cows and sheep, could vary to greater or lesser degrees. This variability is thus considered to be unacceptable, because it cannot be controlled. But according to producers, the practice of including a percentage of cow milk in certain goat cheeses, such as *charollais*, makes production easier and gives the paste a special texture. Incidentally, Italian legislation on DOC cheeses (*Denominazione di origine controllata*) allows for mixed milk.

So which variant must be protected to the detriment of the others, when there are several different production techniques? The rind sausage (*saucisson de couenne*), which is very popular in southeast France and particularly in Dauphiné, is characterized by its variants and the ingredients used (pork rind, head, and fat) and by the percentage of rind, which can vary from twenty to ninety per cent. The distinct geographically circumscribed designations (*andouille, gueuse, murson, saucisse de couenne*) are based on the criteria of well identified techniques and the compositions of the various sausages.

There may also be a second level of variants in the same designation. Thus the *murçon* from La Mure has a number of micro-variants in a restricted area.[4] Given this multifaceted situation, how should the criteria be ordered in protection procedures?

When the producers of *picodon*, a small goat cheese from Drôme and Ardèche, applied for AOC status, the product was mainly made from lactic curds technology.[5] But in one part of Ardèche, it was traditionally made from rennet curds technology. The lactic curd method was chosen and all *picodon* producers had to follow this procedure.

The scale of manufacture also often raises problems. *Rosette* sausage[6] producers are currently discussing whether to adopt geographical protection through a PGI, but production methods between the small artisan producers and big industry differ. Some of these methods have only a slight impact on the integrity of the product while others are more significant and establish a difference. This is the case with the kind of gut used, which influences the ripening process. In small-scale artisan production, the *fuseau*—also called the *rosette*—is used. This natural gut requires a good deal of preparation but the thickness of its skin provides an incomparable finish with direct repercussions—appreciated by connoisseurs—on the flavor. Industrial preparation uses natural reglued gut, thus eliminating the long preparation work and making ripening much faster. On what specifications basis will the producers be able to agree, bearing in mind that industrial production accounts for ninety percent of output?

With products in this category, the question of the options to be selected in order to formalize production methods frequently recurs. Indeed, in the protection scenarios, the situations are often very similar: the specifications impose choices on the grounds of manufacturing techniques concerning both the know-how and the composition of products. These elements initially depended on vernacular knowledge. Controversy often takes place to the background of different conceptions—linked to the technical-economic organization of enterprise—on what the final products should be.

The Status of Knowledge

Technical-scientific knowledge and vernacular knowledge may coexist fairly peacefully, but in certain conditions they may clash, as may be seen when protective measures are introduced.

Cheese productions benefiting from protection come under a system of support and control. The technical advisors appointed to follow them are often ill prepared and analyze practice and knowledge according to their own background, placing the emphasis on "nil defect" and hygiene. Among the features of the Haute-Savoie cheese called *abondance*[7] is a slightly bitter taste due to the draining method. According to the dairy advisor questioned on the subject, this taste is due to a production fault and is not considered to be part of the product's specific character.

In Dombes fish farming is based on very extensive management of ponds. The agricultural fish farming system as it functions today is often perceived as poorly controlled, based on a system of "gathering." This is obviously a pejorative term in the context, downgrading for the producers and reflecting the absence of rational management and expertise about the processes in fish rearing.

In Italy the traditional cherry producers in the Vignola area in Emilia Romagna are said to resist innovation. They are indeed recalcitrant to the introduction of forms of knowledge and practical techniques that change the very foundations of their activity and alter the specific character of their product. This is the case, for example, concerning the introduction of new stocks for grafting and new varieties, and the development of smaller trees implying major changes in management, especially as regards the harvest. The landscape dimension is also of importance here. Thus the empirical knowledge intimately bound up with the very identity of the products is considered a hindrance by the agent responsible for introducing innovations to the orchards and enhancing the product identity.

In France the PGI has given rise to highly controlled application procedures.[8] In this context, producers are commonly reluctant to add their own techniques and knowledge to the specifications, thereby exhibiting a tendency to self-censure; they consider the knowledge with which they are most familiar to be self-evident and therefore out of place in the technical specifications. Instead, they favor a presentation of the history of their product, often using general documents that highlight their fame and historical underpinnings. Thus, it is common for the particular knowledge and expertise that is the basis of the identity and specificity of the product to be left un-stated or to be understated in a selective or watered-down manner. This, moreover, often leads to application reviewers deciding that producer applications are not well grounded! This type of misunderstanding clearly reveals the gap between the two worlds of knowledge.

Another issue is the evaluation of these different kinds of knowledge. Who is really competent to examine the content of the applications for protection? Local products, local varieties are almost unknown, as are the practices and indigenous knowledge associated with them. Over the years the information is lost, because it is not passed through the transmission channels that would have ensured its survival. Moreover, this kind of knowledge was consciously separated from the ordinary teaching circuits, since it was considered as obsolete and of no use in progressive agriculture. There has, therefore, been a paradoxical turnaround; the expert—generally a scientist—is now called upon to focus on skills that for a long time were pushed into the background by official science.

Impossible Reference Models

When the aim is to protect localized traditional knowledge and practice, which by definition are part of a local evolving culture, what reference model must be chosen? Which criteria should be selected to "officialize" that a product is what it claims to be? But what exactly should it be today and how should it be related in terms of technical circumstances and past versions? Contemporary preoccupations concern objects whose production, conservation and consumption contexts have generally completely changed. Who can see clearly? Does an original model still exist?

Let us take the case of taste and its evolution, which must be related to conservation restraints linked to delayed consumption. The *picodon* cheese, for example, was only eaten fresh (i.e., after about a dozen days ripening) at the end of spring, when there was plenty of milk. Most of the production was dried and then stored in a cool place for several months to be consumed at a later date. As required, the cheeses are "started up again" by washing them several times in water with or without some vinegar or alcohol. This washing over several days removes the blue mould from the cheese, softens them and restores the taste, giving them their characteristic "piquant" flavor, conjured up by their name. At present described as the "Dieulefit method," this procedure is included in the specifications, but actually only concerns a very small part of the output.

In general we can observe in many places that the vestiges of a great diversity of practice still have consequences for the variety of taste, identified and sought after by the local consumer. Going back to the example of the murson sausage, the slightly burnt flavor due to the *buclage*,[9] the crackling of the pig's ear and the marrowy part of the rind are tastes and textures that are sought after or rejected according to local manufacturing and consumption practices. But what is the reference version? Moreover, these organoleptic preferences belong to a disappearing rural culture. Preparation methods are often subject to change, so which stage of adaptation should we consider as introducing a "product drift"? Are not connoisseur–consumers still the safest guarantor of a "reference model"?

Similar questions may be raised for *pélardon*, a small goat cheese whose production stretches from the Cévennes to Corbières in southern France. The traditional production method is based on rennet curds, also called "soft curds" (*caillé doux* in French). Since the decline of the 1960s, neocountry people have come to live in this region. They have learned to make this cheese, but following techniques taught in training courses. That is why almost all goat cheese made today uses lactic curds, and this technique has been adopted by the specifications for AOC applications. This raises the very practical question of the relation between the status and distribution of forms of knowledge and the reference model. In the present example, it involves confrontation between local and neolocal players on one hand, as well as agricultural advisers and producers on the other hand.

Relations between Usage and Designation
of Names of Products and Places

The reservation of a place or product name in European protection regulations may lead to debatable exclusions. If the current logic pursued by the players is successful, only the producers who respect the specifications and the control procedures will have the right to sell the Lyon rosette. But such procedures are costly and the small producers—those best placed to guarantee authentic products—are in danger of being deprived of the right to market their products under this name. Commenting on this situation, the big producers reply that in any case the small producers market on a local scale and sell their products to people who already know them. Nonetheless, the genuine problem of collective intellectual property rights persists.

The protection regulations generate various forms of misappropriation. In Ardèche, the local *picodon* producers making the cheese according to "their" method of rennet curds technology, had to "debaptise" their product, when faced with the "legal" lactic curds technology. As for *pélardon*, we still do not know what will happen to the small producers who, in the original cradle of cheese, Les Cévennes, developed their products following the traditional method of rennet curds.

Another question is raised: that of employing geographical names where the regulations are tending to become much stricter. At present there is a debate in France on the use of geographical names: the idea is to prevent producers from using a geographical name for their products, unless they belong to the framework of existing protection procedures.

More generally, legal protection of a name raises many questions concerning tradition, geographical origin, control procedures connected to quality and hygiene. Here the debate brings together both cultural and commercial aspects.

Local Varieties and Breeds

What happens to local varieties, when they are associated with traditional products for which protection is being sought? Several scenarios are possible. When the crop is only one variety, that variety will be designated in the specifications, as, for example, is the case with the Nyons olive, a French AOC, and the *tanche* variety. This might also be the case with many local products subject to European protection.

However, in some so-called traditional systems we often find, especially in fruit production, diversity in terms of varieties. This is the case of the Vignola cherry production, the Ardèche chestnuts, or Norman orchards for cider or perry production.[10] Other plant productions are based genetically on population variety: this is the case for example, with the Espelette pimiento in the Basque Country currently the subject of an AOC application. We are thus dealing with a genetic pool that can vary starting from a recognized local type that is not, however, stabilized. The *coco de Paimpol* bean, in Brittany, another AOC applicant, is based on a local variety type that may vary from one producer to another.

Two levels must therefore be distinguished: on the one hand, the diversity of varieties in a single production system, and on the other, variations within the same variety (population variety).

In all application forms for geographical protection involving plant resources like fruits or vegetables, the priority lies in identifying the genetic material. The procedure is usually associated with the work of observation and selection—i.e., of introducing improvements—required for often heterogeneous and uneven plant material with different disease resistance, and often virus-infected fruit varieties with a certain number of flaws, which, according to the experts, prevent them from qualifying and being used. This happened, for example, to the Puy lentil, and the Moissac grapefruit—both with *Appellation d'origine contrôlée* status in France. Moreover, the sale of seeds is subject to the prior registering of the varieties in the *Official Catalogue of Species and Varieties*, involving the evaluation of the criteria of distinction, homogeneity, and stability (DHS). This procedure takes about two years. Although suitable for the creation of new varieties, this measure runs into problems when dealing with local types or varieties, which by definition, vary within a set. If a continuous connection to a geographical location and a traditional character accorded to the variety type is to be maintained, then it must be developed beginning from this specific character. It is not always easy to establish a formal link between cultivars underlying the product and the compulsory registration system with the Catalogue. One solution may be to authorize—provided good hygiene conditions are applied—this margin of variation within the genetic limits to be established. The question is currently being debated in France because it was raised by the cases of products currently applying for protection: Espelette pimiento, Paimpol coco bean, Cévennes sweet onion and others. Each application is a special case. For example, what is to be done for the Espelette pimiento or the Cévennes sweet onion? Should a local genetic pool be maintained, or some varieties selected from the current population to be made the only authorized marketable variety? Let the producers themselves make the selection? As regards the Paimpol coco beans, the use of the local type has always continued thanks to conservation selection work begun from the best farm varieties. Nonetheless, the producers may use two other varieties, the result of improvement programs, whose agronomic and genetic features resemble very closely the original product but are much more resistant to the usual bean diseases. It seems that the INAO policy is to leave local reproduction to take place within limits acceptable to all sides.

Clearly, local varieties cannot be treated in the same way as new varieties. The key question is knowing if there is a more or less major loss for the genetic pool, linked to establishing the "stabilized" variety assortment or the different forms within the same variety when drafting the specifications.

Given various possible situations, all considerations on this subject must take into account two primary factors: the biological reproduction mode of a plant and its life duration, which in terms of multiplication and production of seeds involves very different technical restraints according to the specific case. Autogamous plants will not be treated in the same way as allogamous, and perennials will not be accorded the same treatment as annuals or biennials. The olive tree is autogamous.[11] It lives to an old age and is propagated by cuttings or grafting: its stability over time is guaranteed and non-problematic. Pimiento and beans are autogamous annuals, they multiply by seeds, and their stability is relatively good. The onion is allogamous and reproduces by seed, and is thus susceptible to the dangers of genetic drift. Other species must be considered in a different light. Potatoes, strawberries or garlic multiply by vegetative reproduction. Their varieties change regularly and require very specialized multiplication work, bearing in mind the related disease risks.

This principle of variety assortment or population variety should be seriously taken into account for each situation, thus creating the conditions for a difficult compromise between maintaining recognized specific features and innovation, circulation, trade and seed control. In this case not only the agrogenetic features of the plants must be considered, but also the local knowledge and practice concerning plant selection, multiplication and reproduction: all these elements are an integral part of the specific character of the final products and must be kept *in situ*. This concept fits in well with the protected designation of origin (PDO) framework, in that both the quality and the features of the product are basically or exclusively due to the geographical environment embracing the natural and human factors and where the production, processing, and preparation take place in a limited geographical area.

There has been talk for some time now of a protected designation of origin for sweet chestnuts in Ardèche. But the laborious procedure has run into many problems, revealing, among other things, how difficult it is to envisage protecting a common plant heritage. How do we deal with the question of local varieties? Given the abundance and diversity of the varieties, which should be chosen and which excluded? Thus while variety assortment made sense in traditional Ardèche society, it is no longer suitable for this context, mainly because the functions of chestnut production have changed. Should local varieties be treated as a living heritage to be preserved and enhanced or as a development potential for orchards engaging in intensive "rational" management? This question is even more interesting since researchers at the French National Institute of Agronomic Research (INRA) are currently working on new varieties corresponding to specific commercial criteria (fruit size, early maturing) which may compete with local varieties, as the vehement arguments of some producers suggest. The creation of a regional chestnut park and the desire to develop a potentially very profitable local resource makes matters even more complicated, since this emblematic tree and fruit are the focus for various expectations.

There are fewer cases as regards local animal breeds: most of them that are not extinct are only present in very small numbers. Moreover, the questions of selection and reproduction are much more complex and the danger of genetic drift omnipresent. The "strain" is often in greater demand than the breed, notably for protected geographical indications, especially in poultry production. Yet certain products may contribute to safeguarding breeds in the framework of geographical protection. Cheese productions encourage this kind of operation: the Abondance and Tarentaise breeds for the abondance and beaufort cheeses—both AOC/PDO—in the French Alps, the Piemontese breed for PDO raschera cheese in the Piemonte region of Italy. Here the breeds have adapted to the pedoclimatic constraints of the mountain region and their milk is very suitable for cheese. Some meats may also contribute to the survival of local breeds in areas where they are still well integrated into the agricultural economy. This is the case of the Bisara pig, which appears in the PGI specifications for two smoked sausages from Trás-os-Montes in Portugal (the Vinhais dry *salpicão* and the Vinhais *chourica de carne or linguica*).

Conclusion

Regulation plays an important role in maintaining the typical features of local agricultural food products and in encouraging their development. Its application is federative and brings coherence in production systems. Left to market laws, these kinds of products would be even more vulnerable. Yet the protection procedures must not contribute to limiting the present diversity by simplifying, stabilizing, formalizing and freezing—all these terms being antinomic to the very notion of diversity.

We might imagine, for many of the local and traditional products, a solution consisting of "leaving things as they are," but the commercial logic underlying our society also plays an influence through hygiene rules that tend to remove any possibility for them of surviving economically, since when they are sold they must respect the health regulations in force.

At the same time, protection, especially AOC/PDO, brings awareness of the specific character of products and in certain cases leads to derogations, linked to the insistence of producers in defending the merits of their methods. Here we might cite the example of using wood in ripening rooms, involving repercussions on the microbiodiversity allowed by this material. In the longer run the ever-present threat of the prohibition of selling raw milk cheeses on an international scale might be warded off thanks to the work of associations defending AOC/PDO cheeses in raising awareness. Is this a question of sacrificing a little to save the essential?

As regards local varieties, the idea of linking variety assembling or population variety to the product is gaining ground in the framework of geographical protection procedures. It means that what constitutes the typical character of production is conserved, but also helps in ensuring the survival of a form of biodiversity. Much less frequent, the procedure integrating local animal breeds is also potentially very interesting, although the erosion has been much more radical than for domestic plants.

By generating new technical, biological, and cultural references and bringing new players onto the scene, regulation has been overlaid on the existing complexity. How and which players can best protect products in this context of variability and diversity? Does this type of protection not primarily concern products already strongly driven by a commercial logic? Given the plethora of small local products, would it not be better to study health rules suitable for them? This would certainly be a sure means of enabling them to survive.

Bourg-en-bresse, 1998

Notes

1. Ressources des terroirs—Cultures, usages, sociétés. Centre national de la recherche scientifique. Eco-anthropologie, FRE 2323. Alimentec, rue Henri de Boissieu, F 01060—Bourg-en-Bresse cedex 09. tel.: 33 (0)4 74 45 52 07, Fax: 33 (0)4 74 45 52 06, email: philippe.marchenay@ethno-terroirs.cnrs.fr, laurence.berard@ethno-terroirs.cnrs.fr. This article was written in 1998; many changes have occurred since then, particularly concerning the protection of geographic origin.
2. Fish breeding in Dombes (department of l'Ain) is based on a original system of exploitation. Fish breeding—especially carp—in water is rotated with cereal crops on dried up pond beds. Each year the ponds are stocked with young fish, emptied, and then fished the following autumn. Every four or five years on average, they are fished in spring and left to dry up for a summer so that they can be used for crops of maize or oats, thus creating a changing countryside.
3. There is also another EC regulation (no. 2078/92 of 30 June 1992) making provisions for methods of farm production compatible with the needs to protect the environment and maintain ecosystems. Oriented towards natural environments, sensitive biotopes and wild species, this also increases in some cases the maintenance of forms of domestic biodiversity: animal breeds, cultivars or local products.
4. The murçon is produced at La Mure on the Matheysine plateau in Isère. But in this case the name has a "ç" and not an "s." The sausage-makers in this area insist very strongly on this spelling, thus distinguishing the product, even though each producer actually has his own recipe.
5. With lactic curds, the aim is acidification, while with rennet curds, the action of the rennet is more important. These technical factors influence the texture and taste of the cheese.
6. A very long big dry sausage (from 70 cm to 1 m), the *rosette* is named after the gut in which it is stuffed: the last section of the pig intestine ending in the rectum. The product requires long maturation. The truncated cone shape of the intestine gives the sausage the look of a bludgeon, while its flavor clearly comes from long maturation. Initially, it was a speciality of the greater Lyon region, which then lent its name to the *rosette de Lyon*.
7. A half cooked pressed paste of unskimmed raw cow milk produced in Haute-Savoie, but originally made in the Val d'Abondance in the upper Chablais.
8. The French government decided to organize the rules intended to protect the link to a place or traditional production method using pre-existing signs. Thus the application for a PGI or CSC depends on the prior acquisition of a label or conformity certificate. Moreover, all "red labels" (Labels rouges) mentioning an origin must now have made a prior application for a PGI.
9. The *buclage* of the pig consists in burning the bristles, usually with straw, immediately after the animal dies. As a traditional practice, it is an alternative to scalding, carried out with boiling water before shaving off the bristles.
10. There are many reasons for this diversity in variety. Different habits, safer harvests, differentiated uses of varieties, staggered production, etc. In Normandy, for example, the specific organoleptic character of ciders and perries is due to a deliberate choice of varieties with complementary required features (sweet, acidic or bitter apples).
11. In autogamous plants, fertilization usually takes place between gametes produced by the same individual. Conversely, allogamy involves natural cross fertilisation, usually between different individuals.

References

Bérard, L. and P. Marchenay. 1994. Ressources des terroirs et diversité bioculturelle; perspectives de recherche. *Journal d'agriculture traditionnelle et de botanique appliquée*, No. spécial "Diversité culturelle, diversité biologique," XXXVI (2): 87–91.

_____ 1995. Lieux, temps et preuve: la construction sociale des produits de terroir. *Terrain*, 24: 153–64.

_____ 1998. Les procédures de patrimonialisation du vivant et leurs conséquences. In *Patrimoine et modernité*, edited by D. POULOT, pp. 159–70. Paris: L'Harmattan.

Bessy, C. and F. Chateaureynaud. 1995. Economie de la perception et qualité des produits. L'exemple des contrefaçons dans le domaine agro-alimentaire. *Cahiers d'économie et sociologie rurales*, 37: 178–99.

Casabianca, F. and E. Valceschini, eds. 1996. *La qualité dans l'agro-alimentaire: émergence d'un champ* de *recherches*, AIP "Construction sociale de la qualité," Paris: INRA.

Chevallier, D., ed. 1991. *Savoir* faire et *pouvoir transmettre. Transmission et apprentissage des savoir-faire et des techniques*, Paris: Editions de la Maison des Sciences de l'Homme.

Communauté Economique Européenne. Règlement (CEE) No. 2081/92 du Conseil du 14 juillet 1992 relatif à la protection des indications géographiques et des appellations d'origine des produits agricoles et des denrées alimentaires. Journal officiel des Communautés européennes, No. L 208/1 et Règlement (CEE) No. 2082/92 du Conseil du 1992 relatif aux attestations de spécificité des produits agricoles et des denrées alimentaires. Journal officiel des Communautés européennes, No. L 208/9.

Creyssel, P. 1991. Agro-alimentaire. Pour une stratégie de normalisation. *Enjeux*, 113: 1–88.

Delbos, G. and P. Jorion. 1984. *La transmission des savoirs*. Paris: Editions de la Maison des Sciences de l'Homme.

Delfosse, C. 1997. Nom de pays et produits de terroir. Enjeux des dénominations géographiques. *Espace géographique*, 26 (3): 220–30.

Goody, J. 1977. *La raison graphique. La domestication de la pensée sauvage*. Paris: Les Editions de Minuit.

_____ 1986. *La logique de l'écriture. Aux origines des sociétés humaines*, Paris: Armand Colin.

Jolivet, G. 1993. Identification de la qualité. Perspectives nationales et européennes. In *Agricultures et sociétés*, edited by Courtet, Berlan-Darqué, Demarne, pp. 168–75. Paris: INRA.

Lenclud, G. 1994. Qu'est-ce que la tradition ? In *Transcrire les mythologies. Tradition, écriture, historicité*, edited by M. Detienne, pp. 25–44. Paris: Albin Michel.

Letablier, M.-T. 1997. *L'art et la matière. Savoirs et ressources locales dans les productions spécifiques*. Noisy-le-Grand: Centre d'études de l'emploi, dossier 11.

Marchenay, P. 1995. Les variétés locales dans le contexte agroalimentaire actuel; mythes et réalités. In *Voyage en alimentation*, edited by N. Eizner, pp. 263–74. Paris: ARF éditions.

Nicolas, F. and E. Valceschini, eds. 1995. *Agro-alimentaire: une économie de la qualité*. Paris: INRA/Economica.

Savoir-Faire. 1991. *Terrain*, no. 16.

Sylvander, B. 1995. Origine géographique et qualité des produits: une approche économique. *Revue de droit rural*, 237: 465–73.

CHAPTER 24

Cultural Research on the Origin and Maintenance of Agricultural Diversity

Stephen Brush

The puzzle of the diversity of life has given rise to numerous domains of human knowledge. In Western science alone, natural history, ethnobiology, evolutionary biology, systematics, and molecular biology address this theme. A more specific discourse has examined the diversity of crops (Darwin 1868; Conklin 1957). Throughout, tension between holism and reductionism has characterized scientific and policy debates. Reductionism is perhaps inevitable, a compelling trend in both theoretical and applied sciences; and one that has dominated research on the origin and nature of crop diversity. This paper explores the limits of reductionism with respect to a current scientific narrative of crop diversity—the idea of genetic erosion.

Conservation Biology

The applied-science conservation biology has emerged at the millennium from our anxiety of an impending collapse in the Earth's biosphere, the menace of extinction (Soulé 1986). Whereas Darwin posed the origin of species as his central question, conservation biology poses the survival of species. Conservation biology is grounded in the theory and mathematics of Island Biogeography (McArthur & Wilson 1967), and it has flourished with the effective use of key symbols and narratives, and familiar fare in popular culture, derived from the structural or master narratives in Western consciousness. The symbols are the "charismatic megafauna" that have represented untamed nature for centuries—wolves, tigers, leviathans. The narratives include stability and diversity, the balance of nature, and the human descent from nature. Conservation biology's vision is patrician, intended to affect public policy and behavior, fashioned of general models and state variables—diversity, energy, ecosystems. In both symbol and narrative, the reductionism of conservation biology aims to save species' diversity by salvaging key fragments of wilderness. The intent of conservation biology is to save a domain for nature so that it can reconquer the Earth's surface if and when human disturbance ceases, whether this be in a century, a millennium, or longer.

Unfortunately social science finds predicting the distant future unfashionable, discredited, and implausible. Thus, conservation biology has no ready-made social science to turn to for estimating the future of most "state variables" that threaten the Earth's biosphere—human population growth, land conversion, the extinction rates of species caught in the path of human expansion, and the emission of climate destabilizing pollutants. Social science potentially has much to offer in understanding all of these variables, but our voice is faint in discourses of biosphere modeling or ecopolitics, diminished not only by disciplinary prejudices but also by uncertainty within social science. Social science's own key narratives, for instance liberal progress or domination and resistance, are contradictory. Even population, a well studied state variable, has widely divergent estimates of when and at what size it will cease growing (Harris 1996). For the other state variables, such as land clearing and use, social science is rather like a primitive form of weather forecasting, informing us that tomorrow's weather will be similar to that of today. The best that social science can say about the near future is that it will be similar to the recent past. We are, therefore, still bereft of ways to address the true time horizon of conservation.

The loss of genetic resources of crops illustrates some of the challenges and conflicts of alloying social science and conservation. This case illustrates the limits of logical models that begin with large state variables but are ultimately determined by human agency and local conditions. Resistance, resiliency, inertia and obstinacy in human affairs are found everywhere, and what happens at the local level is often a chaotic contradiction of general predictions. Local exceptions have long provided grist for anthropologists as critics positioned on the periphery, in opposition to the other disciplines' narratives. However, agnosticism in the face of scientific orthodoxy raises disturbing questions for our discipline. What are we to make of contradictions between global processes and village life? Have we looked in the wrong places or used improper methods? Does local detail obscure larger and longer-term reality?

Genetic Erosion of Crop Resources

Shortly after the rise of modern genetics and the seed industry, crop breeders recognized that their success could well destroy the natural resource base of the industry—the vast stores of genetic material in landraces of crops kept by peasant farmers in centers of crop evolution and diversity. By the 1960s, a narrative of destruction of local crop diversity by global processes—population, technology, commercialization—was firmly established in both crop science and international policy (Frankel and Bennett 1970). This narrative of genetic erosion has two central parts—the production of superior general technology by science and industry and the domination of market relations in all production systems. The rapid diffusion of hybrid crops in the United States and Europe was an early harbinger of a worldwide replacement of local crops. The "Green Revolution" was proof that industrial seeds would bring similar changes to the heartland of genetic diversity in the tropics and subtropics. The response was to collect the genetic resources and store them as "world collections" in gene banks at international agricultural research centers. The narrative of genetic erosion was compelling for several reasons. It evoked the familiar Enlightenment idea of the destruction of nature flowing in the wake of human development. The narrative vouched for the success of agricultural development in less developed countries and rationalized the appropriation of the world's genetic resources into collections controlled by industrial nations.

Anthropologists should quickly perceive a fundamental flaw in the narrative of genetic erosion because it is grounded in an essentialist definition of behavior and culture with a fixed attachment between tradition and crop diversity. The news of the Green Revolution and genetic erosion reached me in Peru, a novice anthropologist researching human ecology and agricultural adaptation in an Andean valley, in the epicenter of genetic diversity of potatoes. Although few industrial inputs were used in the valley, the first Peruvian equivalent of "miracle seed," the *renacimiento* ("renaissance") variety, was found in many of the potato fields, testifying to the Andean farmers' constant vigil for new seed and to the prowess of the *renacimiento* variety. While tendrils of the Green Revolution had established themselves in the village, the impacts seemed neither revolutionary nor erosive. The *renacimiento* variety was just one of many in farmers' inventories of potatoes, and it was not perceived by them as something extraordinary. Here was a local contradiction of the narrative of the Green Revolution: industrial technology in the form of seed was absorbed into the local production system and managed as a local component rather than as the transformative element envisioned by the architects of the Green Revolution.

The fact that farmers in Uchucmarca kept their local potato varieties could not, however, dispel the conclusion that genetic erosion was imminent or inevitable. The hobgoblins of village studies immediately raised up—a single, qualitative observation cannot challenge a narrative of a global process. Anthropologists, looking for the exotic and indigenous, have little trouble finding places where agricultural technology, including crop variety inventories, remains "traditional." However, the concept of genetic erosion needs testing in places experiencing technological change, with new seeds, industrial inputs, and commercial production. I attempted to overcome issues of site specificity and qualitative data by collecting survey data from numerous households and from widely dispersed sites, carried out where new seeds and commercial production were prominent. Nonetheless, my earlier observations were confirmed. Improved varieties easily root themselves in peasant production without displacing local varieties or dramatically reducing their diversity.

While the case of potatoes in Peru challenged the received wisdom about genetic erosion, a tuber crop in a high mountain environment is, perhaps, atypical and too exotic to challenge the narrative. Further research in Peru would not satisfy these challenges, so I initiated research on radically different crops and in different parts of the world. Research in Mexico and Turkey confirmed what I observed in my initial human ecology of Andean agriculture—processes such as technology adoption and commercialization are mediated and transformed at the local level in such a way as to contradict the predictions of the narrative of genetic erosion.

Studies of potatoes, maize, and wheat in centers of domestication and diversity suggest three distinct explanations for the persistence of biological diversity on farms. First, on-farm diversity survives because of environmental advantages of different types of crops and varieties. Thus, one variety or class of varieties will do well at higher altitudes and another in more favorable places. Second, diversity survives because it lowers the risk of crop failure to the household, by providing a form of biological insurance against pests, pathogens, or unfortunate weather. Third, diversity survives because it has a special cultural value, because local varieties, with their aura of social meanings, make good gifts, or because they are prized for taste and quality.

Behaviorism

Challenging the narrative of genetic erosion proved costly to my anthropology. As the genetic erosion narrative pertains to process, so too must the challenge. Ethnobiology initially informed my research, but attention to the structure of lexicons was replaced by behavioral variables of peasant household economy. Culture, an ethereal phenomenon at best, became a residual category in quantitative and ecological analysis of crop variety selection. Confronting a historical process without historical data, avoiding the bias of a single site, and acquiring quantitative data drew me inexorably away from the holistic, descriptive terrain where anthropologists are most comfortable. The most profound problem was approaching a historical process (genetic erosion) with data from a single point in time. Cross sectional analysis aims to surmount this problem by reducing processes (e.g. commercialization or technology adoption) into variables that can be contrasted across households, in relation to a dependent variable (e.g. on-farm diversity). Nevertheless, even our best attempts to capture evolutionary processes in cross-sectional analysis are rather flimsy efforts to contain the chaos of history. The limits of this analysis are reflected in the modest amounts of statistical variability that can be correlated with independent variables. Yet, the absolute lack of critical historical data, either biological or social, afforded no alternative to cross sectional research.

Focus on the household involves its own form of reductionism. Other social units—villages, markets, classes, cultures—play significant roles in provisioning and valuing seed, but their force is mediated through the household. The household is a surrogate to the individual in behavioral analyses of farming systems—a theoretically free agent rationally choosing among alternative production strategies to affect the diversity of crops. Structural and cultural factors—markets, values, class, tastes, community—are only observed through the choices of households.

Three explanations for the persistence of diversity—environmental advantage, risk management, and cultural value—emerged from research on household variety choice. In the behavioral models, these are motivations or utilities deduced from regression models. Anthropologists should find these three explanations to be logically linked into the alloy of ideas, strategies, and behaviors included in culture and social structure, but quantitative behavioral research finds this alloy both methodologically and theoretically cumbersome. It is unlikely that environment, risk, and culture are equally weighted in the collective experience of peasants. Applying Occam's razor, one should logically emerge as pre-eminent. Nevertheless, we continue to show that environment, risk, and culture are synergistic and cannot be ranked.

Stage-wise regression modeling helped to manage the alloy of three distinct explanations, but the loss of ethnographic texture and moment was an outcome of this research strategy. Ironically, success in challenging the narrative has led me back to a problem where anthropological theory and methods are particularly salient. That problem is the one of missing markets.

Anthropology

Explanations deduced from regression analysis of household choice only address one half of the narrative of genetic erosion. Demonstrating that environmental, risk, and cultural value advantage local crop varieties compared to "modern" ones is a testimony of the limits of science and industry to generate universal technology, but it still leaves the puzzle of diversity unsolved. The role of market relations in determining the basis and fate of crop diversity remained unexamined.

Markets, or the lack thereof, weigh heavily on why diversity exists at the household level and why successful efforts by science and industry to breed improved crops may not erode diversity. The rule of comparative advantage pertains as much to households as it does to localities, regions or nations. Households within a village should ascertain that specialization and exchange are advantageous, following the logic explained by Ricardo. Yet, our research on crop variety choice in Peru, Mexico, and Turkey, revealed that peasant households produce more diversity than is necessary or optimal given environmental and risk conditions. Overproduction of diversity may be explained by cultural value of local varieties, especially taste and cooking qualities, but why is it that peasant households have not discovered then the benefits from specialization and exchange, so that not all households need to produce a whole array of varieties? In fact, markets for local varieties at the village level do not seem to operate, and households which consume a particular variety must also grow it. In the parlance of economists (deJanvry et al. 1991), this is a case of "missing markets."

The master narrative on human development in late capitalist society is that market relations will supplant all others in farm economies and production systems. Market hegemony may, indeed, be in store for households everywhere, but there is much to suggest otherwise. Research on crop selection revealed a pattern that has long been familiar to anthropologists. Markets operate in some spheres but not in all, including production. Farmers in all of our study sites were active in the market, especially transactions that were centrifugal to the village, and all types of varieties were marketed, whether local or modern. Nevertheless, consumption of local varieties was not satisfied by local markets. The fact that markets do not work is expressed in the search costs to farmers who attempt find particular varieties for sale or exchange locally and in the risk that they would not be found. The logical response to these costs and risks is to produce diversity at home rather than rely on the market. "Why don't local markets work?" is familiar terrain to anthropologists, who have examined spheres of exchange and embeddedness of economic transactions in the culture of peasant society. The analysis of one narrative, genetic erosion, thus brings us to a much older and profound narrative—the tension between ordering social intercourse according to the presumptions of the market versus non-market principles.

Conclusion

In the proximate future, crop diversity in places like Peru, Mexico, and Turkey would seem to be insulated by the complexity of environment, the limits of science and industry, and the failure of markets to achieve hegemony within village economies. This finding does not excuse us from pursuing conservation, but rather it permits us to improve conservation of crop resources by including farmers as partners in the effort.

A premise of the genetic erosion narrative is that seed can be understood as a utilitarian asset, even though seed is a cultural artifact as well as a production input. The natural place to begin research on crop diversity may be production, but this does not exhaust the social basis of diversity of seed. The crude economics of the genetic erosion narrative, a human analog of Gause's law in ecology, suggests that a diversity should ultimately be supplanted by successful varieties or monocrops. However, a re-appraisal of Gause's law in the face of diversity in nature (Tilman & Pacala 1993) is mirrored in the social science of cultural resistance and resilience. The existence in single places of numerous constraints to survival and unavoidable trade-offs in strategies to meet these constraints result in opportunities for diversity to flourish. Understanding the ecology and fate of crop diversity requires reductionism in both biological and social research, but the limits to reductionism in both sciences are apparent. Our research suggests that the reductionism of marginal utility misrepresents both the complexity and resiliency of local practices (Brush 1992). The search for utility may, in fact, misdirect research on diversity. Boster (1985) makes the point that cognition and perception play critical roles in the rise and maintenance of diversity. Social exchange, cultural aesthetics, word play, and the poetics of identity, robust areas of anthropological research, are likely to inform on the origin and maintenance of crop diversity. One contribution of anthropology to the ecology of diversity is to continue to insist that the complexity of culture is a determinate a factor in the fate of cultural artifacts. While it may be practically and politically impossible to plan conservation without reductionism, understanding diversity cannot rest on utility alone.

References

Boster, J. S. 1985. Selection for perceptual distinctiveness: Evidence from Aguaruna cultivars. *Economic Botany* 39: 310–25.

Brush S. B. 1992. Ethnoecology, biodiversity, and modernization in Andean potato agriculture. *Journal of Ethnobiology* 12: 161–85.

Conklin, H. C. 1957. *Hanunóo agriculture: A report on an integral system of shifting cultivation in the Philippines.* Rome: FAO.

Darwin, C. 1868. *The variation of animals and plants under domestication.* London: J. Murray.

deJanvry, A., M. Fefchamps and E. Sadoulet 1991. Peasant household behavior with missing markets—some paradoxes explained. *Economic Journal* 101: 1400–17.

Frankel, O. H and E. Bennett, eds. 1970. Genetic resources in plants—Their exploration and conservation. IBP Handbook No, 11. Oxford: Blackwell Scientific Pubs.

Harris, J. M. 1996. World agricultural futures: regional sustainability and ecological limits. *Ecological Economics* 17: 95–115.

McArthur, R. H. and E. O. Wilson 1967. *The Theory of Island Biogeography.* Princeton: Princeton University Press.

Soulé, M. E., ed. 1986. *Conservation biology: The science of scarcity and survival.* Sunderland, Massachusetts USA: Sinauer.

Tilman, D. and S. Pacala, 1993. The maintenance of species richness in plant communities. In *Species diversity in ecological communities: Historical and geographical perspectives,* edited by R. E. Ricklefs and D. Schluter. Chicago: The University of Chicago Press.

CHAPTER 25

Activation Practices, History of Environmental Resources, and Conservation

Diego Moreno

"Human Use" of Organisms and Ecosystems?

The definition of the category of conservation proposed for discussion (Papa, in this volume), borrowed from the IUCN, refers to the "human use of organisms and ecosystems." Apart from suggesting that there may exist non-human uses, this definition resurrects the noted opposition between human and natural, which until recent years had characterized environmental research and conservationist policies based on "structural (or 'systemic') ecology" (Moreno 1990). In short, in the proposed definition the term nature has been replaced by "organisms and ecosystems" taken to be synonyms for nature. There is thus a danger of a return to the ideas characterizing the debate on the creation of *geographie humaine*. Today this return can only mean a lack of progress in the concept of ecosystem and its analytical results in many sectors of the environmental sciences (which Papa rightfully acknowledges simply as "nature sciénces") and in particular in the human sciences of history and geography.

A way out of the old dichotomy (Man/Nature) emerged very early in landecology research conducted in Britain in the 1960s. This was the approach of "historical ecology" (often used as the basis for the historical and environmental research carried out in the ethnobotanic and history center in recent years; see, for example, Moreno 1990, 1996, 1997). Thus, for example, to grasp the now clear-cut differences between structural ecology and the approach of historical ecology, the vegetation of a given site is considered by the historical approach as a special kind of artifact (Rackham 1976). Rather than the biogeographers' "bio-climactic stage," vegetation may be seen a stage of production (Metailié 1987), and as such considered the outcome of activation processes to which the environmental resources have been subject in the area in question. The resources (currently present in a given site, area or environmental complex) are thus *historically defined*—conditioned in their own ecology—by the practice (control, production, and activation), and by the forms of knowledge adopted by the previous societies that have settled on the site over time.

The recent development of various sectors of environmental archaeology attests increasingly explicitly to the historical nature and finiteness of environmental resources. There is no primordial natural Eden in the European history of the last 10,000 years (the post-glacial era). This is also true of environmental history outside Europe, as revealed by research into fire history (cf. Pyne 1982). The historical approach or structural approach in ecology gives rise to different models of the history of environmental transformation on different time scales for the observed processes and on different observation scales.

Biodiversity and Activation Processes of Environmental Resources: Problems of Scale of Observation

We are faced with the difficulty of adjusting a problem (and its terminology: biodiversity) that arose from analysis on a global scale, to a local scale. This difficulty also seems to underlie the suggestion made by Brush (1997). On this subject in general as regards a discussion on the problem of biodiversity and the conservation of the European natural heritage, it seems pointless to embark on research that simply compares lists of species, varieties, and products (considered as assets, objects or individuals). Their historical development is therefore reduced to terms of origin, presence, absence, loss or emotion. On a local scale, biodiversity—described, for example, as the floral complexity of a given vegetation—reveals its nature as an "historic product." This effect is obtained and maintained by the harvesters as practice and forms of knowledge (cf. the environmental effects of gathering wild grasses for salads in the eighteenth and twentieth centuries in a Valley near Genoa; see Poggi 1997a, 1997b).

If the general and generalizing models (born on a global observation scale of environmental systems dominated in the 1980s by the structural ecological approach) are to acquire the analytical capacity necessary to inform environmental conservation policies finally aware of the complexity of the historical environmental heritage, these (preliminary) questions of method must be tackled.

Most importantly we must question the issue of the scale of observation. For a study of the historical-social trends in local history, the recent work by Grendi (1996) is of interest. There are a number of immediate consequences from adopting a local perspective in historiography: (a) the problem of time scale of the historical trends (short, medium or long term), (b) the problem of contextualizing social relations—social embedment—in production practice, crop systems, and agrarian structures, which have characterized the local plant and animal forms of production and their historical transformations.

In the second case there has to be a focus on details in the study of new "historiographical objects" in some rural European societies: something similar but perhaps more perspicuous for conservation purposes than "systems of organizing the administrative territory" mentioned by Ortalli (in this volume) or the "social choices of production" suggested by Ambrosoli (also in this volume). The problem may be reformulated in terms of local and social contextualization (i.e., topography) of the practices and forms of knowledge underlying the activation of environmental resources.

Historical Trends and Rural Heritage

Ambrosoli (in this volume) brilliantly avoids hypostatizing the peasant economy and society—i.e., making it immobile in the history of European Society during the *ancien régime*. He identifies possible trends (and connections) between the European agricultures informed by an "agronomy integrated by social groups in the environment" and "an agronomy of selection and choices."

Once we accept this approach—properly referring to forms of knowledge underlying the practice of resource production whose social trend we wish to understand—it is, however, difficult to agree on the fact that in the case of societies with "agronomy integrated in the environment," that the "protection (!) of plant and animal varieties was random."

The work in historical ecology and environmental archaeology now underway in Italy, after the pioneering studies in the English-speaking world (Rackham 1976; and later works), especially those applied to the study of the "products of the land" (Poggi, 1997a) reveal that the *local plant heritage* (this category allows us to deal not only with the domestic flora but also the putatively wild flora) is managed according to definite production strategies. Today they appear to be documented historically more according to the environmental mechanisms introduced by production practices (and their previous effects), than by the sources conventionally referred to by historians of agriculture.

For example, it is very rare to find descriptions or narratives documenting completely or explicitly environmental effects (textual sources), but there are plenty of textual traces, maps and iconography to support the field evidence (ethnobotany and archaeobotany). The historical existence of these "systems forced out" by the development of "selection-based agronomy" in the eighteenth and nineteenth centuries can only be confirmed through this method. At times neither the dialect nor the language of the sources have conserved the ordinary name. This is true of the beech tree pastures created—probably from the Lombard era on—along the spine of the Apennines (*les alpes* and derivative names in medieval toponymy). There is interesting archaeobotanical (palynologic) documentation about them in sites on the Ligurian, Tuscan, and Emilian Apennine areas (cf. Moreno and Davite 1996; Moreno and Poggi 1996a).

Another example is "alder-cropping" in the Eastern Ligurian Apennines (cf. Cervasco et al., forthcoming). In Ambrosoli's language of the history of European agriculture we are faced with a case of the "elimination of symbiontic crops." Once we have topographically localized how much of this practice of the activation of resources is still found in the current ecology of the vegetation in the area of study (the Upper Aveto Valley), we realize that—rather than the elimination of a species or variety (*Alnus incana*)—an entire multiple crop system was eliminated in the nineteenth century.

This kind of environmental and archaeological research, inspired by ethnobotanical methods, has brought to light a rich cultural, environmental, and productive legacy, still passed on today by the farmers of southern Europe. So far this heritage has been "unwittingly" conserved. In practical terms we are talking about environmental resources, practices, forms of knowledge and local plants and animal production. The analysis we mentioned has been applied in an EU research program specifically dedicated to "The products of the land in southern Europe. Ethnological, sensorial and socio-economic characterization of type. Development strategies" (cf. Bérard and Marchenay, in this volume; Moreno and Poggi 1996b). Only by adopting the historical ecology approach and the local history scale of observation will these environmental aspects of the European rural heritage be identified, recovered, and correctly developed. But there is very little time left before the Common Agricultural Policy and its directives wipe out the living parts—literally—of the rural heritage in the residual marginal fringes of local production in Italy. The hygiene-sanitary regulations will be applied to definitively replace local production practice, thus eliminating them from the local market. These resources will thus lose their continuity with the present and the possibility of their economic, cultural, and scientific management.

References

Ambrosoli, M. 2004. Random Conservation and Deliberate Diffusion of Botanical Species: Some Evidence out of the Modern European Agricultural Past [in this volume].

Bérard L. and P. Marchenay. 2004. Diversity, Protection, and Conservation: Local Agricultural Products and Foodstuffs [in this volume].

Bradshaw, R. H. W., 1988. Spatially-precise studies of forest dynamics. In *Vegetation History,* edited by B. Huntley and T. Webb. Antwerp.

Brush, S. in this volume.

Cevasco R., D.Moreno, S. Bertolotto and G. Poggi. Forthcoming. *Historical ecology and post-medieval management practices in Alder woods (Alnus incana (L.) Moench) in the Northern Apennines, Italy. In Advances in forest and woodland history,* edited by Watkins C. University of Nottingham: Proceedings of the British Ecological Society.

Grendi, E. 1993. Storia di una storia locale: perché in Liguria (e in Italia) non abbiamo avuto una *local history? Quaderni Storici.* 82: 141–97.

_____ 1996. *Storia di una storia locale. L'esperienza ligure 1792–1992.* Venice.

Metailie, J. P. 1987. Les sources photographiques et l'histoire du paysage montagnard, l'exemple des pâturages pyrénéens. *Sources, Travaux Historiques,* nos. 9–10: 109–15.

Moreno, D. Ecologia storica. 1989. In *La storiografia italiana degli ultimi vent'anni—Età moderna* edited by L. De Rosa. Bari.

_____ 1990. *Dal documento al terreno. Storia e archeologia dei sistemi agro-silvo-pastorali.* Bologna.

_____ 1990. Past-multiple use of tree-land in the Mediterranean mountains. Experiments on the sweet chestnut culture. *Environmental History Newsletter.* 2: 37–49.

_____ 1992. Història i arqueologia dels recursos mèdio-ambientals. In *Un nou camp d'estudi: història ecològica i història de l'ecologia* edited by R. Garrabou. Barcelona.

_____ 1996. Domestico vs selvatico. Annotazioni su tassonomia e storia locale. *Quaderni storici.* 91: 514–35.

_____ , G. F. Croce, M. A., Guido and C. Montanari. 1993. Pine plantations on ancient grassland: ecological changes in the Mediterranean mountains of Liguria, Italy, during the 19th and 20th centuries. In *Ecological effects of afforestation. Studies in the history and ecology of afforestation in Western Europe,* edited by C. Watkins. CAB International.

_____ and C. Da Vite. 1996. Des saltus aux alpes dans les Apennins du Nord (Italie). Une hypothése sur la phase du haut-moyen-age dans le diagramme pollinique du site de Prato Spilla. In *L'Homme et la Nature au Moyen Age* edited by Colardelle M. Actes du V Congrès International d'Archéologie médievale. Paris.

Moreno D. and G. Poggi. 1996a. Storia delle risorse boschive nelle montagne mediterranee: modelli di interpretazione per le produzioni foraggere in regime consuetudinario. In *L'Uomo e la Foresta (Secc.XII–XVIII)*. Prato.

———— 1996b. Ecologìa historica, caracterizaciòn etnobotànica y valorisaciòn de los "productos de la tierra." *Agricultura y Sociedad,* nos. 80–81: 169–80.

Ortalli, G. 2004. Forms of Knowledge in the Conservation of Natural resources: from the middle Ages to the Venetian "Tribe" [in this volume].

Papa, C. What does it Mean to Conserve Nature? [in this volume].

Poggi, G. 1997a. Pratiche di attivazione: effetti della raccolta tradizionale di vegetali spontanei ed ecologia storica del sito (18°–20° secolo)—(Arbora—Valle T.Recco -Liguria Orientale). *Archeologia Post-Medievale* I.

———— 1997b. Le pratiche di attivazione della copertura vegetale come oggetto geostorico. Dalla cartografia della vegetazione alla cartografia delle risorse vegetali. *Rappresentazioni e pratiche dello spazio in una prospettiva storico-geografica.* Genoa.

Pyne, S. J. 1996. *Fire in America. A Cultural History of Wildland and Rural Fire.* Princeton.

Rackham, O. 1976. *Trees and Woodland in the British Landscape.* London.

———— 1980. *Ancient Woodland. Its history, vegetation and uses in England.* London.

———— 1986. *The History of the Countryside.* London.

———— 1996. Forest history of countries without much forest: Questions of conservation and savanna. In *L'Uomo e la Foresta (Secc.XII–XVIII)*. Prato.

Turner, J. and M. Peglars 1988. Temporaly-precise studies of vegetation history. In *Vegetation History* edited by Huntley B. and Webb T. Dordrecht.

CHAPTER 26

Forms of Knowledge in the Conservation of Natural Resources: From the Middle Ages to the Venetian "Tribe"

Gherardo Ortalli

Anthropocentric Medieval Cultures

To scholars of the Middle Ages and within the history of the environment, the very concept of conservation may appear decidedly anachronistic, if applied in its contemporary meaning. Any talk of a culture of conservation is inevitably fairly ambiguous when used for an age when people never doubted their full right (or even absolute necessity) to be a modifying element of nature, an entity to be used and therefore studied for a number of direct purposes.

In the Middle Ages the idea of preserving the natural heritage as the conservation of an environment whose balances should be interfered with as little as possible or not all, would hardly have been understood—and anyone proposing it would have certainly been judged as bizarre. This kind of attitude was the outcome of several factors expressed at various cultural levels: at a speculative-doctrinal level and at that of daily and living. In other words this held true both in terms of what we might call *Bildung* and *Kultur*.

Considering "high" culture, erudite cosmology and the great world visions, the Middle Ages must be seen as having roots in the Judeo-Christian tradition, which from the outset had a strong anthropocentric connotation (Delort 2001: 62–65). This outlook is found in the early passages of the Old Testament. Man's superiority and control over nature are symbolized by the image of Adam giving names to the animals brought to him by the creator: "to see what he would call them: and whatsoever Adam called every living creature, that *was* the name thereof. And Adam gave names to all cattle, and to the fowl of the air, and to every beast of the field" (*Genesis* 2, 19–20). And for the vegetable world the divine words were also clear: "I have given you every herb bearing seed, which is upon the face of all the earth and every tree in the which is the fruit of a tree yielding seed" (*Genesis* 2, 20).

The Biblical text narrating the pact between God and Noah, who has just disembarked from the Ark after the flood, is even more explicit: "and God blessed Noah and his sons, and said unto them, be fruitful, and multiply, and replenish the earth. And the fear of you and the dread of you shall be on every beast of the earth, and upon every fowl of the air, and upon all the fishes of the sea; into your hand are they delivered. Every moving thing that liveth shall be meat for you; even as the green herb have I given you all things" (*Genesis* 9, 1–3). Although the unity of the cosmos was widely taught by the sources of Christian revelation, there can be no doubt that passages such as these point to a powerful anthropocentrism. Unlike in other civilizations and religions, in Christianity man was created in the image and likeness of God, and as such was not simply part of nature but transcended it. The dialectical relation between man and the natural environment was thus determined by a powerful theological premise: since nature was created for man, there were no limits to its exploitation.

These doctrinal bases paved the way to interpretations of the man-environment relationship whereby Christian anthropocentrism is associated with a consequent arrogance towards nature, considered to be completely subordinate. This topic deserves further study in order to gauge its rather vague schematism, but in the meantime let us dwell on the fact that through these premises, the nascent Christian culture, destined to exercise a hegemonic role, provided an overall interpretation in which the approach to the environment required no notion of respect. The idea of protection was given no place in the organization of forms of knowledge about nature. At the same time, throughout the Middle Ages, the new religion imposed conceptual models intended to systematically strip away the sacred aspects of nature, which had been such a well-known part of the preceding traditional cultures. Eliminating the systems of belief associated with sacred springs and woods, or suppressing the idea of the *genius loci*, meant moving towards a secular interpretation of nature and the only apparent paradox that this secularization was the outcome of a new form of religiosity.

In the Middle Ages conservation can thus only be seen as a kind of human action on the environmental system. The legitimacy of every form of manipulation, transformation, and exploitation was just simply taken for granted. Having said this, however, we must insist on how the attitude towards the nature-system existed on very different epistemological levels. Although the theologian and man of letters could insist on man's supremacy, thus justifying his extreme exploitation of nature, for a long time, at least until the tenth or eleventh century, man—theoretically lord and master through divine investiture—was actually on the defensive. There is thus an initial discrepancy between the doctrinal interpretation and the objective state of the man-environment dialectic. Let us take a closer look at this incongruity.

Theophany and Less Control Over the Environment

During the transition from late Antiquity to the early Middle Ages in our geographical reference areas (i.e., those areas to which the concept of the European Middle Ages is applicable), there was a gradual break-up of the ancient systems of organizing the administrative territory, overwhelmed by the crisis in the late Roman Empire. The structural decline in the control over the environment due to extremely complex new political, economic, and military events (this was the age of the great migrations of peoples) was accentuated by a change in those—for us still relatively obscure—cyclical factors exercising a great influence on overall balances. In particular we find the repercussions of climatic variations and increasingly aggressive epidemics. The traditional Roman world began to suffer from a number of factors: unfavorable climatic conditions, a growth in wooded areas, expanding swamps, variations in rainfall patterns, a dramatic drop in population, reductions in arable land, a return to economic forms of harvesting, and a greater emphasis on hunting and animal-rearing as opposed to arable farming. Incidentally, such processes make this era an unusual phase in the debate on stages of development and the controversial division of human history into phases progressively characterized by hunting, stock raising, and arable farming, according to the theory advanced in 37 BC by Marcus Terentius Varro in *De re rustica* (II, 1, 3–5). But these are issues we cannot explore further here.

Returning to our theme, we see that with the advent of the Middle Ages, more than in the past, nature was felt to be antagonistic and its forces seemed to be threatening and aggressive. Man was almost on the defensive, under pressure from a natural environment that increasingly eluded his control. And it was in this context that the new ideas on the man-environment relation developed in Christian culture. Given these premises, it is hardly surprising that the forms of knowledge pursued by high culture (in centuries when it was monopolized by the clergy) had symbolic, theoretical, and ethico-religious meanings almost completely incompatible with the practical de facto needs associated with the theme of conservation.

In the cultural framework I have tried to describe (with inevitable oversimplification because of its summary form), after the great efforts of patristics, knowledge of nature took the form of theophany (Ortalli 1997: 155–61). The whole world may be seen as the sensible manifestation of God: "every visible and invisible creature maybe be called *theophania,* that is the manifestation of God: *id est divina apparitio"* thus John Scotus Eriugena around 860–65 (*De divisione naturae,* III, 19; Davy 1977: 149–51).

In the late 1960s Lynn Whyte JR explored this natural theophany in a brilliant but highly thought-provoking article, albeit debatable in some of its concluding remarks, unjustly ignored by medieval experts (White 1967): "In the early Church, and always in the Greek East, nature was conceived primarily as a symbolic system through which God speaks to man: the ant is a sermon to sluggards; rising flames are the symbol of the soul's aspiration." Only "in the Latin West by the early thirteenth century natural theology… [take] a very different bent. It was ceasing to be the decoding of the physical symbols of God's communication with man and was becoming the effort to understand God's mind by discovering how his creation operates…It was not until the late eighteenth century that the hypothesis of God became unnecessary to many scientists."

Nature as theophany, then. The beauty and multitude of the stars predicate their marvelous maker; countless drops of rain, sands of the sea, blades of grass in the fields, birds' feathers, the motions of celestial bodies indicating limitless creative knowledge; broad rivers, immense mountains and valleys, the distant sky and the depth of the abyss: all speak of God, whose greatness is infinite. The beauty of all things reveal their splendid creation, the fragrance of perfumes is permeated by the gentle holy spirit. These reflections proposed in *De triplici genere contemplationis* (I, 6–11) by the Premonstratensian and then Carthusian Adam Scotus (who died around 1212/13) neatly illustrate a kind of approach to the natural world which for a long time, in the proper terms, was far from being the exclusive field of aesthetic and exegetic works but also extended into strictly scientific output. The optical studies on the rainbow by Robert Grosseteste, Roger Bacon, and Theodoric of Vriberg or Freiberg, in the forefront of the science of the day, were "an undertaking in religious understanding."

The religious key and the yardstick of allegory (both in the contemplation and interpretation of phenomena) where conducive to the age and widely used. Bestiaries, herbaria, and lapidaries provide excellent examples of this approach to naturalistic knowledge. To form an idea, you only need to dip into the *Phyisiologus*, the fundamental text and in some ways harbinger for all the early medieval "moralized" bestiaries. This work is useful in grasping the change from the great tradition of naturalistic knowledge (especially agronomy) of the Roman tradition to the new early medieval outlook. Take, for example, the chapter on hedgehogs: "they have the shape of a ball and are completely covered in spines. The Physiologist said that the hedgehog climbs up the vine and goes to the grapes, and throws them to the ground before rolling on top of them. The grapes stick to its spines and it takes them to its offspring, leaving the branch bare. You too, therefore, man of faith remain beside the true spiritual vine, if you wish to enter the spiritual press and be preserved in the King's Palace, and if you wish so to come before the holy throne of Christ. How could you let the hedgehog, the evil spirit, climb into your hearts and leave you bare like the bunch on the vine, with no grapes in you?"

Clearly this kind of knowledge had nothing or almost nothing in common with the setting in which we can frame the category of conservation. The hedgehog of the Physiologist was a fictional image for the objective exploitation of natural resources. We must shift the focus then to a series of different notions, of different forms of knowledge. We are talking about those forms of knowledge handed down by the generations through practice. They continued to be pursued throughout the Middle Ages in the daily exploitation of the environment, without, however, raising a real problem of conservation in the face of seemingly infinite resources. Thus, for example, early medieval Europe was full of windmills but studies on hydrodynamics made no progress whatsoever. The moldboard plough replaced the nail plough but no one bothered to record this incredible innovation. The certain increases in grain yields can only be traced with great difficulty through some private contract or rare fiscal related document. The shoulder yoke changed the potential use of animal force, but this was of little interest to those who could read or write and thus document the innovation. From the tenth to the twelfth century technology and practice raced forward at a speed completely ignored by the official depositories of scientific culture. In short, there were very considerable differences in levels and logical incongruities between project and practice, between "ideology" and practical action, and between cosmological models and the concrete world.

The Late Middle Ages and the Problem of Conserving Resources

Despite the context of cultural incongruity described above, the Middle Ages brought a radical change to the modes of exploiting resources. A set of epoch-making phenomena, summed up in the formula "the recovery of the year 1000," led to the de facto issue of conserving resources. And we can get a good idea of this change from documents in the twelfth century. After the long years of the early medieval crisis, Europe entered a phase of rapid development: a period of great forest clearances, a reduction in uncultivated land, improvements in the quality of domestic animals which was returning to premedieval standards (Bökönyi 1974); and farmlands were extended under the pressure of an intense population rise, already consistent in the tenth century. At this point, after long years of vigorous and fairly uncontrolled expansion to the detriment of the natural environment, twelfth-century society, although still bound to natural resources for its basic requirements, gave signs of beginning to adopt specific rules for environmental protection.

Here, too, there is a gap between the epistemological levels, between hegemonic and subordinate cultures or rather (to use other schemas again) between *Bildung* and *Kultur*. In terms of "designing projects," the theme of conserving natural resources was still tenuous. It was obviously not on the medieval agenda. Nonetheless there were a growing number of objective forms of environmental protection. This new climate is clearly indicated by the proliferation of laws in the twelfth century concerning the nascent control of woods, the protection of particularly valuable trees, the obligation to plant new trees, the dangers of polluting rivers and streams, the limits to hunting, and dumping dangerous or toxic materials, and controls on pollutive manufacturing processes. A long series of widespread local measures bears witness to a new awareness. But such measures still reflected the needs of practical knowledge rather than any pressure from new ideologies and new ordering patterns proposed by the institutional depositories of knowledge (Ortalli 1997: 148–54).

Basically there had been a change of attitude towards the environment. The natural heritage was no longer seen as being inexhaustible, unlimited, and eternally renewable, and therefore capable of withstanding totally uncontrolled exploitation. Although not systematic, conservation measures for resources became more frequent. Significantly, they developed side by side with the symptoms suggesting the habitat was deteriorating. Nonetheless this set of uncoordinated measures could in no way be fitted into what we would now call an "environmentalist" perspective. The logic was completely different and any historiographic attempts to reinterpret such measures in a new light seems to me to be decidedly anachronistic. The need to focus on environmental balances grew more intense in the stages of greater population growth. It then diminished after the plague in the mid-fourteenth century only to become more pressing again in the last decades of the fifteenth century when the population almost returned to levels before the great plague. Given this situation, I feel that the undeniable efforts to protect the environment emerged despite the very weak theoretical and conceptual premises. And in any case the almost total lack of an environmental conscience in no way implies there was a lack of awareness about the intrinsic importance of the natural heritage and all the practical consequences. Again the emphasis was very much on pragmatism, without the gap between high culture and practical or popular forms of knowledge being closed.

In short, the medieval experience matured in a situation where "the conservation of resources" followed a logical development that had little in common with "safeguarding the environment." Ultimately, however, the environment objectively began to be protected, mainly because of the population's strict and almost direct dependence on the natural heritage. But again we are dealing with situations and behavior almost diametrically opposed to those of our own age. Our extraordinarily refined knowledge and culture about the environment and its protection, especially in the twentieth century, seems to run into difficulty when it comes to practical action. Conversely, in the Middle Ages the unlimited right to be a dynamic element altering the habitat was never called into doubt, without any significant attempts at theoretical elaboration. There was a much greater willingness to pursue practical protection whenever required by contingent needs immediately perceived by a society with a very direct dependence on natural resources. And this kind of action was on a small but widespread scale, the outcome of diffuse forms of knowledge continually experimented *in situ*.

The Relationship with the Administrative Territory: from the Middle Ages to the Knowledge of the Venetian "Tribe"

Medieval practice maintained a close link between naturalistic forms of knowledge and highly circumscribed local situations. Rather than any grand general theories or developments of the high culture, the specific knowledge of a group or community led to effective actions. This brings us to reflect on biodiversity, but also on the partly related themes of typical products, the building of traditions, and the measures to protect specific local features. These considerations inevitably suggest a comparison with modern situations. From this point of view, looking at typical forms of knowledge (and modes of production), there is a natural tendency to highlight farm production where the logic of the typical product is self-evident. But it is also worthwhile looking at other scenarios, and an interesting field of observation is the highly urbanized Venetian context (Venice is first and foremost a city) also strongly marked by the "interference" of natural elements (lagoon, canals, tides and islands). It is interesting to mention, albeit briefly, the current decisive contrast between traditional widespread forms of knowledge developed in the local context and those produced by general epistemological processes and introduced to the city from outside through procedures with considerable social and economic implications.

Venice has always been characterized by its fragile equilibrium of water and land, built on a refined system of knowledge founded on the continuous and exact measurement of the tide, the relation between salt and fresh water, the influence of rivers and their deposits, the necessary and functional co-existence of islands, *barene* (the flat emerged grassy mud banks in the lagoon only ever covered by very high tide) and *ghebi* (narrow vein-like channels). These forms of knowledge have always played a key role in the survival of the lagoon equilibria. Without them, today the lagoon would not be what it is. The use of natural resources in Venice affected every field, ranging at various times from the opportunities for the salt works to the potential the lagoon waters offered for the art of war, from the action of fishermen to the advantages for farmers on the larger islands, from the resources for all-season lagoon navigation along the length of the upper Adriatic to the favorable conditions for ports, trading, smuggling, etc. In short, this vast extremely urban system of specific cultural features was built on the fulcrum and premise of the lagoon environment and its extraordinary but intractable resources.

These equilibria were administered throughout the centuries by highly differentiated levels of culture and types of knowledge: both theoretical and experimental, and erudite and popular, they were carefully managed and converged in the structures the Venetian State created at various times to keep the lagoon and its resources under control. On this subject, without giving a detailed account of the stages in its long history, we could mention how the lagoon was administered first by the Piovego and Provveditori di Comun, then from the fifteenth century by the Council of Ten and the Senate. In 1501 the magistrature of the three Savi alle Acque was created and then flanked by the Collegio and the Esecutori (there were three from 1521). The Inquistore was created in 1678, while from 1542 a Matematico pubblico, aided by experts, was appointed to give his opinion. The areas of competence of these bodies were distinct but represented overall by the Magistrato alle Acque. They ranged from the technical-scientific field to taxation, with functions of guidance and control, and had extensive judicial powers even in criminal matters (Gasparini 1993: 34–42).

After the fall of the Venetian Republic in 1797, there was obviously still a need for an overall vision, since in 1907 the Magistrato alle Acque was revived as an Italian state body with very wide-ranging powers for an area just under the size of the former Venetian mainland dominions, from Lombardy to Friuli and Trento. The existing testimonies show that even after the Second World War the Magistrato continued to decide on issues concerning the lagoon and "listened to the opinion of nine fisherman," suggests just how far these different forms of knowledge were interwoven. This would seem to be an excellent example of how knowledge produced by research institutes dialogued with empirical information from people with daily experience of life in the lagoon and fully aware of all its vital rhythms. It is an equally interesting example of how this process may be annulled by the new forms of the organization of knowledge introduced by changes in the 1980s, when the state decreed that the basically monopoly management of the lagoon should be placed in the hands of a consortium (the Consorzio Venezia Nuova), bringing together some of the major private operators in Italy in the field of large public works (Mencini 1996: 43–92).

The leap in the logic of the various forms of knowledge may be illustrated in the transition from experimental knowledge, including that of the fishermen, to the mathematical models used by the consortium. Obviously, the example I am citing involves an inevitable and in some ways reductive simplification, but in any case, there is a clear contrast between the pool of knowledge rooted in the experimental approach to daily practice, and another pool of knowledge produced by completely different developments, supported by scientific knowledge imported from outside, by resorting to experts who may even deem getting to know the Venetian situation almost superfluous. Naturally, the outcomes of these different epistemological pathways reflect incredibly different techniques and managerial contexts. Not surprisingly the scientific culture based on the local reality tends to put forward "gentle" solutions of minimal but continuous adjustments to the natural balances aimed at accompanying and guiding the ongoing evolutionary processes. The scientific culture supported by more general capabilities gathered in the private consortium frames the problem in an engineering perspective of large-scale works and incomparably higher (and more costly) levels of technical intervention.

Thus, for example, the return to the traditional measures for the port inlets where the sea enters the lagoon (they were artificially altered this century) is contrasted by a project for a grandiose system of mobile sluices closing the port inlet and regulating the flow of seawater. And instead of re-opening the *casse di colmata* (filled-in areas) and the fish-farming grounds in the lagoon (which now have solid closures instead of the trellis fences allowing for changing tides), a system of barriers and gates protecting the various areas of the city from tidal motion is proposed. These are scientific but also philosophical, ideological, and financial alternatives. And the Venetian "tribe" must reckon with considerable interests and capital outside the effective control of state structures. Instead of continuing with the practice created by a culture founded on knowledge and wisdom acquired *in situ* before being re-elaborated at scientific levels, there is a process of extreme "artificialization" of the whole Venice-system supported by first-order knowledge built up in various places and then transferred into the specific context from outside.

In this situation of crises in the traditional modes of knowledge, and their transmission and application in the face of a very powerful politico-economic system of different forms of knowledge, the impression is that the ongoing processes are not only an arguably inevitable loss of the unique Venetian approach to the development and management of knowledge. It almost seems that the new ways of working on Venice may even transform the overall situation of the city into a kind of "typical product," artificially kept alive, and increasingly tied to specific features constructed abstractly without the contribution of any "tribe" and managed like a kind of commodity to be launched with more or less refined marketing strategies. This is certainly an extraordinarily interesting time, which will soon provide good subject matter for retrospective studies on the sociology, economy, history, and anthropology of Venice.

Session V: Conservation

Edited by Gabriele Iannàccaro

Glauco SANGA observes that the selection and freezing of tradition (see Papa, Bérard and Marchenay, in this volume) is a feature that can be found in every kind of folk knowledge. For instance, when conserving traditional costumes, daily varieties are liable to die away first, while the ones worn on festivities are kept; and then just one of these festive costumes is selected in the end. In the case of Gressoney, the present costume (the only one) was originally the richest festive variety—the one selected by Queen Margherita at the beginning of this century. Simplification seems to be almost a function of diffusion: the larger the area of diffusion, the more unified the types selected, and variants are abandoned. An example can be found in popular songs: lullabies, of private and limited use, have countless variants, while *Bandiera rossa*—a very widespread song of a social kind—was selected as a single type (of course, for political reasons as well).

The real reason is the market, and the increasingly professional status achieved by distribution (more than by production, as songs and popular tales demonstrate, spread throughout the country by storytellers and vagabonds). This market model can be defined the hamburger model: McDonald's theorized that throughout the world the same hamburger should be eaten with the same taste. On this single type very slight local "micro variants" are admitted—variants that do not touch the ground uniformity of the product.

Sanitary legislation hampers many traditional products (cheese, meat products): they recall the hygiene-regulations introduced by the colonial administration against pig breeding in Melanesia, which led to a major crisis in the local society, in which the pig had a fundamental social role.

Ten years ago, in a symposium on *Knowledge Transmission* organized in Rome by Giorgio Raimondo Cardona (who inspired this conference on nature knowledge), Sanga claimed that tradition is a protected kind of transformation. He now fully agrees that the most effective form of conservation is change: dinosaurs are extinct, we are still here, and we are the transformation of a bacterium of four billion years ago.

*

Starting from Benveniste's assumption that Indo-European society is originally triadic (*Institutions indo-européennes*), John TRUMPER points out that it is based on a truth of God established by a priest-caste, a truth of the sword or a truth established by combat, established by a warrior-caste, or military gurus; a truth of men established by and for artisans and those who are not priests or warriors. This constant stretches from ancient Indian society to ancient Celtic society, ranging from East to West and underlying all Indo-European primitive culture. Transposing this into the realm of Indigenous Knowledge: we have priest and warrior castes that are carefully controlling and vetoing information, on the one hand, and on the other an artisan-farmer caste that has the Indigenous Knowledge. When the priest and warrior castes break up owing to conquests, invasions, social or natural upheavals or whatever, even catastrophes of a certain type ("Quid enim cuiquam satis tutum videri potest, si mundus ipse concutitur et partes eius solidissimæ labant?"Nat. Quæst.VI, 1, 4), as may happen, then Indigenous Knowledge comes to the fore even in our long history. Roger I and his relatives and successors, sweeping away Byzantines and Arabs from Italy (1068–1100), had no property laws to appeal to and were forced to consult Indigenous Knowledge, with results documented in various Placiti or Chartæ. A new ruler who takes over property, and who bestows property upon himself, upon his followers or upon the Church etc., has to construct a semblance of juridical possession, and where a lacuna has been created he has, probably to his chagrin, to appeal to Indigenous Knowledge. Such situations are probably what lie behind certain constructs in Common Law, and in the ultimate analysis seem to have been the distant basis of the jury system. However, usually, the other two castes do not seem to cherish having to appeal to the Indigenous Knowledge of the third caste; they prefer to control information and knowledge.

The Academy is an extremely ambiguous sort of entity. Its members are, to a degree, part of the priest-warrior castes, and to a degree, not part of them. They perhaps constitute a contradiction within the caste triad; therefore, as regards Brush's observation "Social science finds predicting the distant future unfashionable, discreditable and implausible," the question is: how can they influence the real priest-priest caste, the real warrior-warrior caste, who are afraid of long-term projections in human planning, and only seem to be interested in conserving their own control of information and knowledge? If humanity and this planet are to survive in any decent way, then what is needed is obviously the potential for long-term projections based on Indigenous Knowledge; a science that is not diagnostic, nor interested in projections, is a pre-science, rather like modern medicine.

It is sometimes possible to trace lost Indigenous Knowledge through toponymy or other ancillary branches of language study. We know from toponymy in the South of Italy that "otters" were once extremely important in the naming of internal and mountain water courses. They are now extinct, which means that such stretches of water must have been bigger and abundant in freshwater fish in at least the Middle Ages. This is no longer true but implies a certain ecology and economy in the past, now absent. Olive-tree cultivation must have been practiced by the Byzantines even in fairly high mountain areas, since, for example, parts of the southern Sila Chain abound in toponymic derivatives of Greek ejlaiva "cultivated olive," "ejlaiwvn "olive groove," and ajgri-elaiva "wild-olive." Is there still the technology to do this? In what sense can mapping out past Indigenous Knowledge in this way, without being archaeological about it, help to project future Indigenous Knowledge?

*

Laura LAURENCICH, anthropologist, director of an eco-museum in Nicaragua, introduced a specialized agriculture (growing vegetables and fruit) alongside the traditional subsistence agriculture. But this is not enough, and natives abandon their fields to flee to the towns and live in miserable slums. What should we do? Of course we should try to conserve culture and tradition, but we must also try to conserve people on their own land, giving them the means and good reasons to remain there. We cannot simply place them in a kind of bell-jar.

*

Pier Giorgio SOLINAS states that paradoxically the most efficient preserver of authenticity is the market. Chianti classico is both the name of a region of Tuscany and of a world-famous wine. The history of Chianti was socially and culturally connected with métayage and nothing survived of that culture but an alimentary product. Now, foreign companies are in possession of the landed estates; for example the well-known Ricasoli wine is produced in a farm owned by Americans, and all around it "agri-tourism" is flourishing. The consumers of this good, which is savory, enological and ethnological at the same time, are tourists traveling by bus from wine farm to wine farm, tasting wine and food and reproducing a version of living in the world of métayage which unfortunately lacks the métayer.

*

Stephen BRUSH maintains that it is impossible to recreate something that never existed in the first place; there is never a force, it is a process and we are living organisms constantly changing. Conservation biology, for example, cannot save biodiversity. Nevertheless, the question of prediction pervades conservation biology, and if we wish to do social science in relation to conservation, some kind of prediction is inevitable. Linguists are equally discomfited by the idea of prediction into any distant time; an article by Michael Crows in 1982, recording the disappearance of languages and the way to preserve, them testifies to this. The best that can be done is to try to predict what is happening. Nobody can expect to save a language by simply recording the lexicon.

*

According to Laurence BÉRARD, the logic of regulation penalizes those producers who do not make the same choice.

As regards "hamburger micro-diversity," European regulations do not consider those local differences that cannot be represented. Some scales of differentiation and scales of difference-protection are necessary.

Other differences result in the production of those goods that are commercialized: for example, food and wine produced by agriculturists are different from typical food and wines, and often "agri-tourism" offers the former.

Regulations also aim to protect product denomination and to control competition: in France, for example, mountain cheese with particular organoleptic characteristics should be differentiated from cheese produced according to fewer restrictions. On the other hand, as most consumers of such products are local consumers, maybe the whole regulation system is hardly necessary.

Moreover, local corporations, regional and tourist-administration boards are interested in these products, and in France they are transforming them into heritage.

*

As regards biodiversity and seeds, Philippe MARCHENAY maintains that the disagreement with agronomists derives from not separating the genetic pool from the genetic lines (which may be "normalized"); there are misunderstandings in relation to the way of creating and managing diversity.

It would be very interesting to anthropologically investigate how the representation of hygiene alters with the different workers involved in food production. The development of an anthropology of negotiation would be equally worth studying, as in this field the role of negotiation is fundamental.

A remarkable point for anthropologists is also the new role of culture in making certain products competitive.

Finally, as those who study this phenomenon are involved in it themselves, they must be cautious in legitimizing some kinds of knowledge rather than others.

*

Diego MORENO states that, as regards parks, instances of biodiversity are often considered mere bioclimatic data, while, on the contrary, landscape projects demand the participation of anthropologists and of those who study biological and cultural conservation since choices are still too often made on the basis of nineteenth-century natural sciences. Anthropologists and historians should not be afraid of progress, of the systems of classification, and of the complexities that biology and environmental sciences are trying to comprehend. However, it must be considered that concepts like that of biodiversity derive from the global-scale environmental studies of the 1970s and 1980s, and they cannot always fit local-scale realities, above all when the local ecosystem is that of *historical ecology* and not of systemic ecology. Park managers and planners should be aware of these mechanisms.

*

Referring to the importance of a concept of diversity that embraces change as underlined by Papa, Tim INGOLD maintains that just as Goody has shown how a focus on thinking rather than thought allows us to bring a more dynamic perspective to bear on understanding cognitive processes it would be useful to focus on processes of diversification rather than on structures of diversity. The imposition of centrally planned regulations specifically designed to conserve biodiversity could have the effect of actually hindering the process of biodiversification, particularly on the local level. On the contrary, biodiversification should always be possible; we should ensure the continuity of variation and change in the organic world.

Referring to a concept of diversity is not the right way to understand difference and its recognition. It is well known that the ecosystems of the subarctic like that of the Saami people in northern Finland, and even more of the Arctic, are characterized by their relatively low biodiversity. There are fewer species, although some of these occur in very large numbers. This, however, does not mean that the Saami inhabit a more impoverished environment, or lead more impoverished lives, or perhaps have a more impoverished body of traditional knowledge, than, say, the Jivaro of the Amazon. Diversity depends upon difference, and difference is what we are interested in. But to represent difference as diversity is to understand difference in a particular way.

Among northern peoples such as the Saami there is just as much difference, and recognition of difference. Rather than being registered in the form of diversity, this difference lies rather in the way in which people are intimately responsive to very fine variations—in the life histories of the animals they deal with, in the particularities of place and the ways these are bound up in the life-histories of persons, and so on. Thus from a purely taxonomic point of view, one reindeer is as good as another, though fine distinctions are made in terms of antler shape, tooth condition, and so on. Yet one must consider that reindeer themselves possess an immense amount of local knowledge, and a pastoral system such as is practised by the Saami practice could not work if this were not the case and if people were not aware of it. Furthermore, the knowledge both of animals and of people is constituted through their own life histories and through their own experiences of relating to the environment and to one another. In short, being perceptually attuned, through a history of relationships, to the recognition of ultra-fine differences is not at all the same as being able to call upon an elaborate taxonomy of types and subtypes.

*

Cristina PAPA objects that assimilating, like Sanga, songs and customs to typical products and McDonald's entails overlooking a relevant difference. As a matter of fact in the circulation of songs, customs, folktales, and languages, intentionality is not involved, while in the merchandizing of a particular product or in the diffusion of McDonald's, intention is predetermined and conscious choices play a decisive role. Preserving is a human activity deriving from choices; subjects, cultures, social classes, and groups decide what to preserve and what not to preserve. There is not any predetermined development of societies; rather, what we find out historically afterwards derives from previous choices that had been made and which could also have been different.

Behind what we preserved of the past—rustic houses but not métayage contracts, wine but not cellar-typologies, agronomists but not the traditional knowledge of métayers—there are particular choices and conditions. What is deliberately preserved is the result of negotiations, that is of a complex system of relationships and powers; therefore it is not up to the ethnographer or to the manager of an ecomuseum to decide what is to be put in a museum or ecomuseum, although their opinions do have some weight.

*

Mauro AMBROSOLI observes that, from the late sixteenth century onwards, every generation has been faced with the problem of agronomic selection, which involves benefits as well as dangers. The typical Italian landscape with trees is indeed a French creation, for Italy's roads were without trees until the end of the eighteenth century. The Italian landscape was made to conform with the naturalistic model of that age, with the added genetic debt of Italo-French plants (which possibly brought about the phylloxera disease).

*

Stephen BRUSH underlines that mentioning hybridization, of cultures and landscapes, is very appropriate, of course, to indigenous knowledge, and very pertinent to an active debate among indigenous people today about ownership of knowledge. According to indigenous ideology, indigenous knowledge is something that can be isolated and treated as a pure entity, but, as Michael Warren recalled, there are ancient currents everywhere in the world that have brought knowledge and biological resources together and mixed them constantly. What is ignored in the debate over the ownership of knowledge is that in order for cultural diversities and for biodiversities to survive, those currents must survive too. The move toward an ownership that has a legitimate political basis to it is an extremely problematic movement and something that is poorly understood. It plays into geopolitics in some ways, and it demands much more debate than has been given.

Index